SCHAUM'S OUTLINE OF

THEORY AND PROBLEMS

OF

BEGINNING
CHEMISTRY

SCHAUM'S OUTLINE OF

THEORY AND PROBLEMS

OF

BEGINNING
CHEMISTRY

DAVID E. GOLDBERG, Ph.D.

Professor of Chemistry
Brooklyn College

SCHAUM'S OUTLINE SERIES

McGRAW-HILL, INC.

New York   St. Louis   San Francisco   Auckland   Bogotá
Caracas   Lisbon   London   Madrid   Mexico City
Milan   Montreal   New Delhi   San Juan   Singapore
Sydney   Tokyo   Toronto

DAVID E. GOLDBERG received his Ph.D. in chemistry from Pennsylvania State University and joined the faculty at Brooklyn College in 1959. His primary interests are chemical and computer science education. He is the author or coauthor of ten books, over 30 journal articles, and numerous booklets for student use.

 This book is printed on recycled paper containing a minimum of 50% total recycled fiber with 10% postconsumer de-inked fiber. Soybean based inks are used on the cover and text.

Formerly published as *Schaum's Outline of theory and problems of Chemistry Foundations*.

Schaum's Outline of Theory and Problems of
BEGINNING CHEMISTRY

4 5 6 7 8 9 10 11 12 13 14 15 16 17 18 19 20 SHP SHP 9 8 7 6 5 4

ISBN 0-07-023679-8

Sponsoring Editor, Jeanne Flagg
Production Supervisor, Fred Schulte
Editing Supervisors, Meg Tobin, Maureen Walker
Cover design by Amy E. Becker.

Library of Congress Cataloging-in-Publication Data

Goldberg, David E. (David Elliott)
    Schaum's outline of theory and problems of chemistry foundations/
David E. Goldberg.
        p.   cm.—(Schaum's outline series)
    ISBN 0-07-023679-8
    1. Chemistry—Outlines, syllabi, etc.   I. Title.
QD41.G65 1991
540'.2'02—dc20   89-78523
                                                                          CIP

To the Student

This book is designed to help you understand chemistry fundamentals. Learning chemistry requires that you master chemical terminology and be able to perform calculations with ease. Toward these ends, many of the examples and problems are formulated to alert you to questions that sound different but are actually the same (see Problem 3.11, for example) or questions that are different but sound very similar (see Problems 4.27 and 5.11, for example). You should not attempt to memorize the solutions to the problems. (There is enough to memorize, without that.) Instead, you must try to understand the concepts involved. You must practice by working many problems, because in addition to the principles, you must get accustomed to the many details involved in solving problems correctly. The key to success in chemistry is working very many problems!

To get the most from this book, use a 5×8 card to cover up the solutions while you are doing the problems. Do not look at the answer first. It is easy to convince yourself that you know how to do a problem by looking at the answer, but generating the answer yourself, as you must do on exams, is not the same. After you have finished, compare your result with the answer given. If the method differs, it does not mean that your method is necessarily incorrect. If your answer is the same, your method is probably correct. Otherwise, try to understand what the difference is, and where you made a mistake, if you did so.

Some of the problems given after the text are very short and/or very easy (see Problems 5.10 and 5.14, for example). They are designed to emphasize a particular point. After you get the correct answer, ask yourself why such a question was asked. Many other problems give analogies to everyday life, to help you understand a chemical principle (see Problems 2.1 with 2.2, 4.14 through 4.17, 5.12 with 5.13, 17.7, and 18.9, for example). The question of how many socks there are in 3 dozen pairs of socks probably will never appear on one of your chemistry exams, but the related question of how many hydrogen atoms there are in 3 moles of hydrogen molecules is apt to be there, and the principle behind this question is almost certain to appear many times throughout the semester.

Make sure you understand the chemical meaning of the terms presented throughout the semester. For example, "significant figures" means something very different in chemical calculations than in economic discussions. Special terms used for the first time in this book will be *italicized*. Whenever you encounter such a term, use it repeatedly until you thoroughly understand its meaning. If necessary, use the Glossary to find the meanings of unfamiliar terms.

Always use the proper units with measurable quantities. It makes quite a bit of difference if your pet is 4 inches tall or 4 feet tall! After Chapter 2, always use the proper number of significant figures in your calculations. Do yourself a favor and use the same symbols and abbreviations for chemical quantities that are used in the text or by your instructor. If you use a different symbol, you might become confused later when that symbol is used for a different quantity.

Some of the problems are stated in parts. After you do the problem by solving the various parts, see if you would know how to solve the same problem if only the last part were asked. Figures such as Figs. 4-5, 8-6, and 11-10 might be of help in figuring out which step of a seemingly complex problem should be done first, and how to proceed from there.

TO THE STUDENT

The order of topics varies in the different courses, and indeed in the same courses given in different institutions. Use the Table of Contents and/or the Index to find the material you wish to study. Not all the courses cover all the material presented in this book. If you see a problem that is totally unfamiliar to you, ask your instructor if that material is to be covered in your course.

DAVID E. GOLDBERG

Preface

This book is designed to help students who have little or no chemistry background do well in their first chemistry course. It can be used effectively in a course preparatory to a general college chemistry course as well as in a course in chemistry for liberal arts students. It should also provide additional assistance to students in the first semester of a chemistry course for nurses and others in the allied health fields. It will even prove to be of value in a high school chemistry course and in a general chemistry course for majors.

The book aims to help the student develop both problem-solving skills and skill in precise reading and interpreting scientific problems and questions. Analogies to everyday life introduce certain types of problems to make the underlying principles less abstract. Many of the problems were devised to clarify particular points often confused by beginning students. To ensure mastery, the book often presents problems in parts, then asks the same question as an entity, to see if the student can do the parts without the aid of the fragmented question. It provides some figures which have proved helpful to a generation of students.

The author gratefully acknowledges the fine critical review of Professor Larry W. Houk and the help of the editors at McGraw-Hill.

David E. Goldberg

Contents

CONTENTS

CONTENTS

Chapter 1

Basic Concepts

1.1 INTRODUCTION

Chemistry is the study of matter and energy and the interaction between them. In this chapter, we learn about the elements, which are the building blocks of all types of matter in the universe, the measurement of matter (and energy) as mass, the properties by which the types of matter can be identified, and a basic classification of matter. The symbols used to represent the elements are also presented, and an arrangement of the elements into classes having similar properties, called a *periodic table*, is introduced. The periodic table is invaluable to the chemist for many types of classification and understanding.

Scientists have gathered so much data that they must have some way of organizing information in a useful form. Toward that end, scientific laws, hypotheses, and theories are used. These forms of generalization are introduced in Sec. 1.7.

1.2 THE ELEMENTS

An *element* is a substance that cannot be broken down into simpler substances by ordinary means. A few more than 100 elements and many combinations of these elements—compounds or mixtures—account for all the materials of the world. Exploration of the moon has provided direct evidence that the earth's satellite is not composed of any different elements than those on earth. Indirect evidence, in the form of light received from the sun and stars, confirms the fact that the same elements make up the entire universe. Helium, from the Greek *helios*, meaning "sun," was discovered in the sun by the characteristic light it emits, before it was discovered on earth.

It is apparent from the wide variety of different materials in the world that there are a great many ways to combine elements. Changing one combination of elements to another is the chief interest of the chemist. It has long been of interest to know the composition of the crust of the earth, the oceans, and the atmosphere, since these are the only sources of raw materials for all the products that humans require. More recently, however, attention has focused on the problem of what to do with the products humans have used and no longer desire. Although elements can change combinations, they cannot be created or destroyed. The iron in a piece of scrap steel might rust, and be changed in form and appearance, but the quantity of iron has not changed. Since there is a limited supply of available iron and since there is a limited capacity to dump unwanted wastes, recycling such materials is very important.

The elements occur in widely varying quantities on earth. The 10 most abundant elements make up 98% of the mass of the crust of the earth. Many elements occur only in traces, and a few are synthetic. Fortunately for humanity, the elements are not distributed uniformly throughout the earth. The distinct properties of the different elements cause them to be concentrated more or less, making them more available as raw materials. For example, sodium and chlorine form salt, which is concentrated in beds by being dissolved in bodies of water which later dry up. Other natural processes are responsible for the distribution of the elements which now exist on earth. It is interesting to note that the different conditions on the moon—for example, the lack of water and air on the surface—might well cause a different sort of distribution of the elements on the earth's satellite.

1

1.3   MATTER AND ENERGY

Chemistry is the study of matter, including its composition, its properties, its structure, the changes which it undergoes, and the laws governing those changes. *Matter* is anything that has mass and occupies space. Any material object, no matter how large or small, is composed of matter. In contrast, light, heat, and sound are forms of energy. *Energy* is the ability to produce change. Whenever a change of any kind occurs, energy is involved, and whenever any form of energy is changed to another form, it is evidence that a change of some kind is occurring or has occurred.

The concept of mass is central to the discussion of matter and energy. The *mass* of an object depends on the quantity of matter in the object. The more mass the object has, the harder it is to set into motion and the harder it is to change its velocity once it is in motion.

Matter and energy are now known to be interconvertible. The quantity of energy producible from a quantity of matter or vice versa is given by Einstein's famous equation

$$E = mc^2$$

where E is the energy, m is the mass of the matter which is converted into energy, and c^2 is a constant—the square of the velocity of light. The constant c^2 is so large,

$$900\,000\,000\,000\,000\,000\,000\ \text{cm}^2/\text{s}^2 \text{ or } 34\,600\,000\,000\ \text{miles}^2/\text{s}^2$$

that tremendous quantities of energy are associated with conversions of minute quantities of matter into energy. The quantity of mass accounted for by the energy contained in a material object is so small that it is not measurable. Hence, the mass of an object is very nearly identical to the quantity of matter in the object. Particles of energy have very small masses despite having no matter whatsoever; that is, all the mass of a particle of light is associated with its energy. Even for the most energetic of light particles, the mass is small. The quantity of mass in any material body corresponding to the energy of the body is so small that it was not even conceived of until Einstein published his theory of relativity in 1905. Thereafter, it had only theoretical significance until World War II, when it was discovered how radioactive processes could be used to transform very small quantities of matter into very large quantities of energy, from which resulted the atomic and hydrogen bombs. Peaceful uses of atomic energy have developed since that time, including the production of a significant fraction of the electric power in many cities.

The mass of an object is directly associated with its weight. The *weight* of a body is the pull on the body by the nearest celestial body. On earth, the weight of a body is the pull of the earth on the body, but on the moon, the weight corresponds to the pull of the moon on the body. The weight of a body is directly proportional to its mass and also depends on the distance of the body from the center of the earth or moon or whatever celestial body the object is near. In contrast, the mass of an object is independent of its position. At any given location, for example on the surface of the earth, the weight of an object is directly proportional to its mass.

When astronauts walk on the moon, they must take care to adjust to the lower gravity on the moon. Their masses are the same no matter where they are, but their weights are about one-sixth as much on the moon as on earth because the moon is so much lighter than the earth. A given push, which would cause an astronaut to jump 1 foot high on earth, would cause him to jump 6 feet on the moon. Since weight and mass are directly proportional on the surface of the earth, chemists have often used the terms interchangeably. The custom formerly was to use the term *weight*, but modern practice tends to use the term *mass* to describe quantities of matter. In this text, the term *mass* will be used, but other chemistry texts might use the term *weight*, and the student must be aware that many instructors still prefer the latter.

The study of chemistry is concerned with the changes that matter undergoes, and therefore chemistry is also concerned with energy. Energy occurs in many forms—heat, light, sound, chemical energy, mechanical energy, electrical energy, and nuclear energy. In general, it is possible to convert each of these forms of energy into others. Except for reactions in which the quantity of matter is changed, as in nuclear reactions, the law of conservation of energy is obeyed:

Energy can be neither created nor destroyed
(in the absence of nuclear reactions).

In fact, many chemical reactions are carried out for the sole purpose of producing energy in a desired form. For example, in the burning of fuels in homes, chemical energy is converted into heat; in the burning of fuels in automobiles, chemical energy is converted into mechanical energy of motion. Reactions occurring in batteries produce electrical energy from the chemical energy stored in the chemicals from which the battery is constructed.

1.4 PROPERTIES

Every material has certain characteristics which distinguish it from other materials and which may be used to establish that two specimens of the same material are indeed the same. Those characteristics that serve to distinguish and identify a specimen of matter are called the *properties* of the specimen. For example, water may be distinguished easily from iron or aluminum, and—although this may appear to be more difficult—iron may readily be distinguished from aluminum by means of the different properties of the metals.

EXAMPLE 1.1. Suggest three ways in which a piece of iron can be distinguished from a piece of aluminum.
1. Iron, but not aluminum, will be attracted by a magnet.
2. If a piece of iron is left in humid air, it will rust. Under the same conditions, aluminum will undergo no appreciable change.
3. If a piece of iron and a piece of aluminum have exactly the same volume, the iron will weigh more than the aluminum.

Physical Properties

The properties related to the *state* (gas, liquid, or solid) or appearance of a material are called *physical properties*. Some commonly known physical properties are listed in Table 1-1. The physical properties of a material can usually be determined without changing the composition of the material. Many physical properties can be measured and described in numerical terms, and comparison of such properties is often the best way to distinguish one material from another.

Table 1-1 Typical Physical and Chemical Properties

Physical Properties	Chemical Properties
Density	Flammability
State at room temperature	Rust resistance
Color	Reactivity
Hardness	Biodegradability
Melting point	
Boiling point	

Chemical Properties

A chemical reaction is a change in which at least one material changes its composition and its set of properties. The characteristic ways in which a specimen of matter undergoes chemical reaction or fails to undergo chemical reaction are called its *chemical properties*. Many examples of chemical properties will be presented in this book. Of the properties of iron listed above, only rusting is a chemical property. Rusting involves a change in composition (from pure iron to iron oxide). The other properties listed do not involve any change in composition of the iron; they are physical changes.

1.5 CLASSIFICATION OF MATTER

In order to study the vast variety of materials that exist in the universe, it is necessary that the study be made in a systematic manner. Therefore, matter is classified according to several different schemes. Matter may be classified as organic or inorganic. It is *organic* if it is a compound of carbon and hydrogen. (A more rigorous definition of organic must wait until Chap. 21.) Otherwise, it is *inorganic*. Another such scheme uses the composition of matter as a basis for classification; other schemes are based on chemical properties of the various classes. For example, substances may be classified as acids, bases, or salts. Each scheme is useful, allowing the study of a vast variety of materials in terms of a given class.

In the method of classification of matter based on composition, a given specimen of material is regarded as either a pure substance or a mixture. An outline of this classification scheme is shown in Table 1-2. The term *pure substance* (or merely substance) refers to a material all parts of which have the same composition and which has a definite and unique set of properties. In contrast, a *mixture* consists of two or more substances and has a somewhat arbitrary composition. The properties of a mixture are not unique, but depend on its composition. The properties of a mixture tend to reflect the properties of the substances of which it is composed; that is, if the composition is changed a little, the properties will change a little.

Table 1-2   Classification of Matter Based on Composition

Substances
Elements
Compounds
Mixtures
Homogeneous mixtures (solutions)
Heterogeneous mixtures (mixtures)

Substances

There are two kinds of substances—elements and compounds. *Elements* are substances that cannot be broken down into simpler substances by ordinary chemical means. Elements cannot be made by the combination of simpler substances. There are slightly more than 100 elements, and every material object in the universe consists of one or more of these elements. Familiar substances which are elements include carbon, aluminum, iron, copper, gold, oxygen, and hydrogen.

Compounds are substances consisting of two or more elements combined in definite proportions by mass to give a material having a definite set of properties different from those of any of its constituent elements. For example, the compound water consists of 88.8% oxygen and 11.2% hydrogen by mass. The physical and chemical properties of water are distinctly different from those of both hydrogen and oxygen. For example, water is a liquid at room temperature and pressure, while the elements of which it is composed are gases under these same conditions. Chemically, water does not burn; hydrogen may burn explosively in oxygen (or air). Any sample of pure water, regardless of its source, has the same composition and the same properties.

There are millions of known compounds, and thousands of new ones are discovered or synthesized each year. Despite such a vast number of compounds, it is possible for the chemist to know certain properties of each one, because compounds can be classified according to their composition and structure, and groups of compounds in each class have some properties in common. For example, organic compounds are generally combustible in oxygen, yielding carbon dioxide and water. So any compound that contains carbon and hydrogen may be predicted by the chemist to be combustible in oxygen.

organic compound + oxygen $\longrightarrow$ carbon dioxide + water + possible other products

Mixtures

There are two kinds of mixtures—homogeneous mixtures and heterogeneous mixtures. *Homogeneous mixtures* are also called *solutions*, and *heterogeneous mixtures* are sometimes called simply mixtures. In heterogeneous mixtures, it is possible to see differences in the sample, merely by looking, although a microscope might be required. In contrast, homogeneous mixtures look the same throughout the sample, even under the best optical microscope.

EXAMPLE 1.2. A teaspoon of salt is added to a cup of warm water. White crystals are seen at the bottom of the cup. Is the mixture homogeneous or heterogeneous? Then the mixture is stirred until the salt crystals disappear. Is the mixture now homogeneous or heterogeneous?

Before stirring, the mixture is heterogeneous; after stirring, the mixture is a solution.

Distinguishing a Mixture from a Compound

Let us imagine an experiment to distinguish a mixture from a compound. Powdered sulfur is yellow in color and it dissolves in carbon disulfide, but it is not attracted by a magnet. Iron filings are black and are attracted by a magnet, but do not dissolve in carbon disulfide. You can mix iron filings and powdered sulfur in any ratio, and get a yellowish-black mixture—the more sulfur, the more yellow the mixture. If you put the mixture in a test tube and hold a magnet alongside the test tube just above the mixture, the iron filings will be attracted, but the sulfur will not. If you pour enough (colorless) carbon disulfide on the mixture, stir, then pour off the resulting yellow liquid, the sulfur has dissolved but the iron has not. The mixture of iron filings and powdered sulfur is described as a mixture because the properties of the combination are still the properties of its components.

If you mix sulfur and iron filings in a certain proportion and then heat the mixture, you can see a red glow spread through the mixture. After it cools, the black solid lump which has been produced, even if crushed into a powder, does not dissolve in carbon disulfide and is not attracted by a magnet. The material has a new set of properties; it is a compound, called iron(II) sulfide. It has a definite composition, and if, for example, you had mixed more iron with the sulfur originally, some iron(II) sulfide and some leftover iron would have resulted. The extra iron would not have become part of the compound.

1.6 REPRESENTATION OF ELEMENTS

Each element has an internationally accepted *symbol* to represent it. A list of the names and symbols of the elements is found on pages 369–371 of this book. Note that symbols for the elements are for the most part merely abbreviations of their names, consisting of either one or two letters. The first letter of the symbol is always written as a capital letter; the second letter, if any, is always written as a lowercase (small) letter. The symbols of a few elements do not suggest their English names, but are derived from the Latin or German names of the elements. The 10 elements whose names do not begin with the same letter as their symbols are listed in Table 1-3. For convenience, on pages 369–371 of this book, these elements are listed twice—once alphabetically by name and again under

Table 1-3   Symbols and Names with Different Initials

Symbol	Name	Symbol	Name
Ag	Silver	Na	Sodium
Au	Gold	Pb	Lead
Fe	Iron	Sb	Antimony
Hg	Mercury	Sn	Tin
K	Potassium	W	Tungsten

the letter which is the first letter of their symbol. It is important to memorize the names and symbols of the most common elements. To facilitate this task, the most familiar elements are listed in Table 1-4.

Table 1-4   Important Elements Whose Names and Symbols Should Be Known

1 H																	2 He
3 Li	4 Be											5 B	6 C	7 N	8 O	9 F	10 Ne
11 Na	12 Mg											13 Al	14 Si	15 P	16 S	17 Cl	18 Ar
19 K	20 Ca	21 Sc	22 Ti	23 V	24 Cr	25 Mn	26 Fe	27 Co	28 Ni	29 Cu	30 Zn	31 Ga	32 Ge	33 As	34 Se	35 Br	36 Kr
37 Rb	38 Sr								46 Pd	47 Ag	48 Cd		50 Sn	51 Sb	52 Te	53 I	54 Xe
55 Cs	56 Ba				74 W				78 Pt	79 Au	80 Hg		82 Pb	83 Bi			86 Rn

	92 U																

The Periodic Table

A convenient way of displaying the elements is in the form of a periodic table, such as is shown on page 372 of this book. The basis for the arrangement of the elements in the periodic table will be discussed at length in Chap. 3. For the present, the periodic table will be regarded as a convenient source of general information about the elements. It will be used repeatedly throughout the book. One example of its use, shown in Fig. 1-1, is to classify the elements as metals or nonmetals. All the

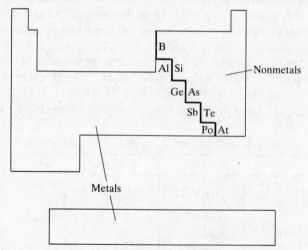

Fig. 1-1   Metals and nonmetals

elements that lie to the left of the stepped line drawn on the periodic table starting to the left of B and descending stepwise to a point between Po and At are metals. The other elements are nonmetals. It is readily seen that the majority of the elements are metals.

The smallest particle of an element that retains the composition of the element is called an *atom*. Details of the nature of atoms are given in Chaps. 3 and 17. The symbol of an element is used to stand for one atom of the element as well as for the element itself.

1.7   LAWS, HYPOTHESES, AND THEORIES

In chemistry, as in all sciences, it is necessary to express ideas in terms having very precise meanings. These meanings are often unlike the meanings of the same words in nonscientific usage. For example, the meaning of the word *property* as used in chemistry is quite different from its meaning in ordinary conversation. Also, in chemical terminology, a concept may be represented by abbreviations, such as symbols or formulas, or by some mathematical expression. Together with the precisely defined terms, such symbols and mathematical expressions constitute a language of chemistry. This language must be learned well. As an aid to recognition of special terms, when such terms are used for the first time in this book, they will be *italicized*.

A statement that generalizes a quantity of experimentally observable phenomena is called a scientific *law*. For example, if a person drops a pencil, it falls downward. This result is predicted by the law of gravity. A generalization that attempts to explain why certain experimental results occur is called a *hypothesis*. When a hypothesis is accepted as true by the scientific community, it is then called a *theory*. One of the most important scientific laws is the law of conservation of mass: During any process (chemical reaction, physical change, or even a nuclear reaction) mass is neither created nor destroyed. Because of the close approximation that the mass of an object is the quantity of matter it contains (excluding the mass corresponding to its energy), the law of conservation of mass can be approximated by the law of conservation of matter: During an ordinary chemical reaction, matter can be neither created nor destroyed.

EXAMPLE 1.3.   When a piece of iron is left in moist air, its surface gradually turns brown and the object gains mass. Explain this phenomenon.

The brown material is iron oxide, rust, formed by a reaction of the iron with the oxygen in the air.

$$\text{iron} + \text{oxygen} \longrightarrow \text{iron oxide}$$

The increase in mass is just the mass of the combined oxygen. When a log burns, the ash which remains is much lighter than the original log, but this is not a contradiction of the law of conservation of matter. In addition to the log, which consists mostly of compounds containing carbon, hydrogen, and oxygen, oxygen from the air is consumed by the reaction. In addition to the ash, carbon dioxide and water vapor are produced by the reaction.

$$\text{log} + \text{oxygen} \longrightarrow \text{ash} + \text{carbon dioxide} + \text{water vapor}$$

The total mass of the ash plus the carbon dioxide plus the water vapor is equal to the total mass of the log plus the oxygen. As always, the law of conservation of matter is obeyed as precisely as chemists can measure. The law of conservation of mass is fundamental to the understanding of chemical reactions. Other laws related to the behavior of matter are equally important, and learning how to apply these laws correctly is a necessary goal of the study of chemistry.

Solved Problems

1.1.   Are elements heterogeneous or homogeneous?

   Ans.   Homogeneous. They look alike throughout the sample because they are alike throughout the sample.

1.2. Are compounds heterogeneous or homogeneous?

Ans. Homogeneous. They look alike throughout the sample because they are alike throughout the sample. Since there is only one substance present, even if it is a combination of elements, it must be alike throughout.

1.3. A generality states that all compounds containing both carbon and hydrogen burn. Do octane and propane burn? Each contains only carbon and hydrogen.

Ans. Yes, both burn. It is easier to learn that all organic compounds burn than to learn a list of hundreds of organic compounds that burn. However, on an examination a question will probably specify one particular organic compound. You must learn a generality and reply with a specific answer.

1.4. Sodium is a very reactive metallic element; for example, it liberates hydrogen gas when treated with water. Chlorine is a yellow-green, choking gas, used in World War I as a poison gas. Contrast these properties with those of the compound of sodium and chlorine—sodium chloride—known as table salt.

Ans. Salt does not react with water to liberate hydrogen, is not reactive, and is not poisonous. It is a white solid and not a silvery metal nor a green gas. In short, it has its own set of properties; it is a compound.

1.5. TNT is a compound of carbon, nitrogen, hydrogen, and oxygen. Carbon occurs in two forms—graphite (the material in "lead pencils") and diamond. Oxygen and nitrogen comprise over 98% of the atmosphere. Hydrogen is an element which reacts explosively with oxygen. Which of the properties of the elements determines the properties of TNT?

Ans. The properties of the elements do not matter. The properties of the compound are quite independent of those of the elements. A compound has its own distinctive set of properties. TNT is most noted for its explosiveness.

1.6. What properties of stainless steel make it more desirable for many purposes than ordinary steel?

Ans. Its resistance to rusting and corrosion.

1.7. What properties of DDT make it useful? What properties make it undesirable?

Ans. DDT's toxicity to insects is its useful property; its toxicity to humans, birds, and other animals makes it undesirable. It is stable, that is, nonbiodegradable (does not decompose spontaneously to simpler substances in the environment). This property makes its use for insects alone more difficult.

1.8. Name an object or an instrument that changes
(*a*) electrical energy to light
(*b*) motion to electrical energy
(*c*) electrical energy to motion
(*d*) chemical energy to heat
(*e*) chemical energy to electrical energy
(*f*) electrical energy to chemical energy

Ans. (*a*) light bulb
(*b*) generator or alternator
(*c*) electric motor
(*d*) gas stove
(*e*) battery
(*f*) rechargeable battery

1.9.   Distinguish clearly between (*a*) mass and matter and (*b*) mass and weight.

   Ans.   (*a*) *Matter* is any kind of material. The mass of an object depends mainly on the matter which it contains. It is affected only very slightly by the energy in it. (*b*) *Weight* is the attraction of the earth on an object. It depends on the mass of the object and its distance to the center of the earth.

1.10.   Distinguish between a theory and a law.

   Ans.   A law tells what happens under a given set of circumstances, while a theory attempts to explain why that behavior occurs.

1.11.   How can you tell if the word "mixture" means any mixture or a heterogeneous mixture?

   Ans.   You can tell from the context. For example, if a problem asks if a sample is a solution or a mixture, the word mixture means heterogeneous mixture. If it asks whether the sample is a compound or a mixture, it means any kind of mixture. Such usage occurs in ordinary English as well as in technical usage. For example, the word *day* has two meanings—one a subdivision of the other. "How many hours are there in a day? What is the opposite of night?"

1.12.   A sample contains 88.8% oxygen and 11.2% hydrogen by mass, and is gaseous and explosive at room temperature. (*a*) Is the sample a compound or a mixture? (*b*) After the sample explodes and cools, it is a liquid. Is the sample now a compound or a mixture? (*c*) Would it be easier to change the percentage of oxygen before or after the explosion?

   Ans.   (*a*) The sample is a mixture. (This compound of hydrogen and oxygen at room temperature—water—is a liquid under these conditions.) (*b*) It is a compound. (*c*) Before the explosion. It is probably easier to add hydrogen or oxygen to the gaseous mixture than to liquid water.

Chapter 2

Mathematical Methods in Chemistry

2.1 INTRODUCTION

Physical sciences, and chemistry in particular, are quantitative in nature. Not only must chemists describe things qualitatively, but they must also measure them quantitatively and compute numeric results from the measurements. The metric system (Sec. 2.2) is a system of units designed to make calculation of measured quantities as easy as possible. Exponential notation (Sec. 2.3) is designed to enable scientists to work with numbers that range from incredibly huge to unbelievably tiny. The factor-label method is introduced in Sec. 2.4 to aid students in deciding how to do certain calculations. The scientist must report the results of the measurements so that any reader will have an appreciation of how accurately the measurements were made. This reporting is done by using the proper number of significant figures (Sec. 2.5). Density calculations are introduced in Sec. 2.6 to enable the student to use all the techniques described thus far.

2.2 METRIC SYSTEM

Scientists measure many different quantities—length, volume, mass (weight), electric current, temperature, pressure, force, magnetic field intensity, radioactivity, and many others. The *metric system* and its recent extension, *Système International d'Unités* (SI), were devised to make measurements and calculations as simple as possible. In this chapter, length, area, volume, and mass will be introduced. Temperature will be introduced in Sec. 2.7 and used extensively in Chap. 11. The quantities to be discussed here are presented in Table 2-1. Their units, abbreviations of the quantities and units, and the legal standards for the quantities are also included.

Table 2-1 Metric Units for Basic Quantities

Quantity	Abbreviation	Basic Unit	Abbreviation	Standard	Comment
Length or distance	l d	meter	m	meter	
Area	A	meter^2	m^2	meter^2	
Volume	V	meter^3 or liter	m^3 L	meter^3	SI unit older metric unit $1\,m^3 = 1\,000\,L$
Mass	m	gram	g	kilogram	$1\,kg = 1\,000\,g$

Length (Distance)

The unit of length, or distance, is the *meter*. Originally conceived of as one ten-millionth of the distance from the north pole to the equator through Paris, the meter is more accurately defined as the distance between two scratches on a platinum-iridium bar kept in Paris. The U.S. standard is the distance between two scratches on a similar bar kept at the National Bureau of Standards in Washington, D.C. The meter is about 10% greater than a yard—39.37 inches to be more precise.

Larger and smaller distances may be measured with units formed by the addition of prefixes to the word meter. The important metric prefixes are listed in Table 2-2. The most important prefixes are *kilo*, *milli*, and *centi*. The prefix *kilo* means 1 000 times the basic unit, no matter to which basic unit it is attached. For example, a kilodollar is $1 000. The prefix *milli* indicates one-thousandth of the basic unit. Thus, 1 millimeter is 0.001 meter; 1 mm = 0.001 m. The prefix *centi* means one-hundredth. A centidollar is one cent; the name for this unit of money comes from the same source as the metric prefix.

Table 2-2   Metric Prefixes

Prefix	Abbreviation	Meaning	Example
mega	M	1 000 000	1 Mm = 1 000 000 m
kilo	**k**	**1 000**	**1 km = 1 000 m**
deci	d	0.1	1 dm = 0.1 m
centi	**c**	**0.01**	**1 cm = 0.01 m**
milli	**m**	**0.001**	**1 mm = 0.001 m**
micro	μ	0.000001	1 μm = 0.000001 m
nano	n	1×10^{-9}	1 nm = 1×10^{-9} m

The metric system was designed to make calculations easier than using the English system in the following ways:

Metric	English
Subdivisions of all dimensions have the same prefixes with the same meanings and the same abbreviations.	Different names for subdivisions.
Subdivisions all differ by powers of 10.	Subdivisions differ by arbitrary factors, rarely powers of 10.
There are no duplicate names with different meanings.	There are duplicate names with different meanings.
The abbreviations are easily recognizable.	The abbreviations are often hard to recognize (e.g., lb for pound and oz for ounce).

Beginning students often regard the metric system as difficult because it is new to them and because they think they must learn all the English-metric conversion factors (Table 2-3). Engineers do have to work in both systems in the United States, but scientists generally do not work in the English system at all. Once you familiarize yourself with the metric system, it is much easier to work with than the English system is.

Table 2-3   Some English-Metric Conversions

	Metric	English
Length	1 meter	39.37 inches
	2.54 cm	1 inch
Volume	1 liter	1.06 U.S. quarts
Mass	1 kilogram	2.2 pounds (avoirdupois)
	28.35 g	1 ounce

EXAMPLE 2.1. (*a*) How many feet are there in 7.50 miles? (*b*) How many meters are there in 7.50 km?

(*a*)
$$7.50 \text{ miles}\left(\frac{5\,280 \text{ feet}}{1 \text{ mile}}\right) = 39\,600 \text{ feet}$$

(*b*)
$$7.50 \text{ km}\left(\frac{1\,000 \text{ m}}{1 \text{ km}}\right) = 7\,500 \text{ m}$$

You can do the calculation of part (*b*) in your head (merely moving the decimal point in 7.50 three places to the right). The calculation of part (*a*) requires a calculator or pencil and paper.

Instructors often require English-metric conversions for two purposes: to familiarize the student with the relative sizes of the metric units in terms of the more familiar English units, and for practice in conversions (see Sec. 2.4). Once you really get into the general chemistry course, the number of English-metric conversions that you do is very small.

One of the main advantages of the metric system is that the same prefixes are used with all quantities, and they always have the same meanings.

EXAMPLE 2.2. The unit of electric power is the watt. What is the meaning of 1 kilowatt?

$$1 \text{ kilowatt} = 1\,000 \text{ watts} \qquad 1 \text{ kW} = 1\,000 \text{ W}$$

Even if you do not recognize the quantity, the prefix always has the same meaning.

EXAMPLE 2.3. How many centimeters are there in 5 m?

Each meter is 100 cm; 5 m is 500 cm.

EXAMPLE 2.4. Since meter is abbreviated m (Table 2-1) and milli is abbreviated m (Table 2-2), how can you tell the difference?

Since milli is a prefix, it must always precede a quantity. If m is used without another letter, or if the m follows another letter, it stands for the unit meter. If m precedes another letter, it stands for the prefix milli.

Area

The extent of a surface is called its *area*. The area of a rectangle (or a square, which is a rectangle with all sides equal) is its length times its width.

$$A = l \times w$$

The dimensions of area are thus the product of the dimensions of two distances. The area of a state or country is usually reported in square miles or square kilometers, for example. If you buy interior paint, you can expect a gallon of paint to cover about 400 ft^2. The units are stated aloud "square feet," but are usually written ft^2. The exponent (the superscript number) means that the unit is multiplied that number of times, just as it does with a number. For example, ft^2 means $\text{ft} \times \text{ft}$.

EXAMPLE 2.5. State aloud the area of Washington, D.C., 87 mile^2.

"87 square miles"

EXAMPLE 2.6. A square is 5.0 m on each side. What is its area?

$$(5.0 \text{ m})^2 = 25 \text{ m}^2$$

Note the difference between "5 meters, squared" and "5 square meters."

$$(5 \text{ m})^2 \qquad \text{and} \qquad 5 \text{ m}^2$$

The former means that the coefficient (5) is also squared; the latter does not.

EXAMPLE 2.7.   A rectangle having an area of $20.0\,m^2$ is $5.00\,m$ long. How wide is it?

$$A = l \times w \qquad w = A/l = 20.0\,m^2/5.00\,m = 4.00\,m$$

Note that the width has a unit of distance (meter).

EXAMPLE 2.8.   What happens to the area of a square when the length of each side is doubled?

Let l = original length of side; then l^2 = original area and $2l$ = new length of side; so $(2l)^2$ = new area. The area has increased from l^2 to $4l^2$; it has increased by a factor of 4. (See Problem 2.14.)

Volume

The SI unit of volume is the cubic meter, m^3. Just as area is derived from length, so can volume be derived from length. Volume is length $\times$ length $\times$ length. Volume also can be regarded as area $\times$ length. The cubic meter is a rather large unit; a cement truck usually can carry between 2 and $3\,m^3$ of cement. Smaller units are dm^3, cm^3, and mm^3; the first two of these are reasonable sizes to be useful in the laboratory.

The older version of the metric system uses the *liter* as the basic unit of volume. It is defined as $1\,dm^3$. Chemists often use this unit in preference to the m^3 because it is about the magnitude of the quantities with which they deal. The student has to know both units, and the relationship between them.

EXAMPLE 2.9.   How many liters are there in $1\,m^3$?

$$1\,m^3 \left(\frac{10\,dm}{1\,m}\right)^3 \left(\frac{1\,L}{1\,dm^3}\right) = 1\,000\,L$$

One cubic meter is $1\,000\,L$.

The liter can have prefixes just as any other unit can. Thus $1\,mL$ is $0.001\,L$, and $1\,kL$ is $1\,000\,L$.

Mass

Mass is a measure of the quantity of material in a sample. We can measure that mass by its inertia —the resistance to change in its motion—or by its weight—the attraction of the sample to the earth. Since weight and mass are directly proportional as long as we stay on the surface of the earth, chemists often use the terms interchangeably. (Physicists do not do that.)

The unit of mass is the *gram*. (Since $1\,g$ is a very small mass, the legal standard of mass in the U.S. is the kilogram.)

Get used to writing the standard abbreviations for units and the proper symbols at the beginning of your study of chemistry, and you will not get mixed up later.

EXAMPLE 2.10.   What is the difference between mg and Mg, two units of mass?

Lowercase m stands for *milli* (Table 2-2), and $1\,mg$ is $0.001\,g$. Capital M stands for *mega*, and $1\,Mg$ is $1\,000\,000\,g$. It is obviously important not to confuse the capital and lowercase m in such cases.

2.3   EXPONENTIAL NUMBERS

The numbers that scientists use range from enormous to extremely tiny. The distances between the stars are literally astronomical—the star nearest to the sun is $23\,500\,000\,000\,000$ miles from it. As another example, the number of atoms of chlorine in $35.5\,g$ of chlorine is $602\,000\,000\,000\,000\,000\,000\,000$, or 602 thousand billion billion. The diameter of one chlorine atom is about $0.000\,000\,01$ cm. To report and work with such large and small numbers, scientists use *exponential notation*. A typical number written in exponential notation looks as follows:

$$\underbrace{2.53}_{\text{Coefficient}} \times \underbrace{\underbrace{10}_{\text{Base}}{}^{\overbrace{4}^{\text{Exponent}}}}_{\text{Exponential part}}$$

The coefficient is merely a decimal number written in the ordinary way. That coefficient is multiplied by the exponential part made up of the base (10) and the exponent. (Ten is the only base which will be used with numbers in exponential form in the general chemistry course.) The exponent tells how many times the base is multiplied by the coefficient:

$$2.53 \times 10^4 = 2.53 \times 10 \times 10 \times 10 \times 10 = 25\,300$$

Since the exponent is 4, four 10s are multiplied by the coefficient.

EXAMPLE 2.11. What is the value of 10^5?

When an exponential is written without an explicit coefficient, a coefficient of 1 is implied:

$$1 \times 10 \times 10 \times 10 \times 10 \times 10 = 100\,000$$

There are five 10s multiplying the implied 1.

EXAMPLE 2.12. What is the value of 10^0? of 10^1?

$10^0 = 1$. There are no 10s multiplying the implied coefficient of 1. $10^1 = 10$. There is one 10 multiplying the implied coefficient of 1.

EXAMPLE 2.13. Write 2.0×10^3 in decimal form.

$$2.0 \times 10 \times 10 \times 10 = 2\,000$$

When scientists write numbers in exponential form, they prefer to write them so that the coefficient has one and only one digit to the left of the decimal point. That notation is called *standard exponential form* or *scientific notation*.

EXAMPLE 2.14. Write $665\,000$ in standard exponential form.

$$665\,000 = 6.65 \times 10 \times 10 \times 10 \times 10 \times 10 = 6.65 \times 10^5$$

The number of 10s is the number of places in $665\,000$ that the decimal point must be moved to the left to get one (nonzero) digit to the left of the decimal point.

Note that the same number $665\,000$ could have been written in exponential form in the following ways:

$$0.665 \times 10 \times 10 \times 10 \times 10 \times 10 \times 10 = 0.665 \times 10^6$$

$$66.5 \times 10 \times 10 \times 10 \times 10 = 66.5 \times 10^4$$

$$665 \times 10 \times 10 \times 10 = 665 \times 10^3$$

$$6\,650 \times 10 \times 10 = 6\,650 \times 10^2$$

$$66\,500 \times 10 = 66\,500 \times 10^1$$

$$665\,000 \times 1 = 665\,000 \times 10^0$$

None of these is in *standard* exponential form, because there are more or fewer than one digit to the left of the decimal point. (The 0 in 0.665 is not counted because it could just as well be omitted.) However, each of them has the same value as 6.65×10^5.

Multiplication of Exponential Numbers

When two numbers in exponential form are multiplied, the coefficients are multiplied and the exponential parts are multiplied separately. The answer is the product of these two answers:

$$(1 \times 10^2) \times (2 \times 10^3) = (1 \times 10 \times 10) \times (2 \times 10 \times 10 \times 10)$$
$$= (1 \times 2) \times (10 \times 10 \times 10 \times 10 \times 10)$$
$$= 2 \times 10^5$$

It is apparent that the number of 10s in the answer—the exponent of the answer—is merely the sum of the exponents of the factors. Hence, the rule for multiplying exponential parts of exponential numbers (having the same base) is to add exponents. This rule saves us from having to write out the 10s each time we multiply.

To multiply exponentials, add the exponents.

EXAMPLE 2.15.   Calculate the product $(2.00 \times 10^6) \times (6.60 \times 10^{13})$.

$$(2.00 \times 10^6) \times (6.60 \times 10^{13}) = (2.00 \times 6.60) \times (10^6 \times 10^{13})$$
$$= 13.2 \times 10^{19} = 1.32 \times 10^{20}$$

Division of Exponential Numbers

To divide exponential numbers, divide the coefficients and divide the exponential parts separately. To divide the exponential parts merely subtract exponents.

$$(6.6 \times 10^5)/(2.0 \times 10^3) = 3.3 \times 10^2$$

We can see that this rule is correct by writing out the dividend and divisor:

$$\frac{6.6 \times 10^5}{2.0 \times 10^3} = \frac{6.6 \times 10 \times 10 \times 10 \times 10 \times 10}{2.0 \times 10 \times 10 \times 10} = 3.3 \times 10 \times 10$$

EXAMPLE 2.16.   Calculate the quotient of $(6.8 \times 10^{102})/(4.0 \times 10^{99})$.

$$(6.8 \times 10^{102})/(4.0 \times 10^{99}) = (6.8/4.0) \times (10^{102-99}) = 1.7 \times 10^3$$

Suppose that we divide a number with a smaller exponent by one with a larger exponent. What will be the result? The answer is obtained by following the same general rule.

EXAMPLE 2.17.   Calculate the quotient of $(6.8 \times 10^3)/(4.0 \times 10^5)$.

$$\frac{6.8 \times 10^3}{4.0 \times 10^5} = \frac{6.8 \times 10 \times 10 \times 10}{4.0 \times 10 \times 10 \times 10 \times 10 \times 10} = 1.7 \times 10^{-2}$$
$$= \frac{1.7}{10 \times 10} = 1.7 \times 10^{-2}$$

The meaning of 10 to a negative exponent is that the coefficient is *divided* by that number of 10s (multiplied together).

You can move the exponential part of a number from numerator to denominator or vice versa merely by changing the sign of the exponent.

$$\frac{1}{10^{-2}} = 10^2 \qquad 10^{-4} = \frac{1}{10^4}$$

EXAMPLE 2.18.   Calculate the value of $(1.60 \times 10^5)/(4.00 \times 10^{-7})$.

$$\frac{1.60 \times 10^5}{4.00 \times 10^{-7}} = \frac{1.60 \times 10^5 \times 10^7}{4.00} = 0.400 \times 10^{12} = 4.00 \times 10^{11}$$

Alternatively,

$$(1.60/4.00) \times (10^5/10^{-7}) = (1.60/4.00) \times (10^{5-(-7)})$$
$$= 0.400 \times 10^{12} = 4.00 \times 10^{11}$$

Addition and Subtraction of Exponential Numbers

When we add or subtract, we always align the decimal points beforehand. In the case of exponential numbers, that means that we add or subtract only numbers having the same exponential parts. The exponent of the answer is the same as the exponent of each of the values added or subtracted.

$$1.5 \times 10^3 + 4.2 \times 10^3 = 5.7 \times 10^3$$

If the exponents of the numbers are not the same, we must adjust them until they are the same before addition or subtraction. Consider

$$1.50 \times 10^3 + 4.2 \times 10^2$$

Since the exponents are not the same, we cannot merely add the coefficients. What we must do is change the coefficient and exponential part of one of the numbers so that the value remains the same but the exponent matches the exponent of the other number:

$$1.50 \times 10^3 = 1.50 \times 10 \times 10 \times 10 = (1.50 \times 10) \times (10 \times 10)$$
$$= 15.0 \times 10^2$$

Then we can add:

$$(15.0 \times 10^2) + (4.2 \times 10^2) = 19.2 \times 10^2 = 1.92 \times 10^3$$

Alternatively, we could have changed the format of the second number:

$$4.2 \times 10^2 = 4.2 \times 10 \times 10 = \frac{4.2 \times 10 \times 10 \times 10}{10} = 0.42 \times 10^3$$

Then addition produces

$$(1.50 \times 10^3) + (0.42 \times 10^3) = 1.92 \times 10^3$$

This answer is the same as the last, but is already in standard exponential form.

EXAMPLE 2.19.   Change 4.00×10^3 to a number with the same value but with an exponent of 2.

$$4.00 \times 10^3 = 40.0 \times 10^2$$

The exponential part of the product is divided by 10; therefore the coefficient is multiplied by 10. Thus, the product is multiplied by $10/10 = 1$, and the value of the product is unchanged. Alternatively, for each place that you move the decimal to the right, subtract 1 from the exponent. For each place that you move the decimal left, add 1 to the exponent. Check yourself to see that you have increased the coefficient and decreased the exponent or vice versa.

Raising an Exponential Number to a Power

To raise an exponential number to a power, raise both the coefficient to the power and the exponential part to the power. To do the latter, multiply the original exponent by the power:

$$(2.0 \times 10^3)^2 = 2.0^2 \times (10^3)^2 = 4.0 \times 10^{2 \times 3} = 4.0 \times 10^6$$

Using an Electronic Calculator

If you use an electronic calculator with exponential capability, note that there is a special key (labeled $\boxed{\text{EE}}$ or $\boxed{\text{EXP}}$) on every such calculator which means "times 10 to the power." If you wish to enter 2×10^3, push the $\boxed{2}$, then the special key, then the $\boxed{3}$. Do NOT push the $\boxed{2}$, then the multiply key, then $\boxed{1}$, then $\boxed{0}$, then the special key, then the $\boxed{3}$. If you do so, your value will be 10 times too large.

EXAMPLE 2.20.　List the keystrokes on an electronic calculator which are necessary to do the following calculation.

$$5.0 \times 10^4 + 2 \times 10^3 =$$

$$\boxed{5}\ \boxed{.}\ \boxed{0}\ \boxed{\text{EXP}}\ \boxed{4}\ \boxed{+}\ \boxed{2}\ \boxed{\text{EXP}}\ \boxed{3}\ \boxed{=}$$

On the electronic calculator, to change the sign of a number you use the $\boxed{+/-}$ key, not the $\boxed{-}$ (minus) key. The $\boxed{+/-}$ key can be used to change the sign of a coefficient of an exponent, depending on when it is pressed. If it is pressed after the $\boxed{\text{EXP}}$ key, it works on the exponent rather than the coefficient.

EXAMPLE 2.21.　List the keystrokes on an electronic calculator which are necessary to do the following calculation.

$$(7.8 \times 10^{-9})/(-5.2 \times 10^{10}) =$$

$$\boxed{7}\ \boxed{.}\ \boxed{8}\ \boxed{\text{EXP}}\ \boxed{9}\ \boxed{+/-}\ \boxed{\div}\ \boxed{5}\ \boxed{.}\ \boxed{2}\ \boxed{+/-}\ \boxed{\text{EXP}}\ \boxed{1}\ \boxed{0}\ \boxed{=}$$

The answer is -1.5×10^{-19}.

2.4　FACTOR-LABEL METHOD

The units of a measurement are an integral part of the measurement. In many ways, they may be treated as algebraic quantities, like x and y in mathematical equations. You must always state the units when making measurements and calculations.

The units are very helpful in suggesting a good approach for solving many problems. For example by considering units in a problem, you can easily decide whether to multiply or divide two quantities to arrive at the answer. The *factor-label method*, also called *dimensional analysis* or the *factor-unit method*, may be used for quantities that are directly proportional to each other. (When one quantity goes up, the other does so in a similar manner. When the number of dimes in a piggy bank goes up, so does the amount in dollars.) Over 75% of the problems in general chemistry can be solved with the factor-label method. Let us look at an example to introduce the factor-label method:
How many dimes are there in 210 dollars? We know that

$$10 \text{ dimes} = 1 \text{ dollar}$$

We may divide both sides of this equation by 10 dimes or by 1 dollar, yielding

$$\frac{10 \text{ dimes}}{10 \text{ dimes}} = \frac{1 \text{ dollar}}{10 \text{ dimes}} \quad \text{or} \quad \frac{10 \text{ dimes}}{1 \text{ dollar}} = \frac{1 \text{ dollar}}{1 \text{ dollar}}$$

Since the numerator and denominator (top and bottom) of the fraction on the left side of the first equation are the same, the ratio is equal to 1. The ratio 1 dollar/10 dimes is therefore equal to 1. By analogous argument, the first ratio of the equation to the right is also equal to 1. That being the case, we can multiply any quantity by either ratio without changing the value of that quantity, because multiplying by 1 does not change the value of anything. We call each ratio a *factor*; the units are the *labels*.

Back to the problem:

$$210 \text{ dollars} \left(\frac{10 \text{ dimes}}{1 \text{ dollar}} \right) =$$

The dollar in the denominator cancels the dollars in the quantity given (the unit, not the number). It does not matter if the units are singular (dollar) or plural (dollars). We multiply the number in the numerator of the ratio and divide by the number in the denominator. That gives us

$$210\ \text{dollars}\left(\frac{10\ \text{dimes}}{1\ \text{dollar}}\right) = 2\,100\ \text{dimes}$$

EXAMPLE 2.22. How many dollars are there in 3 320 dimes?

$$3\,320\ \text{dimes}\left(\frac{1\ \text{dollar}}{10\ \text{dimes}}\right) = 332\ \text{dollars}$$

In this case, the unit dimes canceled. Suppose we had multiplied by the original ratio:

$$3\,320\ \text{dimes}\left(\frac{10\ \text{dimes}}{1\ \text{dollar}}\right) = \frac{33\,200\ \text{dimes}^2}{\text{dollar}}$$

Indeed, this expression has the same value, but the units are unfamiliar, and the answer is useless.

EXAMPLE 2.23. (*a*) What is the mass of 100 mL of mercury? (A 1.00 mL sample has a mass of 13.6 g.) (*b*) What is the volume of 100 g of mercury?

(*a*)
$$100\ \text{mL}\left(\frac{13.6\ \text{g}}{\text{mL}}\right) = 1\,360\ \text{g}$$

(*b*)
$$\frac{100\ \text{g}}{13.6\ \text{g/mL}} = 7.35\ \text{mL}$$

(*a*) When we multiplied the mass per milliliter by the volume, the units mL canceled out, and we were left with grams, which is the unit of mass we wanted. (*b*) When we divided the mass by the mass per milliliter, the units grams canceled out, and we were left with mL, which is the unit of volume we wanted. If we had multiplied in part (*b*) or divided in part (*a*), the units would not have canceled, and we would have obtained a result which was not appropriate to the problem. The units helped us decide whether to multiply or divide.

Instead of dividing by a ratio we can do the same thing by inverting the ratio and multiplying by it.

EXAMPLE 2.24. Change 100 g of mercury to mL, using the data of Example 2.23.

$$100\ \text{g}\left(\frac{1\ \text{mL}}{13.6\ \text{g}}\right) = 7.35\ \text{mL}$$

To use the factor-label method, start with the quantity given (not a rate or ratio). Multiply that quantity by a factor, or more than one factor, until an answer with the desired units is obtained. Compare this method with that of Example 2.23(*b*).

EXAMPLE 2.25. What mass in grams is present in 10.0 mL of sulfuric acid solution, which has a mass of 1.86 g for every 1.00 mL?

Start with the 10.0 mL, not with the ratio of grams to milliliters. Multiply that value by the ratio to get the desired answer:

$$10.0\ \text{mL}\left(\frac{1.86\ \text{g}}{\text{mL}}\right) = 18.6\ \text{g}$$

Percentages can be used as factors. The percentage of something is the number of parts of that thing per hundred parts total. Whatever unit(s) is used for the item in question is also used for the total. For example, the percent by mass of water in a solution is the number of grams of water per hundred grams of solution or the number of kilograms of water per 100 kg of solution. The percent by volume of alcohol in a mixed drink is the number of milliliters of alcohol in 100 mL of the drink, and so forth. If the words "by volume" or some other similar words are not stated, assume percent by mass.

EXAMPLE 2.26. What mass of H_2SO_4 is present in 250 g of solution which is 96.0% H_2SO_4 in water?

$$250\,\text{g solution}\left(\frac{96.0\,\text{g H}_2\text{SO}_4}{100\,\text{g solution}}\right) = 240\,\text{g H}_2\text{SO}_4$$

More than one factor may be needed to solve a particular problem. Either intermediate answers can be calculated and used in the next step, or several factors can be multiplied consecutively.

EXAMPLE 2.27. Calculate the speed in feet per second of an auto going exactly 60 miles per hour.

$$\frac{60\,\text{miles}}{\text{hour}}\left(\frac{5\,280\,\text{feet}}{\text{mile}}\right) = \frac{316\,800\,\text{feet}}{\text{hour}}$$

$$\frac{316\,800\,\text{feet}}{\text{hour}}\left(\frac{1\,\text{hour}}{60\,\text{minutes}}\right) = \frac{5\,280\,\text{feet}}{\text{minute}}$$

$$\frac{5\,280\,\text{feet}}{\text{minute}}\left(\frac{1\,\text{minute}}{60\,\text{seconds}}\right) = \frac{88\,\text{feet}}{\text{second}}$$

Alternatively,

$$\frac{60\,\text{miles}}{\text{hour}}\left(\frac{5\,280\,\text{feet}}{\text{mile}}\right)\left(\frac{1\,\text{hour}}{60\,\text{minutes}}\right)\left(\frac{1\,\text{minute}}{60\,\text{seconds}}\right) = \frac{88\,\text{feet}}{\text{second}}$$

It is usually more reassuring, at least at the beginning, to do such a problem one step at a time. But if you look at the combined solution, you can see that it is easier to do the whole thing at once. In this case, it would have saved us from first multiplying by 60 then dividing by 60.

EXAMPLE 2.28. What is the density of mercury in kg/L if it is 13.6 g/mL?

$$\frac{13.6\,\text{g}}{\text{mL}}\left(\frac{1\,\text{kg}}{1\,000\,\text{g}}\right)\left(\frac{1\,000\,\text{mL}}{\text{L}}\right) = \frac{13.6\,\text{kg}}{\text{L}}$$

EXAMPLE 2.29. Calculate the density of mercury in kg/m^3.

$$\frac{13.6\,\text{kg}}{\text{L}}\left(\frac{1\,000\,\text{L}}{\text{m}^3}\right) = \frac{13\,600\,\text{kg}}{\text{m}^3}$$

Often it is necessary to multiply by a factor raised to a power. Consider the problem of changing $4.00\,\text{m}^3$ to cm^3:

$$4.00\,\text{m}^3\left(\frac{100\,\text{cm}}{\text{m}}\right) =$$

If we multiply by the ratio of 100 cm to 1 m, we will still be left with m^2 (and cm) in our answer. We must multiply by $(100\,\text{cm}/\text{m})$ three times;

$$4.00\,\text{m}^3\left(\frac{100\,\text{cm}}{\text{m}}\right)^3 = 4\,000\,000\,\text{cm}^3$$

$$\left(\frac{100\,\text{cm}}{\text{m}}\right)^3 \quad \text{means} \quad \left(\frac{100\,\text{cm}}{\text{m}}\right)\left(\frac{100\,\text{cm}}{\text{m}}\right)\left(\frac{100\,\text{cm}}{\text{m}}\right)$$

and includes $100^3\,\text{cm}^3$ in the numerator and m^3 in the denominator.

2.5 SIGNIFICANT DIGITS

No matter how accurate the measuring device you use, you can make measurements only to a certain degree of accuracy. For example, would you attempt to measure the length of your shoe (a distance) with an automobile odometer (mileage indicator)? The mileage indicator has tenths of miles (or kilometers) as its smallest scale division, and you can estimate to the nearest hundredth of a mile, or something like 50 feet, but that would be useless to measure the length of a shoe. No matter how you tried, nor how many measurements you made with the odometer, you could not measure such a small distance. In contrast, could you measure the distance from the Empire State Building in New York to the Washington Monument with a 10-cm ruler? You might at first think that it would take a long time, but that it would be possible. However, it would take so many separate measurements, each having some inaccuracy in it, that the final result would be about as bad as measuring the shoe size with the odometer. The conclusion that you should draw from this discussion is that you should use the proper measuring device for each measurement, and that no matter how hard you try, each measuring device has a certain limit to its accuracy, and you cannot measure more accurately with it than its limit of accuracy. In general, you should estimate each measurement to one-tenth the smallest scale division of the instrument that you are using.

Scientists report the accuracy of their measurements by using a certain number of digits. They report all the digits they know for certain plus one extra digit which is an estimate. *Significant figures* or *significant digits* are digits used to report the accuracy of a measurement. (Note the difference between the use of the world significant here and in everyday use, where it indicates "meaningful.") For example, consider the rectangular block pictured in Fig. 2-1. The ruler at the top of the block is divided into centimeters. You can estimate the length of the block to the nearest tenth of a centimeter (millimeter), but you cannot estimate the number of micrometers or even tenths of a millimeter, no matter how you try. You should report the length of the block as 5.4 cm. Using the ruler at the bottom of the block, which has divisions in tenths of centimeters (millimeters), allows you to see for certain that the block is more than 5.4 cm but less than 5.5 cm. You can estimate it as 5.43 cm. Using the extra digit when you report the value allows the person who reads the result to determine that you used the more accurate ruler to make this latter measurement.

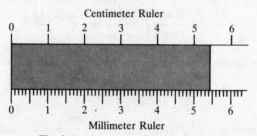

Fig. 2-1   Accuracy of measurement

EXAMPLE 2.30.   Which of the two rulers shown in Fig. 2-1 was used to make each of the following measurements? (*a*) 2.22 cm, (*b*) 3.4 cm, (*c*) 5.35 cm, (*d*) 4.4 cm, and (*e*) 4.40 cm.

The measurements reported in (*a*) and (*c*) can easily be seen to have two decimal places. Since they are reported to the nearest hundredth of a centimeter, they must have been made by the more accurate ruler, the millimeter ruler. The measurements reported in (*b*) and (*d*) were made with the centimeter ruler at the top. In part (*e*), the 0 at the end shows that this measurement was made with the more accurate ruler. Here the distance was measured as more nearly 4.40 cm than 4.41 or 4.39 cm. Thus, the results are estimated to the nearest hundredth of a centimeter, but that value just happens to have a 0 as the estimated digit.

Zeros as Significant Digits

Suppose that we want to report the measurement 2.22 cm in terms of meters. Is our measurement any more or less accurate? No, changing to another set of units does not increase or decrease the accuracy of the measurement. Therefore, we must use the same number of significant digits to report the result. How do we change a number of centimeters to meters?

$$2.22 \, \text{cm} \left(\frac{1 \, \text{m}}{100 \, \text{cm}} \right) = 0.0222 \, \text{m}$$

The zeros in 0.0222 m do not indicate anything about the accuracy with which the measurement was made; they are not significant. (They are important, however.) In a number reported properly, all nonzero digits are significant. Zeros are significant only when they help to indicate the accuracy of the measurement. The following rules are useful to determine when zeros are significant in a properly reported number:

1. All zeros to the left of the first nonzero digit are not significant. The zeros in 0.01 and 007 are not significant (except perhaps to James Bond).
2. All zeros between significant digits are significant. The 0 in 1.01 is significant.
3. All zeros to the right of the decimal point and to the right of the last nonzero digit are significant. The zeros in 1.000 and 3.0 are significant.
4. Zeros to the right of the last nonzero digit in a number with no decimal places are uncertain; they may or may not be significant. The zeros in 100 and 8 000 000 are uncertain. They may be present merely to indicate the magnitude of the number (i.e., locate the decimal point), or they may also indicate something about the accuracy of measurement. [Note: some elementary texts use an overbar to denote the last significant 0 in such numbers ($10\bar{0}$). Other texts use a decimal point at the end of an integer, as in 100., to signify that the zeros are significant. However, these practices are not carried into most regular general chemistry texts or into the chemical literature.] A way to avoid the ambiguity is given in Example 2.32.

EXAMPLE 2.31.   Underline the significant zeros in each of the following measurements, all in meters: (*a*) 1.00, (*b*) 0.0011, (*c*), 0.1010, and (*d*) 10.0.

(*a*) 1.00, (*b*) 0.0011, (*c*) 0.1010, and (*d*) 10.0. In (*a*), the zeros are to the right of the last nonzero digit and to the right of the decimal point (rule 3), so they are significant. In (*b*), the zeros are to the left of the first nonzero digit (rule 1), and so are not significant. In (*c*), the leading 0 is not significant (rule 1), the middle 0 is significant because it lies between two significant 1s (rule 2), and the last 0 is significant because it is to the right of the last nonzero digit and the decimal point (rule 3). In (*d*), the last 0 is significant (rule 3) and the middle 0 is significant because it lies between the significant digits 1 and 0 (rule 2).

EXAMPLE 2.32.   How many significant zeros are there in the number 8 000 000?

The number of significant digits cannot be determined unless more information is given. If there are 8 million people living in New York, and one person moves out, how many are left? The "8 million" people is an estimate, indicating a number nearer to 8 million than to 7 million or 9 million people. If one person moves, the number of people is still nearer to 8 million than to 7 or 9 million, and the population is still properly reported as 8 million.

If you win a lottery and the state deposits $8 000 000 to your account, when you withdraw $1 your balance will be $7 999 999. The accuracy of the bank is much greater than that of the census takers, especially since the census takers update their data only once every 10 years.

To be sure that you know how many significant digits there are in such a number, you can report the number in standard exponential notation. The population of New York would be 8×10^6 people, and the bank account would be $8.000\,000 \times 10^6$ dollars. All digits in standard exponential form are significant.

EXAMPLE 2.33.   Change the following numbers of meters to millimeters. Explain the problem of zeros at the end of a whole number, and how the problem can be solved. (*a*) 3.3 m, (*b*) 3.30 m, and (*c*) 3.300 m.

(*a*)
$$3.3\,\text{m}\left(\frac{1\,000\,\text{mm}}{\text{m}}\right) = 3\,300\,\text{mm} \qquad \text{(two significant digits)}$$

(*b*)
$$3.30\,\text{m}\left(\frac{1\,000\,\text{mm}}{\text{m}}\right) = 3\,300\,\text{mm} \qquad \text{(three significant digits)}$$

(*c*)
$$3.300\,\text{m}\left(\frac{1\,000\,\text{mm}}{\text{m}}\right) = 3\,300\,\text{mm} \qquad \text{(four significant digits)}$$

The magnitudes of the answers are the same, just as the magnitudes of the original values are the same. The numbers of millimeters all look the same, but since we know where the values came from, we know how many significant digits each contains. We can solve the problem by using standard exponential form: (a) 3.3×10^3 mm, (b) 3.30×10^3 mm, and (c) 3.300×10^3 mm.

Significant Digits in Calculations

Special Note on Significant Figures: Electronic calculators do not keep track of significant figures at all. The answers they yield most often have fewer or more digits than the number justified by the measurements. The student must keep track of the significant figures!

Addition and Subtraction

We must report the results of our calculations to the proper number of significant digits. We almost always use our measurements to calculate other quantities and the results of the calculations must indicate to the reader the limit of accuracy with which the actual measurements were made. The rules for significant digits as the result of additions or subtractions with measured quantities are as follows:

We may keep digits only as far to the right as the uncertain digit in the least accurate measurement. For example, suppose you measured a block with the millimeter ruler as 4.33 cm, and another block with the centimeter ruler as 2.1 cm. What is the length of the two blocks together?

$$\begin{array}{r} 2.1 \text{ cm} \\ + 4.33 \text{ cm} \\ \hline 6.43 \text{ cm} \rightarrow 6.4 \text{ cm} \end{array}$$

Since the 1 in the 2.1 cm measurement is uncertain, the 4 in the result is also uncertain. To report 6.43 cm would indicate that we knew the 6.4 for sure and that the 3 was uncertain. Since this is more accurate than our measurements justify, we must *round off* our reported result to 6.4 cm. That result says that we are unsure of the 4 and certain of the 6.

The rule for addition or subtraction can be stated as: Keep digits in the answer only as far to the right as the measurement in which there are digits least far to the right.

digit farthest to the right
in least accurate measurement ─┐

$$\begin{array}{r} 2.1 \text{ cm} \\ + 4.33 \text{ cm} \\ \hline 6.43 \text{ cm} \rightarrow 6.4 \text{ cm} \end{array} \qquad \begin{array}{r} 0.21 \text{ cm} \\ 14.331 \text{ cm} \\ \hline 14.541 \text{ cm} \rightarrow 14.54 \text{ cm} \end{array}$$

last digit retainable ─────────┘

It is not the number of significant digits, but their positions which determine the number of digits in the answer.

Rounding Off

We have seen that we must sometimes reduce the number of digits in our calculated result to indicate the accuracy of the measurements that were made. To reduce the number, we round off numbers other than integer digits using the following rules.

If the first digit which we are to drop is less than 5, we drop the digits without changing the last digit retained.

$$5.433 \rightarrow 5.4$$

If the first digit to be dropped is greater than 5, or is equal to 5 and there are nonzero digits after it to be dropped, the last digit retained is increased by 1:

$$5.46 \quad \rightarrow 5.5$$
$$5.56 \quad \rightarrow 5.6$$
$$5.96 \quad \rightarrow 6.0 \quad \text{(increasing the last digit retained caused a carry)}$$
$$5.5501 \rightarrow 5.6$$

If the first digit to be dropped is a 5, and there are no digits or only zeros after the five, we change the last digit remaining to the nearer even digit. The following numbers are rounded to one decimal place:

$$5.550 \rightarrow 5.6 \qquad 5.55 \rightarrow 5.6$$
$$5.450 \rightarrow 5.4 \qquad 5.45 \rightarrow 5.4$$

For rounding off digits in any whole number place, use the same rules, except that instead of actually dropping digits, replace them with (nonsignificant) zeros. For example,

$$5\,463 \rightarrow 5\,500 \quad \text{(two significant digits)}$$

EXAMPLE 2.34. Round off the following numbers to two significant digits each: (a) 0.0544, (b) 0.544, (c) 5.44, (d) 54.4, and (e) 544.

(a) 0.054, (b) 0.54, (c) 5.4, (d) 54, and (e) 540.

Multiplication and Division

In multiplication and division, different rules apply than in addition and subtraction. It is the *number* of significant digits in each of the factors, rather than their positions, which governs the number of significant digits in the answer. In multiplication and division, the answer retains as many significant digits as there are in the factor with *fewest* significant digits.

EXAMPLE 2.35. Perform each of the following operations to the proper number of significant digits:
(a) $1.50\,\text{cm} \times 2.000\,\text{cm}$
(b) $1.00\,\text{g}/3.00\,\text{cm}^3$
(c) $3.45\,\text{g}/1.15\,\text{cm}^3$

(a) $1.50\,\text{cm} \times 2.000\,\text{cm} = 3.00\,\text{cm}^2$. There are three significant digits in the first factor and four in the second. The answer can retain only three significant digits, equal to the smaller number of significant digits in the factors.

(b) $1.00\,\text{g}/3.00\,\text{cm}^3 = 0.333\,\text{g}/\text{cm}^3$. There are three significant digits, equal to the number of significant digits in each number. Note that the number of decimal places is different in the answer, but in multiplication or division, the number of decimal places is immaterial.

(c) $3.45\,\text{g}/1.15\,\text{cm}^3 = 3.00\,\text{g}/\text{cm}^3$. Since there are three significant digits in each number, there should be three significant digits in the answer. In this case, we had to add zeros, not round off, to get the proper number of significant digits.

When we multiply or divide a measurement by a defined number, rather than by another measurement, we may retain in the answer the number of significant digits that occur in the measurement. For example, if we multiply a number of meters by $1\,000\,\text{mm}/\text{m}$, we may retain the number of significant digits in the number of millimeters that we had in the number of meters. The $1\,000$ is a defined number, not a measurement, and can be regarded as having as many significant digits as needed for any purpose.

EXAMPLE 2.36. How many significant digits should be retained in the answer when we calculate the number of decimeters in 1.465 m?

$$1.465\,\text{m}\left(\frac{10\,\text{dm}}{\text{m}}\right) = 14.65\,\text{dm}$$

The number of significant digits in the answer is 4, equal to the number in the 1.465-m measurement. The 10 dm/m is a definition, and does not limit the number of significant digits to 2.

2.6 DENSITY

Density is a useful property with which to identify substances. *Density* is defined as mass per unit volume:

$$\text{density} = \text{mass}/\text{volume}$$

or in symbols,

$$d = m/V$$

Since it is a quantitative property, it is often more useful for identification than a qualitative property like color or smell. Moreover, density determines whether an object will float in a given liquid. If the object is less dense than the liquid, it will float; if it is more dense, it will sink. It is also useful to discuss density here for practice with the factor-label method of solving problems, and as such, it is often emphasized on early quizzes and examinations.

Density is a ratio—the number of grams per milliliter, for example. In this regard, it is similar to speed.

The word *per* means *divided by*. To get a speed in miles per hour, divide the number of miles by the number of hours.

Say the following speed aloud: 30 miles/hour.

"Thirty miles per hour." (The division symbol is read as the word *per*.)

Distinguish carefully between density and mass, which are often confused in everyday conversation.

EXAMPLE 2.37. Which weighs more, a pound of bricks or a pound of feathers?

Since a pound of each is specified, neither weighs more. But everyone "knows" that bricks are heavier than feathers. The confusion stems from the fact that "heavy" is defined in the dictionary as either "having great mass" or "having high density." Per unit volume, bricks weigh more than feathers; that is, bricks are more dense.

It is relative densities that determine whether an object will float in a liquid. If the object is less dense, it will float (unless it dissolves, of course).

EXAMPLE 2.38. Which has a greater mass: a large wooden desk or a metal needle? Which one will float in water?

The desk has a greater mass. (You can pick up the needle with one finger, but not the desk.) Since the desk is so much larger (greater volume), it displaces more than its own weight of water. Its density is less than that of water, and it will float despite its greater mass. The needle is so small that it does not displace its own weight of water, and thus sinks.

In doing numerical density problems, you may always use the equation $d = m/V$ or the same equation rearranged into the forms $V = m/d$ or $m = dV$. You are often given two of these quantities and asked for the third. You will use the equation $d = m/V$ if you are given mass and volume, but if you are given density and either of the others, you probably should use the factor-label method. That way, you need not manipulate the equation and then substitute; you can solve immediately.

EXAMPLE 2.39 Calculate the density of a 2.0-L body which has a mass of 4.0 kg.

$$d = m/V = (4.0 \,\text{kg})/(2.0 \,\text{L}) = 2.0 \,\text{kg/L}$$

EXAMPLE 2.40. What is the mass of 100.0 mL of lead, which has a density of 11.3 g/mL?

Using the factor-label method:

$$100.0 \,\text{mL} \left(\frac{11.3 \,\text{g}}{\text{mL}} \right) = 1\,130 \,\text{g} = 1.13 \,\text{kg}$$

With the equation:

$$d = m/V$$

$$m = Vd = (100.0\,\text{mL})(11.3\,\text{g/mL}) = 1\,130\,\text{g} = 1.13\,\text{kg}$$

EXAMPLE 2.41. What is the volume of 13.0 g of gold? (density 19.3 g/mL)

$$V = \frac{m}{d} = \frac{13.0\,\text{g}}{19.3\,\text{g/mL}} = 0.674\,\text{mL}$$

or
$$13.0\,\text{g}\left(\frac{1\,\text{mL}}{19.3\,\text{g}}\right) = 0.674\,\text{mL}$$

Use of the equation requires manipulation of the basic equation followed by substitution. The factor-label method gives the same result by merely relying on the units. It also allows combination of the solution with other conversions.

EXAMPLE 2.42. What mass of sulfuric acid is there in 100.0 mL of solution of density 1.85 g/mL which contains 96.0% sulfuric acid by mass?

Note that the *solution* has the density 1.85 g/mL, and that every 100 g of it contains 96.0 g of sulfuric acid.

$$100.0\,\text{mL solution}\left(\frac{1.85\,\text{g solution}}{\text{mL solution}}\right)\left(\frac{96.0\,\text{g acid}}{100\,\text{g solution}}\right) = 178\,\text{g acid}$$

2.7 TEMPERATURE SCALES

Scientists worldwide (and everyone else outside the United States) use the Celsius temperature scale, in which the freezing point of pure water is defined as 0 °C and the normal boiling point of pure water is defined as 100 °C. The *normal boiling point* is the boiling point at 1.00 atm pressure (Chap. 11). The Fahrenheit temperature scale is used principally in the United States. It has the freezing point defined as 32.0 °F and the normal boiling point defined as 212 °F. A comparison of the Celsius and Fahrenheit temperature scales is presented in Fig. 2-2. The temperature differences between the freezing point and normal boiling point on the two scales are 180 °F and 100 °C, respectively. To convert from Fahrenheit temperatures to Celsius temperatures, subtract 32.0 °F and then multiply the result by 100/180 or 5/9.

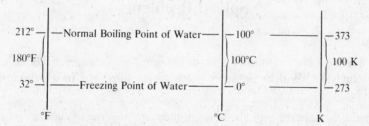

Fig. 2-2   Comparison of Celsius, Fahrenheit, and Kelvin temperature scales

EXAMPLE 2.43. Change 98.6 °F to Celsius.

$$°\text{C} = (98.6\,°\text{F} - 32.0\,°\text{F})5/9 = 37.0\,°\text{C}$$

To convert from Celsius to Fahrenheit, multiply by 9/5 then add 32.0 °F.

EXAMPLE 2.44. Convert 50.0 °C to Fahrenheit.

$$(50.0\,°\text{C})9/5 + 32.0\,°\text{F} = 122\,°\text{F}$$

Kelvin Temperature Scale

The Kelvin temperature scale is defined to have the freezing point of pure water at 273.15 K and the normal boiling point of pure water at 373.15 K. The unit of the Kelvin temperature scale is the *kelvin*. (We do not use "degrees" with Kelvin.) Thus, its temperatures are essentially 273° higher than the same temperatures on the Celsius scale. To convert from Celsius to Kelvin, merely add 273° to the Celsius temperature. To convert in the opposite direction, subtract 273° from the Kelvin temperature to get the Celsius equivalent. (See Fig. 2-2.)

EXAMPLE 2.45.   Convert 100 °C and −20 °C to Kelvin.

$$100\,°C + 273° = 373\,K$$

$$-20\,°C + 273° = 253\,K$$

EXAMPLE 2.46.   Change 50 K and 500 K to Celsius.

$$50\,K - 273° = -223\,°C$$

$$500\,K - 273° = 227\,°C$$

Note that a *change* in temperature in Kelvin is the same as the equivalent change in temperature in Celsius.

EXAMPLE 2.47.   Convert 0 °C and 10 °C to Kelvin. Calculate the change in temperature from 0 °C to 10 °C on both temperature scales.

$$0\,°C + 273° = 273\,K$$

$$10\,°C + 273° = 283\,K$$

The temperature difference on the two scales is

$$
\begin{array}{cc}
10\,°C & 283\,K \\
-\ \ 0\,°C & -273\,K \\
\hline
10\,°C & 10\,K
\end{array}
$$

Solved Problems

METRIC SYSTEM

2.1.   There are dimes in one container and pennies in another. Each container holds the same amount of money. (*a*) Which container holds more coins? (*b*) In which container is each coin more valuable?

   Ans.   The more valuable the coin, the fewer of them are required to make a certain amount of money. The dimes are each more valuable, but there are more pennies.

2.2.   A certain distance is measured in meters and again in centimeters. Which measurement involves (*a*) the larger number? (*b*) the larger unit?

   Ans.   The larger the unit, the fewer of them are required to measure a given distance. There are more centimeters because the meter is larger. Compare this to Problem 2.1.

2.3.   Would 0.20 m^3 of water almost fill a thimble, a jug, or a bathtub?

   Ans.   A bathtub.

2.4. Make the following conversions:
(a) Change 2.5 cm to meters
(b) Change 7.7 kg to grams
(c) Change 3.0 L to milliliters
(d) Change 20.0 kg to milligrams

Ans.

(a)
$$2.5 \, \text{cm}\left(\frac{1 \, \text{m}}{100 \, \text{cm}}\right) = 0.025 \, \text{m}$$

(b)
$$7.7 \, \text{kg}\left(\frac{1\,000 \, \text{g}}{\text{kg}}\right) = 7\,700 \, \text{g}$$

(c)
$$3.0 \, \text{L}\left(\frac{1\,000 \, \text{mL}}{\text{L}}\right) = 3\,000 \, \text{mL}$$

(d)
$$20.0 \, \text{kg}\left(\frac{1\,000 \, \text{g}}{\text{kg}}\right)\left(\frac{1\,000 \, \text{mg}}{\text{g}}\right) = 2.00 \times 10^7 \, \text{mg}$$

2.5. (a) Is there a unit "ounce" that means a weight and another than means a volume? (b) Is there a metric unit that means both weight and volume? (c) What is the difference between an ounce of gold and an ounce of lead?

Ans. (a) Yes. 16 fluid ounces = 1 pint and 16 ounces avoirdupois = 1 pound.
(b) No.
(c) Gold is measured in Troy ounces and lead is measured in avoirdupois ounces.

2.6. What is the difference between a U.S. gallon and an imperial gallon?

Ans. They are two different units despite both having the word gallon in their names. The U.S. gallon is equal to 3.7853 L; the imperial gallon is equal to 4.5460 L.

2.7. Give the volume in liters of each of the following: (a) 1 m³, (b) 1 dm³, (c) 1 cm³, and (d) 1 mm³.

Ans. (a) 1 000 L, (b) 1 L, (c) 0.001 L, and (d) 1×10^{-6} L.

2.8. Give the volume corresponding to each of the following in terms of a dimension involving length cubed: (a) 1 kL, (b) 1 L, (c) 1 mL, and (d) 1 μL.

Ans. (a) 1 m³, (b) 1 dm³, (c) 1 cm³, and (d) 1 mm³.

2.9. Explain in terms of units why area × distance yields volume of a rectangular solid.

Ans. $\text{m}^2 \times \text{m} = \text{m}^3$.

2.10. Give the volume corresponding to each of the following numbers of liters in terms of a dimension involving length cubed: (a) 1 000 L, (b) 1 L, (c) 0.001 L, and (d) 1×10^{-6} L.

Ans. (a) 1 m³, (b) 1 dm³, (c) 1 cm³, and (d) 1 mm³.

2.11. Make the following conversions:
(a) Change 2.5 cm² to m²
(b) Change 20 cm³ to dm³
(c) Change 3.0 L to cm³
(d) Change 7.7 dm³ to L
(e) Change 60 m³ to kL
(f) Change 4.0 m² to cm²
(g) Change 250 mL to cm³

Ans.

(*a*)
$$2.5\,cm^2\left(\frac{1\,m}{100\,cm}\right)^2 = 0.000\,25\,m^2$$

(*b*)
$$20\,cm^3\left(\frac{1\,dm}{10\,cm}\right)^3 = 0.020\,dm^3$$

(*c*)
$$3.0\,L\left(\frac{1\,000\,cm^3}{L}\right) = 3\,000\,cm^3$$

(*d*)
$$7.7\,dm^3\left(\frac{1\,L}{dm^3}\right) = 7.7\,L$$

(*e*)
$$60\,m^3\left(\frac{1\,kL}{m^3}\right) = 60\,kL$$

(*f*)
$$4.0\,m^2\left(\frac{100\,cm}{m}\right)^2 = 40\,000\,cm^2$$

(*g*)
$$250\,mL\left(\frac{1\,cm^3}{mL}\right) = 250\,cm^3$$

2.12. What is wrong with the following problems?
  (*a*) Change 4.5 mL to cm
  (*b*) Change 20 g to mm
  (*c*) Change 40 cm^2 to liters

  Ans. (*a*) You cannot convert a volume to a distance.
     (*b*) You cannot convert a mass to a distance.
     (*c*) You cannot convert an area to a volume.

2.13. (*a*) Change 20.0 miles^2 to square feet. (*b*) Change 20.0 km^2 to square meters.

  Ans.

(*a*)
$$20.0\,miles^2\left(\frac{5\,280\,feet}{mile}\right)^2 = 5.58 \times 10^8\,feet^2$$

(*b*)
$$20.0\,km^2\left(\frac{1\,000\,m}{km}\right)^2 = 2.00 \times 10^7\,m^2$$

2.14. Draw a square 2 cm × 2 cm. In the upper left corner of that square, draw another square 1 cm × 1 cm. How many of the smaller squares will fit into the larger one?

  Ans. Four smaller squares will fit; that is, 4 cm^2 = (2 cm)^2. (See Fig. 2-3.)

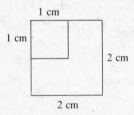

Fig. 2-3

2.15. What is the ratio in areas of a square 1 cm on each side and a square 1 mm on each side?

  Ans. The ratio of centimeters to millimeters is 10 : 1; the ratio of their squares is therefore 100 : 1.

EXPONENTIAL NUMBERS

2.16. What is the meaning of each of the following? (*a*) ten squared, (*b*) ten cubed, and (*c*) ten to the fourth power.

Ans. (*a*) 10^2, (*b*) 10^3, and (*c*) 10^4.

2.17. In the number 7.0×10^3, identify (*a*) the coefficient, (*b*) the exponent, (*c*) the base, and (*d*) the exponential part.

Ans. (*a*) 7.0, (*b*) 3, (*c*) 10, and (*d*) 10^3.

2.18. Which of the following numbers are in standard exponential form?
(*a*) 3.0×10^7 (*f*) 3×10^7
(*b*) 1.23×10^{-4} (*g*) 10×10^4
(*c*) 0.12×10^5 (*h*) 4.4×10^3
(*d*) 1.0×10^{-3} (*i*) 1×10^5
(*e*) 7×10^0

Ans. (*a*), (*b*), (*d*), (*e*), (*f*), (*h*), and (*i*).

2.19. Show by writing out the explicit meaning of the exponential parts that $(2.0 \times 10^3)^2 = 4.0 \times 10^6$.

Ans.

$$
\begin{aligned}
\left(2.0 \times 10^3\right)^2 &= \left(2.0 \times 10^3\right)\left(2.0 \times 10^3\right) \\
&= 2.0 \times 2.0 \times (10 \times 10 \times 10) \times (10 \times 10 \times 10) \\
&= 4.0 \times 10^6
\end{aligned}
$$

2.20. How do you get the exponent of the answer when you
(*a*) Multiply exponentials?
(*b*) Divide exponentials?
(*c*) Add or subtract exponentials?

Ans.   (*a*) Add the exponents.
(*b*) Subtract the exponent of the denominator from that of the numerator.
(*c*) Make sure that the exponents of all the numbers are the same before adding or subtracting. (If they are not convert one or more so that they are the same.) Then use the exponent from each number added or subtracted.

2.21. By expressing the following products as products of 10s, show that to multiply exponentials the exponents must be added.
(*a*) $10^4 \times 10^3$
(*b*) $10^2 \times 10^2$
(*c*) $10^4 \times 10^6$

Ans. (*a*) $(10 \times 10 \times 10 \times 10) \times (10 \times 10 \times 10) = 10^7$
(*b*) $(10 \times 10) \times (10 \times 10) = 10^4$
(*c*) $(10 \times 10 \times 10 \times 10) \times (10 \times 10 \times 10 \times 10 \times 10 \times 10) = 10^{10}$

2.22. By expressing the following quotients as products of 10s, show that to divide exponentials the exponents must be subtracted.
(*a*) $(10^4)/(10^3)$
(*b*) $(10^7)/(10^2)$
(*c*) $(10^4)/(10^2)$
(*d*) $(10^2)/(10^2)$
(*e*) $(10^2)/(10^3)$

Ans.

$$(a) \qquad \frac{10 \times 10 \times 10 \times 10}{10 \times 10 \times 10} = 10^1 = 10$$

$$(b) \qquad \frac{10 \times 10 \times 10 \times 10 \times 10 \times 10 \times 10}{10 \times 10} = 10^5$$

$$(c) \qquad \frac{10 \times 10 \times 10 \times 10}{10 \times 10} = 10^2$$

$$(d) \qquad \frac{10 \times 10}{10 \times 10} = 10^0 = 1$$

$$(e) \qquad \frac{10 \times 10}{10 \times 10 \times 10} = 10^{-1} = \frac{1}{10} = 0.10$$

2.23. Divide the following exponentials:
(a) $10^4/10^6$ (c) $10^7/10^9$
(b) $10^5/10^8$ (d) $10^9/10^{119}$

Ans.   (a) 10^{-2} (c) 10^{-2}
          (b) 10^{-3} (d) 10^{-110}

2.24. Simplify:
(a) $\dfrac{1}{10^{-3}}$ (b) $\dfrac{1}{10^7}$ (c) $\dfrac{10^{-2}}{10^{-2}}$
Ans.   (a) 10^3 (b) 10^{-7} (c) 10^0

2.25. Perform the following calculations:
(a) $(2.20 \times 10^{101})(3.50 \times 10^{22})$
(b) $(2.20 \times 10^{101}) + (3.50 \times 10^{22})$
(c) $(2.20 \times 10^{101})/(3.50 \times 10^{22})$
(d) $(2.20 \times 10^{-101})/(3.50 \times 10^{22})$
(e) $(22.0 \times 10^{-101})/(3.50 \times 10^{-22})$
(f) $(22.0 \times 10^{-101}) - (3.50 \times 10^{-22})$
(g) $(2.20 \times 10^{91}) + (3.50 \times 10^{92})$
(h) $(2.20 \times 10^{101})(3.50 \times 10^{100})$

Ans.   (a) 7.70×10^{123} (e) 6.29×10^{-79}
          (b) 2.20×10^{101} (f) -3.50×10^{-22}
          (c) 6.29×10^{78} (g) 3.72×10^{92}
          (d) 6.29×10^{-124} (h) 7.70×10^{201}

2.26. Convert the following numbers to numbers having the equivalent values but having coefficients with two digits to the left of the decimal point: (a) 3.50×10^5, (b) 6.66×10^{-7}, (c) 1.23×10^0, and (d) 1.23.

Ans.   (a) 35.0×10^4, (b) 66.6×10^{-8}, (c) 12.3×10^{-1}, and (d) 12.3×10^{-1}.

2.27. Convert the following numbers to standard exponential form: (a) 1 423, (b) 0.00403, (c) 0.0003, and (d) 6 000 000.01.

Ans.   (a) 1.423×10^3, (b) 4.03×10^{-3}, (c) 3×10^{-4}, and (d) 6.00000001×10^6.

2.28. Convert the following numbers to decimal form: (*a*) 3.3×10^5, (*b*) 5.06×10^{-2}, and (*c*) 6.9×10^0.

Ans.   (*a*) 330 000, (*b*) 0.0506, and (*c*) 6.9.

2.29. Perform the following calculations:
(*a*) $1.20 \times 10^8 + 9.7 \times 10^7$
(*b*) $1.66 \times 10^{-3} + 6.78 \times 10^{-2}$
(*c*) $7.7 \times 10^{99} + 3.3 \times 10^{99}$
(*d*) $6.02 \times 10^{23} + 6.02 \times 10^{21}$
(*e*) $1.6 \times 10^{-19} + 1.1 \times 10^{-20}$

Ans.   (*a*) 2.17×10^8
(*b*) 6.95×10^{-2}
(*c*) 1.10×10^{100}
(*d*) 6.08×10^{23}
(*e*) 1.7×10^{-19}

2.30. Perform the following calculations:
(*a*) $(1.20 \times 10^8)/(9.7 \times 10^7)$
(*b*) $(1.66 \times 10^{-3})/(6.78 \times 10^{-2})$
(*c*) $(7.7 \times 10^{99})/(3.3 \times 10^{99})$
(*d*) $(6.02 \times 10^{23})/(6.02 \times 10^{21})$
(*e*) $(1.60 \times 10^{-19})/(1.1 \times 10^{-20})$

Ans.   (*a*)   1.2
(*b*) 2.45×10^{-2}
(*c*)   2.3
(*d*) 1.00×10^2
(*e*)   15

2.31. Perform the following calculations:
(*a*) $(1.20 \times 10^8)(9.7 \times 10^7)$
(*b*) $(1.66 \times 10^{-3})(6.78 \times 10^{-2})$
(*c*) $(7.7 \times 10^{99})(3.3 \times 10^{99})$
(*d*) $(6.02 \times 10^{23})(6.02 \times 10^{21})$
(*e*) $(1.6 \times 10^{-19})(1.1 \times 10^{-20})$

Ans.   (*a*) 1.2×10^{16}
(*b*) 1.13×10^{-4}
(*c*) 2.5×10^{199}
(*d*) 3.62×10^{45}
(*e*) 1.8×10^{-39}

2.32. Perform the following calculations:
(*a*) $1.20 \times 10^8 - 9.7 \times 10^7$
(*b*) $1.66 \times 10^{-3} - 6.78 \times 10^{-2}$
(*c*) $7.7 \times 10^{99} - 3.3 \times 10^{99}$
(*d*) $6.02 \times 10^{23} - 6.02 \times 10^{21}$
(*e*) $1.6 \times 10^{-19} - 1.1 \times 10^{-20}$

Ans.   (*a*) 2.3×10^7
(*b*) -6.61×10^{-2}
(*c*) 4.4×10^{99}
(*d*) 5.96×10^{23}
(*e*) 1.5×10^{-19}

2.33. Perform the following calculations:
  (a) $(1.20 \times 10^8 + 9.7 \times 10^7)/(6.04 \times 10^{-3})$
  (b) $(1.66 \times 10^{-3} + 6.78 \times 10^{-2})(1.10 \times 10^4 + 1.3 \times 10^3)$
  (c) $(7.7 \times 10^{99} + 3.3 \times 10^{99}) - (9.4 \times 10^{98})$
  (d) $(6.02 \times 10^{23} + 6.02 \times 10^{21}) + (3.95 \times 10^{22})$
  (e) $(1.6 \times 10^{-19} + 1.1 \times 10^{-20}) - 1$

  Ans. (a) 3.59×10^{10}
  (b) 8.55×10^2
  (c) 1.01×10^{100}
  (d) 6.48×10^{23}
  (e) -1

2.34. Perform the following calculations in your head. Check them if necessary with a calculator.
  (a) $1.23 \times 10^7 \times 10^3$
  (b) $7.65 \times 10^{-3} \times 10^3$
  (c) $5.55 \times 10^{17} \times 10^{-3}$
  (d) $(1.23 \times 10^7)/10^3$
  (e) $(7.65 \times 10^{-3})/10^3$
  (f) $(5.55 \times 10^{17})/10^{-3}$

  Ans. (a) 1.23×10^{10}
  (b) 7.65
  (c) 5.55×10^{14}
  (d) 1.23×10^4
  (e) 7.65×10^{-6}
  (f) 5.55×10^{20}

2.35. (a) Change the following numbers to decimal format and add them. (b) Repeat the calculation using the standard methods for exponential numbers.

$$(1 \times 10^0) + (2.0 \times 10^1)$$

  Ans. (a) $\quad 1 \times 10^0 = 1$
  $2.0 \times 10^1 = 20$
  $\overline{21} = 2.1 \times 10^1$
  (b) $\quad 1 \times 10^0 = 0.1 \times 10^1$
  $2.0 \times 10^1 + 0.1 \times 10^1 = 2.1 \times 10^1$

FACTOR-LABEL METHOD

2.36. (a) Write the reciprocal for the following factor label: 3 miles/hour. (b) Which of these—the reciprocal or the original factor label—is multiplied to change miles to hours?

  Ans. (a) 1 hour/3 miles. (b) The reciprocal:

$$6\,\text{miles}\left(\frac{1\,\text{hour}}{3\,\text{miles}}\right) = 2\,\text{hours}$$

2.37. (a) Write the reciprocal for the following factor label: 4 dollars/pound. (b) Which of these—the reciprocal or the original factor label—is multiplied to change pounds to dollars?

  Ans. (a) 1 pound/4 dollars. (b) The original:

$$2\,\text{pounds}\left(\frac{4\,\text{dollars}}{\text{pound}}\right) = 8\,\text{dollars}$$

2.38. (a) Write the reciprocal for the following factor label: 7.00 dollars/hour. (b) Calculate the number of hours one must work at this rate to earn 100 dollars.

Ans. (*a*) 1 hour/7.00 dollars

(*b*) $100 \text{ dollars} \left(\dfrac{1 \text{ hour}}{7.00 \text{ dollars}} \right) = 14.3 \text{ hours}$

2.39. Calculate the number of cents in 20.33 dollars. (*a*) Do the calculation by first converting the dollars to dimes and then the dimes to cents. (*b*) Repeat with a direct calculation.

Ans.

(*a*)
$$20.33 \text{ dollars} \left(\frac{10 \text{ dimes}}{\text{dollar}} \right) = 203.3 \text{ dimes}$$

$$203.3 \text{ dimes} \left(\frac{10 \text{ cents}}{\text{dime}} \right) = 2033 \text{ cents}$$

(*b*)
$$20.33 \text{ dollars} \left(\frac{10 \text{ dimes}}{\text{dollar}} \right) \left(\frac{10 \text{ cents}}{\text{dime}} \right) = 2033 \text{ cents}$$

or
$$20.33 \text{ dollars} \left(\frac{100 \text{ cents}}{\text{dollar}} \right) = 2033 \text{ cents}$$

2.40. (*a*) Calculate the number of centimeters in 20.33 km. Do the calculation by first converting the kilometers to meters and then the meters to centimeters. (*b*) Repeat with a direct calculation. (*c*) Compare this problem with the last problem.

Ans.

(*a*)
$$20.33 \text{ km} \left(\frac{1000 \text{ m}}{\text{km}} \right) = 20330 \text{ m}$$

$$20330 \text{ m} \left(\frac{100 \text{ cm}}{\text{m}} \right) = 2033000 \text{ cm}$$

(*b*)
$$20.33 \text{ km} \left(\frac{1000 \text{ m}}{\text{km}} \right) \left(\frac{100 \text{ cm}}{\text{m}} \right) = 2033000 \text{ cm}$$

(*c*) The methods are the same.

2.41. Calculate the number of cubic meters in $3.4 \times 10^5 \text{ mm}^3$.

Ans.
$$3.4 \times 10^5 \text{ mm}^3 \left(\frac{1 \text{ m}}{1000 \text{ mm}} \right)^3 = 3.4 \times 10^{-4} \text{ m}^3$$

2.42. Which is bigger, 3.0 L or 3000 mm^3?

Ans. 3.0 L is bigger. It is equal to 3000 cm^3 or 3000000 mm^3.

2.43. Change 3.3 g/cm^3 to kilograms per cubic meter.

Ans.
$$\frac{3.3 \text{ g}}{\text{cm}^3} \left(\frac{100 \text{ cm}}{\text{m}} \right)^3 \left(\frac{1 \text{ kg}}{1000 \text{ g}} \right) = 3.3 \times 10^3 \text{ kg/m}^3$$

2.44. Given that $1 \text{ L} = 1 \text{ dm}^3$, calculate the number of liters in (*a*) 1 m^3. (*b*) 1 cm^3. (*c*) 1 mm^3.

Ans.

(*a*)
$$1 \text{ m}^3 \left(\frac{10 \text{ dm}}{\text{m}} \right)^3 \left(\frac{1 \text{ L}}{\text{dm}^3} \right) = 1000 \text{ L}$$

(*b*)
$$1 \text{ cm}^3 \left(\frac{1 \text{ dm}}{10 \text{ cm}} \right)^3 \left(\frac{1 \text{ L}}{\text{dm}^3} \right) = 0.001 \text{ L}$$

(*c*)
$$1 \text{ mm}^3 \left(\frac{1 \text{ dm}}{100 \text{ mm}} \right)^3 \left(\frac{1 \text{ L}}{\text{dm}^3} \right) = 1 \times 10^{-6} \text{ L}$$

2.45. Commercial sulfuric acid solution is 96.0% H_2SO_4 in water. How many grams of H_2SO_4 is there in 200 g of the commercial solution?

Ans.

$$200\,\text{g solution}\left(\frac{96.0\,\text{g}\,H_2SO_4}{100\,\text{g solution}}\right) = 192\,\text{g}\,H_2SO_4$$

2.46. Pistachio nuts cost \$4.00 per pound. (*a*) How many pounds of nuts can be bought for \$22.00? (*b*) How much does 43.2 pounds of nuts cost?

Ans.

(*a*) $$22.00\,\text{dollars}\left(\frac{1\,\text{pound}}{4.00\,\text{dollars}}\right) = 5.50\,\text{pounds}$$

(*b*) $$43.2\,\text{pounds}\left(\frac{4.00\,\text{dollars}}{\text{pound}}\right) = 172.80\,\text{dollars}$$

2.47. If 42% of a certain class is female, how large is the class if there are 189 females?

Ans.

$$189\,\text{females}\left(\frac{100\,\text{total}}{42\,\text{females}}\right) = 450\,\text{total}$$

2.48. If income tax for a certain student is 15%, how much gross pay does that student need to earn to be able to keep \$150? (Assume no other deductions.)

Ans.   If 15% goes to tax, $100\% - 15\% = 85\%$ is kept.

$$150\,\text{dollars kept}\left(\frac{100\,\text{dollars gross}}{85\,\text{dollars kept}}\right) = 176.47\,\text{dollars gross}$$

SIGNIFICANT DIGITS

2.49. Underline the significant digits in each of the following measurements:
(*a*) 10.0 m
(*b*) 0.010 m
(*c*) 77.01 m
(*d*) 1.0×10^3 m
(*e*) 6.100 m
(*f*) 0.001 m
(*g*) 1.110 m
(*h*) 1.01 m
(*i*) 93.0 m

Ans.   (*a*) <u>10.0</u> m   rules 2 and 3
           (*b*) 0.0<u>10</u> m   rules 1 and 3
           (*c*) <u>77.01</u> m   rule 2
           (*d*) <u>1.0</u> × 10³ m   rule 3
           (*e*) <u>6.100</u> m   rule 3
           (*f*) 0.00<u>1</u> m   rule 1
           (*g*) <u>1.110</u> m   rule 3
           (*h*) <u>1.01</u> m   rule 2
           (*i*) <u>93.0</u> m   rule 3

2.50. What is the difference between the number of significant digits and the number of decimal places in a measurement?

Ans. The number of significant digits is the number of digits that reflect the accuracy of the measurement. The number of decimal places is the number of digits after the decimal point. The two have little to do with each other, so do not get them confused.

2.51. Calculate the answer to each of the following expressions to the correct number of significant digits.

   (*a*) 2.20 cm × 1.1 cm =
   (*b*) 2.20 cm − 1.1 cm =
   (*c*) (2.20 g)/(1.1 cm^3) =
   (*d*) 1.034 cm + 6.6 cm − 4.01 cm =
   (*e*) 69.0 cm + 0.002 cm
   (*f*) 9.66 cm + 3.44 cm − 0.01 cm =
   (*g*) 7.77 cm + 0.506 cm =

   Ans. (*a*) 2.4 cm^2
           (*b*) 1.1 cm
           (*c*) 2.0 g/cm^3
           (*d*) 3.6 cm
           (*e*) 69.0 cm
           (*f*) 13.09 cm
           (*g*) 8.28 cm

2.52. Calculate the answer to each of the following expressions to the correct number of significant digits.

   (*a*) (1.23 cm + 6.6 cm) × 5.00 cm =
   (*b*) 5.89 cm × (1.0 cm − 7.633 cm) =
   (*c*) (5.79 cm × 5.5 cm) − 6.02 cm^2 =
   (*d*) 7 001 cm − 1.00 cm =
   (*e*) 7.0 × 10^5 cm × 5.55 cm × 10^3 cm =
   (*f*) 6.6 × 10^{-3} cm − 5.5 × 10^{-2} cm =
   (*g*) (2.00 cm × 3.35 cm) + (4.21 cm^2 − 1.23 cm^2) =

   Ans. (*a*) 39 cm^2
           (*b*) − 39 cm^2
           (*c*) 26 cm^2
           (*d*) 7.000 × 10^3 cm
           (*e*) 3.9 × 10^9 cm^3
           (*f*) − 4.8 × 10^{-2} cm
           (*g*) 9.68 cm^2

2.53. To the correct number of significant digits, calculate the number of

   (*a*) millimeters in 2.0 m
   (*b*) kilograms in 23 g
   (*c*) millimeters in 23.32 cm
   (*d*) centimeters in 0.0010 m

   Ans. (*a*) 2.0 × 10^3 mm   (two significant digits)
           (*b*) 0.023 kg   (two significant digits)
           (*c*) 233.2 mm   (four significant digits)
           (*d*) 0.10 cm   (two significant digits)

2.54. A beaker plus contents has a combined mass of 120.2 g. The beaker has a mass of 119.0 g. If the density of the contents is 2.05 g/mL, what is the volume of the contents to the correct number of significant digits?

   Ans.

$$(120.2\,\text{g} - 119.0\,\text{g})\left(\frac{1\,\text{mL}}{2.05\,\text{g}}\right) = 0.59\,\text{mL}$$

There are only two significant digits since the subtraction yields an answer with two significant digits. That answer limits the final volume to two significant digits.

2.55. To how many decimal places should you report the reading of (a) a 50-mL buret (graduated in 0.1 mL)? (b) an analytic balance (calibrated to 0.001 g)?

   Ans.   (a) 2. (b) 4. (One estimated digit in each.)

2.56. Round off each of the following calculated results to the nearest integer (whole number): (a) 2.500 cm, (b) 3.500 cm, (c) 3.5 cm, (d) 2.501 cm, (e) 43.95 cm, and (f) 1.445 cm.

   Ans.   (a) 2 cm, (b) 4 cm, (c) 4 cm, (d) 3 cm, (e) 44 cm, and (f) 1 cm.

2.57. Calculate each answer to the correct number of significant digits:
   (a) $(2.46\,g)/(1.23\,mL) =$
   (b) $(246\,g)/(1.23\,L) =$
   (c) $(393\,g)/(2\,L) =$

   Ans.   (a) $(2.46\,g)/(1.23\,mL) = 2.00\,g/mL$
             (b) $(246\,g)/(1.23\,L) = 2.00 \times 10^2\,g/L$
             (c) $(393\,g)/(2\,L) = 200\,g/L = 2 \times 10^2\,g/L$

   Three significant digits are required in (a) and in (b); one significant digit is required in (c).

2.58. Perform the following calculations to the proper number of significant digits. All measurements are in centimeters.
   (a) $1.20 \times 10^8 - 9.7 \times 10^7$
   (b) $1.66 \times 10^{-3} - 6.78 \times 10^{-2}$
   (c) $7.7 \times 10^{99} - 3.3 \times 10^{99}$
   (d) $6.02 \times 10^{23} - 6.02 \times 10^{21}$
   (e) $1.6 \times 10^{-19} - 1.1 \times 10^{-20}$

   Ans.   (a) $2.3 \times 10^7\,cm$
             (b) $-6.61 \times 10^{-2}\,cm$
             (c) $4.4 \times 10^{99}\,cm$
             (d) $5.96 \times 10^{23}\,cm$
             (e) $1.5 \times 10^{-19}\,cm$

2.59. Explain why every 0 in the coefficient is significant in a number properly expressed in standard exponential notation.

   Ans.   No zeros are needed for the sole purpose of determining the magnitude of the number. (The exponential part of the number does that.) The only other reason a 0 would be present is that it is significant.

2.60. Add, expressing the answer in the proper number of significant figures:
   (a) $1\,000.0\,cm + 0.01\,cm$
   (b) $1.0000 \times 10^3\,cm + 1 \times 10^{-2}\,cm$

   Ans.   (a)   1 000.0 cm
                      0.01 cm
                 ‾‾‾‾‾‾‾‾‾‾‾‾
                 1 000.01 cm → 1 000.0 cm

   The fraction added is so small relative to the first number that it does not make any difference within the number of significant figures allowed.

(b) This part is exactly the same as part (a). The same numbers, expressed with the same number of significant digits, are used. Be aware that when a small number is added to a much larger one, the small number may not affect the value of the larger within the accuracy of its measurement.

2.61. Perform the following calculations to the proper number of significant digits:
  (a) $(1.20 \times 10^8 \, g + 9.7 \times 10^7 \, g)/(6.04 \times 10^3 \, L)$
  (b) $(1.66 \times 10^{-3} \, m + 6.78 \times 10^{-2} \, m)(1.1 \times 10^4 \, m + 1.3 \times 10^3 \, m)$
  (c) $(7.7 \times 10^{99} \, m + 3.3 \times 10^{99} \, m) - (9.4 \times 10^8 \, m)$
  (d) $(6.02 \times 10^{23} \, atoms + 6.02 \times 10^{21} \, atoms) + (3.95 \times 10^{22} \, atoms)$
  (e) $(1.6 \times 10^{-19} \, cm + 1.1 \times 10^{-20} \, cm) - 1 \, cm$

  Ans. (a) $3.59 \times 10^4 \, g/L$ (The sum has three significant digits.)
       (b) $8.3 \times 10^2 \, m^2$ (The second sum limits the answer to two significant digits.)
       (c) $1.10 \times 10^{100} \, m$ (The sum has three significant digits, and the number subtracted is too small to affect the sum at all.)
       (d) $6.48 \times 10^{23} \, atoms$
       (e) $-1 \, cm$ (The sum is too small to affect -1 cm at all.)

2.62. Determine the answers to the following expression to the correct number of significant digits:
  (a) $18.00 \, mL(1.50 \, g/mL) =$
  (b) $0.170 \, m - 2.2 \, mm =$
  (c) $161 \, cm + 1.53 \, cm =$
  (d) $\dfrac{3.00 \, g}{2.00 \, mL} =$

  Ans. (a) $18.00 \, mL(1.50 \, g/mL) = 27.0 \, g$ (three significant digits)
       (b) $0.170 \, m - 2.2 \, mm = 0.170 \, m - 0.0022 \, m = 0.168 \, m$ (three decimal places)
       (c) $161 \, cm + 1.53 \, cm = 162.53 \, cm \rightarrow 163 \, cm$
       (d) $\dfrac{3.00 \, g}{2.00 \, mL} = 1.50 \, g/mL$ (three significant digits)

2.63. The radius of a circle is 3.00 cm. (a) What is the diameter of the circle? (b) What is the area of the circle?

$$d = 2r \qquad A = \pi r^2$$

  Ans. (a) The diameter is twice the radius:

$$d = 2r = 2(3.00 \, cm) = 6.00 \, cm$$

The 2 here is a definition, not a measurement, and does not limit the accuracy of the answer.

  (b) $A = \pi r^2 = (3.14159)(3.00 \, cm)^2 = 28.3 \, cm^2$

We can use the value of π to as many significant digits as we wish. It is not a measurement.

DENSITY

2.64. Calculate the density of a solution of which 12.5 g occupies 9.6 mL.

  Ans.

$$d = m/V = (12.5 \, g)/(9.6 \, mL) = 1.3 \, g/mL$$

2.65. The density of air is about 1.25 g/L. The density of water is about 1.00 g/mL. Which is more dense?

  Ans. Water is more dense, 1000 g/L. Be sure that when you compare two densities, you use the same units for both.

2.66. The density of lead is 11.34 g/mL. (a) What volume is occupied by 500.0 g of lead? (b) What is the mass of 500.0 mL of lead?

Ans.

(a)
$$500.0 \, \text{g} \left(\frac{1 \, \text{mL}}{11.34 \, \text{g}} \right) = 44.09 \, \text{mL}$$

(b)
$$500.0 \, \text{mL} \left(\frac{11.34 \, \text{g}}{\text{mL}} \right) = 5\,670 \, \text{g} = 5.670 \, \text{kg}$$

2.67. The density of gold is 19.3 g/mL, that of lead is 11.34 g/mL, and that of liquid mercury is 13.6 g/mL. Will gold float in liquid mercury? Will lead?

Ans. Lead will float in mercury, because it is less dense than mercury, but gold will not float because it is more dense.

2.68. Convert a density of 1.0 g/mL to (a) kg/L. (b) kg/m³.

Ans.

(a)
$$\frac{1.0 \, \text{g}}{\text{mL}} \left(\frac{1 \, \text{kg}}{10^3 \, \text{g}} \right) \left(\frac{10^3 \, \text{mL}}{\text{L}} \right) = \frac{1.0 \, \text{kg}}{\text{L}}$$

(b)
$$\frac{1.0 \, \text{kg}}{\text{L}} \left(\frac{10^3 \, \text{L}}{\text{m}^3} \right) = \frac{1.0 \times 10^3 \, \text{kg}}{\text{m}^3}$$

2.69. Calculate the density of a board 3.0 m long, 3.0 cm thick, and 1.0 dm wide, which has a mass of 8.0 kg. Will the board float in water (density 1.00 g/mL)?

Ans. The volume of the board is length × width × thickness.

$$3.0 \, \text{m} \left(\frac{100 \, \text{cm}}{\text{m}} \right) = 300 \, \text{cm long}$$

$$1.0 \, \text{dm} \left(\frac{10 \, \text{cm}}{\text{dm}} \right) = 10 \, \text{cm wide}$$

$$V = 300 \, \text{cm} \times 10 \, \text{cm} \times 3.0 \, \text{cm} = 9.0 \times 10^3 \, \text{cm}^3 = 9.0 \, \text{L}$$

The density is the mass divided by the volume:

$$(8.0 \, \text{kg}) / (9.0 \, \text{L}) = 0.89 \, \text{kg/L}$$

The board will float in water, since it is less dense. The density of water is 1.00 kg/L (see the preceding problem).

TEMPERATURE SCALES

2.70. Convert −40 °F to Celsius.

Ans.

$$°\text{C} = (°\text{F} - 32°)5/9 = (-40° - 32°)5/9 = -40 \, °\text{C}$$

2.71. Change the following temperatures in Celsius to Kelvin. (a) 50 °C, (b) −50 °C, and (c) −272 °C.

Ans. (a) 50 °C + 273° = 323 K
    (b) −50 °C + 273° = 223 K
    (c) −272 °C + 273° = 1 K

2.72. Change the following temperatures in Kelvin to Celsius. (a) 277 K, (b) 1 000 K, and (c) 333 K.

Ans. (a) 4 °C, (b) 727 °C, and (c) 60 °C.

Supplementary Problems

2.73. Change 60.0 km/hour to meters per second.

Ans.

$$\frac{60.0\,\text{km}}{\text{hour}}\left(\frac{1\,000\,\text{m}}{\text{km}}\right)\left(\frac{1\,\text{hour}}{60\,\text{min}}\right)\left(\frac{1\,\text{min}}{60\,\text{s}}\right) = 16.7\,\text{m/s}$$

2.74. Match the value on the left with each corresponding value on the right. (More than one value might correspond for each.)

1 mL	1 dm^3
	1 cm^3
1 L	1 mm^3
	1 m^3
1 kL	10^3 L
	10^{-3} L
1 μL	10^0 L
	10^{-6} L

Ans.

$$1\,\text{mL} = 1\,\text{cm}^3 = 10^{-3}\,\text{L}$$

$$1\,\text{L} = 1\,\text{dm}^3 = 10^0\,\text{L}$$

$$1\,\text{kL} = 1\,\text{m}^3 = 10^3\,\text{L}$$

$$1\,\mu\text{L} = 1\,\text{mm}^3 = 10^{-6}\,\text{L}$$

2.75. Make the following English-metric conversions:
(*a*) Change 2.5 inches to cm
(*b*) Change 15 kg to pounds
(*c*) Change 3.00 L to fluid ounces
(*d*) Change 7.7 pounds to grams
(*e*) Change 60.0 miles per hour to meters per second
(*f*) Change 4.00 inch^2 to cm^2
(*g*) Change 250.0 mL to quarts (U.S.)
(*h*) Change 20.0 feet to meters
(*i*) Change 25 kg to ounces
(*j*) Change 19 km to miles

Ans.

(*a*)
$$2.5\,\text{inches}\left(\frac{2.54\,\text{cm}}{\text{inch}}\right) = 6.4\,\text{cm}$$

(*b*)
$$15\,\text{kg}\left(\frac{2.2\,\text{pounds}}{\text{kg}}\right) = 33\,\text{pounds}$$

(*c*)
$$3.00\,\text{L}\left(\frac{1.06\,\text{quarts}}{\text{L}}\right)\left(\frac{32\,\text{ounces}}{\text{quart}}\right) = 102\,\text{ounces}$$

(*d*)
$$7.7\,\text{pounds}\left(\frac{1.0\,\text{kg}}{2.2\,\text{pounds}}\right)\left(\frac{1\,000\,\text{g}}{\text{kg}}\right) = 3\,500\,\text{g}$$

(*e*)
$$\frac{60.0\,\text{miles}}{\text{hour}}\left(\frac{1\,760\,\text{yards}}{\text{mile}}\right)\left(\frac{36\,\text{inches}}{\text{yard}}\right)\left(\frac{1\,\text{m}}{39.37\,\text{inches}}\right)\left(\frac{1\,\text{hour}}{3\,600\,\text{s}}\right) = \frac{26.8\,\text{m}}{\text{s}}$$

(f) $$4.00 \text{ inch}^2 \left(\frac{2.54 \text{ cm}}{\text{inch}} \right)^2 = 25.8 \text{ cm}^2$$

(g) $$250.0 \text{ mL} \left(\frac{1 \text{ L}}{1\,000 \text{ mL}} \right) \left(\frac{1.06 \text{ quart}}{\text{L}} \right) = 0.265 \text{ quart}$$

(h) $$20.0 \text{ feet} \left(\frac{12 \text{ inches}}{\text{foot}} \right) \left(\frac{2.54 \text{ cm}}{\text{inch}} \right) \left(\frac{1 \text{ m}}{100 \text{ cm}} \right) = 6.10 \text{ m}$$

(i) $$25 \text{ kg} \left(\frac{2.2 \text{ pounds}}{\text{kg}} \right) \left(\frac{16 \text{ ounces}}{\text{pound}} \right) = 880 \text{ ounces}$$

(j) $$19 \text{ km} \left(\frac{0.621 \text{ mile}}{\text{km}} \right) = 12 \text{ miles}$$

2.76. Express, in standard exponential notation, the number of
(a) milliliters in 3.0 L
(b) grams in 4.0 kg
(c) centimeters in 5.0 m
(d) liters in 6.0 mL
(e) grams in 7.0 ng
(f) meters in 9.0 dm
(g) kilograms in 1.0 mg

Ans.
(a) 3.0×10^3 mL
(b) 4.0×10^3 g
(c) 5.0×10^2 cm
(d) 6.0×10^{-3} L
(e) 7.0×10^{-9} g
(f) 9.0×10^{-1} m
(g) 1.0×10^{-6} kg

2.77. Express, in standard exponential notation, the number of
(a) cubic centimeters in 1.0 L
(b) milliliters in 1.0 m^3
(c) liters in 1.0 m^3
(d) cubic meters in 1.0 L
(e) cubic meters in 1.0 cm^3
(f) cubic meters in 1.0 mm^3
(g) cubic meters in 1.0 mL
(h) liters in 1.0 mL
(i) kiloliters in 1.0 m^3

Ans.
(a) $1.0 \times 10^3 \text{ cm}^3$
(b) 1.0×10^6 mL
(c) 1.0×10^3 L
(d) $1.0 \times 10^{-3} \text{ m}^3$
(e) $1.0 \times 10^{-6} \text{ m}^3$
(f) $1.0 \times 10^{-9} \text{ m}^3$
(g) $1.0 \times 10^{-6} \text{ m}^3$
(h) 1.0×10^{-3} L
(i) 1.0 kL

2.78. What exponential number best represents each of the following metric prefixes?
(a) kilo
(b) milli
(c) centi
(d) micro
(e) nano
(f) deci

Ans. (a) 10^3
(b) 10^{-3}
(c) 10^{-2}
(d) 10^{-6}
(e) 10^{-9}
(f) 10^{-1}

2.79. Round off the atomic weights (found in the periodic table) of the first 20 elements to three significant digits each.

Ans.

H	1.01	C	12.0	Na	23.0	S	32.1
He	4.00	N	14.0	Mg	24.3	Cl	35.5
Li	6.94	O	16.0	Al	27.0	Ar	39.9
Be	9.01	F	19.0	Si	28.1	K	39.1
B	10.8	Ne	20.2	P	31.0	Ca	40.1

2.80. What is wrong with the following problem? Change 100 g to milliliters.

Ans. You cannot change a mass to a volume (unless you have a value for density).

2.81. A car was driven 50.0 miles per hour for 1.00 hour, then 30.0 miles per hour for 1.00 hour. Calculate its average speed.

Ans. The average speed is the total distance divided by the total time. The total time is 2.00 hours. The total distance is given by

$$1.00\,\text{hour}\left(\frac{50.0\,\text{mile}}{\text{hour}}\right) + 1.00\,\text{hour}\left(\frac{30.0\,\text{miles}}{\text{hour}}\right) = 80.0\,\text{miles}$$

$$\text{average speed} = (80.0\,\text{miles})/(2.00\,\text{hours}) = 40.0\,\text{miles/hour}$$

2.82. A car was driven 50.0 miles per hour for 1.00 mile, then 30.0 miles per hour for 1.00 mile. Calculate its average speed.

Ans. Note the difference between this problem and the last. Here the total distance is 2.00 miles, and the total time given by

$$1.00\,\text{mile}\left(\frac{1\,\text{hour}}{50.0\,\text{miles}}\right) + 1.00\,\text{mile}\left(\frac{1\,\text{hour}}{30.0\,\text{miles}}\right) = 0.0533\,\text{hour}$$

$$\text{average speed} = 2.00\,\text{miles}/0.0533\,\text{hour} = 37.5\,\text{miles/hour}$$

2.83. The density of an alloy (a mixture) containing 60.0% gold plus other metals is $14.2\,\text{g/cm}^3$. If gold costs $500.00 per ounce avoirdupois, what is the value of the gold in a 13.2-cm^3 piece of jewelry?

Ans.

$$13.2\,\text{cm}^3\left(\frac{14.2\,\text{g alloy}}{\text{cm}^3}\right)\left(\frac{60.0\,\text{g gold}}{100.0\,\text{g alloy}}\right)\left(\frac{1\,\text{ounce gold}}{28.35\,\text{g gold}}\right)\left(\frac{500.00\,\text{dollars}}{\text{ounce gold}}\right) = 1\,980\,\text{dollars}$$

2.84. A solution with density 1.13 g/mL contains 30.0 g of sugar per 100 mL of solution. What mass of solution will contain 500.0 g of sugar? What mass of water is required to make this solution?

Ans.

$$500.0\,\text{g sugar}\left(\frac{100\,\text{mL solution}}{30.0\,\text{g sugar}}\right)\left(\frac{1.13\,\text{g solution}}{\text{mL solution}}\right) = 1\,880\,\text{g solution}$$

$$1\,880\,\text{g solution} - 500\,\text{g sugar} = 1\,380\,\text{g water}$$

2.85. How far can Joe drive his car on $23.00 if the car goes 12.5 miles per gallon and gasoline costs $1.11 per gallon?

Ans.

$$23.00 \text{ dollars}\left(\frac{1 \text{ gallon}}{1.11 \text{ dollars}}\right)\left(\frac{12.5 \text{ miles}}{\text{gallon}}\right) = 259 \text{ miles}$$

2.86. How far can Joe drive his car on $23.00 if the car goes 12.5 miles per gallon and gasoline costs $1.11 per gallon, with a 4.00% discount for cash?

Ans. If there is a 4.00% discount, the amount to be paid is 96.0% of list price.

$$23.00 \text{ dollars paid}\left(\frac{1.00 \text{ dollar list}}{0.960 \text{ dollar paid}}\right)\left(\frac{1 \text{ gallon}}{1.11 \text{ dollars}}\right)\left(\frac{12.5 \text{ miles}}{\text{gallon}}\right)$$

$$= 270 \text{ miles}$$

2.87. Calculate the density of sulfuric acid, 1.86 g/mL, in kilograms per liter.

Ans.

$$\frac{1.86 \text{ g}}{\text{mL}}\left(\frac{1 \text{ kg}}{1\,000 \text{ g}}\right)\left(\frac{1\,000 \text{ mL}}{\text{L}}\right) = \frac{1.86 \text{ kg}}{\text{L}}$$

2.88. Calculate the density of 200.0 g rectangular block with length 7.00 cm, width 3.00 cm, and thickness 2.00 cm.

Ans. 4.76 g/cm^3.

2.89. A gold alloy chain has a density of 15.0 g/cm^3. The cross-sectional area of the chain is 3.00 mm^2. What length of chain has a mass of 10.00 g?

Ans.

$$V = 10.00 \text{ g}\left(\frac{1 \text{ cm}^3}{15.0 \text{ g}}\right) = 0.667 \text{ cm}^3\left(\frac{10 \text{ mm}}{\text{cm}}\right)^3 = 667 \text{ mm}^3$$

$$\text{length} = \frac{V}{A} = \frac{667 \text{ mm}^3}{3.00 \text{ mm}^2} = 222 \text{ mm} = 22.2 \text{ cm}$$

2.90. How much does the chain of the last problem cost if the gold alloy costs $450/ounce? (1 ounce = 28.35 g)

Ans.

$$10.00 \text{ g}\left(\frac{1 \text{ ounce}}{28.35 \text{ g}}\right)\left(\frac{\$450.00}{\text{ounce}}\right) = \$158.70$$

2.91. Using the data of Problem 2.67, determine whether a cube of metal 2.0 cm on each side is mercury, gold, or lead. The mass of the cube is 154 g.

Ans. The volume of a cube is equal to its length cubed, or $(2.0 \text{ cm})^3 = 8.0 \text{ cm}^3$.

$$d = m/V = (154 \text{ g})/(8.0 \text{ cm}^3) = 19 \text{ g/cm}^3$$

The metal is gold, because the calculated density is closer to that of gold than to that of either of the other metals.

2.92. Commercial sulfuric acid solution is 96.0% H_2SO_4 in water and has a density of 1.86 g/mL. How many grams of H_2SO_4 is there in 200.0 mL of the commercial solution?

Ans.

$$200.0 \, \text{mL solution} \left(\frac{1.86 \, \text{g solution}}{\text{mL solution}} \right) \left(\frac{96.0 \, \text{g H}_2\text{SO}_4}{100 \, \text{g solution}} \right) = 357 \, \text{g H}_2\text{SO}_4$$

         ↑   ↑   ↑

   quantity given   density of solution   percent

2.93. An ashtray is made of 50.0 g of glass and 50.0 g of iron. The density of the glass is 2.50 g/mL and that of the iron is 7.86 g/mL. What is the density of the ashtray?

Ans. The density of the ashtray is its mass divided by its volume. Its mass is 100.0 g. Its volume is the total of the volumes of the glass and iron:

$$50.0 \, \text{g glass} \left(\frac{1 \, \text{mL}}{2.50 \, \text{g}} \right) = 20.0 \, \text{mL glass}$$

$$50.0 \, \text{g iron} \left(\frac{1 \, \text{mL}}{7.86 \, \text{g}} \right) = 6.36 \, \text{mL iron}$$

The total volume is 26.4 mL, and the density is

$$d = m/V = (100.0 \, \text{g})/(26.4 \, \text{mL}) = 3.79 \, \text{g/mL}$$

Chapter 3

Atoms and Atomic Weights

3.1 INTRODUCTION

In this chapter we will discuss:
1. Dalton's postulates regarding the existence of the atom and the laws on which those postulates are based.
2. Atomic weights, their uses and limitations.
3. The structure of the atom.
4. The existence of isotopes.
5. The periodic table, which for now is presented only enough to introduce the concepts of periodic groups or families and the numbers of electrons in the outermost electron shells.

3.2 THE ATOMIC THEORY

In 1804, John Dalton proposed the existence of atoms. He not only postulated that atoms exist, as had ancient Greek philosophers, but he also attributed to the atoms certain properties. His postulates were as follows:
1. Elements are composed of indivisible particles, called atoms.
2. Atoms of a given element all have the same mass, and the mass of an atom of a given element is different from the mass of an atom of any other element.
3. When atoms combine to form compounds, the atoms of one element combine with those of the other element(s) to form molecules.
4. Atoms of two or more elements may combine in different ratios to form different compounds.
5. The most common ratio of atoms is 1:1, and where more than one compound of two or more elements exist, the most stable is the one with 1:1 ratio of atoms. This postulate is incorrect.

Dalton's postulates created great activity among chemists, who sought to prove or disprove them. The fifth postulate was very quickly shown to be incorrect, and the first three have had to be modified in the light of later knowledge. However, the first four postulates were close enough to the truth to lay the foundations for a basic understanding of mass relationships in chemical compounds and chemical reactions.

Dalton's postulates were based on three laws which had been developed shortly before he proposed his theory.

The law of conservation of mass states that mass is neither created nor destroyed in a chemical reaction.

The law of definite composition states that every chemical compound is made up of elements in a definite ratio by mass.

The law of multiple proportions states that when two different compounds are formed from the same elements, the ratio of masses of one element in the two compounds for a given mass of any other element is a small whole number.

Dalton argued that these laws are entirely reasonable if the elements are composed of atoms. For example, the reason that mass is neither gained nor lost in a chemical reaction is that the atoms merely change partners with each other; they do not appear or disappear. The constant composition of compounds stems from the fact that the compounds consist of a definite ratio of atoms, each with a definite mass. The law of multiple proportions is due to the fact that different numbers of atoms of

44

one element can react with a given number of atoms of a second element, and since the atoms must combine in whole number ratios, the ratio of masses must also be in whole numbers.

3.3   ATOMIC WEIGHTS

Once Dalton's hypotheses had been proposed, the next logical step was to determine the *relative* masses of the atoms of the elements. Since there was no way at that time to determine the mass of an individual atom, the relative masses were the best information available. That is, one could tell that an atom of one element had a mass twice as great as an atom of a different element (or $\frac{15}{4}$ times as much, or 17.3 times as much, etc.). How could even these relative masses be determined? They could be determined by taking equal (large) numbers of atoms of two elements and by determining the ratio of masses of these collections of atoms.

For example, a large number of carbon atoms has a total mass of 12.0 g, and an equal number of oxygen atoms has a total mass of 16.0 g. Since the number of atoms of each kind is equal, the ratio of masses of one carbon atom to one oxygen atom is 12.0 to 16.0. How can one be sure that there are an equal number of carbon and oxygen atoms? One ensures equal numbers by using a compound of carbon and oxygen in which there are an equal number of atoms of the two elements (i.e., carbon monoxide).

A great deal of difficulty was encountered at first, because Dalton's fifth postulate gave an incorrect ratio of numbers of atoms in many cases. Such a large number of incorrect results were obtained that it soon became apparent that the fifth postulate was not correct. It was not until some 50 years later than an experimental method was devised to determine the atomic ratios in compounds, at which time the scale of relative atomic weights was determined in almost the present form. These relative weights are called the *atomic weights*.

The atomic weight of the lightest element, hydrogen, was originally taken to be 1 atomic mass unit (amu). The modern values of the atomic weights are based on the most common kind of carbon atom, called "carbon-twelve" and written ^{12}C, as the standard. The mass of ^{12}C is measured in the modern *mass spectrometer*, and ^{12}C is *defined* to have an atomic weight of exactly 12 amu. On this scale hydrogen has an atomic weight of 1.008 amu.

Different names are used for the atomic weight by different authors and different abbreviations are used for it. The term *dalton* is used by some, in honor of John Dalton, and these authors use the abbreviation D. Other authors use the name *atomic mass unit*. The abbreviation u rather than amu is sometimes encountered.

The atomic weight of an element is the relative mass of an average atom of the element compared with ^{12}C, which has an atomic weight of exactly 12. Thus, since a sulfur atom has a mass $\frac{8}{3}$ times that of a carbon atom, the atomic weight of sulfur is

$$12 \text{ amu} \times \tfrac{8}{3} = 32 \text{ amu}$$

A complete list of the modern values of the atomic weights of the elements is given in Table 3-1.

3.4   ATOMIC STRUCTURE

From 50 years to 100 years after Dalton proposed his theory, various discoveries showed that the atom is not indivisible, but really is composed of parts. Natural radioactivity and the interaction of electricity with matter are two different types of evidence for this subatomic structure. The most important subatomic particles are listed in Table 3-2, along with their most important properties. The protons and neutrons occur in a very tiny *nucleus* (plural, nuclei). The electrons occur outside the nucleus.

There are two types of electric charges that occur in nature—positive and negative. Charges of these two types are opposite one another, and cancel the effect of the other. Bodies with opposite charge types attract each other; those with the same charge type repel each other. If a body has equal

Table 3-1 The Elements

Element	Symbol	Atomic Number	Atomic Weight
Actinium	Ac	89	(227)
Aluminum	Al	13	26.9815
Americium	Am	95	(243)
Antimony	Sb	51	121.75
Argon	Ar	18	39.948
Arsenic	As	33	74.9216
Astatine	At	85	(210)
Silver	Ag	47	107.868
Gold	Au	79	196.9665
Barium	Ba	56	137.34
Berkelium	Bk	97	(249)
Beryllium	Be	4	9.01218
Bismuth	Bi	83	208.9806
Boron	B	5	10.81
Bromine	Br	35	79.904
Cadmium	Cd	48	112.40
Calcium	Ca	20	40.08
Californium	Cf	98	(251)
Carbon	C	6	12.011
Cerium	Ce	58	140.12
Cesium	Cs	55	132.9055
Chlorine	Cl	17	35.453
Chromium	Cr	24	51.996
Cobalt	Co	27	58.9332
Copper	Cu	29	63.546
Curium	Cm	96	(247)
Dysprosium	Dy	66	162.50
Einsteinium	Es	99	(254)
Erbium	Er	68	167.26
Europium	Eu	63	151.96
Fermium	Fm	100	(253)
Fluorine	F	9	18.9984
Francium	Fr	87	(223)
Iron	Fe	26	55.847
Gadolinium	Gd	64	157.25
Gallium	Ga	31	69.72
Germanium	Ge	32	72.59
Gold	Au	79	196.9665
Hafnium	Hf	72	178.49
Helium	He	2	4.00260
Holmium	Ho	67	164.9303
Hydrogen	H	1	1.0080
Mercury	Hg	80	200.59
Indium	In	49	114.82
Iodine	I	53	126.9045
Iridium	Ir	77	192.22
Iron	Fe	26	55.847
Krypton	Kr	36	83.80
Potassium	K	19	39.102
Lanthanum	La	57	138.9055
Lawrencium	Lr	103	(257)
Lead	Pb	82	207.2
Lithium	Li	3	6.941
Lutetium	Lu	71	174.97
Magnesium	Mg	12	24.305
Manganese	Mn	25	54.9380
Mendelevium	Md	101	(256)

Table 3-1 The Elements (*Continued*)

Element	Symbol	Atomic Number	Atomic Weight
Mercury	Hg	80	200.59
Molybdenum	Mo	42	95.94
Neodymium	Nd	60	144.24
Neon	Ne	10	20.179
Neptunium	Np	93	237.0482
Nickel	Ni	28	58.71
Niobium	Nb	41	92.9064
Nitrogen	N	7	14.0067
Nobelium	No	102	(254)
Sodium	Na	11	22.9898
Osmium	Os	76	190.2
Oxygen	O	8	15.9994
Palladium	Pd	46	106.4
Phosphorus	P	15	30.9738
Platinum	Pt	78	195.09
Plutonium	Pu	94	(242)
Polonium	Po	84	(210)
Potassium	K	19	39.102
Praseodymium	Pr	59	140.9077
Promethium	Pm	61	(145)
Protactinium	Pa	91	231.0359
Lead	Pb	82	207.2
Radium	Ra	88	226.0254
Radon	Rn	86	(222)
Rhenium	Re	75	186.2
Rhodium	Rh	45	102.9055
Rubidium	Rb	37	85.4678
Ruthenium	Ru	44	101.07
Samarium	Sm	62	150.4
Scandium	Sc	21	44.9559
Selenium	Se	34	78.96
Silicon	Si	14	28.086
Silver	Ag	47	107.868
Sodium	Na	11	22.9898
Strontium	Sr	38	87.62
Sulfur	S	16	32.06
Antimony	Sb	51	121.75
Tin	Sn	50	118.69
Tantalum	Ta	73	180.9479
Technetium	Tc	43	98.9062
Tellurium	Te	52	127.60
Terbium	Tb	65	158.9254
Thallium	Tl	81	204.37
Thorium	Th	90	232.0381
Thulium	Tm	69	168.9342
Tin	Sn	50	118.69
Titanium	Ti	22	47.90
Tungsten	W	74	183.85
Uranium	U	92	238.029
Vanadium	V	23	50.9414
Tungsten	W	74	183.85
Xenon	Xe	54	131.30
Ytterbium	Yb	70	173.04
Yttrium	Y	39	88.9059
Zinc	Zn	30	65.37
Zirconium	Zr	40	91.22

numbers of charges of the two types, it has no net charge, and is said to be *neutral*. The charge on the electron is a fundamental unit of electric charge (equal to 1.6×10^{-19} C), and is given the symbol *e*.

EXAMPLE 3.1.   Using the data of Table 3-2, find the charge on a nucleus which contains (*a*) 6 protons and 6 neutrons and (*b*) 6 protons and 8 neutrons.

$$(a) \quad 6(+1) + 6(0) = +6 \qquad (b) \quad 6(+1) + 8(0) = +6$$

Both nuclei have the same charge. Although the nuclei have different numbers of neutrons, the neutrons have no charges, so they do not affect the charge on the nucleus.

Table 3-2   Subatomic Particles

	Charge (*e*)	Mass (amu)	Location
Proton	+1	1.00728	In nucleus
Neutron	0	1.00894	In nucleus
Electron	−1	0.0005414	Outside nucleus

EXAMPLE 3.2.   To the nearest integer, calculate the mass (in amu) of a nucleus which contains (*a*) 6 protons and 6 neutrons and (*b*) 6 protons and 8 neutrons.
(*a*)   6(1 amu) + 6(1 amu) = 12 amu
(*b*)   6(1 amu) + 8(1 amu) = 14 amu

The nuclei differ in mass but not in charge.

Uncombined atoms as a whole are electrically neutral.

EXAMPLE 3.3.   Refer to Table 3-2 and deduce which two of the types of subatomic particles in an uncombined atom occur in equal numbers.

The number of positive charges must equal the number of negative charges, since the atom has a net charge of 0. The number of positive charges, as shown in the table, is equal to the number of protons. The number of negative charges, also from the table, is equal to the number of electrons. Therefore, in an uncombined atom, the number of protons must equal the number of electrons.

The number of protons in the nucleus determines the chemical properties of the element. That number is called the *atomic number* of the element. Each element has a different atomic number. An element may be identified by giving its name or its atomic number. Atomic numbers may be specified by use of a subscript before the symbol of the element. For example, carbon may be designated $_6$C. The subscript is really unnecessary, since all carbon atoms have atomic number 6, but it is sometimes useful to include it. Atomic numbers are listed in the periodic table and in Table 3-1.

EXAMPLE 3.4.   (*a*) What is the charge on a neon nucleus? (*b*) What is the charge on a neon atom?

(*a*) +10, equal to the atomic number of neon (from Table 3-1). (*b*) 0 (all uncombined atoms have net charge of 0). Note that these questions sound very much alike, but are very different. You must read questions in chemistry very carefully.

3.5 ISOTOPES

Atoms having the same number of protons but different numbers of neutrons are called *isotopes* of each other. The number of neutrons does not affect the chemical properties of the atoms appreciably, so all isotopes of a given element have essentially the same chemical properties. Different isotopes have different masses and different nuclear properties, however.

The sum of the number of protons plus number of neutrons in the isotope is called the *mass number* of the isotope. Isotopes are usually distinguished from each other by their mass numbers, given as a superscript before the chemical symbol for the element. Carbon-twelve is an isotope of carbon with a symbol ^{12}C.

EXAMPLE 3.5. (*a*) What is the sum of the number of protons and the number of neutrons in ^{12}C? (*b*) What is the number of protons in ^{12}C? (*c*) What is the number of neutrons in ^{12}C?

(*a*) 12, its mass number. (*b*) 6, its atomic number, given in the periodic table. (*c*) $12 - 6 = 6$. Note that the mass number for most elements is *not* given in the periodic table.

EXAMPLE 3.6. Choose the integer quantities from the following list: (*a*) atomic number, (*b*) atomic weight, and (*c*) mass number.

Atomic number and mass number are integer quantities; atomic weight is not in general equal to an integer.

EXAMPLE 3.7. Choose the quantities which appear in the periodic table from the following list: (*a*) atomic number, (*b*) atomic weight, and (*c*) mass number.

Atomic number and atomic weight appear in the periodic table. The mass numbers of only a few elements (which do not occur naturally) appear there in parentheses.

The atomic weight of an element is the weighted average of the masses of the individual isotopes of the element.

EXAMPLE 3.8. ^{79}Br comprises 50.54% of naturally occurring bromine, and ^{81}Br comprises 49.46%. The mass of ^{79}Br is 78.9183 amu and the mass of ^{81}Br is 80.9163 amu. What is the atomic weight of bromine?

The *weighted average* is the sum of the mass of each isotope times its fraction present:

$$78.9183 \text{ amu}(0.5054) + 80.9163 \text{ amu}(0.4946) = 79.91 \text{ amu}$$

3.6 THE PERIODIC TABLE

The *periodic table* is an extremely useful tabulation of the elements. It is constructed in a manner such that each vertical column contains elements which are chemically similar. The elements in the columns are called *groups* or *families*. (Elements in some of the groups are very similar to each other. Elements in others of the groups are less similar. For example, the elements of the first group resemble each other more than the elements of the fourth group from the end, headed by N.) Each row in the table is called a *period* (Fig. 3-1).

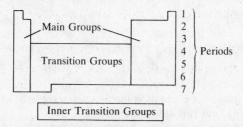

Fig. 3-1   The periodic table

There are three distinct areas of the periodic table—the *main group elements, the transition group elements,* and *the inner transition group elements.* We will focus our attention at first on the main group elements, whose properties are easiest to learn and to understand.

The periods and the groups are identified differently. The periods are labeled from 1 to 7. Some reference is made to period numbers. The groups are referred to extensively by number. Unfortunately, the groups have been labeled in three different ways:

Classical   Main groups are labeled IA through VIIA plus 0. Transition groups are labeled IB through VIII (although not in that order).

Amended   Main groups and transition groups are labeled IA through VIII and then IB through VIIB plus 0.

Modern   Groups are labeled with Arabic numerals from 1 through 18.

Although the modern designation is seemingly simpler, it does not point up some of the relationships that the older designations do. In this book, the classical system will be followed, with the modern number often included in parentheses. The designations are tabulated in Table 3-3. You should check to see which designation your text uses, and use that designation yourself.

Table 3-3 Periodic Group Designations

Classical	IA	IIA	IIIB	$\cdots$	VIII	IB	IIB	IIIA	$\cdots$	VIIA	0
Amended	IA	IIA	IIIA	$\cdots$	VIII	IB	IIB	IIIB	$\cdots$	VIIB	0
Modern	1	2	3	$\cdots$	8 9 10	11	12	13	$\cdots$	17	18

Note that three columns are collected as one group (VIII) in the two older designations.

EXAMPLE 3.9. (a) What is the relationship between the *numbers* of the first eight columns in all three designations? (b) of the last eight columns?

(a) They are all the same, 1 through 8. (b) The modern designation is 10 higher for each of these groups except the last, where it is 18 higher.

Several important groups are given names. The Group IA (1) metals (not including hydrogen) are called the *alkali metals*. Group IIA (2) elements are known as the *alkaline earth metals*. The Group VIIA (17) elements are called the *halogens*. The Group IB (11) metals are known as the *coinage metals*. These names lessen the need for using group numbers, and thereby the confusion from the different systems.

The electrons in atoms are arranged in shells. (A more detailed account of electronic structure will be presented in Chap. 17.) The maximum number of electrons that can fit in any shell n is given by

$$\text{maximum number} = 2n^2$$

Since there are only about 100 electrons total in even the biggest atoms, it can easily be seen that the shells numbered 5 or higher never get filled with electrons. Another important limitation is that the outermost shell, called the *valence shell*, can never have more than eight electrons in it. The number of electrons in the valence shell is a periodic property.

Shell number	1	2	3	4	5	6	7
Maximum number of electrons	2	8	18	32	50	72	98
Maximum number as outermost shell	2	8	8	8	8	8	8

EXAMPLE 3.10. (a) How many electrons could fit in the first seven shells of an atom if the shells filled to their capacity in numeric order? (b) Why does this not happen?

(a) A total of 280 (2 + 8 + 18 + 32 + 50 + 72 + 98) electrons could be held in this fictional atom. (b) Filling one atom with 280 electrons does not happen because the shells do not fill to capacity before the next ones start to fill and also because even the biggest atom has only a few more than 100 electrons total.

EXAMPLE 3.11. (a) What is the maximum number of electrons in the third shell of an atom in which there are electrons in the fourth shell? (b) What is the maximum number of electrons in the third shell of an atom in which there are no electrons in the fourth shell?

(*a*) 18 is the maximum number in the third shell. (*b*) 8 is the maximum number if there are no electrons in higher shells. (There can be no electrons in higher shells if there are none in the fourth shell.)

EXAMPLE 3.12. What is the maximum number of electrons in the first shell when it is the outermost shell? when it is not the outermost shell?

The maximum number is 2. It does not matter if it is the outermost shell or not, 2 is the maximum number of electrons in the first shell.

EXAMPLE 3:13. Arrange the 11 electrons of sodium into shells.

$$2 \qquad 8 \qquad 1$$

The first two electrons fill the first shell, and the next eight fill the second shell. That leaves the one electron left in the third shell.

The number of outermost electrons is crucial to the chemical bonding of the atom. (See Chap. 5.) For main group elements, the number of outermost electrons is equal to the classical group number, except that it is 2 for helium and 8 for the other group 0 (18) elements. (It is equal to the modern group number minus 10 except for helium and the first two groups.)

Electron dot diagrams use the chemical symbol to represent the nucleus plus the inner electrons and a dot to represent each valence electron. Such diagrams will be extremely useful in Chap. 5.

EXAMPLE 3.14. Write the electron dot symbols for the elements Li through Ne.

$$\text{Li}\cdot \qquad \text{Be}: \qquad \cdot\text{B}: \qquad \cdot\dot{\text{C}}: \qquad \cdot\dot{\text{N}}: \qquad :\dot{\text{O}}: \qquad :\ddot{\text{F}}: \qquad :\ddot{\text{Ne}}:$$

Solved Problems

THE ATOMIC THEORY

3.1. Why was Dalton's contribution different from that of the ancient Greeks who postulated the existence of atoms?

Ans.   Dalton based his postulates on experimental evidence; the Greeks did not.

3.2. Could Dalton have done his work 100 years earlier?

Ans.   No; the laws on which his theory was based had not been established until shortly before his work.

3.3. Two oxides of iron have compositions as follows:

Compound 1	77.7% Fe	22.3% O
Compound 2	69.9% Fe	30.1% O

Show that these compounds obey the law of multiple proportions.

Ans.   The ratio of the mass of one element in one compound to the mass of that element in the other compound—for a fixed mass of the other element—must be in a ratio of small whole numbers. Let us calculate the mass of oxygen in the two compounds per gram of iron. That is, the fixed mass will be 1.00 g Fe.

$$1.00\,\text{g Fe}\left(\frac{22.3\,\text{g O}}{77.7\,\text{g Fe}}\right) = 0.287\,\text{g O}$$

$$1.00\,\text{g Fe}\left(\frac{30.1\,\text{g O}}{69.9\,\text{g Fe}}\right) = 0.431\,\text{g O}$$

Is the ratio of the masses of oxygen a ratio of small integers?

$$\frac{0.431\,g}{0.287\,g} = \frac{1.50}{1} = \frac{3}{2}$$

The ratio of 1.50:1 is equal to the ratio 3:2, and the law of multiple proportions is satisfied. Note that the ratio of masses of iron to oxygen is not necessarily a ratio of small integers; the ratio of mass of oxygen in one compound to mass of oxygen in the other compound is what must be in the small integer ratio.

3.4. A sample of water purified from an iceberg in the Arctic Ocean contains 88.8% oxygen by mass. (*a*) What is the percent of oxygen by mass in water purified from a tropical forest? (*b*) How can you predict that percentage?

   Ans.   (*a*) 88.8%. (*b*) The law of definite proportions states that all water, no matter what its source, contains the same percentage of its constituent elements.

3.5. If a 20-pound log burns into 3 ounces of ashes, how is the law of conservation of mass obeyed?

   Ans.   The log plus oxygen has a certain mass. The ashes plus the carbon dioxide and water (and perhaps a few other compounds) must have a combined mass which totals the same as the combined mass of the log and oxygen. The law does not state that the total mass before and after the reaction must be the mass of the solids only.

3.6. What happens to a postulate of a scientific theory if it leads to incorrect results?

   Ans.   It is changed or abandoned.

3.7. Chemists do not use the law of multiple proportions in their everyday work. Why was this law introduced in this book?

   Ans.   It was introduced to show that Dalton's atomic theory was based on experimental data.

ATOMIC WEIGHTS

3.8. The average man in a certain class weighs 180 pounds. The average woman in that class weighs 120 pounds. The male/female ratio is 3:2. What is the average weight of a person in the class?

   Ans.

$$\left(\frac{3}{5}\right)(180\,\text{pounds}) + \left(\frac{2}{5}\right)(120\,\text{pounds}) = 156\,\text{pounds}$$

3.9. What is the difference, if any, between (*a*) atomic weight and atomic mass? (*b*) atomic weight unit and atomic mass unit?

   Ans.   (*a*) Atomic weight and atomic mass are synonymous, as are (*b*) atomic weight unit and atomic mass unit. (Note that the atomic mass unit is the unit of atomic mass or atomic weight.)

3.10. Determine the total weight of all the men in your chemistry class, and the total weight of all the women (by using an anonymous questionnaire). (*a*) Attempt to determine the ratio of the average weight of each man to that of each woman without counting the number of each. (*b*) Attempt the same determination at a dated party.

   Ans.   (*a*) You cannot determine the ratio of averages without knowing something about the numbers present. (*b*) At a dated party, where there is apt to be a 1:1 ratio of men to women, the ratio of the averages is equal to the ratio of the total weights.

ATOMIC STRUCTURE

3.11. (*a*) How many protons are there in the nucleus of a chlorine atom? (*b*) How many protons are there in the chlorine atom?

    Ans. (*a*) and (*b*) 17 (all the protons are in the nucleus, so it is not necessary to specify the nucleus). Here are two questions which sound different, but really are the same. Again, you must read the questions carefully. You must understand the concepts and the terms and you must not try to memorize.

3.12. Explain on the basis of the information in Table 3-2 (*a*) why the nucleus is positively charged. (*b*) why the nucleus contains most of the mass of the atom. (*c*) why electrons are attracted to the nucleus. (*d*) why the atom may be considered mostly empty space.

    Ans. (*a*) The only positively charged particles, the protons, are located there. (*b*) The protons and neutrons, both massive compared to the electrons, are located there. (*c*) The nucleus is positively charged and they are negatively charged. (*d*) The nucleus is tiny compared to the atom as a whole, and it contains most of the mass of the atom. Thus, the remainder of the atom contains little mass, and may be thought of as mostly empty space.

3.13. If uncombined atoms as a whole are neutral, how can they be made up of charged particles?

    Ans. The number of positive charges and negative charges (protons and electrons) happen to be equal, and the effect of the positive charges cancels the effect of the negative charges.

ISOTOPES

3.14. Complete the following table (for uncombined atoms):

	Symbol	Atomic Number	Mass Number	Number of Protons	Number of Electrons	Number of Neutrons
(*a*)	^{14}C					
(*b*)		17	35			
(*c*)				92		143
(*d*)					16	18
(*e*)		35				44

  Ans.

	Symbol	Atomic Number	Mass Number	Number of Protons	Number of Electrons	Number of Neutrons
(*a*)	^{14}C	6	14	6	6	8
(*b*)	^{35}Cl	17	35	17	17	18
(*c*)	^{235}U	92	235	92	92	143
(*d*)	^{34}S	16	34	16	16	18
(*e*)	^{79}Br	35	79	35	35	44

    (*a*) Atomic number is determined from the identity of the element, and is equal to the number of protons and to the number of electrons. Its value is given in the periodic table. The mass number is given, and the number of neutrons is equal to the mass number minus the atomic number. (*b*) The atomic number and the number of protons are always the same, and these are the same as the number of electrons in uncombined atoms. (*c*) The sum of the number of protons and neutrons equals the mass number. In general, if we use Z for the atomic number, A for the mass number, and n for the number of neutrons, we get the simple equation:

$$A = Z + n$$

Once we are given two of these, we may calculate the third. Also,

$$Z = \text{number of protons} = \text{number of electrons in uncombined atom}$$

3.15. Can you guess from data in the periodic table what the most important isotope of bromine is? With the additional information that ^{80}Br does not occur naturally, how can you amend that guess?

Ans. The atomic weight of Br is about 80, so one might be tempted to guess that ^{80}Br is the most important isotope. Knowing that it is not, one then can guess that an about-equal mixture of ^{79}Br and ^{81}Br occurs, which is correct.

3.16. The number of which subatomic particle is different in atoms of two isotopes?

Ans. The number of neutrons differs.

3.17. Consider the results of Problem 3.16 and the fact that ^{12}C and ^{14}C have very similar chemical properties. Which is more important in determining chemical properties, nuclear charge or nuclear mass?

Ans. Since the charge is the same and the mass differs, the chemical properties are due to the charge.

3.18. What is the net charge on an atom which contains 17 protons, 18 neutrons, and 18 electrons?

Ans. The charge is -1. There are 18 electrons, each with -1 charge for a total of -18, and there are 17 protons each with a $+1$ charge, for a total of $+17$. It does not matter how many neutrons there are, since they have no charge. $+17 + (-18) = -1$. If the initial letter of the particle stands for the number of that particle, then

$$\text{charge} = (+1)p + (-1)e + (0)n = (+1)(17) + (-1)(18) + (0)(18) = -1$$

3.19. What is the mass number of an atom with 17 protons, 18 neutrons, and 17 electrons?

Ans. The mass number is the number of protons plus neutrons; the number of electrons is immaterial.
$$17 + 18 = 35$$

3.20. Two uncombined atoms are isotopes of each other. One has 24 electrons. How many electrons does the other have?

Ans. The second isotope also has 24 electrons. The only difference between isotopes is the number of neutrons.

3.21. Complete the following table (for uncombined atoms):

	Symbol	Atomic Number	Mass Number	Number of Protons	Number of Electrons	Number of Neutrons
(a)	^{51}V					
(b)		56	137			
(c)			56	26		
(d)			31		15	
(e)		38				52

Ans.

	Symbol	Atomic Number	Mass Number	Number of Protons	Number of Electrons	Number of Neutrons
(a)	^{51}V	23	51	23	23	28
(b)	^{137}Ba	56	137	56	56	81
(c)	^{56}Fe	26	56	26	26	30
(d)	^{31}P	15	31	15	15	16
(e)	^{90}Sr	38	90	38	38	52

3.22. What subscripts could be used in the symbols in column 1 of the answer to Problem 3.14?

Ans.   The atomic numbers can be used as subscripts to the left of the chemical symbols.

3.23. (*a*) Which columns of the answer to Problem 3.14 are always the same as other columns for uncombined atoms? (*b*) Which column is the sum of two others? (*c*) Which column is the same as the superscripts of the symbols?

Ans.   (*a*)   Atomic number, number of protons, and number of electrons.
     (*b*)   Mass number = number of protons + number of neutrons.
     (*c*)   The superscripts in the first column are repeated in the third column.

THE PERIODIC TABLE

3.24. How are the 19 electrons of the potassium atom arranged in shells?

Ans.

$$2 \qquad 8 \qquad 8 \qquad 1$$

The first shell of any atom holds a maximum of two electrons, and the second shell holds a maximum of eight. Thus, the first two electrons of potassium fill the first shell, and the next eight fill the second shell. The outermost shell of any atom can hold at most eight electrons. In potassium, there are nine electrons left, which would fit into the third shell if it were not the outermost shell. However, if we put the nine electrons into the third shell, it would be the outermost shell. Therefore, we put 8 of the remaining electrons in that shell. That leaves the one electron left in the fourth shell.

3.25. How many electrons are there in the outermost shell of each of the following elements? (*a*) Na, (*b*) Mg, (*c*) Al, (*d*) Si, (*e*) P, (*f*) S, (*g*) Cl, and (*h*) Ar.

Ans.   (*a*) 1, (*b*) 2, (*c*) 3, (*d*) 4, (*e*) 5, (*f*) 6, (*g*) 7, and (*h*) 8.

3.26. Write the electron dot symbol for each atom in Problem 3.25.
Ans.

$$\text{Na·} \qquad \text{Mg:} \qquad \text{:Al·} \qquad \text{:Si·} \qquad \text{:}\overset{\cdot}{\text{P}}\text{·} \qquad \text{:}\overset{\cdot}{\text{S}}\text{:} \qquad \text{:}\overset{\cdot}{\text{Cl}}\text{:} \qquad \text{:}\overset{\cdot\cdot}{\text{Ar}}\text{:}$$

3.27. How many electrons are there in the outermost shell of (*a*) H and (*b*) He?

Ans.   (*a*) One. There is only one electron in the whole atom. The shell can hold two electrons, but in the free hydrogen atom, there is only one electron present. (*b*) Two.

Supplementary Problems

3.28. Distinguish clearly between (*a*) neutron and nucleus, (*b*) mass number and atomic weight, (*c*) atomic number and mass number, (*d*) atomic number and atomic weight, (*e*) atomic weight and atomic mass unit, and (*f*) atomic mass and atomic mass unit.

Ans.   (*a*) The nucleus is a distinct part of the atom. Neutrons are subatomic particles which, along with protons, are located in the nucleus. (*b*) Mass number refers to individual isotopes. It is the sum of the numbers of protons and neutrons. Atomic weight refers to the naturally occurring mixture of isotopes, and is the relative mass of the average atom compared to ^{12}C. (*f*) Atomic mass is the same as atomic weight [see (*b*)]. Atomic mass unit is the unit of atomic weight.

3.29. Complete the following table (for uncombined atoms):

	Symbol	Atomic Number	Mass Number	Number of Protons	Number of Electrons	Number of Neutrons
(a)	^{13}C					
(b)		11				12
(c)			61		27	
(d)				35		46
(e)			234			144
(f)					38	52

Ans.

	Symbol	Atomic Number	Mass Number	Number of Protons	Number of Electrons	Number of Neutrons
(a)	^{13}C	6	13	6	6	7
(b)	^{23}Na	11	23	11	11	12
(c)	^{61}Co	27	61	27	27	34
(d)	^{81}Br	35	81	35	35	46
(e)	^{234}Th	90	234	90	90	144
(f)	^{90}Sr	38	90	38	38	52

3.30. Does hydrogen peroxide, H_2O_2, have the same composition as water, H_2O? Does your answer violate the law of definite proportions? Explain briefly.

Ans. The compounds have different ratios of hydrogen to oxygen atoms, and thus different mass ratios. The law of definite proportions applies to each compound individually, not to the two different compounds. H_2O and H_2O_2 each follow the law of definite proportions (and together they also follow the law of multiple proportions).

3.31. Show that the compounds copper(I) oxide and copper(II) oxide obey the law of multiple proportions.

	% Copper	% Oxygen
Copper(I) oxide	88.82	11.18
Copper(II) oxide	79.89	20.11

Does the law require that the ratio (mass of copper)/(mass of oxygen) be integral for each of these compounds?

Ans. The ratio of mass of oxygen to mass of copper is not necessarily integral. Take a constant mass of copper, for example, 1.000 g Cu. From the ratios given, there are

$$\text{copper(I) oxide} \qquad 1.000\,\text{g Cu}\left(\frac{11.18\,\text{g O}}{88.82\,\text{g Cu}}\right) = 0.1259\,\text{g O}$$

$$\text{copper(II) oxide} \qquad 1.000\,\text{g Cu}\left(\frac{20.11\,\text{g O}}{79.89\,\text{g Cu}}\right) = 0.2517\,\text{g O}$$

The ratio of masses of O (per 1.000 g Cu) in the two compounds is

$$\frac{0.2517}{0.1259} = \frac{1.999}{1.000} = \frac{2}{1}$$

The ratio of masses of oxygen in the two compounds (for a given mass of Cu) is the ratio of small integers, as required by the law of multiple proportions. The ratio of mass of copper to mass of oxygen is not integral.

3.32. (a) Compare the number of seats in Yankee Stadium during a midweek afternoon game to the number during a weekend doubleheader. (b) Compare the number of locations available for electrons in the third shell of a hydrogen atom and the third shell of a uranium atom.

Ans.   (*a*) The number of seats is the same at both times. (The number occupied probably is different.) (*b*) The number of locations is the same. In uranium, all 18 locations are filled, whereas in hydrogen they are empty.

3.33.   Which of the following familiar metals are main group elements and which are transition metals? (*a*) Al, (*b*) Fe, (*c*) Au, (*d*) Ag, (*e*) Na, and (*f*) Cu.

Ans.   (*a*) and (*e*) are main group elements.

3.34.   In which section (main group, transition group, inner transition group) are the nonmetals found?

Ans.   Main group.

3.35.   In which atom is it more difficult to predict the number of valence electrons—Fe or Na?

Ans.   Fe (Na is a main group element, with valence electrons equal to its group number).

Chapter 4

Formulas and Formula Weights

4.1 INTRODUCTION

Atoms and their symbols were introduced in Chap. 3 and 1. In this chapter, the representation of compounds by their formulas will be developed. The formula for a compound (Sec. 4.3) contains much information of use to the chemist. We will learn how to calculate the number of atoms of each element in a formula unit of a compound. Since atoms are so tiny, we will learn to use large groups of atoms—moles of atoms—to ease our calculations. We will learn to calculate the percent by mass of each element in the compound. We will learn how to calculate the simplest formula from percent composition data, and to calculate molecular formulas from simplest formulas and molecular weights. The procedure for writing formulas from names or from knowledge of the elements involved will be presented in Chaps. 5, 6, and 13.

4.2 MOLECULES AND FORMULA UNITS

Some elements combine by covalent bonding (Chap. 5) into units called *molecules*, which have relatively few atoms in each. Other elements combine by gaining or losing electrons to form *ions*, which are charged atoms (or groups of atoms). The ions are attracted to each other, a type of bonding called ionic bonding (Chap. 5), and form combinations containing millions or more of ions of each element. To identify these ionic compounds, the simplest formula is generally used. (For now, you can tell that a compound is ionic if it contains at least one metal atom or the NH_4 group. If not, it is most likely bonded in molecules. For more on bonding, see Chap. 5.)

Formulas for ionic compounds represent one *formula unit*. However, molecules too have formulas, and thus formula units. Even uncombined atoms of an element have formulas. Thus, formula units may refer to uncombined atoms, molecules, or atoms combined in an ionic compound:
one formula unit may be

    one atom of uncombined element, e.g., Hg

    one molecule of a covalently bonded compound, e.g., CO

    one simple unit of an ionic compound, e.g., NaCl

EXAMPLE 4.1. What is the formula unit of the element lithium? the compound nitrogen monoxide? the compound lithium chloride?

The formula unit of lithium is Li, one atom of lithium. The formula unit of nitrogen monoxide is NO, one molecule of nitrogen monoxide. The formula unit of lithium chloride is LiCl.

The word *molecule* or the general term *formula unit* may be applied to one unit of CO. The word *atom* or the term *formula unit* may be applied to one unit of uncombined Hg. However, there is no special name for one unit of NaCl. *Formula unit* is the best designation. (Some instructors and some texts refer to "molecules" of NaCl, and especially to "molecular weight" of NaCl, because the calculations done on formula units do not depend on the type of bonding involved. However, strictly speaking, the terms *molecule* and *molecular weight* should be reserved for substances bonded into molecules.)

4.3 CHEMICAL FORMULAS

Writing a formula implies that the atoms in the formula are bonded together in some way. The relative numbers of atoms of the elements in a compound are shown in a chemical formula by writing the symbols of the elements followed by appropriate *subscripts* to denote how many atoms of each element there are in the formula unit. A subscript *following* the symbol gives the number of atoms of that element per formula unit. If there is no subscript, *one* atom per formula unit is implied. For example, the formula H_2SO_4 describes a molecule containing two atoms of hydrogen and four atoms of oxygen, along with one atom of sulfur. Sometimes groups of atoms which are bonded together within a molecule or within an ionic compound are grouped together in the formula in parentheses. The number of such groups is indicated by a subscript following the closing parenthesis. For example, the 2 in

$$(NH_4)_2SO_4$$

states that there are two NH_4 groups present per formula unit. There is only one SO_4 group; therefore parentheses are not necessary around it.

EXAMPLE 4.2. How many H atoms and how many S atoms are there per formula unit in $(NH_4)_2SO_3$?

There are two NH_4 groups, each containing four H atoms, for a total of eight H atoms per formula unit. There is only one S atom; the "3" defines the number of O atoms.

In summary, chemical formulas yield the following information:
1. which elements are present
2. the ratio of the number of atoms of each element to the number of atoms of each other element
3. the number of atoms of each element per formula unit of compound
4. the fact that all the atoms represented are bonded together in some way.

You cannot tell from a formula how many atoms of each element are present in a given sample of substance, because there might be a little or a lot of the substance present. The formula tells the *ratio* of atoms of each element to all the others, and the ratio of atoms of each element to formula units as a whole.

EXAMPLE 4.3. (*a*) Can you tell how many ears and how many noses were present at the last Super Bowl football game? Can you guess how many ears there were per nose? how many ears per person? (*b*) Can you tell how many hydrogen and oxygen atoms there are present in a sample of pure water? Can you tell how many hydrogen atoms there are per oxygen atom? per water molecule?

(*a*) Since the problem does not give the number of people at the game, it is impossible to tell the number of ears or noses from the information given. The ratio of ears to noses and ears to people are both likely to be $2:1$. (*b*) Since the problem does not give the quantity of water, it is impossible to tell the number of hydrogen atoms or oxygen atoms from the information given. The ratios of hydrogen atoms to oxygen atoms and hydrogen atoms to water molecules are both $2:1$.

Note that a pair of hydrogen atoms bonded together is a hydrogen molecule. Seven elements, when uncombined with other elements, form *diatomic* molecules. These elements are hydrogen, nitrogen, oxygen, fluorine, chlorine, bromine, and iodine. They are easy to remember because the last six form a large "7" in the periodic table:

N O F
       Cl
       Br
       I

4.4   FORMULA WEIGHTS

The *formula weight* of a compound is the sum of the atomic weights of all the atoms (not merely each kind of atom) in the formula. Thus, in the same way a symbol is used to represent an element, a formula is used to represent a compound and also one unit of the compound. The formula weight of the compound or the mass of 1 mol of the compound is easily determined on the basis of the formula (Sec. 4.5). Note that just as *formula unit* may refer to uncombined atoms, molecules, or atoms combined in an ionic compound, the term *formula weight* may refer to the atomic weight of an atom, the molecular weight of a molecule, or the formula weight of a formula unit of an ionic compound.

Use at least three significant digits in formula weight calculations.

EXAMPLE 4.4.   What is the formula weight of $MgSO_4$?

$$\text{atomic weight of magnesium} = 24.31\,\text{amu}$$

$$\text{atomic weight of sulfur} = 32.06\,\text{amu}$$

$$4 \times \text{atomic weight of oxygen} = 4 \times 16.00\,\text{amu} = 64.00\,\text{amu}$$

$$\text{formula weight} = \text{total} = 120.37\,\text{amu}$$

4.5   THE MOLE

Atoms and molecules are incredibly small. For over a hundred years after Dalton postulated their existence, no one was able to work with just one atom or molecule. (In recent times, with special apparatus, it has been possible to see the effects of individual atoms and molecules, but this subject will be developed later.) Just as the dozen is used as a convenient number of items in everyday life, the *mole* may be best thought of as a number of items. The mole is 6.02×10^{23} items, a number called *Avogadro's number*. This is a very large number: six hundred two thousand billion billion! The entire earth has a mass of 6×10^{24} kg. Thus, the earth has only 10 times as many kilograms as a mole of carbon has atoms. One can have a mole of any item, but it makes little sense to speak of moles of anything but the tiniest of particles, like atoms and molecules. It might seem unusual to give a name to a number, but remember we do the same thing in everyday life; a dozen is the name for 12 items. Just as a grocer finds selling eggs by the dozen more convenient than selling them individually, the chemist finds calculations more convenient with moles. The number of formula units (i.e., the number of atoms, of molecules of molecular elements or compounds, or of formula units of ionic compounds) can be converted to moles of the same substance, and vice versa, using Avogadro's number (Fig. 4-1).

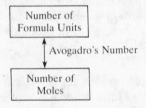

Fig. 4-1   Avogadro's number conversions

Mole is abbreviated mol. Do not use *m* or *M* for mole; these symbols are used for other quantities related to moles, and so you will be confused if you use either of them. Note: A mole is referred to by some authors as a "gram molecular weight" because a mole of molecules has a mass in grams equal to its molecular weight. In this terminology, a "gram atomic weight" is a mole of atoms, and a "gram formula weight" is a mole of formula units.

The formula weight of a substance is equal to its number of grams per mole. Avogadro's number is the number of atomic mass units in 1 g. It is defined in that manner so that the atomic weight of an element (in amu) is numerically equal to the number of grams of the element per mole. Consider helium, with atomic weight 4.0:

$$\frac{4.0\,\text{amu}}{\text{He atom}} = \frac{4.0\,\text{amu}}{\text{He atom}}\left(\frac{6.02\times10^{23}\ \text{He atoms}}{\text{mol}}\right)\left(\frac{1\,\text{g}}{6.02\times10^{23}\,\text{amu}}\right) = \frac{4.0\,\text{g}}{\text{mol}}$$

Avogadro's number appears both in the numerator and the denominator of this expression; the values reduce to a factor of 1 (they cancel out) and the numeric value in grams per mole is equal to the numeric value of the atomic weight in amu per atom.

$$\frac{4.0\,\text{amu}}{\text{He atom}} = \frac{4.0\,\text{g}}{\text{mol He atoms}}$$

A similar argument leads to the conclusion that the formula weight of any element or compound is equal to the number of grams per mole of the element or compound.

EXAMPLE 4.5.   How many feet tall is a stack of a dozen shoe boxes, each 4 inches tall?

$$\frac{4\,\text{inches}}{\text{box}} = \frac{4\,\text{inches}}{\text{box}}\left(\frac{12\,\text{boxes}}{\text{dozen}}\right)\left(\frac{1\,\text{foot}}{12\,\text{inches}}\right) = \frac{4\,\text{feet}}{\text{dozen}}$$

The same number, but in different units, is obtained because the number of inches in 1 foot is the same as the number in 1 dozen. Compare this process to the one above for grams per mole.

Changing grams to moles and moles to grams is perhaps the most important calculation you will have to make all year (Fig. 4-2). Some authors use the term *molar mass* for the mass of 1 mol of any substance. The units are typically grams per mole.

Mass	← Formula Weight →	Number of Moles

Fig. 4-2   Formula weight conversions
The mass of a substance can be converted to moles, and vice versa, with the formula weight.

EXAMPLE 4.6.   Calculate the mass of 1.00 mol of CCl_4.

The formula weight of CCl_4 is given by

$$C = 12.0\,\text{amu}$$
$$4\,Cl = 4\times35.5\,\text{amu} = 142\,\text{amu}$$
$$CCl_4 = \text{total} = 154\,\text{amu}$$

Thus, the mass of 1.00 mol of CCl_4 is 154 g.

EXAMPLE 4.7.   Calculate the mass of 2.50 mol NaCl.

The formula weight of NaCl is given by

Na	23.0 amu
Cl	35.5 amu
NaCl	58.5 amu

NaCl has a formula weight of 58.5 amu.

$$2.50 \text{ mol NaCl} \left(\frac{58.5 \text{ g NaCl}}{\text{mol NaCl}} \right) = 146 \text{ g NaCl}$$

EXAMPLE 4.8.   Calculate the mass in grams of 1 carbon atom.

First we can calculate the number of moles of carbon, using Avogadro's number:

$$1 \text{ C atom} \left(\frac{1 \text{ mol C}}{6.02 \times 10^{23} \text{ C atoms}} \right) = 1.66 \times 10^{-24} \text{ mol C}$$

Then we calculate the mass from the number of moles and the formula (atomic) weight:

$$1.66 \times 10^{-24} \text{ mol C} \left(\frac{12.0 \text{ g}}{\text{mol C}} \right) = 1.99 \times 10^{-23} \text{ g}$$

Alternately, we can combine these expressions into one:

$$1 \text{ C atom} \left(\frac{1 \text{ mol C}}{6.02 \times 10^{23} \text{ C atoms}} \right) \left(\frac{12.0 \text{ g}}{\text{mol C}} \right) = 1.99 \times 10^{-23} \text{ g}$$

Still another solution method:

$$1 \text{ C atom} \left(\frac{12.0 \text{ amu}}{\text{C atom}} \right) \left(\frac{1.00 \text{ g}}{6.02 \times 10^{23} \text{ amu}} \right) = 1.99 \times 10^{-23} \text{ g}$$

How can we count such a large number of items as Avogadro's number? One way, which we also can use in everyday life, is to weigh a small number and the entire quantity. We can count the small number, and the ratio of the number of the small portion to the number of the entire quantity is equal to the ratio of their masses.

EXAMPLE 4.9.   A TV show requires contestants to guess the number of grains of rice in a gallon container. The closest contestant after 4 weeks will win a big prize. How could you prepare for such a contest, without actually counting the grains in a gallon of rice?

One way to get a good estimate is to count 100 grains of rice and weigh that sample. Weigh a gallon of rice. Then the ratio of number of grains of rice in the small sample (100) to number of grains of rice in the large sample (which is the unknown) is equal to the ratio of masses. Suppose the 100 grains of rice weighed 0.012 pound and the gallon of rice weighed 8.00 pounds. The unknown number of grains in a gallon x can be calculated easily:

$$\frac{x \text{ grains in gallon}}{100 \text{ grains}} = \frac{8.80 \text{ pounds}}{0.012 \text{ pound}}$$

$$x = 73\,000$$

Why not weigh just one grain? The calculation would be simpler, but weighing just one grain might be impossible with the balances available.

The situation is similar in counting atoms, but much more difficult. Individual atoms cannot be seen to be counted, nor can they be weighed in the ordinary manner. Still, if the mass of 1 atom can be determined (in amu, for example) the number of atoms in a mole can be calculated. Historically, what chemists have done in effect is to weigh given numbers of atoms of different elements; they have gotten the ratio of the weights of individual atoms from the ratio of the weight of the different elements and the relative numbers of atoms of the elements. The actual number of atoms in a mole is of surprisingly little consequence in everyday chemical work (but important on chemistry examinations).

The number of moles of each element in a mole of compound is stated in the chemical formula. Hence, the formula can be used to convert the number of moles of the compound to the number of moles of its component elements, and vice versa (Fig. 4-3).

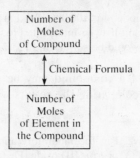

Fig. 4-3 Chemical formula as mole ratio
The number of moles of each element in a compound and the number of moles of the compound as a whole are related by the subscript of that element in the chemical formula.

4.6 PERCENT COMPOSITION OF COMPOUNDS

The term *percent* means the quantity or number of units out of one hundred units total. Percentage is computed by finding the fraction of the total quantity represented by the quantity under discussion, and multiplying by 100%. For example, if a group of 12 persons includes 9 females, the percent females in the group is

$$\left(\frac{9}{12}\right) \times 100\% = 75\%$$

In other words, if there were 100 persons in the group and the ratio of females to males were the same, 75 of the group would be female. Percentage is a familiar concept to anyone who pays sales taxes.

The concept of percentage is often used to describe the composition of compounds. If the formula of a compound is known, the percent by mass of an element in the compound is determined by computing the fraction of the formula weight which is made up of that element, and multiplying that fraction by 100%. Thus, an element X with atomic weight 40.0 amu in a compound XY of formula weight 99.0 amu will be present in

$$\left(\frac{40.0 \text{ amu}}{99.0 \text{ amu}}\right) \times 100\% = 40.4\%$$

The formula gives the number of moles of atoms of each element in each mole of the compound; then the number of grams of each element in the number of grams of compound in 1 mol of the compound is fairly easily computed. Knowing the mass of 1 mol of the compound and the mass of each element in that quantity of compound allows calculations of the percent by mass of each element.

EXAMPLE 4.10.   Calculate the percent composition of $MgSO_4$; that is, the percent by mass of each element in the compound.

One mole of the compound contains 1 mol of magnesium, 1 mol of sulfur, and 4 mol of oxygen atoms. The formula weight is 120.37 amu; hence, there is 120.37 g/mol $MgSO_4$.

One mole of magnesium has a mass of 24.31 g. The percent magnesium is therefore

$$\% \text{ Mg} = \left(\frac{24.31 \text{ g Mg}}{120.37 \text{ g MgSO}_4}\right) \times 100\% = 20.20\% \text{ Mg}$$

One mole of sulfur has a mass of 32.06 g. The percent sulfur is given by

$$\% \text{ S} = \left(\frac{32.06 \text{ g S}}{120.37 \text{ g MgSO}_4}\right) \times 100\% = 26.63\% \text{ S}$$

Four moles of oxygen has a mass of 4×16.00 g. The percent oxygen is given by

$$\% \text{ O} = \left(\frac{4 \times 16.00 \text{ g O}}{120.37 \text{ g MgSO}_4}\right) \times 100\% = 53.17\% \text{ O}$$

The total of all the percentages in the compound is 100.00%. Within the accuracy of the calculation, the total of all the percentages must be 100%. This result may be interpreted to mean that if there were 100.00 g of $MgSO_4$, then 20.20 g would be magnesium, 26.63 g would be sulfur, and 53.17 g would be oxygen.

In laboratory work, the identity of a compound may be established by determining its percent composition experimentally and then comparing the results with the percent composition calculated from its formula.

EXAMPLE 4.11.   A compound was analyzed in the laboratory and found to contain 77.7% iron and 22.3% oxygen. Is the compound iron(II) oxide, FeO, or iron(III) oxide, Fe_2O_3?

The percent composition of iron(II) oxide is determined by dividing the mass of 1 mol of iron and the mass of 1 mol of oxygen by the mass of 1 mol of iron(II) oxide, and multiplying by 100%:

$$\% \text{ Fe} = \left(\frac{55.85 \text{ g Fe}}{71.85 \text{ g oxide}} \right) \times 100.0\% = 77.73\% \text{ Fe}$$

$$\% \text{ O} = \left(\frac{16.00 \text{ g O}}{71.85 \text{ g oxide}} \right) \times 100.0\% = 22.27\% \text{ O}$$

These are the percentages reported, so the compound is FeO. Fe_2O_3 contains a smaller percentage of iron.

4.7   EMPIRICAL FORMULAS

The formula of a compound gives the relative number of atoms of the different elements present. It also gives the relative number of moles of the different elements present. As was shown in Sec. 4.6, the percent by mass of each element in a compound may be computed from its formula. Conversely, if the formula is not known, it may be deduced from the experimentally determined composition. This procedure is possible because once the relative masses of the elements are found, the relative numbers of moles of each may be determined. Formulas derived in this manner are called *empirical formulas* or simplest formulas. In solving a problem in which percent composition is given, any sized sample may be considered, since the percentage of each element does not depend on the size of the sample. The most convenient size to consider is 100 g, for with that size sample, the percentage of each element is equal to the same number of grams.

EXAMPLE 4.12.   What is the empirical formula of a compound which contains 40.0% sulfur and 60.0% oxygen, by mass?

In each 100.0 g of the substance, there will be 40.0 g of sulfur and 60.0 g of oxygen. The number of moles of each element in this sample is given by

$$40.0 \text{ g S} \left(\frac{1 \text{ mol S}}{32.06 \text{ g S}} \right) = 1.25 \text{ mol S}$$

$$60.0 \text{ g O} \left(\frac{1 \text{ mol O}}{16.00 \text{ g O}} \right) = 3.75 \text{ mol O}$$

Therefore, the ratio of moles of O to moles of S

$$\frac{3.75 \text{ mol O}}{1.25 \text{ mol S}} = \frac{3 \text{ mol O}}{1 \text{ mol S}}$$

is 3 to 1. The ratio of 3 mol of oxygen to 1 mol of sulfur corresponds to the formula SO_3.

If more than two elements are present, divide all the numbers of moles by the smallest to attempt to get an integral ratio. Even after this step, it might be necessary to multiply every result by a small integer to get integral ratios, corresponding to the empirical formula.

EXAMPLE 4.13. Determine the empirical formula of a compound containing 29.1% Na, 40.5% S, and 30.4% O.

$$29.1 \text{ g Na} \left(\frac{1 \text{ mol Na}}{23.0 \text{ g Na}} \right) = 1.27 \text{ mol Na}$$

$$40.5 \text{ g S} \left(\frac{1 \text{ mol S}}{32.06 \text{ g S}} \right) = 1.26 \text{ mol S}$$

$$30.4 \text{ g O} \left(\frac{1 \text{ mol O}}{16.0 \text{ g O}} \right) = 1.90 \text{ mol O}$$

$$\frac{1.27 \text{ mol Na}}{1.26} = 1.01 \text{ mol Na}$$

$$\frac{1.26 \text{ mol S}}{1.26} = 1.00 \text{ mol S}$$

$$\frac{1.90 \text{ mol O}}{1.26} = 1.51 \text{ mol O}$$

Multiplying each value by 2 yields 2 mol Na, 2 mol S, and 3 mol O, corresponding to $Na_2S_2O_3$.

EXAMPLE 4.14. Determine the empirical formula of a sample of a compound which contains 79.59 g Fe and 30.40 g O.

Since the numbers of grams are given, the step changing percent to grams is not necessary.

$$79.59 \text{ g Fe} \left(\frac{1 \text{ mol Fe}}{55.85 \text{ g Fe}} \right) = 1.425 \text{ mol Fe}$$

$$30.40 \text{ g O} \left(\frac{1 \text{ mol O}}{16.00 \text{ g O}} \right) = 1.900 \text{ mol O}$$

Dividing each by 1.425 yields 1.000 mol Fe for every 1.333 mol O. This ratio is still not integral. Multiplying these values by 3 yields integers. (We can round off when a value is within 1 or 2% of an integer, but not more.)

$$3.000 \text{ mol Fe} \quad \text{and} \quad 3.999 \text{ or } 4 \text{ mol O}$$

The formula is Fe_3O_4.

4.8 MOLECULAR FORMULAS

Formulas describe the composition of compounds. Empirical formulas give the mole ratio of the various elements. However, sometimes different compounds have the same ratio of moles of atoms of the same elements. For example, acetylene, C_2H_2, and benzene, C_6H_6, each have 1:1 ratios of moles of carbon atoms to moles of hydrogen atoms. Such compounds have the same percent compositions. However, they do not have the same number of atoms in each molecule. The molecular formula is a formula which gives all the information that the empirical formula gives (the mole ratios of the various elements) plus the information of how many atoms are in each molecule. In order to deduce molecular formulas from experimental data, the percent composition plus the molecular weight are usually determined. The molecular weight may be determined in several ways, one of which will be described in Chap. 11.

It is apparent that the compounds C_2H_2 and C_6H_6 have different molecular weights. That of C_2H_2 is 26 amu; that of C_6H_6 is 78 amu. It is easy to determine the molecular weight from the molecular formula, but how can the molecular formula be determined from the empirical formula and the molecular weight? The following steps are used, with benzene having a molecular weight of 78 and an empirical formula of CH serving as an example.

EXAMPLE 4.15.

Step	Example
1. Determine the formula weight corresponding to 1 empirical formula.	CH has a formula weight of $12 + 1 = 13$
2. Divide the molecular weight by the empirical formula weight.	molecular weight = 78 empirical weight = 13 $\frac{78}{13} = 6$
3. Multiply the number of atoms of each element in the empirical formula by the whole number found in step 2.	$(CH)_6 = C_6H_6$

EXAMPLE 4.16. A compound contains 85.7% carbon and 14.3% hydrogen, and has a molecular weight of 56.0 amu. What is its molecular formula?

The first step is to determine the empirical formula from the percent composition data.

$$85.7 \text{ g C}\left(\frac{1 \text{ mol C}}{12.0 \text{ g C}}\right) = 7.14 \text{ mol C}$$

$$14.3 \text{ g H}\left(\frac{1 \text{ mol H}}{1.008 \text{ g H}}\right) = 14.2 \text{ mol H}$$

The empirical formula is CH_2.

Now the molecular formula is determined using the steps outlined above:

1. Weight of $CH_2 = 14.0$ amu
2. Number of units = $(56.0 \text{ amu})/(14.0 \text{ amu}) = 4$
3. Molecular formula = $(CH_2)_4 = C_4H_8$

Solved Problems

MOLECULES AND FORMULA UNITS

4.1. Why is the term *molecular weight* inappropriate for NaCl?

> *Ans.* NaCl is an ionic compound; it does not form molecules and so does not have a molecular weight. The formula weight of NaCl is 58.5 amu, calculated and used in exactly the same manner as a molecular weight for a molecular compound would be calculated and used.

4.2. The simplest type of base contains OH^- ions. Which of the following compounds is more apt to be a base, CH_3OH or NaOH?

> *Ans.* NaOH is ionic, and is a base. CH_3OH is covalently bonded, and is not a base.

4.3. Which of the following compounds occur in molecules?
(a) C_6H_{12}, (b) $C_6H_{12}O_6$, (c) CH_4O, (d) $COCl_2$, (e) $CoCl_2$, (f) CO, (g) NH_4Cl, and (h) $NiCl_2$.

> *Ans.* All but (e), (g), and (h) form molecules; (e), (g), and (h) are ionic.

CHEMICAL FORMULAS

4.4. How many atoms of each element are there in a formula unit of each of the following elements or compounds?

(a) S_8

(b) Na_2SO_3

(c) $(NH_4)_2S$

(d) H_2O_2

(e) Cl_2

(f) H_3PO_4

(g) $Hg_2(NO_3)_2$

(h) $(NH_4)_2HPO_4$

Ans. (*a*) 8 S atoms	(*e*) 2 Cl atoms
(*b*) 2 Na, 1 S, 3 O atoms	(*f*) 3 H, 1 P, 4 O atoms
(*c*) 2 N, 8 H, 1 S atoms	(*g*) 2 Hg, 2 N, 6 O atoms
(*d*) 2 H, 2 O atoms	(*h*) 2 N, 9 H, 1P, 4 O atoms

4.5. Which seven elements form diatomic molecules when they are uncombined with other elements?

Ans. H_2, N_2, O_2, F_2, Cl_2, Br_2, and I_2. To help you remember these elements, note that the last six of these form a pattern resembling "7" in the periodic table.

FORMULA WEIGHTS

4.6. What is the formula weight of each of the items in Problem 4.4?

Ans. (*a*) 8×32.06 amu = 256.5 amu
  (*b*) 2 Na: 2×23.0 amu = 46.0 amu
         S: 1×32.0 amu = 32.0 amu
        3 O: 3×16.0 amu = 48.0 amu
        total   = 126.0 amu
  (*c*) 68.0 amu
  (*d*) 34.0 amu
  (*e*) 71.0 amu
  (*f*) 98.0 amu
  (*g*) 525.0 amu
  (*h*) 132.0 amu

4.7. What is the difference between the atomic weight of fluorine and the molecular weight of fluorine?

Ans. The atomic weight of fluorine is 19.0 amu, as seen in the periodic table or a table of atomic weights. The molecular weight of fluorine, corresponding to F_2, is twice that value, 38.0 amu.

4.8. The standard for atomic weights is ^{12}C, at exactly 12 amu. What is the standard for formula weights?

Ans. ^{12}C, the same standard, is used for formula weights.

4.9. Calculate the formula weight of each of the following compounds: (*a*) NaCl, (*b*) NaCN, (*c*) CCl_4, (*d*) $Mg(CN)_2$, and (*e*) LiOH.

Ans. (*a*) 58.5 amu, (*b*) 49.0 amu, (*c*) 154 amu, (*d*) 76.3 amu, and (*e*) 23.9 amu.

4.10. What mass in amu is in 1 formula unit of $Al_2(SO_4)_3$?

Ans. There are 2 aluminum atoms, 3 sulfur atoms, and 12 oxygen atoms in the formula unit.

$$2 \times \text{atomic weight of aluminum} = 54.0 \text{ amu}$$

$$3 \times \text{atomic weight of sulfur} = 96.2 \text{ amu}$$

$$12 \times \text{atomic weight of oxygen} = 192.0 \text{ amu}$$

$$\text{formula weight} = \text{total} = 342.2 \text{ amu}$$

The formula represents 342.2 amu of aluminum sulfate.

THE MOLE

4.11. How many O atoms are there in 1 molecule of H_2O_2? How many moles of O atoms are there in 1 mol of H_2O_2?

Ans. There are two O atoms per molecule of H_2O_2, and 2 mol O atoms per mol H_2O_2. The chemical formula provides both these ratios.

4.12. What mass in grams is there in 1.00 mol of $Al_2(SO_4)_3$?

Ans. There are 2 aluminum atoms, 3 sulfur atoms, and 12 oxygen atoms in each formula unit.

$$2 \times \text{atomic weight of aluminum} = 54.0 \, \text{amu}$$
$$3 \times \text{atomic weight of sulfur} = 96.2 \, \text{amu}$$
$$12 \times \text{atomic weight of oxygen} = 192.0 \, \text{amu}$$
$$\text{formula weight} = \text{total} = 342.2 \, \text{amu}$$

The 1.00 mol represents 342.2 g of aluminum sulfate.

4.13. How many atoms of Na are there in 1.00 mol Na? What is the mass of 1.00 mol Na?

Ans. There are 6.02×10^{23} atoms in 1.00 mol Na (Avogadro's number). There is 23.0 g of Na in 1.00 mol Na (equal to the atomic weight in grams). This problem requires use of two of the most important conversion factors involving moles. Note which one is used with masses and which one is used with numbers of atoms (or molecules of formula units). With *numbers* of atoms, molecules, or formula units, use Avogadro's *number*; with mass or *weight* use the formula *weight*.

Problems 4.14 through 4.17 are easier to do when both parts are asked together. Note the differences among the parts labeled (*a*) and also among the parts labeled (*b*). On examinations, you are likely to be asked only one such problem at a time, so you must read the problems carefully and recognize the difference between similar-sounding problems.

4.14. (*a*) How many socks are there in exactly 72 dozen socks? (*b*) How many hydrogen atoms are there in 72 mol of hydrogen atoms?

Ans.

(*a*) $$72 \, \text{dozen socks} \left(\frac{12 \, \text{socks}}{\text{dozen socks}} \right) = 864 \, \text{socks}$$

(*b*) $$72 \, \text{mol H} \left(\frac{6.02 \times 10^{23} \, \text{H}}{\text{mol H}} \right) = 4.3 \times 10^{25} \, \text{H}$$

4.15. (*a*) How many pairs of socks are there in exactly 72 dozen pairs of socks? (*b*) How many hydrogen molecules are there in 72 mol of hydrogen molecules?

Ans.

(*a*) $$72 \, \text{dozen pair socks} \left(\frac{12 \, \text{pair socks}}{\text{dozen pair socks}} \right) = 864 \, \text{pair socks}$$

(*b*) $$72 \, \text{mol H}_2 \left(\frac{6.02 \times 10^{23} \, \text{H}_2}{\text{mol H}_2} \right) = 4.3 \times 10^{25} \, \text{H}_2$$

4.16. (*a*) How many pairs of socks can be made with exactly 72 dozen (identical) socks? (*b*) How many hydrogen molecules can be made with 72 mol of hydrogen atoms?

Ans.

$$(a) \qquad 72 \text{ dozen socks} \left(\frac{1 \text{ dozen pair socks}}{2 \text{ dozen socks}} \right) \left(\frac{12 \text{ pair socks}}{\text{dozen pair socks}} \right) = 432 \text{ pair socks}$$

(from definition of a pair)   (from definition of a dozen)

$$(b) \qquad 72 \text{ mol H} \left(\frac{1 \text{ mol H}_2}{2 \text{ mol H}} \right) \left(\frac{6.02 \times 10^{23} \text{ H}_2}{\text{mol H}_2} \right) = 2.2 \times 10^{25} \text{ H}_2$$

(from definition of a molecule)   (from definition of a mole)

4.17. (*a*) How many socks are there in exactly 72 dozen pairs of socks? (*b*) How many hydrogen atoms are there in 72 mol of hydrogen molecules?

Ans.

$$(a) \qquad 72 \text{ dozen pair socks} \left(\frac{12 \text{ pair socks}}{\text{dozen pair socks}} \right) \left(\frac{2 \text{ socks}}{\text{pair socks}} \right) = 1\,728 \text{ socks}$$

$$(b) \qquad 72 \text{ mol H}_2 \left(\frac{6.02 \times 10^{23} \text{ H}_2}{\text{mol H}_2} \right) \left(\frac{2 \text{ H}}{\text{H}_2} \right) = 8.7 \times 10^{25} \text{ H}$$

4.18. Each of Problems 4.14 through 4.17 is different. How important is it to read the problem carefully?

Ans.   It is obvious that the wording of these problems is only a little different in each case but the answers are very different. One of the prime requirements for successful completion of general chemistry is the ability to read the problems and interpret them correctly. You must not try to speed-read a chemistry text, and especially not problems on exams.

4.19. (*a*) Which contains more pieces of fruit: a dozen cherries or a dozen watermelons? Which weighs more? (*b*) Which contains more atoms: a mole of helium or a mole of uranium? Which weighs more?

Ans.   (*a*) Both have the same number of fruits (12), but since each watermelon weighs more than a cherry, the dozen watermelons weigh more than the dozen cherries. (*b*) Both have the same number of atoms (6.02×10^{23}), but since uranium has a greater atomic weight (see the periodic table), the mole of uranium weighs more.

4.20. What is the difference between a mole of hydrogen atoms and a mole of hydrogen molecules?

Ans.   The hydrogen molecules contain two hydrogen atoms each; hence the mole of hydrogen molecules contains twice as many atoms as the mole of hydrogen atoms. A mole of hydrogen molecules contains 2 mol H atoms.

4.21. Show that the ratio of the number of moles of two elements in a compound is equal to the ratio of the number of atoms of the two elements.

Ans.

$$\frac{x \text{ mol A}}{\text{mol B}} = \frac{x \text{ mol A} \left(\dfrac{6.02 \times 10^{23} \text{ atoms A}}{\text{mol A}} \right)}{\text{mol B} \left(\dfrac{6.02 \times 10^{23} \text{ atoms B}}{\text{mol B}} \right)} = \frac{x \text{ atoms A}}{\text{atom B}}$$

4.22. How many moles of each substance is there in each of the following masses of substances?
  (a) 101 g Na
  (b) 55.2 g $(NH_4)_2SO_3$
  (c) 222 g NaOH
  (d) 1.04 g NaCl
  (e) 19.0 g F_2
  (f) 44.4 g NaH_2PO_4

Ans.

(a)
$$101 \text{ g Na} \left(\frac{1 \text{ mol Na}}{23.0 \text{ g Na}} \right) = 4.39 \text{ mol Na}$$

(b)
$$55.2 \text{ g } (NH_4)_2SO_3 \left(\frac{1 \text{ mol } (NH_4)_2SO_3}{116 \text{ g } (NH_4)_2SO_3} \right) = 0.476 \text{ mol } (NH_4)_2SO_3$$

(c)
$$222 \text{ g NaOH} \left(\frac{1 \text{ mol NaOH}}{40.0 \text{ g NaOH}} \right) = 5.55 \text{ mol NaOH}$$

(d)
$$1.04 \text{ g NaCl} \left(\frac{1 \text{ mol NaCl}}{58.5 \text{ g NaCl}} \right) = 0.0178 \text{ mol NaCl}$$

(e)
$$19.0 \text{ g } F_2 \left(\frac{1 \text{ mol } F_2}{38.0 \text{ g } F_2} \right) = 0.500 \text{ mol } F_2$$

(f)
$$44.4 \text{ g } NaH_2PO_4 \left(\frac{1 \text{ mol } NaH_2PO_4}{120 \text{ g } NaH_2PO_4} \right) = 0.370 \text{ mol } NaH_2PO_4$$

4.23. What is the mass of each of the following:
  (a) 1.25 mol NaI
  (b) 2.42 mol $NaNO_3$
  (c) 4.02 mol $Ba(NO_2)_2$
  (d) 0.042 mol C_8H_{18}

Ans.

(a)
$$1.25 \text{ mol NaI} \left(\frac{150 \text{ g NaI}}{\text{mol NaI}} \right) = 188 \text{ g NaI}$$

(b)
$$2.42 \text{ mol } NaNO_3 \left(\frac{85.0 \text{ g } NaNO_3}{\text{mol } NaNO_3} \right) = 206 \text{ g } NaNO_3$$

(c)
$$4.02 \text{ mol } Ba(NO_2)_2 \left(\frac{229 \text{ g } Ba(NO_2)_2}{\text{mol } Ba(NO_2)_2} \right) = 921 \text{ g } Ba(NO_2)_2$$

(d)
$$0.042 \text{ mol } C_8H_{18} \left(\frac{114 \text{ g } C_8H_{18}}{\text{mol } C_8H_{18}} \right) = 4.8 \text{ g } C_8H_{18}$$

4.24. (a) Define the unit millimole (mmol). (b) How many sodium atoms are there in 1 mmol of Na?
  Ans. (a) 1 mmol = 0.001 mol. (b) 6.02×10^{20} Na atoms.

4.25. Without doing actual calculations, determine which of the following have masses greater than 1 mg:

 (a) 1 CO_2 molecule
 (b) 1 mol CO_2
 (c) 1 amu
 (d) 6.02×10^{23} amu
 (e) 6.02×10^{23} H atoms
 (f) $\dfrac{1}{(6.02 \times 10^{23})}$ mol H atoms

Ans.   (b), (d), and (e) have masses greater than 1 mg.

4.26. (a) Calculate the formula weight of H_2SO_4. (b) Calculate the number of grams in 1.00 mol of H_2SO_4. (c) Calculate the number of grams in 2.00 mol of H_2SO_4. (d) Calculate the number of grams in 0.473 mol of H_2SO_4.

Ans.   (a)   2 H   2.02 amu
             S   32.06 amu
            4 O   64.00 amu
            total 98.08 amu

   (b) 98.08 g

   (c) $2.00 \, \text{mol} \left(\dfrac{98.08 \, \text{g}}{\text{mol}} \right) = 196 \, \text{g}$

   (d) $0.473 \, \text{mol} \left(\dfrac{98.08 \, \text{g}}{\text{mol}} \right) = 46.4 \, \text{g}$

4.27. (a) Determine the number of moles of ammonia in 100 g of ammonia, NH_3. (b) Determine the number of moles of nitrogen atoms in 100 g of ammonia. (c) Determine the number of moles of hydrogen atoms in 100 g of ammonia.

Ans.

   (a) $100 \, \text{g NH}_3 \left(\dfrac{1 \, \text{mol NH}_3}{17.0 \, \text{g NH}_3} \right) = 5.88 \, \text{mol NH}_3$

   (b) $5.88 \, \text{mol NH}_3 \left(\dfrac{1 \, \text{mol N}}{\text{mol NH}_3} \right) = 5.88 \, \text{mol N}$

   (c) $5.88 \, \text{mol NH}_3 \left(\dfrac{3 \, \text{mol H}}{\text{mol NH}_3} \right) = 17.6 \, \text{mol H}$

4.28. Determine the number of mercury atoms in 14.7 g of Hg_2Cl_2.

Ans.

   $14.7 \, \text{g Hg}_2\text{Cl}_2 \left(\dfrac{1 \, \text{mol Hg}_2\text{Cl}_2}{472 \, \text{g Hg}_2\text{Cl}_2} \right) \left(\dfrac{2 \, \text{mol Hg}}{\text{mol Hg}_2\text{Cl}_2} \right) \left(\dfrac{6.02 \times 10^{23} \, \text{Hg atoms}}{\text{mol Hg}} \right)$

   $= 3.75 \times 10^{22} \, \text{atoms}$

4.29. How many hydrogen atoms are there in 1.00 mol of ammonia, NH_3?

Ans.

   $1 \, \text{mol NH}_3 \left(\dfrac{6.02 \times 10^{23} \, \text{NH}_3 \, \text{molecules}}{\text{mol NH}_3} \right) \left(\dfrac{3 \, \text{H atoms}}{\text{NH}_3 \, \text{molecule}} \right) = 1.81 \times 10^{24} \, \text{H atoms}$

4.30. (*a*) Create one factor that will change 300 g of calcium carbonate to a number of formula units of calcium carbonate. (*b*) Is it advisable to learn and use such a factor?

Ans.

(*a*)
$$\frac{1 \text{ mol } CaCO_3}{100 \text{ g } CaCO_3} \left(\frac{6.02 \times 10^{23} \text{ units } CaCO_3}{\text{mol } CaCO_3} \right) = \frac{6.02 \times 10^{23} \text{ units } CaCO_3}{100 \text{ g } CaCO_3}$$

(*b*) It is possible to use such a conversion factor, but it is advisable while you are learning to use the factors involved with moles to use as few different ones as possible. That way, you have to remember fewer. Also, in each conversion you will change either the unit (mass → moles) or the chemical ($CaCO_3$ → O atoms) in a factor, but not both. Many texts do use such combined factors, however.

4.31. How many moles of water can be made with 7.2 mol H atoms (plus enough O atoms)?

Ans.

$$7.2 \text{ mol } H \left(\frac{1 \text{ mol } H_2O}{2 \text{ mol } H} \right) = 3.6 \text{ mol } H_2O$$

4.32. How many moles of Na are there in 3.0 mol Na_2SO_4?

Ans.

$$3.0 \text{ mol } Na_2SO_4 \left(\frac{2 \text{ mol } Na}{\text{mol } Na_2SO_4} \right) = 6.0 \text{ mol } Na$$

4.33. (*a*) If a dozen pairs of socks weighs 9 ounces, how much would the same socks weigh if they were unpaired? (*b*) What is the mass of 1 mol O_2? What is the mass of the same number of oxygen atoms unbonded to each other?

Ans. (*a*) The socks would still weigh 9 ounces; unpairing them makes no difference in their mass.

(*b*) The mass of 1 mol O_2 is 32 g; the mass of 2 mol O is also 32 g. The mass does not depend on whether they are bonded or not.

4.34. How many molecules are there in 1.00 mol Cl_2? How many atoms are there? What is the mass of 1.00 mol Cl_2?

Ans. There are 6.02×10^{23} molecules in 1.00 mol Cl_2 (Avogadro's number). Since there are two atoms per molecule, there are

$$6.02 \times 10^{23} \text{ molecules} \left(\frac{2 \text{ atoms}}{\text{molecule}} \right) = 1.20 \times 10^{24} \text{ atoms } Cl \text{ in } 1.00 \text{ mol } Cl_2$$

The mass is that of 1.00 mol Cl_2 or 2.00 mol Cl:

$$1.00 \text{ mol } Cl_2 \left(\frac{2 \times 35.5 \text{ g } Cl_2}{\text{mol } Cl_2} \right) = 71.0 \text{ g } Cl_2$$

$$2.00 \text{ mol } Cl \left(\frac{35.5 \text{ g } Cl}{\text{mol } Cl} \right) = 71.0 \text{ g } Cl$$

The *mass* is the same, no matter whether we focus on the atoms or molecules. (Compare the mass of 1 dozen pairs of socks rolled together and that of the same socks unpaired. Would the two masses differ? If so, which would be greater? See Problem 4.33.)

4.35. Make a table showing the number of molecules and the mass of each of the following: (*a*) 1.00 mol Cl_2, (*b*) 2.00 mol Cl_2, and (*c*) 0.703 mol Cl_2.

Ans.

		Number of molecules	Mass	
(*a*)	1.00 mol Cl_2	6.02×10^{23}	71.0 g	(see Problem 4.34)
(*b*)	2.00 mol Cl_2	1.20×10^{24}	142 g	
(*c*)	0.703 mol Cl_2	4.23×10^{23}	49.9 g	

Once you know the mass or number of molecules in 1 mol, you merely have to multiply to get the mass or number of molecules in any other given number of moles. Sometimes the calculation is easy enough to do in your head.

4.36. What mass of oxygen is combined with 1.25×10^{24} atoms of sulfur in Na_2SO_4?

Ans.

$$1.25 \times 10^{24} \text{ atoms S}\left(\frac{1 \text{ mol S}}{6.02 \times 10^{23} \text{ atoms S}}\right)\left(\frac{4 \text{ mol O}}{\text{mol S}}\right)\left(\frac{16.0 \text{ g O}}{\text{mol O}}\right) = 133 \text{ g O}$$

4.37. How can you measure the thickness of a sheet of notebook paper with a 10-cm ruler?

Ans. One way is to measure the combined thickness of many sheets, and divide that distance by the number of sheets. For example, if 500 sheets is 5.05 cm thick, then each sheet is (5.05 cm)/500 = 0.0101 cm thick. Note that it is impossible to measure 0.0101 cm with a centimeter ruler, but we can accomplish the same purpose indirectly. (See Problem 4.38.)

4.38. How can you measure the mass of a carbon atom? Compare this problem with Problem 4.37.

Ans. We want to measure the mass of a large number of carbon atoms, and divide the total mass by the number of atoms. However, we have the additional problem here, compared with counting sheets of paper, that atoms are too small to count. We can "count" them by combining them with a known number of atoms of another element. For example, to count a number of carbon atoms, combine them with a known number of oxygen atoms to form CO, in which the ratio of atoms of carbon to oxygen is 1:1.

4.39. A 208-g sample of a "new" element Z reacts with 32.0 g of O_2. Assuming the atoms of Z react in a 1:1 ratio with oxygen *molecules*, calculate the atomic weight of Z.

Ans. Since the 208 g of Z reacts with 1.00 mol of O_2 in a 1:1 mole ratio, there must be 1.00 mol of Z in 208 g. Therefore, 208 g is 1.00 mol, and the atomic weight of Z is 208 amu.

4.40. To form a certain compound, 48.0 g of oxygen reacts with 178 g of tin. What is the formula of the compound?

Ans. The formula is the mole ratio:

$$48.0 \text{ g O atoms}\left(\frac{1 \text{ mol O atoms}}{16.0 \text{ g O}}\right) = 3.00 \text{ mol O atoms}$$

$$178 \text{ g Sn atoms}\left(\frac{1 \text{ mol Sn atoms}}{118.7 \text{ g Sn}}\right) = 1.50 \text{ mol Sn atoms}$$

The mole ratio is 1.50 mol Sn/3.00 mol O = 1 mol Sn/2 mol O. The formula is SnO_2.

4.41. If 48.0 g O_2 were used in Problem 4.40, how would the problem change?

Ans. The answer is the same; 48.0 g of O_2 is 48.0 g of O atoms (bonded in pairs). See Problem 4.33.

4.42. How many hydrogen atoms are there in 4.0 mol NH_3?

Ans.

$$4.0 \text{ mol NH}_3\left(\frac{6.02 \times 10^{23} \text{ molecules NH}_3}{\text{mol NH}_3}\right)\left(\frac{3 \text{ H atoms}}{\text{molecule NH}_3}\right) = 7.2 \times 10^{24} \text{ H atoms}$$

4.43. How many S atoms are there in $6.00\,mol\,S_8$?

Ans.

$$6.00\,mol\,S_8\left(\frac{6.02\times10^{23}\,\text{molecules }S_8}{mol\,S_8}\right)\left(\frac{8\,\text{atoms }S}{\text{molecule }S_8}\right)$$

$$= 2.89\times10^{25}\,S\,\text{atoms}$$

PERCENT COMPOSITION OF COMPOUNDS

4.44. A 100-g sample of water has a percent composition of 88.8% oxygen and 11.2% hydrogen. (*a*) What is the percent composition of a 5.00-g sample of water? (*b*) Calculate the number of grams of oxygen in a 5.00-g sample of water.

Ans. (*a*) 88.8% O and 11.2% H. The percent composition does not depend on the sample size.

  (*b*) $5.00\,g\,H_2O\left(\dfrac{88.8\,g\,O}{100\,g\,H_2O}\right) = 4.44\,g\,O.$

4.45. Calculate the percent composition of DDT ($C_{14}H_9Cl_5$).

Ans.

$$
\begin{array}{lll}
\text{C:} & 14\times12.01 = & 168.1\,\text{amu} \\
\text{H:} & 9\times1.008 = & 9.07\,\text{amu} \\
\text{Cl:} & 5\times35.45 = & 177.2\,\text{amu} \\
\text{Total} & & = 354.4\,\text{amu}
\end{array}
$$

The percent carbon is found by dividing the mass of carbon in one molecule by the weight of the molecule and multiplying the quotient by 100%:

$$\%\,C = \left(\frac{168.1\,\text{amu C}}{354.4\,\text{amu total}}\right)\times100\% = 47.43\%\,C$$

$$\%\,H = \left(\frac{9.07\,\text{amu H}}{354.4\,\text{amu total}}\right)\times100\% = 2.56\%\,H$$

$$\%\,Cl = \left(\frac{177.2\,\text{amu Cl}}{354.4\,\text{amu total}}\right)\times100\% = 50.00\%\,Cl$$

The percentages add up to 99.99%. (The answer is correct within the accuracy of the number of significant figures used.)

4.46. Calculate the percent composition of each of the following: (*a*) C_4H_8 and (*b*) C_6H_{12}.

Ans. (*a*)
$$
\begin{array}{lll}
\text{C:} & 4\times12.0 = & 48.0\,\text{amu} \\
\text{H:} & 8\times1.0 \;\; = & 8.0\,\text{amu} \\
\text{Total} & & = 56.0\,\text{amu}
\end{array}
$$

The percent carbon is found by dividing the mass of carbon in one molecule by the weight of the molecule and multiplying the quotient by 100%:

$$\%\,C = \left(\frac{48.0\,\text{amu C}}{56.0\,\text{amu total}}\right)\times100\% = 85.7\%\,C$$

$$\%\,H = \left(\frac{8.0\,\text{amu H}}{56.0\,\text{amu total}}\right)\times100\% = 14.3\%\,H$$

The two percentages add up to 100.0%.

(b) C: $6 \times 12.0 = 72.0$ amu
     H: $12 \times 1.0 = 12.0$ amu
             Total $= 84.0$ amu

$$\% \, C = \left(\frac{72.0 \, \text{amu C}}{84.0 \, \text{amu total}} \right) \times 100\% = 85.7\% \, C$$

$$\% \, H = \left(\frac{12.0 \, \text{amu H}}{84.0 \, \text{amu total}} \right) \times 100\% = 14.3\% \, H$$

The percentages are the same as those in part (a). That result might have been expected. Since the ratio of atoms of carbon to atoms of hydrogen is the same (1:2) in both compounds, the ratio of masses also ought to be the same, and their percent by mass ought to be the same. From another viewpoint, this result means that the two compounds cannot be distinguished from each other by their percent compositions alone.

4.47. A detective analyzes a drug and finds that it contains 80.22% carbon and 9.62% hydrogen. Could the drug be pure tetrahydrocannabinol ($C_{21}H_{30}O_2$)?

Ans.

21 C: $21 \times 12.01 = 252.2$ amu

30 H: $30 \times 1.008 = 30.24$ amu

2 O: $2 \times 16.00 = 32.00$ amu

formula weight $= 314.4$ amu

$$\% \, C = \left(\frac{252.2 \, \text{amu}}{314.4 \, \text{amu}} \right) \times 100\% = 80.22\% \, C$$

$$\% \, H = \left(\frac{30.24 \, \text{amu}}{314.4 \, \text{amu}} \right) \times 100\% = 9.618\% \, H$$

Since the percentages are the same, the drug *could* be tetrahydrocannabinol. (It is not proven to be, however. If the percent composition were different, it would be proven *not* to be pure tetrahydrocannabinol.)

4.48. A certain mixture of salt (NaCl) and sugar ($C_{12}H_{22}O_{11}$) contains 50.0% chlorine by mass. Calculate the percentage of salt in the mixture.

Ans.   The percent chlorine in NaCl is given by

$$\left(\frac{35.5 \, \text{g Cl}}{58.5 \, \text{g NaCl}} \right) \times 100\% = 60.7\% \, Cl$$

The percent sodium chloride in the sample is given by

$$\frac{50.0 \, \text{g Cl}}{100 \, \text{g sample}} = \frac{50.0 \, \text{g Cl} \left(\dfrac{100 \, \text{g NaCl}}{60.7 \, \text{g Cl}} \right)}{100 \, \text{g sample}} = \frac{82.4 \, \text{g NaCl}}{100 \, \text{g sample}} = 82.4\% \, NaCl$$

When using percentages, be careful to distinguish percentage of what in what!

EMPIRICAL FORMULAS

4.49. (a) Write a formula for a molecule with 4 phosphorus atoms and 10 oxygen atoms per molecule. (b) What is the empirical formula of this compound?

Ans.   (a) P_4O_{10}. (b) P_2O_5.

4.50. Which of the formulas in Problem 4.4 obviously are not empirical formulas?

Ans.   (a), (d), (e), (g), (h). The formulas in the first four of these parts can be divided by a small integer to give a simpler formula, so these cannot be empirical formulas. (They must have at least some covalent bonds.) In part (h), the hydrogen atoms are not all grouped together, so this also is not merely an empirical formula.

4.51. Which do we use to calculate the empirical formula of an oxide, the atomic weight of oxygen (16 amu) or the molecular weight of oxygen (32 amu)?

Ans.   The atomic weight. We are solving for a formula, which is a ratio of atoms. This type of problem has nothing to do with oxygen gas, O_2.

4.52. Calculate the empirical formula of a compound consisting of 85.7% C and 14.3% H.

Ans.   Assume that 100.0 g of the compound is analyzed. Since the same percentages are present no matter what the sample size, we can consider any size sample we wish, and considering 100 g makes the calculations easier. The number of grams of the elements are then 85.7 g C and 14.3 g H.

$$85.7 \text{ g C}\left(\frac{1 \text{ mol C}}{12.0 \text{ g C}}\right) = 7.14 \text{ mol C}$$

$$14.3 \text{ g H}\left(\frac{1 \text{ mol H}}{1.008 \text{ g H}}\right) = 14.2 \text{ mol H}$$

The empirical formula (or any formula) must be in the ratio of small integers. Thus, we attempt to get the ratio of moles of carbon to moles of hydrogen into an integer ratio; we divide all the numbers of moles by the smallest number of moles:

$$\frac{7.14 \text{ mol C}}{7.14} = 1.00 \text{ mol C}$$

$$\frac{14.2 \text{ mol H}}{7.14} = 1.99 \text{ mol H}$$

We can round off when a value is within 1 or 2% of an integer, but not more. The ratio of moles of C to moles of H is 1:2, so the empirical formula is CH_2.

4.53. Calculate the empirical formula of a compound containing 72.36% Fe and the rest oxygen.

Ans.   The oxygen must be 27.64%, to total 100.00%.

$$72.36 \text{ g Fe}\left(\frac{1 \text{ mol Fe}}{55.85 \text{ g Fe}}\right) = 1.296 \text{ mol Fe}$$

$$27.64 \text{ g O}\left(\frac{1 \text{ mol O}}{16.00 \text{ g O}}\right) = 1.728 \text{ mol O}$$

Dividing both of these numbers by the smaller yields

$$\frac{1.296 \text{ mol Fe}}{1.296} = 1.000 \text{ mol Fe}$$

$$\frac{1.728 \text{ mol O}}{1.296} = 1.333 \text{ mol O}$$

This is still not a whole number ratio, since 1.333 is much too far from an integer to round off. Since 1.333 is about $1\frac{1}{3}$, multiply *both* numbers of moles by 3:

$$3.000 \text{ mol Fe} \quad \text{and} \quad 3.999 \text{ mol O}$$

This is close enough to an integer ratio, so the empirical formula is Fe_3O_4.

4.54. If each of the following mole ratios is obtained in an empirical formula problem, what should it be multiplied by to get an integer ratio?
(a) 1.50/1
(b) 1.33/1
(c) 1.25/1
(d) 1.75/1
(e) 1.67/1

Ans.   (a) 2, to get 3/2
(b) 3, to get 4/3
(c) 4, to get 5/4
(d) 4, to get 7/4
(e) 3, to get 5/3

4.55. Determine the empirical formula of a compound which has a percent composition Mg: 20.2%, S: 26.6%, O: 53.2%.

Ans.   In a 100-g sample, there are

$$20.2 \text{ g Mg} \left(\frac{1 \text{ mol Mg}}{24.3 \text{ g Mg}} \right) = 0.831 \text{ mol Mg}$$

$$26.6 \text{ g S} \left(\frac{1 \text{ mol S}}{32.0 \text{ g S}} \right) = 0.831 \text{ mol S}$$

$$53.2 \text{ g O} \left(\frac{1 \text{ mol O}}{16.0 \text{ g O}} \right) = 3.325 \text{ mol O}$$

To get integer mole ratios, divide by the smallest, 0.831 mol:

$$\frac{0.831 \text{ mol Mg}}{0.831} = 1.00 \text{ mol Mg}$$

$$\frac{0.831 \text{ mol S}}{0.831} = 1.00 \text{ mol S}$$

$$\frac{3.325 \text{ mol O}}{0.831} = 4.00 \text{ mol O}$$

The mole ratio is 1 mol Mg to 1 mol S to 4 mol O; the formula is $MgSO_4$.

4.56. Calculate the empirical formula for each of the following compounds: (a) Na, 32.4%; S, 22.5%; O, 45.0% and (b) C, 80.0%; H, 20.0%.

Ans.

		Dividing by the Smallest Yields
(a) $32.4 \text{ g Na} \left(\frac{1 \text{ mol Na}}{23.0 \text{ g Na}} \right) = 1.41 \text{ mol Na}$		2.0
$22.5 \text{ g S} \left(\frac{1 \text{ mol S}}{32.06 \text{ g S}} \right) = 0.702 \text{ mol S}$		1.0
$45.0 \text{ g O} \left(\frac{1 \text{ mol O}}{16.00 \text{ g O}} \right) = 2.81 \text{ mol O}$		4.0

The empirical formula is Na_2SO_4.

Dividing by the
smallest Yields

(*b*)

$$80.0 \, \text{g C} \left(\frac{1 \, \text{mol C}}{12.01 \, \text{g C}} \right) = 6.66 \, \text{mol C}$$ 1.0

$$20.0 \, \text{g H} \left(\frac{1 \, \text{mol H}}{1.008 \, \text{g H}} \right) = 19.8 \, \text{mol H}$$ 3.0

The empirical formula is CH_3.

MOLECULAR FORMULAS

4.57. Explain why we cannot calculate a molecular formula for a compound of sulfur, sodium, and oxygen?

Ans.   The compound is ionic; it does not form molecules.

4.58. List five possible molecular formulas for a compound with empirical formula CH_2.

Ans. C_2H_4, C_3H_6, C_4H_8, C_5H_{10}, C_6H_{12}, and any other formula with a C to H ratio of $1:2$.

4.59. Which one of the following could possibly be defined as "the ratio of moles of each of the given elements to moles of each of the others"? (*a*) Empirical formula, (*b*) molecular formula, or (*c*) percent composition by mass.

Ans.   Choice (*a*). This is a useful definition of empirical formula. The molecular formula gives the ratio of moles of each element to moles of the compound, plus the information given by the empirical formula. The percent composition does not deal with moles, but is a ratio of masses.

4.60. A compound consists of 85.7% C and 14.3% H. Its molecular weight is 42.0 amu. (*a*) Calculate its empirical formula. (*b*) Calculate its empirical formula weight. (*c*) Calculate the number of empirical formula units in one molecule. (*d*) Calculate its molecular formula.

Ans.   (*a*)   The empirical formula is calculated to be CH_2, as presented in Problem 4.52.
           (*b*)   The empirical formula weight is 14.0 amu, corresponding to 1 C and 2 H atoms.
           (*c*)   There are

$$\frac{42.0 \, \text{amu/molecule}}{14.0 \, \text{amu/empirical formula unit}} = \frac{3 \, \text{empirical formula units}}{\text{molecule}}$$

           (*d*)   The molecular formula is $(CH_2)_3$ or C_3H_6.

4.61. The percent composition of a certain compound is 92.26% C and 7.74% H. Its molecular weight is 78.0 amu. (*a*) Determine its empirical formula. (*b*) Determine its molecular formula.

Ans.

(*a*)

$$92.26 \, \text{g C} \left(\frac{1 \, \text{mol C}}{12.0 \, \text{g C}} \right) = 7.69 \, \text{mol C}$$

$$7.74 \, \text{g H} \left(\frac{1 \, \text{mol H}}{1.008 \, \text{g H}} \right) = 7.68 \, \text{mol H}$$

The ratio is $1:1$, and the empirical formula is CH.
(*b*)   The empirical formula weight is 13.0 amu. There are

$$\frac{78.0 \, \text{g/mol}}{13.0 \, \text{g/empirical formula unit}} = \frac{6 \, \text{empirical formula units}}{\text{mol}}$$

The molecular formula is C_6H_6.

Supplementary Problems

4.62. Define or identify each of the following:

Molecule	Avogadro's number
Ion	Percent
Formula	Empirical formula
Formula unit	Molecular formula
Formula weight	Molar mass
Mole	Empirical formula weight
Molecular weight	Molecular mass

4.63. Combine Figures 4-1, 4-2, and 4-3 into one figure. List all the conversions possible using the combined figure.

    Ans.   The figure is presented as Fig. 4-4. One can convert from mass to moles, moles of component elements, or number of formula units. Additionally, one can convert from number of formula units to moles, to moles of component elements, or to mass. Also from moles of component elements to moles of compound, number of formula units of compound, or mass of compound. Finally, from moles of compound to number of formula units, mass, or number of moles of component elements.

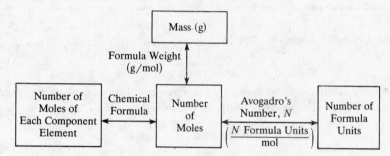

Fig. 4-4   Conversions involving moles

4.64. A certain fertilizer is advertised to contain 12.0% K_2O. What percentage of the fertilizer is potassium?

    Ans.   9.96% K.

4.65. A compound consists of 85.7% and 14.3% H. Its molecular weight is 42.0 amu. Calculate its molecular formula.

    Ans.   This problem is exactly the same as Problem 4.60. The steps are the same even though they are not specified in the statement of this problem.

4.66. How is molecular weight related to formula weight?

    Ans.   They are the same for compounds which form molecules.

4.67. Does the term *atomic weight* refer to uncombined atoms, atoms bonded in compounds, or both?

    Ans.   Both.

4.68. (*a*) Calculate the percent composition of C_2H_4. (*b*) Calculate the percent composition of C_4H_8. (*c*) Compare the results and explain the reason for these results.

    Ans.   (*a, b*) There is 85.7% C and 14.3% H in each. (*c*) They are the same because they have the same ratio of moles of elements.

4.69. A compound has a molecular weight of 46.0 amu and its percent composition is 69.6% oxygen and 30.4% nitrogen. What is its molecular formula?

Ans. NO_2.

4.70. What mass of oxygen is contained in 250 g $CaCO_3$?

Ans.

$$250 \text{ g CaCO}_3 \left(\frac{1 \text{ mol CaCO}_3}{100 \text{ g CaCO}_3} \right) \left(\frac{3 \text{ mol O}}{\text{mol CaCO}_3} \right) \left(\frac{16.0 \text{ g O}}{\text{mol O}} \right) = 120 \text{ g O}$$

You can also do this problem using percent composition (Sec. 4.6).

4.71. Determine the molecular formula of a compound with molecular weight between 50 and 60 which contains 88.8% C and 11.2% H.

Ans.

$$88.8 \text{ g C} \left(\frac{1 \text{ mol C}}{12.0 \text{ g C}} \right) = 7.40 \text{ mol C}$$

$$11.2 \text{ g H} \left(\frac{1 \text{ mol H}}{1.008 \text{ g H}} \right) = 11.1 \text{ mol H}$$

Dividing both numbers of moles by 7.40 yields 1.00 mol C and 1.50 mol H. Multiplying both of these by 2 yields the empirical formula C_2H_3. The empirical formula weight is thus 27.0 amu. The number of empirical formula units in a mole can be calculated by using 55 amu for the molecular weight. The number must be an integer.

$$\frac{55 \text{ amu/molecule}}{27.0 \text{ amu/empirical formula unit}} = \frac{2 \text{ empirical formula units}}{\text{molecule}}$$

If we had used 50 amu or 60 amu, the answer would still have been closer to the integer 2 than to any other integer. The molecular formula is thus C_4H_6.

4.72. Calculate the empirical formula of each of the following compounds:

    (*a*) 27.93% Fe   24.05% S   48.01% O
    (*b*) 36.0% Al   64.0% S
    (*c*) 72.0% Mn   28.0% O
    (*d*) 71.05% Co   28.95% O
    (*e*) 12.2% H   87.8% C
    (*f*) 36.77% Fe   21.10% S   42.13% O

Ans.   (*a*) $Fe_2S_3O_{12}$, (*b*) Al_2S_3, (*c*) Mn_3O_4, (*d*) Co_2O_3, (*e*) C_3H_5, (*f*) $FeSO_4$.

4.73. To Fig. 4-4 (Problem 4.63), add a box for the mass of the substance in atomic mass units, and also add the factor labels with which that box can be connected to two others.

Ans.   The result is shown in Fig. 4-5.

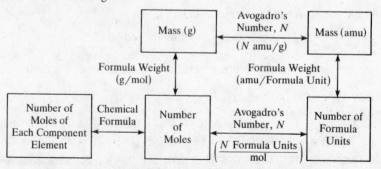

Fig. 4-5   Conversions involving moles and formula units

Chapter 5

Chemical Bonding

5.1 INTRODUCTION

In Chap. 4 we learned to write formulas and interpret formulas for compounds, and to calculate formulas from experimental data. In this chapter, we learn why compounds have the formulas they have; for example, why sodium chloride is $NaCl$ and not $NaCl_2$.

The vast bulk of materials found in nature are compounds or mixtures of compounds rather than free elements. On or near the earth's surface, the nonmetallic elements oxygen, nitrogen, sulfur, and carbon are sometimes found in the uncombined state, and the noble gases are always found in nature uncombined. Also, the metals copper, silver, mercury, and gold sometimes occur in the free state. It is thought that, except for the noble gases, these elements have been liberated from their compounds somewhat recently (compared with the age of the earth) by geological or biological processes. It is a rule of nature that the state which is most probably encountered corresponds to the state of lowest energy. For example, water flows downhill under the influence of gravity and iron rusts when exposed to air. Since compounds are encountered more often than free elements, then, it can be inferred that the combined state must be the state of low energy compared with the state of the corresponding free elements. Indeed, those elements that do occur naturally as free elements must possess some characteristics that correspond to a relatively low energy state.

In this chapter, some aspects of chemical bonding will be discussed. It will be shown that chemical combination corresponds to the tendency of atoms to assume the most stable electron configuration possible.

5.2 THE OCTET RULE

The elements helium, neon, argon, krypton, xenon, and radon—known as the *noble gases*—almost always have monatomic molecules. Their atoms are not combined with atoms of other elements or with other atoms like themselves. Prior to 1962, no compounds of these elements were known. (Since 1962, some compounds of krypton, xenon, and radon have been prepared.) Why are these elements so stable, while the elements with atomic numbers 1 less or 1 more are so reactive? The answer lies in the electronic structures of their atoms. The electrons in atoms are arranged in shells, as described in Sec. 3.6. (A more detailed account of electronic structure will be presented in Chap. 17.)

EXAMPLE 5.1. (*a*) Arrange the 11 electrons of sodium into shells. (*b*) Arrange the 10 electrons of neon into shells.

		\multicolumn{3}{c}{Shell Number}		
		1	2	3
(*a*)	Na	2	8	1
(*b*)	Ne	2	8	

The first two electrons fill the first shell and the next eight electrons fill the second shell. That leaves the one electron left in sodium for the third shell.

The charge on the nucleus and the number of electrons in the valence shell determine the chemical properties of the atom. The electron configurations of the noble gases (except for that of helium) correspond to a valence shell containing eight electrons—a very stable configuration called an *octet*. Atoms of other main group elements tend to react with other atoms in various ways to achieve the octet, as is discussed in the next sections. The tendency to achieve an octet of electrons in the outermost shell is called the *octet rule*. If the outermost shell is the first shell, that is, if there is only one shell occupied, then the maximum number of electrons is two. A configuration of two electrons in the first shell, with no other shells occupied by electrons, is stable and therefore is also said to obey the octet rule.

5.3   IONS

The electron configuration of a sodium atom is

$$\text{Na} \quad 2 \quad 8 \quad 1$$

It is readily seen that if a sodium atom were to lose one electron, the resulting species would have the configuration

$$\text{Na}^+ \quad 2 \quad 8 \quad 0 \quad \text{or more simply} \quad \text{Na}^+ \quad 2 \quad 8$$

The nucleus of a sodium atom contains 11 protons, and if there are only 10 electrons surrounding the nucleus after the atom has lost one electron, the atom will have a net charge of 1+. An atom (or group of atoms) that contains a net charge is called an *ion*. In chemical notation, an ion is represented by the symbol of the atom with the charge indicated as a superscript to the right. Thus, the sodium ion is written Na^+. Na^+ has the same configuration of electrons as a neon atom has (see Example 5.1b). Ions that have the electron configurations of noble gases are stable. Note the very important differences between the sodium ion and a neon atom—the different nuclear charges and the net 1+ charge on Na^+. Therefore the Na^+ ion is not as stable as the Ne atom.

EXAMPLE 5.2.   What is the electron configuration of a fluoride ion, obtained by adding an electron to a fluorine atom?

The electron configuration of a fluorine atom is

$$\text{F} \quad 2 \quad 7$$

Upon gaining an additional electron, the fluorine atom achieves the electron configuration of neon:

$$\text{F}^- \quad 2 \quad 8$$

However, since the fluorine atom contains 9 protons in its nucleus and now contains 10 electrons outside the nucleus, it has a net negative charge; it is an ion. The ion is designated F^- and named the *fluoride ion*.

Compounds—even ionic compounds—have no net charge. In the compound sodium fluoride, there are sodium ions and fluoride ions; the oppositely charged ions attract each other and form a regular geometric arrangement, as shown in Fig. 5-1. This attraction is called an *ionic bond*. There are equal numbers of Na^+ ions and F^- ions, and the compound is electrically neutral. It would be

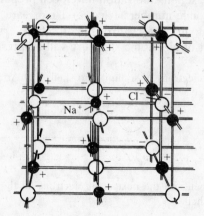

Fig. 5-1   Ball-and-stick model of the sodium chloride structure

The lines are not covalent bonds (Sec. 5.5), but only indications of the positions of the ions. Sodium fluoride is only one compound having the sodium chloride structure.

inaccurate to speak of a molecule of solid sodium fluoride or of a bond between a specific sodium ion and a specific fluoride ion. The substance NaF is extremely stable because of (1) the stable electron configurations of the ions and (2) the attractions between the oppositely charged ions.

The electron configurations of ions of many elements, especially main group elements, can be predicted by assuming that the gain or loss of electrons by an atom results in a configuration analogous to that of a noble gas—a "noble gas" configuration, which contains an octet of electrons in the valence shell. Not all the ions that could be predicted with this rule actually form. For example, few monatomic ions have charges of 4+, and no monatomic ions have charges of 4−. Group IVA atoms tend to bond in another way (Sec. 5.5).

EXAMPLE 5.3.   Predict the charge on a calcium ion and that on a bromide ion, and deduce the formula of calcium bromide.

The electron configuration of a calcium atom is

$$\text{Ca} \qquad 2 \qquad 8 \qquad 8 \qquad 2$$

By losing two electrons, calcium attains the electron configuration of argon and thereby acquires a charge of 2+ .

$$\text{Ca}^{2+} \qquad 2 \qquad 8 \qquad 8$$

A bromine atom has the configuration

$$\text{Br} \qquad 2 \qquad 8 \qquad 18 \qquad 7$$

By gaining one electron, the bromine atom attains the electron configuration of krypton and also attains a charge of 1−. The two ions expected are therefore Ca^{2+} and Br^-. Since calcium bromide as a whole cannot have any net charge, there must be two bromide ions for each calcium ion; hence, the formula is CaBr_2.

The ionic nature of these compounds can be shown by experiments in which the charged ions are made to carry an electric current. If a compound that consists of ions is dissolved in water and the solution is placed between electrodes in an apparatus like that shown in Fig. 5-2, the solution will conduct electricity when the electrodes are connected to the terminals of a battery. Each type of ion moves toward the electrode having a charge opposite to that of the ion. Positively charged ions are called *cations*, and negatively charged ions are called *anions*. Thus, cations migrate to the negative electrode, and anions migrate to the positive electrode. In order to have conduction of electricity, the ions must be free to move. In the solid state, an ionic compound will not conduct. However, if the compound is heated until it melts or if it is dissolved in water, the resulting liquid will conduct electricity because in the liquid state the ions are free to move.

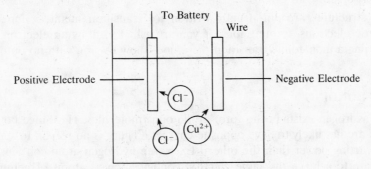

Fig. 5-2   Conducting by ions in solution

5.4   ELECTRON DOT NOTATION

To represent the formation of bonds between atoms, it is convenient to use a system known as *electron dot notation*. In this notation, the symbol for an element is used to represent the nucleus of an atom of the element plus all of the electrons except those in the outermost (valence) shell. The outermost electrons are represented by dots (or circles or crosses). For example, the dot notation for the first 10 elements in the periodic table is as follows:

$$\text{H}\cdot \quad \text{He}: \quad \text{Li}\cdot \quad \text{Be}: \quad :\text{B}\cdot \quad :\overset{\cdot}{\text{C}}\cdot \quad :\overset{\cdot}{\text{N}}\cdot \quad :\overset{\cdot\cdot}{\text{O}}: \quad :\overset{\cdot\cdot}{\text{F}}: \quad :\overset{\cdot\cdot}{\text{Ne}}:$$

Using electron dot notation, the production of sodium fluoride, magnesium fluoride, and magnesium oxide may be pictured as follows: A sodium atom and a fluorine atom react in a 1:1 ratio, since sodium has one electron to lose from its outermost shell and fluorine requires one more electron to complete its outermost shell.

$$\text{Na}\cdot + \cdot\overset{\cdot\cdot}{\underset{\cdot\cdot}{\text{F}}}: \longrightarrow \text{Na}^+ + :\overset{\cdot\cdot}{\underset{\cdot\cdot}{\text{F}}}:^-$$

To lose its entire outermost shell, a magnesium atom must lose two electrons. Since each fluorine atom needs only one electron to complete its octet, it takes two fluorine atoms to react with one magnesium atom:

$$\text{Mg}: \quad + \quad \begin{matrix} \cdot\overset{\cdot\cdot}{\underset{\cdot\cdot}{\text{F}}}: \\ \\ \cdot\overset{\cdot\cdot}{\underset{\cdot\cdot}{\text{F}}}: \end{matrix} \quad \longrightarrow \text{Mg}^{2+} \quad + \quad \begin{matrix} :\overset{\cdot\cdot}{\underset{\cdot\cdot}{\text{F}}}:^- \\ \\ :\overset{\cdot\cdot}{\underset{\cdot\cdot}{\text{F}}}:^- \end{matrix}$$

The magnesium atom has two electrons in its outermost shell. Each oxygen atom has six electrons in its outermost shell, and requires two more electrons to attain its octet. Each oxygen atom therefore requires one magnesium atom from which to obtain the two electrons, and magnesium and oxygen react in a 1:1 ratio.

$$\text{Mg}: \quad + \quad :\overset{\cdot\cdot}{\underset{\cdot\cdot}{\text{O}}}: \longrightarrow \text{Mg}^{2+} \quad + \quad :\overset{\cdot\cdot}{\underset{\cdot\cdot}{\text{O}}}:^{2-}$$

EXAMPLE 5.4. With the aid of the periodic table, use electron dot notation to determine the formula of the ionic compound formed between potassium and sulfur.

Potassium is in group IA and sulfur is in group VIA, and so each potassium atom has one outermost electron and each sulfur atom has six. Therefore, it takes two potassium atoms to supply the two electrons needed by one sulfur atom.

$$\begin{matrix} \text{K}\cdot \\ \\ \text{K}\cdot \end{matrix} \quad + \quad :\overset{\cdot\cdot}{\underset{\cdot}{\text{S}}}: \longrightarrow \begin{matrix} \text{K}^+ \\ \\ \text{K}^+ \end{matrix} \quad + \quad :\overset{\cdot\cdot}{\underset{\cdot\cdot}{\text{S}}}:^{2-}$$

The formula of potassium sulfide is K_2S.

Electron dot structures are not usually written for transition metal or inner transition metal atoms. They do lose electrons, forming ions. If you are asked to draw an electron dot diagram for a compound containing a monatomic transition metal ion, show the ion with no outermost electrons.

5.5 COVALENT BONDING

The element hydrogen exists in the form of diatomic molecules, H_2. Since both hydrogen atoms are identical, they are not likely to have opposite charges. (There is no reason to suppose that one has more electron-attracting power than the other.) Each free hydrogen atom contains a single electron, and if the atoms are to achieve the same electron configuration as atoms of helium, they must each acquire a second electron. If the two hydrogen atoms are allowed to come sufficiently close to each other, their two electrons will effectively belong to both atoms. The positively charged hydrogen nuclei are attracted to the pair of electrons, and effectively, a bond is formed. The bond formed from the sharing of a pair of electrons (or more than one pair) between two atoms is called a *covalent bond*. The hydrogen molecule is more stable than two separate hydrogen atoms. By sharing a pair of electrons, each hydrogen atom acquires a configuration analogous to that of a helium atom. Other pairs of nonmetallic atoms share electrons in the same way.

The formation of covalent bonds between atoms can be conventionally depicted by means of the electron dot notation. The formation of some covalent bonds is shown in this manner below:

$$H\cdot + \cdot H \longrightarrow H:H$$

$$:\ddot{C}l\cdot + \cdot\ddot{C}l: \longrightarrow :\ddot{C}l:\ddot{C}l:$$

$$H\cdot + \cdot\ddot{C}l: \longrightarrow H:\ddot{C}l:$$

$$\cdot\dot{C}\cdot + 4\cdot\ddot{C}l: \longrightarrow \begin{matrix} :\ddot{C}l: \\ :\ddot{C}l:C:\ddot{C}l: \\ :\ddot{C}l: \end{matrix}$$

In these examples, it can be seen that the carbon and chlorine atoms can achieve octets of electrons by sharing pairs of electrons with other atoms. Hydrogen atoms attain "duets" of electrons because the first shell is complete when it contains two electrons. We note from Sec. 5.4 that main group cations generally lose all their valence electrons, and then have none left in their valence shell.

Sometimes it is necessary for two atoms to share more than one pair of electrons to attain octets. For example, the nitrogen molecule, N_2, can be represented as follows:

$$:N:::N:$$

Three pairs of electrons must be shared in order that each nitrogen atom has an octet of electrons. The formation of strong covalent bonds between nitrogen atoms in N_2 is responsible for the relative inertness of nitrogen gas.

Every group of electrons shared between two atoms constitutes a covalent bond. When one pair of electrons is involved, the bond is called a *single bond*. When two pairs of electrons unite two atoms, the bond is called a *double bond*. Three pairs of electrons shared between two atoms constitute a *triple bond*. Examples of these types of bonds are given below:

$$H:H \qquad\qquad :\ddot{O}::C::\ddot{O}: \qquad\qquad :N:::N: \qquad\qquad :\ddot{F}:\underset{\cdot\cdot}{\overset{\cdot\cdot}{S}}:\ddot{F}:$$

Single bond   Double bonds   Triple bond   Two single bonds

Some small points might be noted now that double bonds have been introduced. Hydrogen atoms rarely bond to oxygen atoms which are double-bonded to some other atom. Halogen atoms rarely form double bonds.

There are many exceptions to the so-called octet rule, but only a few will be encountered within the scope of this book. Such cases will be discussed in detail as they are encountered.

The constituent atoms in polyatomic ions are also linked by covalent bonds. In these cases, the net charge on the ion is determined by the total number of electrons and the total number of protons. For example, the ammonium ion, NH_4^+, formed from five atoms, contains one fewer electron than the number of protons. A nitrogen atom plus 4 hydrogen atoms contains a total of 11 protons and 11 electrons, but the ion has only 10 electrons, 8 of which are valence electrons.

$$\left[\begin{matrix} H \\ \overset{\cdot\cdot}{H:N:H} \\ H \end{matrix}\right]^+$$

Similarly, the hydroxide ion contains one valence electron more than the total in the two individual atoms—oxygen and hydrogen.

$$:\ddot{O}:H^-$$

Covalent Bonding of More Than Two Atoms

Writing electron dot diagrams for molecules or ions containing only two atoms is relatively easy. When *several* atoms are to be represented as being linked together by means of covalent bonds, the following procedure may be used to determine precisely the total number of electrons which must be shared among the atoms. The procedure, useful for compounds in which the atoms obey the octet rule, will be illustrated using sulfur dioxide as an example.

	Steps		Example	
1.	Determine the number of valence electrons available.		S	6
			2 O	12
		Total		18
2.	Determine the number of electrons necessary to satisfy the octet (or duet) rule with no electron sharing.		S	8
			2 O	16
		Total		24
3.	The difference between the numbers obtained in steps 2 and 1 is the number of bonding electrons.	Required		24
		Available		− 18
		To be shared		6
4.	Place the atoms as symmetrically as possible. (Note that a hydrogen atom cannot be bonded to more than one atom, since it is capable of sharing only two electrons.)		O S O	
5.	Place the number of electrons to be shared between the atoms, a pair at a time, at first one pair between each pair of atoms. Use as many pairs as remain to make double or triple bonds.		O::S:O	
6.	Add the remainder of the available electrons to complete the octets (or duets) of all the atoms. There should be just enough if the molecule or ion follows the octet rule.		$\ddot{O}::\ddot{S}:\ddot{O}:$	

EXAMPLE 5.5. Draw electron dot diagrams showing the bonding in the following compounds: (a) CO_2, (b) HCN, (c) SiH_4, and (d) BaF_2.

(a)

	C	O	Total
Electrons needed	8	2 × 8	24
Available	4	2 × 6	16
To be shared			8

$$O::C::O$$

Add the rest of the electrons:

$$:\ddot{O}::C::\ddot{O}:$$

(b)

	H	C	N	Total
Electrons needed	2	8	8	18
Available	1	4	5	10
To be shared				8

$$H:C:::N$$

Add the rest of the electrons:

$$H:C:::N:$$

(c)

	Si	H	Total
Electrons needed	8	4×2	16
Available	4	4×1	8
To be shared			8

$$\begin{array}{c} H \\ H\!:\!\overset{..}{\underset{}{Si}}\!:\!H \\ H \end{array}$$

(d)

	Ba	F	Total
Electrons needed	0	2×8	16
Available	2	2×7	16
To be shared			0

$$Ba^{2+} \qquad 2 : \overset{..}{\underset{..}{F}} : ^{-}$$

No covalent bonds exist in BaF_2; there is no electron sharing.

In polyatomic *ions*, more or fewer electrons are available than the number that come from the valence shells of the atoms in the ion. This gain or loss of electrons provides the charge.

EXAMPLE 5.6. Draw an electron dot diagram for CO_3^{2-}.
1. Valence electrons available

C	4	
3 O	18	
charge	2	(2− charge means 2 extra electrons)
Total	24	

2. Valence electrons required

C	8
3 O	24
Total	32

3.

Required	32
Available	−24
To be shared	8

5. With shared electrons $\quad \left[O::C:O \atop \overset{..}{O} \right]^{2-}$

6. With all electrons $\quad \left[: \overset{..}{\underset{..}{O}} :: C : \overset{..}{\underset{..}{O}} : \atop : \overset{..}{\underset{..}{O}} : \right]^{2-}$

The Scope of the Octet Rule

It must be emphasized that the octet rule does not describe the electron configuration of all compounds. The very existence of any compounds of the noble gases is evidence that the octet rule does not apply in all cases. Other examples of compounds that do not obey the octet rule are BF_3, PF_5, and SF_6. But the octet rule does summarize, systematize, and explain the bonding in so many compounds that it is well worth learning and understanding. Compounds in which atoms attain the configuration of helium (the duet) are considered to obey the octet rule, despite the fact that they achieve only the duet characteristic of the complete first shell of electrons.

5.6 DISTINCTION BETWEEN IONIC AND COVALENT BONDING

The word *bonding* applies to any situation in which two or more atoms are held together in such close proximity that they form a characteristic species which has distinct properties and which can be represented by a chemical formula. In compounds consisting of ions, bonding results from the attractions between the oppositely charged ions. In such compounds in the solid state, each ion is surrounded on all sides by ions of the opposite charge. (For example, see Fig. 5-1.) In a solid ionic compound, it is incorrect to speak of a bond between specific pairs of ions.

In contrast, covalent bonding involves the sharing of electron pairs between two specific atoms, and it is possible to speak of a definite bond. For example, in molecules of H_2 and CCl_4 there are one and four covalent bonds per molecule, respectively.

Polyatomic ions, such as OH^-, NO_3^-, and NH_4^+ possess covalent bonds as well as an overall charge.

$$:\!\ddot{O}\!:\!H^- \qquad \left[:\!\ddot{O}\!:\!:\!N\!:\!\ddot{O}\!: \atop :\!\ddot{O}\!: \right]^- \qquad \left[{H \atop H\!:\!\ddot{N}\!:\!H \atop H} \right]^+$$

The charges on polyatomic ions cause ionic bonding between these groups of atoms and oppositely charged ions. In writing electron dot structures, the distinction between ionic and covalent bonds must be clearly indicated. For example, an electron dot diagram for the compound NH_4NO_3 would be

$$\left[{H \atop H\!:\!\ddot{N}\!:\!H \atop H} \right]^+ \qquad \left[:\!\ddot{O}\!:\!:\!N\!:\!\ddot{O}\!: \atop :\!\ddot{O}\!: \right]^-$$

5.7 PREDICTING THE NATURE OF BONDING IN COMPOUNDS

Electronegativity

Electronegativity is a qualitative measure of the ability of an atom to attract electrons involved in covalent bonds. Atoms with higher electronegativities have greater electron-attracting ability. Selected values of electronegativity are given in Table 5-1. The greater the electronegativity difference between a pair of elements, the more likely they are to form an ionic compound; the lower the difference in electronegativity, the more likely that, if they form a compound, the compound will be covalent.

Table 5-1   Selected Electronegativities

H 2.1							He
Li 1.0	Be 1.5	B 2.0	C 2.5	N 3.0	O 3.5	F 4.0	Ne
Na 0.9	Mg 1.2	Al 1.5	Si 1.8	P 2.1	S 2.5	Cl 3.0	Ar
K 0.8	Ca 1.0	Ga 1.6	Ge 1.8	As 2.0	Se 2.4	Br 2.8	Kr
Rb 0.8	Sr 1.0	In 1.7	Sn 1.8	Sb 1.9	Te 2.1	I 2.5	Xe
Cs 0.7	Ba 0.9	Tl 1.8	Pb 1.9	Bi 1.9	Po 2.0	At 2.2	Rn

You need not memorize values of electronegativity (although those of the second-period elements are very easy to learn). You may generalize and state that the greater the separation in the periodic table, the greater the electronegativity difference. Also, in general electronegativity increases to the right and upward in the periodic table. Compounds are generally named and formulas are written for them with the less electronegative element first. (Some hydrogen compounds are exceptions. The symbol for hydrogen is written first only for acids. Hydrogen combined with a halogen atom or with a polyatomic anion forms an acid. In NH_3, the H is written last despite its lower electronegativity because NH_3 is not an acid.) You may not even have to consider electronegativity in making deductions about chemical systems—for example, in naming compounds (Chap. 6). In fact, the following generalizations about bonding can be made without reference to electronegativity.

Most binary compounds (compounds of two elements) of metals and nonmetals are essentially ionic. All compounds involving only nonmetals are essentially covalent except for compounds containing the NH_4^+ ion.

Practically all tertiary compounds (compounds of three elements) contain covalent bonds. If one or more of the elements is a metal, there is likely to be ionic as well as covalent bonding involved in the compound.

Formation of Ions in Solution

When some molecules containing only covalent bonds are dissolved in water, they react with the water to produce ions in solution. For example, pure hydrogen chloride, HCl, and pure ammonia, NH_3, consist of molecules containing only covalent bonds. When cooled to sufficiently low temperatures ($-33\,°C$ for NH_3, $-85\,°C$ for HCl) these substances condense to liquids. However, the liquids do not conduct electricity, since they are still covalent and contain no ions. In contrast, when HCl is dissolved in water, the resulting solution conducts electricity well. Aqueous solutions of ammonia also conduct, but poorly. In these cases, the following reactions occur to the indicated extent to yield ions:

$$H\!:\!\ddot{\underset{\cdot\cdot}{C}l}\!: + H\!:\!\ddot{\underset{\cdot\cdot}{O}}\!:\!H \longrightarrow \left[H\!:\!\overset{\displaystyle H}{\underset{\cdot\cdot}{O}}\!:\!H \right]^+ + :\!\ddot{\underset{\cdot\cdot}{C}l}\!:^- \qquad (100\%)$$

$$H\!:\!\ddot{N}\!:\!H \atop \underset{\displaystyle H}{} + H\!:\!\ddot{\underset{\cdot\cdot}{O}}\!:\!H \longrightarrow \left[H\!:\!\overset{\displaystyle H}{N}\!:\!H \atop \underset{\displaystyle H}{} \right]^+ + :\!\ddot{\underset{\cdot\cdot}{O}}\!:\!H^- \qquad (\text{approximately } 1\%)$$

H_3O^+ is often abbreviated H^+.

Solved Problems

THE OCTET RULE

5.1. Arrange the electrons in each of the following atoms in shells: (a) Li, (b) Na, (c) K, and (d) Rb.

Ans.

		Shell Number				
		1	2	3	4	5
(a)	Li	2	1			
(b)	Na	2	8	1		
(c)	K	2	8	8	1	
(d)	Rb	2	8	18	8	1

5.2. Arrange the electrons in each of the following atoms in shells: (*a*) F, (*b*) Cl, and (*c*) Br.

Ans.

		Shell Number				
		1	2	3	4	5
(*a*)	F	2	7			
(*b*)	Cl	2	8	7		
(*c*)	Br	2	8	18	7	

5.3. How many electrons are there in the outermost shell of each of the following? (*a*) Mg, (*b*) Si, (*c*) P, (*d*) Br, and (*e*) Kr.

Ans.   (*a*) 2, (*b*) 4, (*c*) 5, (*d*) 7, and (*e*) 8.

5.4. Which electron shell is most important to the bonding for each of the following atoms? (*a*) H, (*b*) Li, (*c*) Na, (*d*) K, (*e*) Rb, and (*f*) Cs.

Ans.   In each case the outermost shell is the most important. (*a*) 1, (*b*) 2, (*c*) 3, (*d*) 4, (*e*) 5, and (*f*) 6.

5.5. Explain why uncombined atoms of all of the elements in a given main group of the periodic table will be represented by a similar electron dot notation.

Ans.   They all have the same number of valence electrons.

5.6. How many electrons can fit into an atom in which the outermost shell is the (*a*) first shell? (*b*) second shell? (*c*) third shell?

Ans.   (*a*) 2. The first shell is filled. (*b*) 10. The first and second shells are both filled. (*c*) 18. The first and second shells are filled, and there are eight electrons in the third shell (the maximum number before the fourth shell starts filling).

5.7. Compare the number of electrons in Problem 5.6 with the atomic numbers of the first three noble gases.

Ans.   They are the same—2, 10, and 18.

5.8. How does lithium achieve the "octet" configuration?

Ans.   Li loses an electron, leaving it with the electron configuration of He. A configuration of two electrons in its outermost shell corresponds to the octet because the outermost shell is the first shell, which can hold only two electrons.

5.9. Which elements acquire the electron configuration of helium by covalent bonding?

Ans.   Only hydrogen. Lithium and beryllium are metals, which tend to lose electrons (and form ionic bonds) rather than share. The resulting configuration of two electrons in the first shell, with no other shells occupied, is stable, and therefore is also said to satisfy the octet rule. Second-period elements of higher atomic number tend to acquire the electron configuration of neon. If the outermost shell of an atom is the first shell, the maximum number of electrons in the atom is 2.

IONS

5.10. What is the difference between NO_2 and NO_2^-?

Ans.   The first is a compound and the second is an ion—a part of a compound.

5.11. (*a*) What is the charge on the sodium ion? (*b*) What is the charge on the sodium atom? (*c*) What is the charge on the sodium nucleus?

Ans.   (*a*) 1+, (*b*) 0, and (*c*) 11+. Note how important it is to read the questions carefully.

5.12. You have been given a stack of $3.00 gift certificates for a store which gives no change for these certificates. What is the minimum number of $2.00 items that you can buy without wasting money? How many certificates will you use?

Ans.   You can buy three items (for $6) with two certificates (worth $6).

5.13. What is the formula of the compound of aluminum and sulfur?

Ans. Al_2S_3. Aluminum has three electrons in the valence shell of each atom, and sulfur requires two. This problem involves the same reasoning as the last problem.

5.14. When a sodium atom loses an electron to form Na^+, how many electrons are there in what is now the outermost shell? in the valence shell?

Ans.   There are eight electrons in the second shell, which is now the outermost shell, since the one electron in the third shell has been lost. There are now zero electrons in the valence shell.

5.15. Arrange the electrons in each of the following ions in shells: (a) Mg^{2+}, (b) Ca^{2+}, and (c) O^{2-}.

Ans.

		\multicolumn{4}{c}{Shell Number}			
		1	2	3	4
(a)	Mg^{2+}	2	8	0	
(b)	Ca^{2+}	2	8	8	0
(c)	O^{2-}	2	8		

5.16. Write the formulas for the compounds of (a) Na^+ and ClO_2^- and (b) Mg^{2+} and ClO^-. Explain why one of the formulas requires parentheses.

Ans. $NaClO_2$ and $Mg(ClO)_2$. The parentheses mean two ClO^- ions; no parentheses means two O atoms in one ClO_2^- ion.

5.17. Write formulas for the compounds formed by the following pairs of ions: (a) Na^+ and Cl^-, (b) Na^+ and S^{2-}, (c) Ba^{2+} and S^{2-}, (d) Al^{3+} and S^{2-}, (e) Mg^{2+} and N^{3-}, and (f) Co^{2+} and ClO^-.

Ans.   (a) NaCl, (b) Na_2S, (c) BaS, (d) Al_2S_3, (e) Mg_3N_2, and (f) $Co(ClO)_2$.

5.18. Write formulas for the compounds formed by the reaction of (a) sodium and sulfur, (b) barium and bromine, (c) aluminum and oxygen, (d) lithium and nitrogen, (e) magnesium and nitrogen, (f) aluminum and fluorine, and (g) magnesium and sulfur.

Ans.   (a) Na_2S, (b) $BaBr_2$, (c) Al_2O_3, (d) Li_3N, (e) Mg_3N_2, (f) AlF_3, and (g) MgS.

5.19. What ions are present in each of the following compounds?

(a) $FeCl_2$
(b) Cu_2S
(c) CuO

(d) $Ba(ClO)_2$
(e) $(NH_4)_3PO_4$
(f) $CaCO_3$

Ans.   (a) Fe^{2+} and Cl^-
(b) Cu^+ and S^{2-}
(c) Cu^{2+} and O^{2-}

(d) Ba^{2+} and ClO^-
(e) NH_4^+ and PO_4^{3-}
(f) Ca^{2+} and CO_3^{2-}

5.20. Complete the following table by writing the formula of the compound formed by the cation at the left and the anion at the top. NH_4Br is given as an example.

	Br^-	BrO^-	BrO_3^-	SO_3^{2-}	AsO_4^{3-}
NH_4^+	NH_4Br				
K^+					
Fe^{2+}					
Cr^{3+}					

Ans.

	Br^-	BrO^-	BrO_3^-	SO_3^{2-}	AsO_4^{3-}
NH_4^+	NH_4Br	NH_4BrO	NH_4BrO_3	$(NH_4)_2SO_3$	$(NH_4)_3AsO_4$
K^+	KBr	$KBrO$	$KBrO_3$	K_2SO_3	K_3AsO_4
Fe^{2+}	$FeBr_2$	$Fe(BrO)_2$	$Fe(BrO_3)_2$	$FeSO_3$	$Fe_3(AsO_4)_2$
Cr^{3+}	$CrBr_3$	$Cr(BrO)_3$	$Cr(BrO_3)_3$	$Cr_2(SO_3)_3$	$CrAsO_4$

5.21. In 2.50 mol $Ba(ClO_3)_2$, (*a*) how many moles of barium ions are present? (*b*) How many moles of oxygen atoms are present? (*c*) How many moles of chlorate ions are present?

Ans. (*a*) 2.50 mol Ba^{2+} ions
   (*b*) 15.0 mol O atoms
   (*c*) 5.00 mol ClO_3^- ions

ELECTRON DOT NOTATION

5.22. How many electrons are "required" in the valence shell of atoms of each of the following elements in their compounds? (*a*) S, (*b*) H, (*c*) O, and (*d*) Mg.

Ans. (*a*) 8, (*b*) 2, (*c*) 8, and (*d*) 0.

5.23. Draw an electron dot diagram for each of the following:
   (*a*) Sr
   (*b*) As
   (*c*) Br
   (*d*) Br^-

   (*e*) Se
   (*f*) P^{3-}
   (*g*) Rn
   (*h*) Al^{3+}

Ans. (*a*) Sr:

   (*b*) $:\overset{\cdot}{A}s\cdot$

   (*c*) $:\overset{\cdot}{Br}:$

   (*d*) $:\overset{\cdot\cdot}{Br}:^-$

   (*e*) $:\overset{\cdot}{Se}:$

   (*f*) $:\overset{\cdot\cdot}{P}:^{3-}$

   (*g*) $:\overset{\cdot}{Rn}:$

   (*h*) Al^{3+} (All valence electrons have been lost.)

5.24. Draw an electron dot diagram for Li_3N.

Ans. $3\,Li^+$ $:\overset{\cdot\cdot}{N}:^{3-}$

5.25. Draw electron dot diagrams for sodium atoms and sulfur atoms (*a*) before they react with each other and (*b*) after they react with each other.

Ans. (*a*) $Na\cdot$ $:\overset{\cdot}{S}:$ $Na\cdot$

   (*b*) Na^+ $:\overset{\cdot\cdot}{S}:^{2-}$ Na^+

COVALENT BONDING

5.26. Draw electron dot diagrams for CH_4 and CH_3OH.

Ans.

$$
\begin{array}{cc}
\text{H} & \text{H} \\
\text{H:}\overset{..}{\underset{..}{\text{C}}}\text{:H} \qquad & \text{H:}\overset{..}{\underset{..}{\text{C}}}\text{:}\overset{..}{\underset{..}{\text{O}}}\text{:H} \\
\text{H} & \text{H}
\end{array}
$$

5.27. Identify the octet of carbon by encircling the electrons that satisfy the octet rule for carbon in Example 5.5(*a*). Put rectangles around the electrons that satisfy the octet for the oxygen atoms.

Ans.

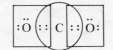

5.28. Identify the bonding electrons and the nonbonding electrons on the sulfur atom in the electron dot diagram of SO_2.

Ans.

$$
\begin{array}{c}
\text{nonbonding} \\
\mid \ \ \ \ \ \ \ \mid \\
\text{:O::S:O:} \\
\mid \ \ \ \ \mid \\
\text{bonding}
\end{array}
$$

5.29. Draw an electron dot diagram for $COCl_2$.

Ans.

$$
\begin{array}{c}
\text{:}\overset{..}{\underset{..}{\text{Cl}}}\text{:}\overset{..}{\text{C}}\text{:}\overset{..}{\underset{..}{\text{Cl}}}\text{:} \\
\text{:O:}
\end{array}
$$

5.30. Draw electron dot diagrams for oxygen atoms and sulfur atoms (*a*) before they react with each other and (*b*) after they react to form SO_2.

Ans.   (*a*) $\overset{:\dot{\text{O}}:}{\underset{:\dot{\text{O}}:}{\text{:}\dot{\text{S}}\text{:}}}$ (*b*) $\text{:}\overset{..}{\text{O}}\text{::}\overset{..}{\text{S}}\text{:}\overset{..}{\underset{..}{\text{O}}}\text{:}$

5.31. The gaseous elements hydrogen, nitrogen, and fluorine exist as diatomic molecules when they are not combined with other elements. Draw an electron dot structure for each molecule.

Ans.

$$
\text{H:H} \qquad\qquad \text{:N:::N:} \qquad\qquad \text{:}\overset{..}{\underset{..}{\text{F}}}\text{:}\overset{..}{\underset{..}{\text{F}}}\text{:}
$$

5.32. Explain why hydrogen atoms cannot form double bonds.

Ans.   They cannot hold more than two electrons in their valence shell, because it is the first shell. A double bond includes four electrons.

5.33. Draw electron dot diagrams for (*a*) NaCl, (*b*) SF_2, (*c*) PH_3, and (*d*) NH_4Br.

Ans.   (*a*) Na^+ $\text{:}\overset{..}{\underset{..}{\text{Cl}}}\text{:}^-$

(*b*) $\text{:}\overset{..}{\underset{..}{\text{F}}}\text{:}\overset{..}{\underset{..}{\text{S}}}\text{:}\overset{..}{\underset{..}{\text{F}}}\text{:}$

(*c*) $\overset{\text{H:}\overset{..}{\text{P}}\text{:H}}{\text{H}}$

(*d*) $\left[\overset{\text{H}}{\underset{\text{H}}{\text{H:}\overset{..}{\text{N}}\text{:H}}}\right]^+$ $\text{:}\overset{..}{\underset{..}{\text{Br}}}\text{:}^-$

In part (*a*), for example, do not draw the positive ion too near the negative ion; they are bonded by ionic bonding and do not share electrons.

5.34. Draw an electron dot diagram for each of the following: (*a*) SO_3, (*b*) SO_3^{2-}, (*c*) K_2SO_3, and (*d*) H_2SO_3.

Ans. (*a*) :Ö::S:Ö: (with :Ö: above S)

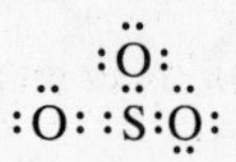

  (*b*) [:Ö:S:Ö:]$^{2-}$ (with :Ö: above S)

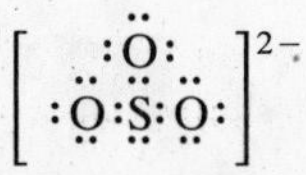

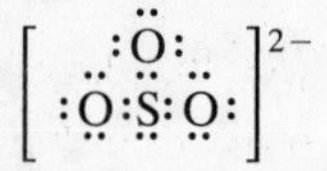

  (*c*) $2K^+$ [:Ö:S:Ö:]$^{2-}$ (with :Ö: above S)

  (*d*) H:Ö:S:Ö:H (with :Ö: above S)

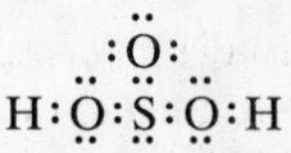

In (*a*), a double bond is needed to make the octet of sulfur. In (*b*), the extra pair of electrons makes the set of atoms an ion and eliminates the need for a double bond. In (*c*), that same ion is present, along with the two potassium ions to balance the charge. In (*d*), two hydrogen atoms are covalently bonded to oxygen atoms to complete the compound.

5.35. Draw electron dot structures for each of the following molecules: (*a*) CO, (*b*) CO_2, (*c*) HCN, (*d*) N_2O (an unsymmetrical molecule, with the two nitrogen atoms adjacent to each other).

Ans. (*a*) :C:::O:

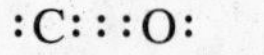

  (*b*) :Ö::C::Ö:

  (*c*) H:C:::N:

  (*d*) :N::N::Ö:

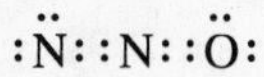

5.36. Draw the electron dot diagram for ammonium sulfide, $(NH_4)_2S$.

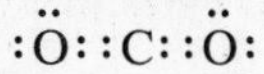

Ans.

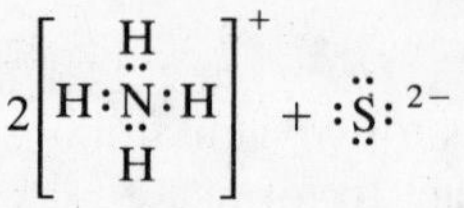

$$2\begin{bmatrix} \text{H} \\ \text{H:N:H} \\ \text{H} \end{bmatrix}^+ + :\ddot{\underset{..}{S}}:^{2-}$$

5.37. Draw electron dot diagrams for the following: (*a*) PCl_3, (*b*) H_2O, (*c*) H_2O_2, (*d*) ClO_3^-, and (*e*) NH_3.

Ans. (*a*) :Cl:P:Cl:
               :Cl:

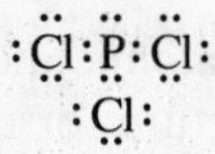

  (*b*) H:Ö:H

  (*c*) H:Ö:Ö:H

  (*d*) :Ö:Cl:Ö:⁻
               :Ö:

  (*e*) H:N:H
            H

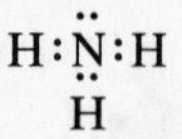

DISTINCTION BETWEEN IONIC AND COVALENT BONDING

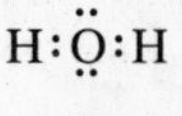

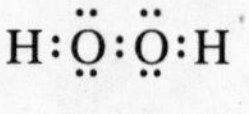

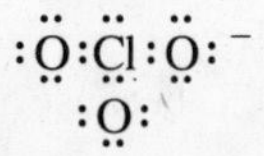

5.38. Draw an electron dot diagram for (*a*) Na_2SO_4 and (*b*) H_2SO_4. What is the major difference between them?

Ans. (*a*) $2Na^+$ [:Ö:S:Ö:]$^{2-}$ (with :Ö: above and :Ö: below S)
                         :Ö:

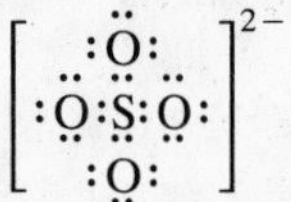

  (*b*) H:Ö:S:Ö:H (with :Ö: above and :Ö: below S)
            :Ö:

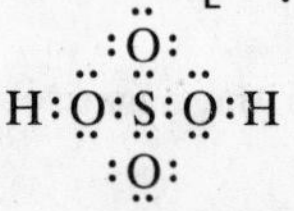

In the sodium salt there is ionic bonding as well as covalent bonding; in the hydrogen compound, there is only covalent bonding.

5.39. Describe the bonding of the chlorine atoms in each of the following substances: (a) Cl_2, (b) SCl_2, and (c) $MgCl_2$.

　　Ans.　(a) The chlorine atoms are bonded to each other with a covalent bond. (b) The chlorine atoms are both bonded to the sulfur atom with covalent bonds. (c) The chlorine atoms are changed to Cl^- ions, and are bonded to the magnesium ion by ionic bonds.

5.40. What type of bonding is present in each of the following compounds? (a) $MgCl_2$, (b) SCl_2, and (c) $Mg(ClO)_2$.

　　Ans.　(a) Ionic, (b) covalent, and (c) both ionic and covalent.

PREDICTING THE NATURE OF BONDING IN COMPOUNDS

5.41. Which element of each of the following pairs has the higher electronegativity? Consult Table 5-1 only after writing down your answer. (a) Mg and Cl, (b) S and O, and (c) Cl and O.

　　Ans.　(a) Cl. It lies farther to the right in the periodic table. (b) O. It lies farther up in the periodic table. (c) O. It lies farther up in the periodic table, and is more electronegative even though it lies one group to the left. (See Table 5-1.)

5.42. Which element is named first in the compound of each of the following pairs? (a) As and S, (b) As and Br, and (c) S and I.

　　Ans.　(a) As, since it lies to the left of S and below it. (b) As, since it lies left of Br. (c) S, since it lies left of I, despite the fact that it is above I.

Supplementary Problems

5.43. (a) How many electrons are there in the outermost shell of an atom of phosphorus? (b) How many *additional* electrons is it necessary for an atom of phosphorus to share in order to attain an octet configuration? (c) How many additional electrons is it necessary for an atom of chlorine to share in order to attain an octet configuration? (d) Write the formula for a compound of phosphorus and chlorine. (e) Draw an electron dot structure showing the arrangement of electrons in a molecule of the compound.

　　Ans.　(a)　5. (It is in periodic group VA.)
　　　　　(b)　3　$(8 - 5 = 3)$
　　　　　(c)　1　$(8 - 7 = 1)$
　　　　　(d)　PCl_3
　　　　　(e)　　$\ddot{\underset{..}{Cl}} : \underset{..}{P} : \ddot{\underset{..}{Cl}}$
　　　　　　　　　$: \underset{..}{Cl} :$

5.44. Phosphorus forms two covalent compounds with chlorine, PCl_3 and PCl_5. Discuss these compounds in terms of the octet rule.

　　Ans.　PCl_3 obeys the octet rule; PCl_5 does not. PCl_5 has to bond five chlorine atoms around the phosphorus atom, each with a pair of electrons, for a total of 10 electrons around phosphorus.

5.45. Draw an electron dot diagram for NO. Explain why it cannot follow the octet rule.

　　Ans.　$: \underset{.}{N} : : \ddot{O} :$
　　　　　There is an odd number of electrons in NO; there is no way that there can be eight around each atom.

5.46. Which of the following compounds involve covalent bonding? Which involve electron sharing? (*a*) $MgCl_2$, (*b*) SCl_2, and (*c*) $(NH_4)_2S$.

    Ans. SCl_2 and $(NH_4)_2S$ involve covalent bonding, and therefore by definition, electron sharing. $(NH_4)_2S$ also exhibits ionic bonding.

5.47. Draw an electron dot diagram for NH_4HS.

    Ans.

$$\left[\begin{array}{c} H \\ H\!:\!\overset{..}{N}\!:\!H \\ H \end{array} \right]^{+} \qquad H\!:\!\overset{..}{\underset{..}{S}}\!:^{-}$$

5.48. Write a formula for a binary compound formed between each of the following pairs of elements.

(*a*)	Na	F	(*h*)	Mg	N
(*b*)	K	S	(*i*)	C	Cl
(*c*)	Mg	Cl	(*j*)	Cl	Ca
(*d*)	Li	S	(*k*)	O	Mg
(*e*)	Li	N	(*l*)	P	Cl
(*f*)	Al	Br	(*m*)	S	F
(*g*)	Al	O			

    Ans.

(*a*)	NaF	(*h*)	Mg_3N_2
(*b*)	K_2S	(*i*)	CCl_4
(*c*)	$MgCl_2$	(*j*)	$CaCl_2$ (The metal ion is written first.)
(*d*)	Li_2S	(*k*)	MgO
(*e*)	Li_3N	(*l*)	PCl_3 (or PCl_5)
(*f*)	$AlBr_3$	(*m*)	SF_2 (or SF_4 or SF_6)
(*g*)	Al_2O_3		

5.49. Write the formula for the compound formed by the combination of each of the following pairs of elements. State whether the compound is ionic or covalent. (*a*) Mg and Br; (*b*) Ca and O; (*c*) Si and F; and (*d*) Br and Cl.

    Ans.

(*a*)	$MgBr_2$	ionic
(*b*)	CaO	ionic
(*c*)	SiF_4	covalent
(*d*)	BrCl	covalent

5.50. Distinguish between each of the following pairs: (*a*) an ion and an ionic bond. (*b*) an ion and a free atom. (*c*) a covalent bond and an ionic bond. (*d*) a triple bond and three single bonds on the same atom. (*e*) a polyatomic molecule and a polyatomic ion.

    Ans.   (*a*) An ion is a charged atom or group of atoms; an ionic bond is the attraction between ions.

           (*b*) An ion is charged and a free atom is uncharged.

           (*c*) A covalent bond involves sharing of electrons; an ionic bond involves electron transfer and as a result the formation of ions.

           (*d*) Although both involve three pairs of electrons, the triple bond has all three pairs of electrons between two atoms and three single bonds have each pair of electrons between different pairs of atoms.

$$:N:::N: \qquad\qquad H\!:\!\overset{..}{N}\!:\!H$$
$$ H$$

               triple bond   three single bonds

           (*e*) Both have more than one atom. The polyatomic ion is charged, and is only part of a compound; the polyatomic molecule is uncharged and represents a complete compound.

5.51. BF_3 and PF_5 are nonoctet rule compounds. Draw an electron dot diagram for each.

Ans.

$$:\overset{\displaystyle ..}{\underset{\displaystyle ..}{F}}:B:\overset{\displaystyle ..}{\underset{\displaystyle ..}{F}}: \qquad \begin{matrix} :\overset{..}{F}. & .\overset{..}{F}: \\ & P. \\ :\overset{..}{F}. & .\overset{..}{F}: \\ & :\overset{..}{F}: \end{matrix}$$

5.52. How many electrons remain in the valence shell after the loss of electrons by the neutral atom to form each of the following ions: (*a*) Pb^{2+} (*b*) Pb^{4+}

Ans.   (*a*) The Pb^{2+} still has two electrons in its sixth shell. (*b*) The Pb^{4+} has no electrons left in that shell.

5.53. Complete the following table by writing the formula of the compound formed by the cation at the left and the anion at the top.

	IO_2^-	IO_3^-
Fe^{2+}		
Fe^{3+}		

Ans.

	IO_2^-	IO_3^-
Fe^{2+}	$Fe(IO_2)_2$	$Fe(IO_3)_2$
Fe^{3+}	$Fe(IO_2)_3$	$Fe(IO_3)_3$

5.54. Draw an electron dot diagram for HNO_2. Note: Double-bonded oxygen atoms do not bond hydrogen atoms as a rule.

Ans.

$$H:\overset{\displaystyle ..}{\underset{\displaystyle ..}{O}}:N::\overset{\displaystyle ..}{\underset{\displaystyle ..}{O}}:$$

The hydrogen is bonded to the single-bonded oxygen atom.

5.55. Draw electron dot diagrams for H_3PO_3 and H_3PO_2. Four atoms are bonded directly to the phosphorus atom in each case.

Ans.

$$\begin{matrix} & :\overset{..}{\underset{..}{O}}: & \\ H:\overset{..}{\underset{..}{O}}:P:\overset{..}{\underset{..}{O}}:H \\ & H & \end{matrix} \qquad \begin{matrix} & H & \\ H:\overset{..}{\underset{..}{O}}:P:\overset{..}{\underset{..}{O}}: \\ & H & \end{matrix}$$

5.56. In 2.50 mol $NaClO_3$, (*a*) how many moles of sodium ions are present? (*b*) How many moles of chlorate ions are present?

Ans.   (*a*) 2.50 mol Na^+ ions. (*b*) 2.50 mol ClO_3^- ions. The subscript "3" refers to the number of O atoms per anion, not to the number of anions.

Chapter 6

Inorganic Nomenclature

6.1 INTRODUCTION

Naming and writing formulas for inorganic compounds are extremely important skills. For example, a physician might prescribe barium sulfate for a patient in preparation for a stomach x-ray. If barium sulfite or barium sulfide is given instead, the patient might die from barium poisoning. Such a seemingly small difference in the name makes a very big difference in the properties! (Barium sulfate is too insoluble to be toxic.)

There is a vast variety of inorganic compounds, and the compounds are named according to varying systems of nomenclature. The first job to do when you wish to name a compound is to determine which class it is in. Rules for the major classes will be given here. Compounds that are rarely encountered in general chemistry courses will not be covered. An outline of the classes that will be presented is given in Table 6-1, and rules for naming compounds in the different classes are illustrated in Fig. 6-1. Rules for writing formulas from names will be presented.

Table 6-1 Nomenclature Divisions for Inorganic Compounds

Binary nonmetal-nonmetal compounds
Ionic compounds
Monatomic cations with constant charges
Monatomic cations with variable charges
Polyatomic cations
Monatomic anions
Oxyanions
Varying numbers of oxygen atoms
Special anions
Inorganic acids
Acid Salts
Hydrates

6.2 BINARY COMPOUNDS OF NONMETALS

The first compounds to be discussed will be compounds of two nonmetals. These binary compounds are named with the element to the left or below in the periodic table named first. The other element is then named, with its ending changed to -*ide* and a prefix added to denote the number of atoms of that element present. If one of the elements is to the left and the other below, the one to the left is named first unless that element is oxygen or fluorine, in which case it is named last. The same order of elements is used in writing formulas for these compounds. (The element with the lower electronegativity is usually named first; refer to Table 5-1.) The prefixes are presented in Table 6-2. The first six prefixes are the most important to memorize.

The systematic names presented for binary nonmetal-nonmetal compounds are not used for the hydrogen compounds of group III, IV, and V elements or for water. These compounds have common names which are used instead. Water and ammonia (NH_3) are the most important compounds in this class.

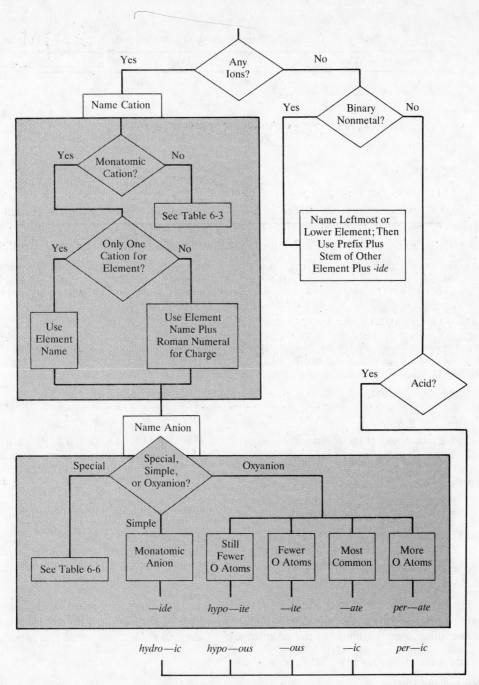

Fig. 6-1 Outline of nomenclature rules

EXAMPLE 6.1. Name and write the formula for a compound containing two atoms of oxygen and one atom of nitrogen in each molecule.

This compound is a compound of two nonmetals. The nitrogen is named first, since it lies to the left of oxygen in the periodic table. Then the oxygen is named, with its ending changed to *-ide* and a prefix denoting the number of oxygen atoms present.

$$\text{nitrogen dioxide} \qquad NO_2$$

EXAMPLE 6.2. Name (*a*) CO and (*b*) CO_2.

Both of these compounds are compounds of two nonmetals. The carbon is named first, since it lies to the left of oxygen in the periodic table. Then the oxygen is named, with its ending changed to *-ide* and a prefix denoting the number of oxygen atoms present. (*a*) Carbon monoxide and (*b*) carbon dioxide.

Table 6-2   Prefixes for Nonmetal-Nonmetal Compounds

Number of Atoms	Prefix
1	mono (or mon before names starting with a or o)
2	di
3	tri
4	tetra (or tetr before names starting with a or o)
5	penta (or pent before names starting with a or o)
6	hexa
7	hepta
8	octa
9	nona
10	deca

EXAMPLE 6.3.   Write the formulas for (*a*) sulfur dioxide and (*b*) sulfur trioxide.

(*a*) SO_2 and (*b*) SO_3. Sulfur is named first, since it lies below oxygen in the periodic table. (The element that is first in the name is also first in the formula.) The prefixes tell how many atoms of the second element there are in each molecule.

EXAMPLE 6.4.   Name P_2O_5.

Phosphorus lies to the left and below oxygen in the periodic table, so it is named first. According to the rules, the name is phosphorus pentoxide, and professional chemists call this compound phosphorus pentoxide. (Some texts for beginning students call the compound diphosphorus pentoxide, to aid students in writing the formula from the name.)

EXAMPLE 6.5.   Name (*a*) ClO_2 and (*b*) NCl_3.

(*a*) Chlorine dioxide and (*b*) nitrogen trichloride. In ClO_2, O is above Cl but to its left in the periodic table. In NCl_3, N is above Cl and to its left. When this situation arises with oxygen or fluorine, that element is written last; any other such pair has the element to the left written first. That is the reason for the order of naming in ClO_2 and NCl_3.

EXAMPLE 6.6.   Name H_2O.

Water. The common name is used.

6.3   NAMING IONIC COMPOUNDS

Ionic compounds are composed of *cations* (positive ions) and *anions* (negative ions). The cation is always named first, and then the anion is named. The name of the cation does not depend on the nature of the anion, and the name of the anion does not depend on the nature of the cation.

6.3.1   Naming the Cation

Naming of the positive ion depends on whether the cation is *monatomic* (has one atom). If not, the special names given in Sec. 6.3.2 are used. If the cation is monatomic, the name depends on whether the element forms more than one positive ion in its compounds. For example, sodium forms only one positive ion in all its compounds—Na^+. Iron forms two positive ions—Fe^{2+} and Fe^{3+}. Cations of elements that form only one type of ion in all their compounds need not be further identified in the name. Thus, Na^+ may simply be called the sodium ion. Cations of metals that occur with two or more different charges must be further identified. $Fe(NO_3)_2$ and $Fe(NO_3)_3$ occur with Fe^{2+} and Fe^{3+} ions, respectively. If we just call the ion the "iron ion," we will not know which one it is. Therefore, for monatomic cations, we use a Roman numeral in parentheses attached to the name to tell the charge on such ion. (Actually, oxidation numbers are used for this purpose, but if you have not yet studied oxidation numbers—Chap. 13—follow the rules given next.) Thus, Fe^{2+} is called the iron(II) ion, and Fe^{3+} is called the iron(III) ion.

The elements that form only one cation are the alkali metals (group IA), the alkaline earth metals (group IIA), zinc, cadmium, aluminum, and most often silver. The charge on the ions that these elements form in their compounds is always equal to their periodic table group number (or group number minus 10 in the newest labeling system in the periodic table).

EXAMPLE 6.7.   Name $CoCl_2$ and $BaCl_2$.

Since Co is not among the elements that always form ions of the same charge in all of its compounds, the charge must be stated. The name is cobalt(II) chloride. Since Ba is an alkaline earth element, the charge in its compounds is always 2+, so there is no need to mention the charge in the name. The compound is barium chloride.

EXAMPLE 6.8.   Write formulas for (a) copper(I) oxide and (b) copper(II) oxide.

(a) Cu_2O and (b) CuO. Note carefully that the Roman numerals in the names mean one thing—the charge on the ion—and the Arabic-numeral subscripts in the formulas mean another—number of atoms. Here the copper(I) has a charge of 1+, and therefore two copper(I) ions are required to balance the charge on one oxide ion. The copper(II) ion has a charge of 2+, and therefore one such ion is sufficient to balance the 2− charge on the oxide ion.

EXAMPLE 6.9.   Name NCl_3.

Nitrogen trichloride. Although this is a binary compound of two nonmetals, it can be named with Roman numeral designations. It is indeed possible to call this nitrogen(III) chloride in the most modern usage, but most chemists do not do that yet.

6.3.2   Polyatomic Cations

Several cations that consist of more than one atom are important in general chemistry. There are few enough of these important ions to learn them individually. They are presented in Table 6-3. There are several ions like uranyl ion, also of limited importance in general chemistry. More will be said about these ions in the sections on oxidation states (Chap. 13).

Table 6-3   Several Polyatomic Cations

$NH_4{}^+$	Ammonium ion	Very important
$Hg_2{}^{2+}$	Mercury(I) ion	Somewhat important
	or mercurous ion	
$UO_2{}^{2+}$	Uranyl ion	Not too important

6.3.3   Classical Nomenclature System

An older system for naming cations of elements having more than one possible cation uses the ending -ic for the ion with the higher charge and the ending -ous for the ion with the lower charge. In this system, the Latin names for some of the elements were used instead of the English names. Thus, this system is harder in two ways than the *Stock system*, described above—you must know whether a particular ion has a related ion of higher or lower charge and you must know the Latin root. You should study this section if your text or your instructor uses this system; otherwise you may omit it. The names for common ions in this system are given in Table 6-4. Note that transition metals except the coinage metals do not have ions with charges of 1+. (This system was also used for nonmetal-nonmetal compounds, and still exists in the designation of nitrogen oxides. N_2O is called nitrous oxide and NO is called nitric oxide.)

Table 6-4   Names of Cations in Classical System

Transition Metals			
Vanadous	V^{2+}	Vanadic	V^{3+}
Chromous	Cr^{2+}	Chromic	Cr^{3+}
Manganous	Mn^{2+}	Manganic	Mn^{3+}
Ferrous	Fe^{2+}	Ferric	Fe^{3+}
Cobaltous	Co^{2+}	Cobaltic	Co^{3+}
Nickelous	Ni^{2+}	Nickelic	Ni^{4+}
Cuprous	Cu^{+}	Cupric	Cu^{2+}
Argentous	Ag^{+}	Argentic	Ag^{2+} (rare)
Aurous	Au^{+}	Auric	Au^{3+}
Mercurous	Hg_2^{2+}	Mercuric	Hg^{2+}
Palladous	Pd^{2+}	Palladic	Pd^{4+}
Platinous	Pt^{2+}	Platinic	Pt^{4+}
Main Group Metals			
Stannous	Sn^{2+}	Stannic	Sn^{4+}
Plumbous	Pb^{2+}	Plumbic	Pb^{4+}
Inner Transition Element			
Cerous	Ce^{3+}	Ceric	Ce^{4+}

6.3.4   Naming Anions

Common anions may be grouped as follows: monatomic anions, *oxyanions*, and special anions. There are special endings for the first two groups; the third group is small enough to be memorized.

Monatomic Anions

If the anion is monatomic (has one atom), the name of the element is amended by changing the ending to *-ide*. Note that this ending is also used for binary nonmetal-nonmetal compounds. All monatomic anions have names ending in *-ide*, but there are a few anions that consist of more than one atom which also end in *-ide*—the most important of these are OH^- and CN^-. OH^- is called the hydroxide ion and CN^- is called the cyanide ion.

The charge on every monatomic anion is equal to the group number minus 8 (or 18, if the most modern periodic table group numbering system is used).

EXAMPLE 6.10.   What is the charge on (*a*) the oxide ion, (*b*) the nitride ion, and (*c*) the bromide ion?
(*a*)   Oxygen is in group VI (16), and so the charge is $6 - 8 = -2$ (or $16 - 18 = -2$).
(*b*)   Nitrogen is in group V (15), and so the charge is $5 - 8 = -3$.
(*c*)   Bromine is in group VII (17), and so the charge is $7 - 8 = -1$.

Oxyanions

Oxyanions consist of an atom of an element plus some number of atoms of oxygen covalently bonded to it. The name of the anion is given by the name of the element with its ending changed to either *-ate* or *-ite*. In some cases, it is also necessary to add the prefix *per-* or *hypo-* to distinguish all the possible oxyanions from one another. For example, there are four oxyanions of chlorine, which are named as follows:

$$ClO_4^-\qquad \text{perchlorate ion}$$
$$ClO_3^-\qquad \text{chlorate ion}$$
$$ClO_2^-\qquad \text{chlorite ion}$$
$$ClO^-\qquad \text{hypochlorite ion}$$

One may think of the *-ite* ending as meaning "one fewer oxygen atom." The *per-* and *hypo-* prefixes then mean "one more oxygen atom" and "still one fewer oxygen atom," respectively. Thus, perchlorate means one more oxygen atom than chlorate has. Hypochlorite means still one fewer oxygen atom than chlorite has. Note that all four oxyanions have the same central atom (Cl) and the same charge (1−). The only difference in their constitutions is the number of oxygen atoms.

Other elements have similar sets of oxyanions, but not all have four different oxyanions. You should learn the names of the seven ions ending in *-ate* for the most common elements. These are the most important oxyanions. (Use the rules given above for remembering the others.) These ions are presented in Table 6-5. Note that the ones with central atoms in odd periodic groups have odd charges and those in even periodic groups have even charges.

Note that not all the possible oxyanions of these elements exist. If the name and formula are not given in Table 6-5, the ion is not known. If you learn the seven ions that end in *-ate* plus the meaning of the ending *-ite* and the prefixes, you will be able to write formulas for 20 oxyanions. You may double this number of names by learning an additional rule in Sec. 6.4. Note from Table 6-5 that for each central element, all ions present have the same charge.

Table 6-5 The Most Common Oxyanions

ClO_4^-	Perchlorate	ClO_3^-	Chlorate	ClO_2^-	Chlorite	ClO^-	Hypochlorite
BrO_4^-	Perbromate	BrO_3^-	Bromate	BrO_2^-	Bromite	BrO^-	Hypobromite
IO_4^-	Periodate	IO_3^-	Iodate	IO_2^-	Iodite	IO^-	Hypoiodite
		NO_3^-	Nitrate	NO_2^-	Nitrite		
		PO_4^{3-}	Phosphate	PO_3^{3-}	Phosphite	PO_2^{3-}	Hypophosphite
		SO_4^{2-}	Sulfate	SO_3^{2-}	Sulfite		
		CO_3^{2-}	Carbonate				

EXAMPLE 6.11. Name the following ions without consulting Table 6-5: (*a*) SO_3^{2-}, (*b*) ClO^-, and (*c*) IO_4^-.

(*a*) Remembering that sulfate is SO_4^{2-}, we note that this ion has one fewer oxygen atom. It must be the sulfite ion. (*b*) Remembering that chlorate is ClO_3^-, we note that this ion has two fewer oxygen atoms. Chlorite would have one fewer; this ion must be the hypochlorite ion. (*c*) Remembering that iodate is IO_3^-, we note that this ion has one more oxygen atom. It must be the periodate ion.

Special Anions

There are a few anions that seem rather unusual but are often used in general chemistry. Perhaps the best way to remember them is to memorize them, but some hints to help do that will be given below. The special anions are given in Table 6-6.

Table 6-6 Special Anions

CrO_4^{2-}	Chromate
$Cr_2O_7^{2-}$	Dichromate
MnO_4^-	Permanganate
$C_2H_3O_2^-$	Acetate
CN^-	Cyanide
OH^-	Hydroxide
O_2^{2-}	Peroxide

Note that chromate and permanganate have central atoms in periodic groups with the same numbers as those of sulfate and perchlorate, and they have analogous formulas to these ions. Dichromate is related to chromate by having one more Cr atom and three more O atoms. The cyanide and hydroxide ions have already been discussed. The acetate ion is really the ion of an organic acid, which is why it is so unusual in this grouping. The peroxide ion has two oxygen atoms and a total charge of 2−.

6.3.5   Putting the Names of the Ions Together

Now that we know how to name the cations and anions, we merely have to put the two names together to get the names of ionic compounds. The cation is named first and the anion is named next. The number of cations and anions per formula unit need not be included in the name of the compound because anions have characteristic charges, and the charge on the cation has already been established by its name. There are as many cations and anions as needed to get a neutral compound with the lowest possible subscripts.

EXAMPLE 6.12.   Name (a) $Ba(NO_3)_2$ and (b) $Co(ClO_3)_2$.

(a) The cation is the barium ion. The anion is the nitrate ion. The compound is barium nitrate. Note that we do not state anything to indicate the presence of two nitrate anions; that we can deduce from the fact that the barium ion has a 2+ charge and nitrate has a 1− charge. (b) The cation is cobalt(II). We know that it is cobalt(II) because its charge must balance two chlorate ion charges, each 1−. The compound is cobalt(II) chlorate.

6.3.6   Writing Formulas for Ionic Compounds

Formulas of many ionic compounds are easily written by consideration of the charges on their ions. In order to do this, the charges on the ions must be memorized. The charges on most common ions are given in this section. There are both positive and negative ions. The formula of a compound in written so that the ions are combined in ratios such that there is a net charge of 0 on the compound as a whole. For example, to write the formula for sodium chloride, one sodium ion (having a charge of 1+) is combined with one chloride ion (having a charge of 1−); hence, the formula is NaCl. The algebraic total of the charges is 0. To write the formula of aluminum chloride, an aluminum ion (having a charge of 3+) is combined with three chloride ions; the resulting formula is $AlCl_3$. Again, the total algebraic sum of the charges is 0. The formula for ammonium sulfate includes two ammonium ions (each with a 1+ charge) and one sulfate ion (with a charge of 2−), written as $(NH_4)_2SO_4$. As a further example, the formula for aluminum oxide requires the combination of aluminum ions with oxygen ions (with a charge of 2−) in such a manner that the net charge is 0. The formula is Al_2O_3, corresponding to $2 \times (3+)$ for aluminum plus $3 \times (2-)$ for oxygen, a total of 0.

EXAMPLE 6.13.   Write the formulas for (a) copper(II) perchlorate and (b) sodium sulfite.

(a) The copper(II) ion is Cu^{2+}; perchlorate is ClO_4^-, having one more oxygen atom than chlorate and a single minus charge. The formula is therefore $Cu(ClO_4)_2$. It takes two perchlorate ions, each with a single negative charge, to balance one copper(II) ion. (b) The sodium ion is Na^+. (Sodium forms no other positive ion.) Sulfite is SO_3^{2-}, having one fewer oxygen atom than sulfate and a 2− charge. It takes two sodium ions to balance the charge on one sulfite ion, and so the formula is Na_2SO_3.

EXAMPLE 6.14.   Write the formulas for (a) lithium sulfide, (b) lithium iodide, (c) aluminum sulfate, and (d) magnesium chlorate.

(a) Li_2S, (b) LiI, (c) $Al_2(SO_4)_3$, and (d) $Mg(ClO_3)_2$.

6.4   NAMING INORGANIC ACIDS

The anions described in the preceding sections may be formed by reaction of the corresponding acids with hydroxides:

$$HCl + NaOH \longrightarrow NaCl + H_2O$$

$$H_2SO_4 + 2\,NaOH \longrightarrow Na_2SO_4 + 2\,H_2O$$

$$H_3PO_4 + 3\,NaOH \longrightarrow Na_3PO_4 + 3\,H_2O$$

The salts formed by these reactions consist of cations and anions. The cation in each case is Na^+, and the anions are Cl^-, SO_4^{2-}, and PO_4^{3-}, respectively. In these examples, the chloride ion, sulfate ion, and phosphate ion are formed from their parent acids. Thus the acids and anions are related, and so are their names.

When they are pure, acids are not ionic. When we put them into water solution, seven become fully ionized. These are called *strong acids*, and include HCl, $HClO_3$, $HClO_4$, HBr, HI, HNO_3, and H_2SO_4. The others ionize at least to some extent, and are called *weak acids*. Strong acids react completely with water to form ions and weak acids react to some extent to form ions, but both types react completely with hydroxides to form ions. Formulas for acids conventionally are written with the hydrogen atoms which can ionize first. Different names for some acids are given when the compound is pure and when it is dissolved in water. For example, HCl is called hydrogen chloride when it is in the gas phase, but in water it ionizes to give hydrogen ions and chloride ions, and is called hydrochloric acid. The names for all the acids corresponding to the anions in Table 6-5 can be deduced by the following simple rule. Note that the number of hydrogen atoms in the acid is the same as the number of negative charges on the anion.

> *Replace the -ate ending of an anion with "-ic acid" or replace the -ite ending with "-ous acid."*

This rule does not change if the anion has a prefix *per-* or *hypo-*; if the anion has such a prefix, so does the acid. If not, the acid does not either.

> *If the anion ends in -ide, add the prefix hydro- and change the ending to "-ic acid."*

Ion Ending	Acid Name Components
-ate	-ic acid
-ite	-ous acid
-ide	hydro—ic acid

EXAMPLE 6.15.   Name the following as acids: (*a*) HBr, (*b*) HNO_2, (*c*) H_3PO_4, (*d*) HIO, and (*e*) $HClO_4$.

(*a*) HBr is related to Br^-, the bromide ion. The *-ide* ending is changed to "*-ic* acid," and the prefix *hydro-* is added. The name is hydrobromic acid.

(*b*) HNO_2 is related to NO_2^-, the nitrite ion. The *-ite* ending is changed to "*-ous* acid." The name is nitrous acid.

(*c*) H_3PO_4 is related to PO_4^{3-}, the phosphate ion. The *-ate* ending is changed to "*-ic* acid," and so the name is phosphoric acid.

(*d*) HIO is related to IO^-, the hypoiodite ion. The prefix *hypo-* is not changed, but the *-ite* ending is changed to "*-ous* acid." The name is hypoiodous acid.

(*e*) $HClO_4$ is related to ClO_4^-, the perchlorate ion. The prefix *per-* is not changed, but the ending is changed to "*-ic* acid." The name is perchloric acid.

EXAMPLE 6.16.   Write formulas for the following acids: (*a*) nitric acid, (*b*) hydroiodic acid, (*c*) hypophosphorous acid, and (*d*) perbromic acid.

(*a*)   Nitric acid is related to the nitrate ion, NO_3^-. The acid has one hydrogen ion, corresponding to the one negative charge on nitrate ion. The formula is HNO_3.

(*b*)   The acid is related to the iodide ion, I^-. The acid has one hydrogen atom, because the ion has one negative charge. The formula is HI.

(*c*)   The acid is related to hypophosphite, PO_2^{3-}. The acid has three hydrogen atoms, corresponding to the three negative charges on the ion. Its formula is H_3PO_2.

(*d*) $HBrO_4$.

6.5   ACID SALTS

In Sec. 6.4 the reactions of hydroxides and acids were presented. It is possible for an acid with more than one ionizable hydrogen atom (with more than one hydrogen written first in the formula) to react with fewer hydroxide ions, and to form a product with some ionizable hydrogen atoms left:

$$H_2SO_4 + NaOH \longrightarrow NaHSO_4 + H_2O \qquad (Na^+ \text{ and } HSO_4^-)$$

$$H_3PO_4 + NaOH \longrightarrow NaH_2PO_4 + H_2O \qquad (Na^+ \text{ and } H_2PO_4^-)$$

$$H_3PO_4 + 2NaOH \longrightarrow Na_2HPO_4 + 2H_2O \qquad (Na^+ \text{ and } HPO_4^{2-})$$

The products formed are called *acid salts*, and the anions contain at least one ionizable hydrogen atom and at least one negative charge. The sum of the negative charges plus hydrogen atoms equals the original number of hydrogen atoms in the parent acid and also the number of negative charges in the normal anion. For example, HSO_4^- contains one hydrogen atom plus one negative charge, for a total of two. That is the number of hydrogen atoms in H_2SO_4 and the number of negative charges in SO_4^{2-}.

The anions of acid salts are named with the word hydrogen placed before the name of the normal anion. Thus, HSO_4^- is the hydrogen sulfate ion. To denote two atoms, the prefix *di-* is used. HPO_4^{2-} is the hydrogen phosphate ion, while $H_2PO_4^-$ is the dihydrogen phosphate ion. In an older naming system, the prefix *bi-* was used instead of the word hydrogen. Thus, HCO_3^- was called the bicarbonate ion instead of the more modern name, hydrogen carbonate ion.

EXAMPLE 6.17. Name HS^-.

Hydrogen sulfide ion.

EXAMPLE 6.18. What are the formulas for the dihydrogen phosphite ion and sodium dihydrogen phosphite?

$H_2PO_3^-$ and NaH_2PO_3. Note that the two hydrogen atoms plus the one charge total 3, equal to the number of hydrogen atoms in phosphorous acid. The one charge is balanced by the sodium ion in the complete compound.

6.6 HYDRATES

Some stable ionic compounds are capable of bonding to a certain number of molecules of water per formula unit. Thus, copper(II) sulfate forms the stable $CuSO_4 \cdot 5H_2O$, with five molecules of water per $CuSO_4$ unit. This type of compound is called a *hydrate*. The name of the compound is the name of the *anhydrous* (without water) compound with a designation for the number of water molecules appended. Thus, $CuSO_4 \cdot 5H_2O$ is called copper(II) sulfate pentahydrate. The 5, written on line, multiplies everything after it until the next centered dot or the end of the formula. Thus, included in $CuSO_4 \cdot 5H_2O$ there are ten H atoms and nine O atoms (five from the water and four in the sulfate ion).

Solved Problems

INTRODUCTION

6.1. Explain why the following three compounds, with such similar-looking formulas, have such different names:

$$\begin{array}{ll} AlCl_3 & \text{Aluminum chloride} \\ CoCl_3 & \text{Cobalt(III) chloride} \\ PCl_3 & \text{Phosphorus trichloride} \end{array}$$

Ans. The three compounds belong to different nomenclature classes. Aluminum in its compounds always forms 3+ ions, and thus there is no need to state 3+ in the name. Cobalt forms 2+ and 3+ ions, and we need to designate which of these exists in this compound. PCl_3 is a binary nonmetal-nonmetal compound, using a prefix to denote the number of chlorine atoms.

BINARY COMPOUNDS OF NONMETALS

6.2. Name the following compounds: (a) CO, (b) SO_2, (c) SO_3, (d) CCl_4, (e) PCl_5, and (f) SF_6.

    Ans. (a) Carbon monoxide, (b) sulfur dioxide, (c) sulfur trioxide, (d) carbon tetrachloride, (e) phosphorus pentachloride, and (f) sulfur hexafluoride.

6.3. Name the following compounds: (a) P_2S_5 and (b) NCl_3.

    Ans. (a) Phosphorus pentasulfide and (b) nitrogen trichloride.

6.4. Name the following: (a) a compound with three oxygen atoms and one bromine atom per molecule; (b) a compound with one iodine atom and three chlorine atoms per molecule; and (c) a compound with one sulfur atom and four iodine atoms per molecule.

    Ans. (a) Bromine trioxide, (b) iodine trichloride, and (c) sulfur tetraiodide.

6.5. Write formulas for each of the following compounds: (a) carbon tetrabromide, (b) sulfur dibromide, and (c) sulfur hexafluoride.

    Ans. (a) CBr_4, (b) SBr_2, and (c) SF_6.

6.6. Name (a) NH_3 and (b) H_2O.

    Ans. (a) Ammonia and (b) water.

6.7. Write formulas for each of the following compounds: (a) carbon tetrachloride, (b) carbon disulfide, and (c) carbon monoxide.

    Ans. (a) CCl_4, (b) CS_2, and (c) CO.

6.8. Write formulas for each of the following compounds: (a) nitrogen monoxide, (b) nitrogen tribromide, and (c) nitrogen triiodide.

    Ans. (a) NO, (b) NBr_3, and (c) NI_3.

NAMING IONIC COMPOUNDS

6.9. Name NH_3 and $NH_4{}^+$.

    Ans. Ammonia and ammonium ion. Note carefully the difference between these names and formulas.

6.10. How can a beginning student recognize an ionic compound?

    Ans. If the compound contains the $NH_4{}^+$ ion or a metal atom, it is most likely ionic. (It might have *internal* covalent bonds.) Otherwise, it is covalent, with no ions present.

6.11. In naming $SO_4{}^{2-}$, professional chemists might say "sulfate" or "sulfate ion," but in naming Na^+, they always say "sodium ion." Explain the difference.

    Ans. "Sulfate" is always an ion; "sodium" might refer to the element, the atom, or the ion, and so a distinction must be made.

6.12. Name the following ions: (a) Cu^+, (b) Cr^{2+}, (c) Pd^{2+}, and (d) $Hg_2{}^{2+}$.

    Ans. (a) Copper(I) ion, (b) chromium(II) ion, (c) palladium(II) ion, and (d) mercury(I) ion.

6.13. In which of the following periodic groups are located the metals that form ions of only one charge? (a) IA, (b) IIA, (c) IIIA, (d) VIA, (e) VIIA, (f) IB, (g) VIII, and (h) IIB.

    Ans. (a) All the alkali metals (but not hydrogen). (b) All the alkaline earth metals. (c) Aluminum ion. (f) Silver ion. (h) Zinc and cadmium ions.

6.14. Name the following compounds: (*a*) $NaClO_2$ and (*b*) $Ba(ClO)_2$.

    Ans.   (*a*) Sodium chlorite and (*b*) barium hypochlorite. Note that parentheses enclose the ClO^- ions, because there is a subscript to show that there are two of them. In (*a*), there is only one anion, which contains two oxygen atoms.

6.15. Explain one way to remember why mercury(I) ion has the Roman numeral I in it.

    Ans.   The average charge on the two Hg atoms in Hg_2^{2+} is 1+.

6.16. Explain why the formula for mercury(I) ion is Hg_2^{2+} rather than Hg^+.

    Ans.   The actual formula implies that the two mercury atoms are covalently bonded together.

6.17. Name the following ions: (*a*) Fe^{3+}, (*b*) I^-, (*c*) BrO_2^-, and (*d*) NO_2^-.

    Ans.   (*a*) Iron(III) ion, (*b*) iodide ion, (*c*) bromite ion, and (*d*) nitrite ion.

6.18. Name the following ions: (*a*) PO_3^{3-}, (*b*) S^{2-}, and (*c*) SO_4^{2-}.

    Ans.   (*a*) Phosphite ion, (*b*) sulfide ion, and (*c*) sulfate ion.

6.19. Write formulas for each of the following compounds: (*a*) sodium bromide, (*b*) sodium bromate, (*c*) sodium bromite, (*d*) sodium hypobromite, and (*e*) sodium perbromate.

    Ans.   (*a*) NaBr, (*b*) $NaBrO_3$, (*c*) $NaBrO_2$, (*d*) NaBrO, and (*e*) $NaBrO_4$.

6.20. What is the difference between ClO_3 and ClO_3^-?

    Ans.   The first is a compound, chlorine trioxide; the second is an ion, the chlorate ion. The chlorate ion has one extra electron, as shown by the minus sign.

6.21. (*a*) Write the formula for sodium sulfide, sodium sulfate, and sodium sulfite. (*b*) How many elements are implied in the compound by the *-ide* ending? by the *-ite* ending? by the *-ate* ending? (*c*) Name a particular element that is implied by the *-ate* ending.

    Ans.   (*a*) Na_2S, Na_2SO_4, and Na_2SO_3.
                (*b*)   *-ide*: at least two (a monatomic anion); *-ate* and *-ite*: at least three.
                (*c*)   There is oxygen plus another element in the anion.

6.22. Name the following compounds: (*a*) FeF_2, (*b*) FeF_3, (*c*) FeO, (*d*) Fe_2O_3, and (*e*) $FePO_4$.

    Ans.   (*a*) Iron(II) fluoride, (*b*) iron(III) fluoride, (*c*) iron(II) oxide, (*d*) iron(III) oxide, and (*e*) iron(III) phosphate.

6.23. Name the following compounds: (*a*) NaCl, (*b*) Na_2SO_4, and (*c*) Na_3PO_4.

    Ans.   (*a*) Sodium chloride, (*b*) sodium sulfate, and (*c*) sodium phosphate. The only indication that there are one, two, and three sodium ions in the compounds is in the knowledge of the charges on the ions.

6.24. Name the following compounds: (*a*) NH_4Cl, (*b*) $(NH_4)_2S$, and (*c*) $(NH_4)_3PO_4$.

    Ans.   (*a*) Ammonium chloride, (*b*) ammonium sulfide, and (*c*) ammonium phosphate. Note that the last two formulas require parentheses around the ammonium ion, but that the first one does not since there is only one ammonium ion per formula unit.

6.25. Name the following compounds: (a) $Ba_3(PO_4)_2$ and (b) $Co_2(SO_4)_3$.

Ans.   (a) Barium phosphate and (b) cobalt(III) sulfate. We recognize that cobalt has a 3+ charge because two cobalt ions are needed to balance three sulfate ions, each of which has a 2− charge.

6.26. Fill in the table with the formula of the compound whose cation is named at the left and whose anion is named at the column head.

	Chloride	Sulfate	Phosphate	Hydroxide
Ammonium				
Sodium				
Cobalt(II)				
Iron(III)				

Ans.

	Chloride	Sulfate	Phosphate	Hydroxide
Ammonium	NH_4Cl	$(NH_4)_2SO_4$	$(NH_4)_3PO_4$	NH_4OH (unstable)
Sodium	$NaCl$	Na_2SO_4	Na_3PO_4	$NaOH$
Cobalt(II)	$CoCl_2$	$CoSO_4$	$Co_3(PO_4)_2$	$Co(OH)_2$
Iron(III)	$FeCl_3$	$Fe_2(SO_4)_3$	$FePO_4$	$Fe(OH)_3$

6.27. Name the following ions: (a) BrO^-, (b) NO_3^-, and (c) SO_3^{2-}.

Ans.   (a) Hypobromite ion, (b) nitrate ion, and (c) sulfite ion.

6.28. Name the following ions: (a) $C_2H_3O_2^-$, (b) $Cr_2O_7^{2-}$, (c) O_2^{2-}, (d) MnO_4^-, and (e) CrO_4^{2-}.

Ans.   (a) Acetate ion, (b) dichromate ion, (c) peroxide ion, (d) permanganate ion, and (e) chromate ion.

6.29. Name the following ions: (a) H^+, (b) OH^-, (c) CN^-, (d) NH_4^+, and (e) K^+.

Ans.   (a) Hydrogen ion, (b) hydroxide ion, (c) cyanide ion, (d) ammonium ion, and (e) potassium ion.

6.30. Write formulas for each of the following compounds: (a) sodium cyanide, (b) sodium hydroxide, and (c) sodium peroxide.

Ans.   (a) $NaCN$, (b) $NaOH$, and (c) Na_2O_2 (from Table 6-6).

6.31. What is the difference between the names phosphorus and phosphorous?

Ans.   The first is the name of the element; the second is the name of the acid with fewer oxygen atoms than phosphoric acid—H_3PO_3, phosphorous acid.

6.32. Write formulas for each of the following compounds: (a) barium bromide, (b) copper(II) bromate, and (c) cobalt(III) fluoride.

Ans.   (a) $BaBr_2$, (b) $Cu(BrO_3)_2$, and (c) CoF_3.

6.33. Name and write formulas for each of the following: (a) the compound of sodium and bromine; (b) the compound of magnesium and bromine; and (c) the compound of aluminum and bromine.

Ans.   (a) Sodium bromide, $NaBr$; (b) magnesium bromide, $MgBr_2$; and (c) aluminum bromide, $AlBr_3$. The bromide ion has a 1− charge. The sodium, magnesium, and aluminum ions have charges of 1+, 2+, and 3+, respectively, and the subscripts given are the smallest possible to just balance the charges.

6.34. Name and write formulas for each of the following: (*a*) the compound of sodium and sulfur; (*b*) the compound of magnesium and sulfur; and (*c*) the compound of aluminum and sulfur.

Ans. (*a*) Sodium sulfide, Na_2S; (*b*) magnesium sulfide, MgS; and (*c*) aluminum sulfide, Al_2S_3. The sulfide ion has a 2− charge. The sodium, magnesium, and aluminum ions have charges of 1+, 2+, and 3+, respectively, and the subscripts given are the smallest possible to just balance the charges.

6.35. Name and write formulas for each of the following: (*a*) two compounds of copper and bromine; (*b*) two compounds of nickel and fluorine; and (*c*) two compounds of chromium and bromine. (If necessary, see Table 6-4 for data.)

Ans. (*a*) Copper(I) bromide and copper(II) bromide, $CuBr$ and $CuBr_2$; (*b*) nickel(II) fluoride and nickel(IV) fluoride, NiF_2 and NiF_4; and (*c*) chromium(II) bromide and chromium(III) bromide, $CrBr_2$ and $CrBr_3$.

6.36. Name and write formulas for each of the following: (*a*) two compounds of iron and bromine; (*b*) two compounds of palladium and bromine; and (*c*) two compounds of mercury and bromine.

Ans. (*a*) Iron(II) bromide and iron(III) bromide, $FeBr_2$ and $FeBr_3$; (*b*) palladium(II) bromide and palladium(IV) bromide, $PdBr_2$ and $PdBr_4$; and (*c*) mercury(I) bromide and mercury(II) bromide, Hg_2Br_2 and $HgBr_2$. (The charges on the cations can be obtained from Table 6-4.)

6.37. Name and write formulas for each of the following: (*a*) the compound of bromine and sodium, (*b*) the compound of bromine and magnesium, and (*c*) the compound of bromine and aluminum.

Ans. This answer is exactly the same as that of Problem 6.33. The metal is named first even if it is given last in the statement of the problem.

6.38. What are the charges on the following ions? (*a*) cyanide, (*b*) barium ion, (*c*) sulfide, (*d*) nitride, and (*e*) chloride.

Ans. (*a*) 1−, (*b*) 2+, (*c*) 2−, (*d*) 3−, and (*e*) 1−.

6.39. Using the periodic table if necessary, write formulas for the following compounds: (*a*) hydrogen bromide, (*b*) magnesium chloride, (*c*) barium sulfide, (*d*) aluminum fluoride, (*e*) beryllium bromide, (*f*) barium selenide, and (*g*) sodium iodide.

Ans. (*a*) HBr, (*b*) $MgCl_2$, (*c*) BaS, (*d*) AlF_3, (*e*) $BeBr_2$, (*f*) $BaSe$, and (*g*) NaI.

6.40. Write the formula for each of the following compounds: (*a*) hydrogen iodide, (*b*) calcium chloride, (*c*) lithium oxide, (*d*) silver nitrate, (*e*) iron(II) sulfide, (*f*) aluminum chloride, (*g*) ammonium sulfate, (*h*) zinc carbonate, (*i*) iron(III) oxide, (*j*) sodium phosphate, (*k*) iron(II) acetate, (*l*) ammonium cyanide, and (*m*) copper(II) chloride.

Ans. (*a*) HI, (*b*) $CaCl_2$, (*c*) Li_2O, (*d*) $AgNO_3$, (*e*) FeS, (*f*) $AlCl_3$, (*g*) $(NH_4)_2SO_4$, (*h*) $ZnCO_3$, (*i*) Fe_2O_3, (*j*) Na_3PO_4, (*k*) $Fe(C_2H_3O_2)_2$, (*l*) NH_4CN, and (*m*) $CuCl_2$.

NAMING INORGANIC ACIDS

6.41. Name the following acids: (*a*) HCl, (*b*) $HClO$, (*c*) $HClO_2$, (*d*) $HClO_3$, and (*e*) $HClO_4$.

Ans. (*a*) Hydrochloric acid, (*b*) hypochlorous acid, (*c*) chlorous acid, (*d*) chloric acid, and (*e*) perchloric acid.

6.42. Name the following compound: (*a*) H_2S (pure) and (*b*) H_2S (as an acid).

Ans. (*a*) Hydrogen sulfide and (*b*) hydrosulfuric acid.

6.43. Write formulas for each of the following acids: (*a*) hydrochloric acid, (*b*) phosphorous acid, and (*c*) hypochlorous acid.

Ans. (*a*) HCl, (*b*) H_3PO_3, and (*c*) $HClO$.

6.44. Name HI in two ways.

Ans. Hydrogen iodide and hydroiodic acid.

6.45. How many ionizable hydrogen atoms are there in $HC_2H_3O_2$?

Ans. One. Ionizable hydrogen atoms are written first. (The other three hydrogen atoms will not react with sodium hydroxide.)

ACID SALTS

6.46. Write formulas for each of the following compounds: (*a*) ammonium hydrogen sulfite, (*b*) sodium hydrogen sulfide, and (*c*) iron(II) bicarbonate.

Ans. (*a*) NH_4HSO_3, (*b*) $NaHS$, and (*c*) $Fe(HCO_3)_2$.

6.47. Write formulas for each of the following compounds: (*a*) calcium hydrogen carbonate, (*b*) disodium hydrogen phosphate, (*c*) sodium dihydrogen phosphate, and (*d*) calcium dihydrogen phosphate.

Ans. (*a*) $Ca(HCO_3)_2$ (the single charge on HCO_3^- is not sufficient to balance the 2+ charge on Ca; two anions are necessary), (*b*) Na_2HPO_4, (*c*) NaH_2PO_4, and (*d*) $Ca(H_2PO_4)_2$. (The same prefix is used to denote two atoms in (*b*), (*c*), and (*d*) as in nonmetal-nonmetal compounds.)

6.48. Write formulas for each of the following compounds: (*a*) sodium hydrogen phosphate, (*b*) sodium dihydrogen phosphate, and (*c*) magnesium hydrogen phosphate.

Ans. (*a*) Na_2HPO_4, (*b*) NaH_2PO_4, and (*c*) $MgHPO_4$. (The charges must balance in each case.)

HYDRATES

6.49. Name the following compound, and state how many hydrogen atoms it contains per formula unit:

$$Na_2SO_3 \cdot 7H_2O$$

Ans. Sodium sulfite heptahydrate. It contains fourteen H atoms per formula unit.

6.50. Name $CuSO_4$ and $CuSO_4 \cdot 5H_2O$.

Ans. Copper(II) sulfate and copper(II) sulfate pentahydrate.

Supplementary Problems

6.51. Name (*a*) H^+, (*b*) H^-, and (*c*) NaH.

Ans. (*a*) Hydrogen ion, (*b*) hydride ion, and (*c*) sodium hydride.

6.52. Write formulas for each of the following compounds: (*a*) ammonium cyanide, (*b*) mercury(I) nitrate, and (*c*) uranyl sulfate.

    Ans.   (*a*) NH_4CN (from Tables 6-3 and 6-6), (*b*) $Hg_2(NO_3)_2$ (from Table 6-3) and (*c*) UO_2SO_4 (from Table 6-3).

6.53. Write a formula for each of the following compounds: (*a*) phosphorus pentasulfide, (*b*) iodine heptafluoride, and (*c*) nitrogen dioxide.

    Ans.   (*a*) P_2S_5 (analogous to P_2O_5), (*b*) IF_7 (one of the few times the prefix *hepta-* is used), and (*c*) NO_2.

6.54. State the meaning of each of the following terms: (*a*) *per-*, (*b*) *hypo-*, (*c*) *hydro-*, (*d*) hydrogen (as part of an ion).

    Ans.   (*a*) *Per-* means "more oxygen atoms." For example, the perchlorate ion (ClO_4^-) has more oxygen atoms than does the chlorate ion (ClO_3^-). (*b*) *Hypo-* means "fewer oxygen atoms." For example, the hypochlorite ion (ClO^-) has fewer oxygen atoms than does the chlorite ion (ClO_2^-). (*c*) *Hydro-* means "no oxygen atoms." For example, hydrochloric acid (HCl) has no oxygen atoms, in contrast to chloric acid ($HClO_3$). (*d*) Hydrogen signifies an acid salt, such as NaHS—sodium hydrogen sulfide.

6.55. What relationship is there in the meaning of the prefix *per-* when used with an oxyanion and when used in peroxide?

    Ans.   In both cases, it means "more oxygen atoms."

6.56. What are the differences among the following questions? (*a*) What is the formula of the compound of sulfur and barium? (*b*) What is the formula of the compound of Ba^{2+} and S^{2-}? (*c*) What is the formula for barium sulfide?

    Ans.   In each case, the answer is BaS. Part (*b*) gives the ions and their charges, and so is perhaps easiest to answer. Part (*a*) gives the elements, so it is necessary to know that periodic group IIA elements always form 2+ ions in all their compounds and that sulfur forms a 2− ion in its compounds with metals. It is also necessary to remember that the metal is named first. In part (*c*), the fact that there is only one compound of these two elements is deduced by the fact that the barium is stated with no Roman numeral, and that sulfide is a specific ion with a specific (2−) charge.

6.57. Name (*a*) Cu_2S and (*b*) CuS.

    Ans.   (*a*) Copper(I) sulfide and (*b*) copper(II) sulfide.

6.58. Distinguish between SO_3 and SO_3^{2-}.

    Ans. SO_3 is a compound, and SO_3^{2-} is an ion—part of a compound.

6.59. Complete the following table by writing the formula of the compound formed by the cation at the left and the anion at the top. NH_4Cl is given as an example.

	Cl^-	ClO^-	ClO_4^-	SO_4^{2-}	PO_4^{3-}
NH_4^+	NH_4Cl				
K^+					
Ba^{2+}					
Al^{3+}					

    Ans.

	Cl^-	ClO^-	ClO_4^-	SO_4^{2-}	PO_4^{3-}
NH_4^+	NH_4Cl	NH_4ClO	NH_4ClO_4	$(NH_4)_2SO_4$	$(NH_4)_3PO_4$
K^+	KCl	KClO	$KClO_4$	K_2SO_4	K_3PO_4
Ba^{2+}	$BaCl_2$	$Ba(ClO)_2$	$Ba(ClO_4)_2$	$BaSO_4$	$Ba_3(PO_4)_2$
Al^{3+}	$AlCl_3$	$Al(ClO)_3$	$Al(ClO_4)_3$	$Al_2(SO_4)_3$	$AlPO_4$

6.60. From the data of Table 6-4, determine which charge is most common for first transition series elements (those elements in period 4 among the transition groups).

Ans. The 2+ charge is seen to be most common, occurring with every element in the first transition series which is listed. (Only Sc has no 2+ ion.)

6.61. Name the compounds of Problem 6.59.

Ans.

	Cl^-	ClO^-	ClO_4^-	SO_4^{2-}	PO_4^{3-}
NH_4^+	Ammonium chloride	Ammonium hypochlorite	Ammonium perchlorate	Ammonium sulfate	Ammonium phosphate
K^+	Potassium chloride	Potassium hypochlorite	Potassium perchlorate	Potassium sulfate	Potassium phosphate
Ba^{2+}	Barium chloride	Barium hypochlorite	Barium perchlorate	Barium sulfate	Barium phosphate
Al^{3+}	Aluminum chloride	Aluminum hypochlorite	Aluminum perchlorate	Aluminum sulfate	Aluminum phosphate

6.62. What is the difference in the rules for remembering charges on monatomic anions and oxyanions?

Ans. Monatomic anions have charges equal to the group number minus 8. The oxyanions have even charges for even-group central atoms and odd charges for odd-group central atoms, but the rule does not give a simple way to determine the charge definitely for oxyanions as it does for monatomic anions.

6.63. Use the periodic table relationships to write formulas for (*a*) the selenate ion and (*b*) the arsenate ion.

Ans. Selenium is below sulfur in the periodic table, and arsenic is below phosphorus. We write formulas analogous to those for sulfate and phosphate: (*a*) SeO_4^{2-} and (*b*) AsO_4^{3-}.

Chapter 7

Chemical Equations

7.1 INTRODUCTION

A chemical reaction is described by means of a shorthand notation called a *chemical equation*. One or more substances, called *reactants* or *reagents*, are allowed to react to form one or more other substances, called *products*. Instead of using words, equations are written using the formulas for the substances involved. For example, a reaction used to prepared oxygen may be described in words as follows:

mercury(II) oxide when heated yields oxygen gas plus mercury

Using the formulas for the substances involved, the process could be written

$$HgO \xrightarrow{\text{heat}} O_2 + Hg \qquad \text{(unfinished)}$$

A chemical equation describes a chemical reaction in many ways as an empirical formula describes a chemical compound. The equation describes not only which substances react, but the relative number of moles of each undergoing reaction and the relative number of moles of each product formed. Note especially that it is the mole ratios in which the substances react, not how much is present, that the equation describes. In order to show the quantitative relationships, the equation must be *balanced*. That is, it must have the same number of atoms of each element used up and produced (except for special equations that describe nuclear reactions). The law of conservation of mass is thus obeyed, and also the "law of conservation of atoms." *Coefficients* are used before the formulas for elements and compounds to tell how many formula units of that substance are involved in the reaction. A coefficient does not imply any chemical bonding between units of the substance it is placed before. The number of atoms involved in each formula unit is multiplied by the coefficient to get the total number of atoms of each element involved. Later, when equations with individual ions are written (Chap. 9), the net charge on each side of the equation, as well as the numbers of atoms of each element, must be the same to have a balanced equation. The absence of a coefficient in a *balanced* equation implies a coefficient of 1.

The balanced equation for the decomposition of HgO is

$$2\,HgO \xrightarrow{\text{heat}} O_2 + 2\,Hg$$
$$\underbrace{\hspace{4cm}}_{\text{coefficients}}$$

EXAMPLE 7.1. Draw a ball-and-stick diagram depicting the reaction represented by the following equation:

$$2\,H_2 + O_2 \longrightarrow 2\,H_2O$$

The figure is shown as Fig. 7.1.

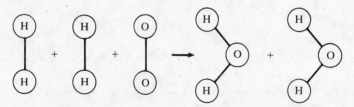

Fig. 7-1   The reaction of hydrogen and oxygen

EXAMPLE 7.2. Interpret the following equation:

$$2\,Na + Cl_2 \longrightarrow 2\,NaCl$$

114

The equation states that elementary sodium reacts with elementary chlorine to produce sodium chloride, table salt. (The fact that chlorine is one of the seven elements that occur in diatomic molecules when not combined with other elements is indicated.) The numbers before the Na and NaCl are coefficients, stating how many formula units of these substances are involved. If there is no coefficient in a balanced equation, a coefficient of 1 is implied, and so the absence of a coefficient before the Cl_2 implies one Cl_2 molecule. The equation thus states that when the two reagents react, they do so in a ratio of two atoms of sodium to one molecule of chlorine, to form two formula units of sodium chloride. In addition, it states that when the two reagents react, they do so in a ratio of 2 mol of sodium to 1 mol of chlorine molecules, to form 2 mol of sodium chloride. The ratios of moles of each reactant and product to every other reactant or product are implied:

$$\frac{2\,\text{mol Na}}{1\,\text{mol Cl}_2} \qquad \frac{2\,\text{mol Na}}{2\,\text{mol NaCl}} \qquad \frac{1\,\text{mol Cl}_2}{2\,\text{mol NaCl}}$$

$$\frac{1\,\text{mol Cl}_2}{2\,\text{mol Na}} \qquad \frac{2\,\text{mol NaCl}}{2\,\text{mol Na}} \qquad \frac{2\,\text{mol NaCl}}{1\,\text{mol Cl}_2}$$

EXAMPLE 7.3. How many ratios are implied for a reaction in which four reactants plus products are involved?

Twelve. (Each of the four coefficients has a ratio with the other three, and three 4's equals 12.)

7.2 BALANCING SIMPLE EQUATIONS

If you know the reactants and products of a chemical reaction, you should be able to write an equation for the reaction and balance it. In writing the equation, first write down the correct formulas for all reactants and products. After they are written down, only then start to balance the equation. Do not balance the equation by changing the formulas of the substances involved. For simple equations, you should balance the equation "by inspection." (Balancing oxidation-reduction equations will be presented in Chap. 13.) The following rules will help you to balance simple equations.

1. Before the equation is balanced, there are no coefficients for any reactant or product; after the equation is balanced, the absence of a coefficient implies a coefficient of 1. To prevent ambiguity while you are balancing the equation, place a question mark before each substance, and replace the question mark as you figure out each real coefficient. After you get some practice, you will not need to use the question marks.
2. Assume a coefficient of 1 for the most complicated substance in the equation. Since you are getting ratios, you can assume any value for one of the substances. Then work from this substance to figure out the coefficients of the others, one at a time.
3. If an element appears in more than one reactant or more than one product, leave that element for last, if possible.
4. Optionally, if a polyatomic ion is involved which does not change during the reaction, you may treat the whole thing as one unit, instead of considering the atoms which make it up.
5. After you have provided coefficients for all the substances, if any fractions are present, multiply every coefficient by a small integer to clear the fractions.
6. Always check to see that you have the same number of atoms of each element on the two sides of the equation after you finish.

EXAMPLE 7.4. Phosphoric acid, H_3PO_4, reacts with sodium hydroxide to produce sodium phosphate and water. Write a balanced chemical equation for the reaction.

Step 1: Write down the formulas for reactants and products. Add question marks.

$$? \, H_3PO_4 + ? \, NaOH \longrightarrow ? \, Na_3PO_4 + ? \, H_2O$$

Step 2: Assume a coefficient of 1 for a complicated reactant or product.

$$1 \, H_3PO_4 + ? \, NaOH \longrightarrow ? \, Na_3PO_4 + ? \, H_2O$$

Balance the other substances from that one.

$$1\,H_3PO_4 + ?\,NaOH \longrightarrow 1\,Na_3PO_4 + ?\,H_2O$$

$$1\,H_3PO_4 + 3\,NaOH \longrightarrow 1\,Na_3PO_4 + ?\,H_2O$$

$$1\,H_3PO_4 + 3\,NaOH \longrightarrow 1\,Na_3PO_4 + 3\,H_2O$$

$$H_3PO_4 + 3\,NaOH \longrightarrow Na_3PO_4 + 3\,H_2O$$

Since the coefficient of H_3PO_4 is 1, there is one phosphorus atom on the left of the equation. Phosphorus appears in only one product, and so that product must have a coefficient of 1. The one Na_3PO_4 has three Na atoms in it, and so there must be three Na atoms on the left; the NaOH gets a coefficient of 3. There are three H atoms in H_3PO_4 and three more in three NaOH, and so three water molecules are produced. The oxygen atoms are balanced, with seven on each side. We drop the coefficients of 1 to finish our equation.

Step 5 is not necessary.

Step 6: Check: We find 6 H atoms, 1 P atom, 3 Na atoms, and 7 O atoms on each side. Alternatively (step 4), we count 6 H atoms, 1 PO_4 group, 3 Na atoms, and 3 other O atoms on each side of the equation.

EXAMPLE 7.5. Zinc metal reacts with HCl to produce $ZnCl_2$ and hydrogen gas. Write a balanced equation for the process.

Step 1:

$$?\,Zn + ?\,HCl \longrightarrow ?\,ZnCl_2 + ?\,H_2$$

We note that hydrogen is one of the seven elements that form diatomic molecules when in the elementary state.

Step 2:

$$?\,Zn + ?\,HCl \longrightarrow 1\,ZnCl_2 + ?\,H_2$$

$$1\,Zn + 2\,HCl \longrightarrow 1\,ZnCl_2 + ?\,H_2$$

$$1\,Zn + 2\,HCl \longrightarrow 1\,ZnCl_2 + 1\,H_2$$

$$Zn + 2\,HCl \longrightarrow ZnCl_2 + H_2$$

Step 6: There are one Zn atom, two H atoms, and two Cl atoms on each side of the equation.

EXAMPLE 7.6. Balance the following equation:

$$Cu(NO_3)_2 + NaI \longrightarrow CuI + I_2 + NaNO_3$$

Step 1:

$$?\,Cu(NO_3)_2 + ?\,NaI \longrightarrow ?\,CuI + ?\,I_2 + ?\,NaNO_3$$

Step 2:

$$1\,Cu(NO_3)_2 + ?\,NaI \longrightarrow ?\,CuI + ?\,I_2 + ?\,NaNO_3$$

$$1\,Cu(NO_3)_2 + ?\,NaI \longrightarrow 1\,CuI + ?\,I_2 + 2\,NaNO_3$$

$$1\,Cu(NO_3)_2 + 2\,NaI \longrightarrow 1\,CuI + ?\,I_2 + 2\,NaNO_3$$

$$1\,Cu(NO_3)_2 + 2\,NaI \longrightarrow 1\,CuI + \tfrac{1}{2}\,I_2 + 2\,NaNO_3$$

Step 5:

$$2\,Cu(NO_3)_2 + 4\,NaI \longrightarrow 2\,CuI + 1\,I_2 + 4\,NaNO_3$$

$$\text{or } 2\,Cu(NO_3)_2 + 4\,NaI \longrightarrow 2\,CuI + I_2 + 4\,NaNO_3$$

7.3 PREDICTING THE PRODUCTS OF A REACTION

Before you can balance a chemical equation, you have to know the formulas for all the reactants and products. If the names are given for these substances, you have to know how to write formulas from the names (Chap. 6). If reactants only are given, you have to know how to predict the products from the reactants. This latter topic is the subject of this section.

To simplify the discussion, we will classify simple chemical reactions into five types:

Type 1: combination reactions
Type 2: decomposition reactions
Type 3: substitution reactions
Type 4: double-substitution reactions
Type 5: combustion reactions

More complex oxidation-reduction reactions will be discussed in Chap. 13.

Combination Reactions

A combination reaction is a reaction of two reactants to produce one product. The simplest combination reactions are the reactions of two elements to form a compound. After all, if two elements are treated with each other, they can either react or not. There is no other possibility, since neither can decompose. In most reactions like this, there will be a reaction. The main problem is to write the formula of the one product correctly, and then balance the equation. In this process, first determine the formulas of the products from the rules of chemical combination (Chap. 5). Only when the formulas of the reactants and products have all been written down do you balance the equation by adjusting the coefficients.

EXAMPLE 7.7. Complete and balance the following equations:

$$Na + Cl_2 \longrightarrow$$

$$Mg + O_2 \longrightarrow$$

$$Li + S \longrightarrow$$

The products are determined first (from electron dot structures, if necessary) to be NaCl, MgO, and Li_2S, respectively. These are placed to the right of the respective arrows, and the equations are then balanced.

$$2\,Na + Cl_2 \longrightarrow 2\,NaCl$$

$$2\,Mg + O_2 \longrightarrow 2\,MgO$$

$$2\,Li + S \longrightarrow Li_2S$$

EXAMPLE 7.8. Write a complete, balanced equation for the reaction of each of the following pairs of elements: (a) magnesium and sulfur, (b) potassium and bromine, and (c) aluminum and oxygen.

(a) The reactants are Mg and S. The Mg can lose two electrons [it is in periodic group IIA (2)], and sulfur can gain two electrons [it is in periodic group VIA (16)]. The ratio of magnesium to sulfur is thus 1:1, and the compound which will be formed is MgS:

$$Mg + S \longrightarrow MgS$$

The equation is already balanced.

(b) The reactants are K and Br_2. (In its elementary form, bromine is stable as diatomic molecules.) The combination of a group IA (1) metal and a group VIIA (17) nonmetal produces a salt with a 1:1 ratio of atoms: KBr.

$$K + Br_2 \longrightarrow KBr \quad \text{(unbalanced)}$$

$$2\,K + Br_2 \longrightarrow 2\,KBr$$

(c)
$$Al + O_2 \longrightarrow Al_2O_3 \quad \text{(unbalanced)}$$

$$4\,Al + 3\,O_2 \longrightarrow 2\,Al_2O_3$$

It is possible for an element and a compound of that element or for two compounds containing a common element to react by combination. The most common type in general chemistry is the reaction of a metal oxide with a nonmetal oxide to produce a salt with an oxyanion. For example,

$$MgO + SO_3 \longrightarrow MgSO_4$$

Decomposition Reactions

The second type of simple reaction is *decomposition*. This reaction is also easy to recognize. Typically, only one reactant is given. A type of energy, such as heat or electricity, may also be indicated. This reactant decomposes to its elements, to an element and a simpler compound, or to two simpler compounds. *Binary compounds* may yield two elements or an element and a simpler compound. *Ternary* (three-element) *compounds* may yield an element and a compound or two simpler compounds. These possibilities are shown in Fig. 7.2.

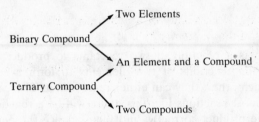

Fig. 7-2 Decomposition possibilities

A *catalyst* is a substance that speeds up a chemical reaction without undergoing a permanent change in its own composition. Catalysts are often but not always noted above or below the arrow in the chemical equation. Since a small quantity of catalyst is sufficient to cause a large quantity of reaction, the amount of catalyst need not be specified; it is not balanced like the reactants and products. In this manner, the equation for a common laboratory preparation of oxygen is written

$$2\,KClO_3 \xrightarrow{\ MnO_2\ } 2\,KCl + 3\,O_2$$

EXAMPLE 7.9. Write a complete, balanced equation for the reaction that occurs when (*a*) HgO is heated, (*b*) H_2O is electrolyzed, and (*c*) $KClO_3$ is heated in the presence of the catalyst MnO_2.

(*a*) With only one reactant, what can happen? No simpler compound of Hg and O is evident, and the compound decomposes to its elements:

$$HgO \longrightarrow Hg + O_2 \qquad (unbalanced)$$

Remember that oxygen occurs in diatomic molecules when it is uncombined.

$$2\,HgO \longrightarrow 2\,Hg + O_2$$

(*b*) Note that in most of these cases, energy of some type is added to make the compound decompose.

$$2\,H_2O \longrightarrow 2\,H_2 + O_2$$

(*c*) $$2\,KClO_3 \longrightarrow 2\,KCl + 3\,O_2$$

A ternary compound does not yield three elements; this one yields an element and a simpler compound. The MnO_2 may be indicated over or below the arrow, or it may be omitted entirely.

Substitution or Replacement Reactions

Elements have varying abilities to combine. Among the most reactive metals are the alkali metals and the alkaline earth metals. On the opposite end of the scale of reactivities, among the least active metals or the most stable metals are silver and gold, prized for their lack of reactivity. *Reactive* means the opposite of *stable*, but means the same as *active*.

When a free element reacts with a compound of different elements, the free element will replace one of the elements in the compound if the free element is more reactive than the element it replaces. In general, a free metal will replace the metal in the compound, or a free nonmetal will replace the nonmetal in the compound. A new compound and a new free element are produced. As usual, the formulas of the products are written on the bases presented in Chap. 5. The formula of a product does not depend on the formula of the reacting element or compound. For example, consider the reactions of sodium with iron(II) chloride and fluorine with sodium chloride:

$$2\,Na + FeCl_2 \longrightarrow 2\,NaCl + Fe$$

$$F_2 + 2\,NaCl \longrightarrow 2\,NaF + Cl_2$$

Sodium, a metal, replaces iron, another metal. Fluorine, a nonmetal, replaces chlorine, another nonmetal. (In some high-temperature reactions, a nonmetal can displace a relatively inactive metal from its compounds.) The formulas for F_2, NaCl, NaF, and Cl_2 are written on the basis of the rules of chemical bonding (Chap. 5).

You can easily recognize the possibility of a substitution reaction because you are given a free element and a compound of different elements.

EXAMPLE 7.10. Look only at the reactants in the following equations. Tell which of the reactions represent substitution reactions.

(*a*) $$4\,Li + O_2 \longrightarrow 2\,Li_2O$$

(*b*) $$2\,Li + MgO \longrightarrow Li_2O + Mg$$

(*c*) $$2\,KClO_3 \longrightarrow 2\,KCl + 3\,O_2$$

(*d*) $$Cl_2 + 2\,FeCl_2 \longrightarrow 2\,FeCl_3$$

(*b*) only. (*a*) is a combination, (*c*) is a decomposition, and (*d*) is a combination. Note in (*d*) that chlorine is added to a compound of chlorine.

If the free element is less active than the corresponding element in the compound, no reaction will take place. A short list of metals in order of their reactivities and an even shorter list of nonmetals are presented in Table 7-1. The metals in the list range from very active to very stable; the nonmetals listed range from very active to fairly active. A more comprehensive list, a table of standard reduction potentials, is presented in general chemistry textbooks.

Table 7-1   Relative Reactivities of Some Metals and Nonmetals

	Metals	Nonmetals	
Most Active Metals	Alkali and Alkaline	F	Most Active Nonmetals
	Earth Metals	O	
	Al	Cl	
	Zn	Br	
	Fe	I	Less Active Nonmetals
	Pb		
	H		
	Cu		
Less Active Metals	Ag		
	Au		

EXAMPLE 7.11. Complete and balance the following equations. If no reaction occurs, indicate that fact by writing "NR."

(*a*) $$NaCl + Fe \longrightarrow$$

(*b*) $$NaF + Cl_2 \longrightarrow$$

(*a*) $$NaCl + Fe \longrightarrow NR$$

(*b*) $$NaF + Cl_2 \longrightarrow NR$$

In each of these cases, the free element is less active than the corresponding element in the compound, and cannot replace that element from its compound.

In substitution reactions, hydrogen in its compounds with nonmetals often acts like a metal; hence, it is listed among the metals in Table 7.1.

EXAMPLE 7.12. Complete and balance the following equation. If no reaction occurs, indicate that fact by writing "NR."

$$Zn + HCl \longrightarrow$$

$$Zn + 2HCl \longrightarrow ZnCl_2 + H_2$$

Zinc is more reactive than hydrogen (Table 7-1), and replaces it from its compounds. Note that free hydrogen is in the form H_2 (Sect. 4.3).

In substitution reactions with acids, metals that can form two different ions in their compounds generally form the one with the lower charge. For example, iron can form Fe^{2+} and Fe^{3+}. In its reaction with HCl, $FeCl_2$ is formed. In contrast, in combination with the free element, the higher-charged ion is often formed if sufficient nonmetal is available.

$$2Fe + 3Cl_2 \longrightarrow 2FeCl_3$$

See Table 6-4 for possible charges on some common metal ions.

Double-Substitution or Double-Replacement Reactions

Double-substitution or *double-replacement reactions*, also called *double-decomposition reactions* or *metathesis reactions*, involve two ionic compounds, most often in aqueous solution. In this type of reaction, the cations simply swap anions. The reaction proceeds if a solid or a covalent compound is formed from ions in solutions. All gases at room temperature are covalent. Some reactions of ionic solids plus ions in solution also occur. Otherwise, no reaction takes place. For example,

$$AgNO_3 + NaCl \longrightarrow AgCl + NaNO_3$$

$$HCl(aq) + NaOH \longrightarrow NaCl + H_2O$$

$$CaCO_3(s) + 2HCl \longrightarrow CaCl_2 + CO_2 + H_2O$$

In the first reaction, two ionic compounds in water are mixed. The AgCl formed by the swapping of anions is insoluble, causing the reaction to proceed. The solid AgCl formed from solution is an example of a *precipitate*. In the second reaction, a covalent compound, H_2O, is formed from its ions in solution, H^+ and OH^-, causing the reaction to proceed. In the third reaction, a solid reacts with the acid in solution to produce two covalent compounds.

Since it is useful to know what state each reagent is in, we often designate the state in the equation. The modern practice is to add to the formula the designation in parentheses: (s) for solid, (l) for liquid, (g) for gas, and (aq) for aqueous solution. Thus, a reaction of silver nitrate with sodium chloride in aqueous solution, yielding solid silver chloride and aqueous sodium nitrate, may be written as

$$AgNO_3(aq) + NaCl(aq) \longrightarrow AgCl(s) + NaNO_3(aq)$$

Older practices used an underline or an arrow pointing down to denote precipitate and an overbar or arrow pointing up to denote gas.

$$AgNO_3 + NaCl \longrightarrow AgCl\downarrow + NaNO_3$$

$$AgNO_3 + NaCl \longrightarrow \underline{AgCl} + NaNO_3$$

Just as with replacement reactions, double-replacement reactions may or may not proceed. They need a driving force. The driving force in replacement reactions is reactivity; here it is insolubility or covalence. In order for you to be able to predict if a double-replacement reaction will proceed, you must know some solubilities of ionic compounds. A short list of solubilities is given in Table 7-2.

Table 7-2   Some Solubility Classes

Soluble	Insoluble
Chlorates	$BaSO_4$
Acetates	Most sulfides
Nitrates	Most oxides
Alkali metal salts	Most carbonates
Ammonium salts	Most phosphates
Chlorides, except for.........	$AgCl$, $PbCl_2$, Hg_2Cl_2, $CuCl$

EXAMPLE 7.13.   Complete and balance the following equation. If no reaction occurs, indicate that fact by writing "NR."

$$NaCl + KNO_3 \longrightarrow$$

$$NaCl + KNO_3 \longrightarrow NR$$

If a double-substitution reaction took place, $NaNO_3$ and KCl would be produced. However, both of these are soluble and ionic; hence, there is no driving force and therefore no reaction.

In double-replacement reactions, the charges on the metal ions (and indeed on nonmetal ions if they do not form covalent compounds) generally remain the same throughout the reaction.

EXAMPLE 7.14.   Complete and balance the following equations. If no reaction occurs, indicate that fact by writing "NR."

$$FeCl_3 + AgNO_3 \longrightarrow$$
$$FeCl_2 + AgNO_3 \longrightarrow$$

$$FeCl_3 + 3\,AgNO_3 \longrightarrow Fe(NO_3)_3 + 3\,AgCl$$
$$FeCl_2 + 2\,AgNO_3 \longrightarrow Fe(NO_3)_2 + 2\,AgCl$$

If you start with Fe^{3+} (first equation), you wind up with Fe^{3+}. If you start with Fe^{2+} (second equation), you wind up with Fe^{2+}.

NH_4OH and H_2CO_3 are unstable. If one of these products were expected as a product of a reaction, either NH_3 plus H_2O or CO_2 plus H_2O would be obtained instead:

$$NH_4OH \longrightarrow NH_3 + H_2O$$
$$H_2CO_3 \longrightarrow CO_2 + H_2O$$

Combustion Reactions

Reactions of elements and compounds with oxygen are so prevalent that they may be considered a separate type of reaction. Compounds of carbon, hydrogen, oxygen, sulfur, nitrogen, and other elements may be burned. Of greatest importance, if a reactant contains carbon, carbon monoxide or carbon dioxide will be produced, depending upon how much oxygen is available. Reactants containing hydrogen always produce water on burning. Of less importance, NO and SO_2 are other products of burning in oxygen. (To produce SO_3 requires a catalyst in a combustion reaction with O_2.)

EXAMPLE 7.15.   Complete and balance the following equations:

(a) $$C_2H_4 + O_2 \text{ (limited amount)} \longrightarrow$$
(b) $$C_2H_4 + O_2 \text{ (excess amount)} \longrightarrow$$

(a) $$C_2H_4 + 2\,O_2 \longrightarrow 2\,CO + 2\,H_2O$$
(b) $$C_2H_4 + 3\,O_2 \longrightarrow 2\,CO_2 + 2\,H_2O$$

If sufficient O_2 is available (3 mol O_2 per mole C_2H_4), CO_2 is the product. In both cases, H_2O is produced.

Acids and Bases

Generally, acids and bases react according to the rules for replacement and double-replacement reactions given above. They are so important, however, that a special nomenclature has developed for acids and their reactions. Acids were introduced in Sec. 6.4. They may be identified in their formulas by having the H representing hydrogen written first, and in their names by the presence of the word acid. An acid will react with a base to form a *salt* and water. The process is called *neutralization*. Neutralization reactions will be used as examples in Sec. 10.5, on titration.

$$HBr + NaOH \longrightarrow NaBr + H_2O$$
$$|$$
$$\text{a salt}$$

The driving force for such reactions is the formation of water, a covalent compound.

As pure compounds, acids are covalent. When placed in water, they react with the water to form ions; it is said that they *ionize*. If they react 100% with the water, they are said to be *strong acids*. The seven common strong acids are listed in Table 7-3. All the rest are *weak*; that is, the rest ionize only a few percent, and largely stay in their covalent forms. Both strong and weak acids react 100% with metal hydroxides. All soluble metal hydroxides are ionic in water.

Table 7-3   The Seven Common Strong Acids

HCl, $HClO_3$, $HClO_4$, HBr, HI, HNO_3, H_2SO_4 (first proton only)

Solved Problems

INTRODUCTION

7.1. How many oxygen atoms are there in each of the following, perhaps as part of a balanced chemical equation? (*a*) $7\,H_2O$, (*b*) $3\,Ba(NO_3)_2$, (*c*) $4\,CuSO_4 \cdot 5H_2O$, and (*d*) $2\,UO_2(ClO_3)_2$.

Ans.   (*a*) 7, (*b*) 18, (*c*) 36, and (*d*) 16.

BALANCING SIMPLE EQUATIONS

7.2. Balance the following equation:

$$C + Cu_2O \xrightarrow{\text{heat}} CO + Cu$$

Ans.

$$C + Cu_2O \longrightarrow CO + 2\,Cu$$

7.3. Balance the following equation:

$$NH_4Cl + NaOH \longrightarrow NH_3 + H_2O + NaCl$$

Ans.

$$?\,NH_4Cl + ?\,NaOH \longrightarrow ?\,NH_3 + ?\,H_2O + ?\,NaCl$$
$$1\,NH_4Cl + ?\,NaOH \longrightarrow ?\,NH_3 + ?\,H_2O + ?\,NaCl$$
$$1\,NH_4Cl + ?\,NaOH \longrightarrow 1\,NH_3 + ?\,H_2O + 1\,NaCl$$
$$1\,NH_4Cl + 1\,NaOH \longrightarrow 1\,NH_3 + ?\,H_2O + 1\,NaCl$$
$$1\,NH_4Cl + 1\,NaOH \longrightarrow 1\,NH_3 + 1\,H_2O + 1\,NaCl$$
$$NH_4Cl + NaOH \longrightarrow NH_3 + H_2O + NaCl$$

7.4.   Balance the following equation:

$$La(HCO_3)_3 + H_2SO_4 \longrightarrow La_2(SO_4)_3 + H_2O + CO_2$$

Ans.

$$? La(HCO_3)_3 + ? H_2SO_4 \longrightarrow ? La_2(SO_4)_3 + ? H_2O + ? CO_2$$

$$? La(HCO_3)_3 + ? H_2SO_4 \longrightarrow 1 La_2(SO_4)_3 + ? H_2O + ? CO_2$$

$$2 La(HCO_3)_3 + 3 H_2SO_4 \longrightarrow 1 La_2(SO_4)_3 + ? H_2O + ? CO_2$$

$$2 La(HCO_3)_3 + 3 H_2SO_4 \longrightarrow 1 La_2(SO_4)_3 + 6 H_2O + 6 CO_2$$

$$2 La(HCO_3)_3 + 3 H_2SO_4 \longrightarrow La_2(SO_4)_3 + 6 H_2O + 6 CO_2$$

7.5.   Balance the following equation:

$$NaOH + NH_4Cl + AgCl \longrightarrow Ag(NH_3)_2Cl + NaCl + H_2O$$

Ans.

$$? NaOH + ? NH_4Cl + ? AgCl \longrightarrow ? Ag(NH_3)_2Cl + ? NaCl + ? H_2O$$

$$? NaOH + ? NH_4Cl + ? AgCl \longrightarrow 1 Ag(NH_3)_2Cl + ? NaCl + ? H_2O$$

$$? NaOH + 2 NH_4Cl + 1 AgCl \longrightarrow 1 Ag(NH_3)_2Cl + ? NaCl + ? H_2O$$

$$? NaOH + 2 NH_4Cl + 1 AgCl \longrightarrow 1 Ag(NH_3)_2Cl + 2 NaCl + ? H_2O$$

$$2 NaOH + 2 NH_4Cl + 1 AgCl \longrightarrow 1 Ag(NH_3)_2Cl + 2 NaCl + ? H_2O$$

$$2 NaOH + 2 NH_4Cl + 1 AgCl \longrightarrow 1 Ag(NH_3)_2Cl + 2 NaCl + 2 H_2O$$

$$2 NaOH + 2 NH_4Cl + AgCl \longrightarrow Ag(NH_3)_2Cl + 2 NaCl + 2 H_2O$$

7.6.   Balance the following chemical equations:

(*a*) $$NaOH + H_3PO_4 \longrightarrow Na_3PO_4 + H_2O$$

(*b*) $$Ba(OH)_2 + H_3PO_4 \longrightarrow Ba_3(PO_4)_2 + H_2O$$

(*c*) $$NCl_3 + H_2O \longrightarrow HClO + NH_3$$

(*d*) $$Fe + HCl \longrightarrow FeCl_2 + H_2$$

(*e*) $$BaCO_3 + HClO_3 \longrightarrow Ba(ClO_3)_2 + CO_2 + H_2O$$

(*f*) $$NaC_2H_3O_2 + HCl \longrightarrow NaCl + HC_2H_3O_2$$

(*g*) $$HCl + Zn \longrightarrow ZnCl_2 + H_2$$

(*h*) $$HCl + NaHCO_3 \longrightarrow NaCl + H_2O + CO_2$$

Ans.

(*a*) $$3 NaOH + H_3PO_4 \longrightarrow Na_3PO_4 + 3 H_2O$$

(*b*) $$3 Ba(OH_2) + 2 H_3PO_4 \longrightarrow Ba_3(PO_4)_2 + 6 H_2O$$

(*c*) $$NCl_3 + 3 H_2O \longrightarrow 3 HClO + NH_3$$

(*d*) $$Fe + 2 HCl \longrightarrow FeCl_2 + H_2$$

(*e*) $$BaCO_3 + 2 HClO_3 \longrightarrow Ba(ClO_3)_2 + CO_2 + H_2O$$

(*f*) $$NaC_2H_3O_2 + HCl \longrightarrow NaCl + HC_2H_3O_2$$

(*g*) $$2 HCl + Zn \longrightarrow ZnCl_2 + H_2$$

(*h*) $$HCl + NaHCO_3 \longrightarrow NaCl + H_2O + CO_2$$

7.7.   Write balanced equations for each of the following reactions:

(*a*)   Lithium plus oxygen yields lithium oxide

(*b*)   Mercury(II) oxide when heated yields mercury and oxygen

   (c)   Carbon plus oxygen yields carbon monoxide

   (d)   Carbon plus oxygen yields carbon dioxide

   (e)   Methane (CH_4) plus oxygen yields carbon dioxide plus water

   (f)   Ethane (C_2H_6) plus oxygen yields carbon dioxide plus water

   (g)   Ethylene (C_2H_4) plus oxygen yields carbon dioxide plus water

Ans.

   (a)
$$4\,Li + O_2 \longrightarrow 2\,Li_2O$$

   (b)
$$2\,HgO \xrightarrow{\text{heat}} 2\,Hg + O_2$$

   (c)
$$2\,C + O_2 \longrightarrow 2\,CO$$

   (d)
$$C + O_2 \longrightarrow CO_2$$

   (e)
$$CH_4 + 2\,O_2 \longrightarrow CO_2 + 2\,H_2O$$

   (f)
$$2\,C_2H_6 + 7\,O_2 \longrightarrow 4\,CO_2 + 6\,H_2O$$

   (g)
$$C_2H_4 + 3\,O_2 \longrightarrow 2\,CO_2 + 2\,H_2O$$

7.8.   Write balanced chemical equations for the following reactions:

   (a)   Sodium plus bromine yields sodium bromide

   (b)   Phosphorus trichloride plus water yields phosphorous acid plus hydrogen chloride

   (c)   Sodium hydroxide plus sulfuric acid yields sodium sulfate plus water

   (d)   Ethane, C_2H_6, plus oxygen yields carbon monoxide plus water

   (e)   Octane, C_8H_{18}, plus oxygen yields carbon dioxide plus water

   (f)   Zinc plus hydrobromic acid yields zinc bromide plus hydrogen

   (g)   Potassium chlorate when heated yields potassium chloride plus oxygen

   (h)   Copper plus silver nitrate yields copper(II) nitrate plus silver

   (i)   Calcium hydrogen carbonate plus heat yields calcium carbonate plus carbon dioxide plus water

   (j)   Calcium hydrogen carbonate plus hydrobromic acid yields calcium bromide plus carbon dioxide plus water

   (k)   Lead(II) nitrate plus sodium chromate yields lead(II) chromate plus sodium nitrate

   (l)   Barium hydroxide plus nitric acid yields barium nitrate plus water

   (m) Copper(II) sulfate plus water yields copper(II) sulfate pentahydrate

   (n)   Copper(II) nitrate plus sodium iodide yields copper(I) iodide plus iodine plus sodium nitrate

Ans.

   (a) $2\,Na + Br_2 \longrightarrow 2\,NaBr$ (Remember that free bromine is Br_2.)

   (b) $PCl_3 + 3\,H_2O \longrightarrow H_3PO_3 + 3\,HCl$

   (c) $2\,NaOH + H_2SO_4 \longrightarrow Na_2SO_4 + 2\,H_2O$

   (d) $2\,C_2H_6 + 5\,O_2 \longrightarrow 4\,CO + 6\,H_2O$

   (e) $2\,C_8H_{18} + 25\,O_2 \longrightarrow 16\,CO_2 + 18\,H_2O$

   (f) $Zn + 2\,HBr \longrightarrow ZnBr_2 + H_2$

   (g) $2\,KClO_3 \longrightarrow 2\,KCl + 3\,O_2$ (Remember that free oxygen is O_2.)

   (h) $Cu + 2\,AgNO_3 \longrightarrow Cu(NO_3)_2 + 2\,Ag$

   (i) $Ca(HCO_3)_2 \longrightarrow CaCO_3 + H_2O + CO_2$

   (j) $Ca(HCO_3)_2 + 2\,HBr \longrightarrow CaBr_2 + 2\,H_2O + 2\,CO_2$

   (k) $Pb(NO_3)_2 + Na_2CrO_4 \longrightarrow PbCrO_4 + 2\,NaNO_3$

   (l) $Ba(OH)_2 + 2\,HNO_3 \longrightarrow Ba(NO_3)_2 + 2\,H_2O$

   (m) $CuSO_4 + 5\,H_2O \longrightarrow CuSO_4 \cdot 5\,H_2O$

   (n) $2\,Cu(NO_3)_2 + 4\,NaI \longrightarrow 2\,CuI + I_2 + 4\,NaNO_3$

7.9.   Balance the following equation:

$$Ba(ClO_3)_2 + Na_2SO_4 \longrightarrow BaSO_4 + NaClO_3$$

Ans.

$$Ba(ClO_3)_2 + Na_2SO_4 \longrightarrow BaSO_4 + 2\,NaClO_3$$

The oxygen atoms need not be considered individually if the ClO_3^- and SO_4^{2-} ions are considered as groups. (See step 4, Sec. 7.2.)

7.10.   Balance the following equation:

$$Fe + FeCl_3 \longrightarrow FeCl_2$$

Ans.   Balance the Cl first, since the Fe appears in two places in the reactants. Here, the Fe happens to be balanced automatically.

$$Fe + 2\,FeCl_3 \longrightarrow 3\,FeCl_2$$

7.11.   Write balanced chemical equations for the following reactions: (*a*) Hydrogen fluoride is produced by the reaction of hydrogen and fluorine. (*b*) Hydrogen combines with fluorine to yield hydrogen fluoride. (*c*) Fluorine reacts with hydrogen to give hydrogen fluoride.

Ans.   (*a*), (*b*), and (*c*).

$$H_2 + F_2 \longrightarrow 2\,HF$$

7.12.   Write balanced chemical equations for the following reactions: (*a*) Hydrogen fluoride is produced by the reaction of hydrochloric acid and sodium fluoride. (*b*) Hydrochloric acid combines with sodium fluoride to yield hydrogen fluoride. (*c*) Sodium fluoride reacts with hydrochloric acid to give hydrofluoric acid.

Ans.   (*a*), (*b*), and (*c*).

$$HCl + NaF \longrightarrow NaCl + HF$$

7.13.   Balance the following chemical equations:

(*a*) $$Na_2CO_3 + HClO_3 \longrightarrow NaClO_3 + CO_2 + H_2O$$

(*b*) $$Ca(HCO_3)_2 + HCl \longrightarrow CaCl_2 + CO_2 + H_2O$$

(*c*) $$H_2S + O_2 \longrightarrow H_2O + SO_2$$

(*d*) $$BiCl_3 + H_2O \longrightarrow BiOCl + HCl$$

(*e*) $$NH_3 + O_2 \longrightarrow NO + H_2O$$

(*f*) $$Cu_2S + O_2 \longrightarrow Cu + SO_2$$

(*g*) $$H_2O_2 \longrightarrow H_2O + O_2$$

Ans.

(*a*) $$Na_2CO_3 + 2\,HClO_3 \longrightarrow 2\,NaClO_3 + CO_2 + H_2O$$

(*b*) $$Ca(HCO_3)_2 + 2\,HCl \longrightarrow CaCl_2 + 2\,CO_2 + 2\,H_2O$$

(*c*) $$2\,H_2S + 3\,O_2 \longrightarrow 2\,H_2O + 2\,SO_2$$

(*d*) $$BiCl_3 + H_2O \longrightarrow BiOCl + 2\,HCl$$

(*e*) $$4\,NH_3 + 5\,O_2 \longrightarrow 4\,NO + 6\,H_2O$$

(*f*) $$Cu_2S + O_2 \longrightarrow 2\,Cu + SO_2$$

(*g*) $$2\,H_2O_2 \longrightarrow 2\,H_2O + O_2$$

7.14. Balance the following chemical equations:

(a) $H_2S + CuCl_2 \longrightarrow HCl + CuS$

(b) $Hg_2Cl_2 + NH_3 \longrightarrow HgNH_2Cl + Hg + NH_4Cl$

(c) $Pb(NO_3)_2 + KI \longrightarrow PbI_2 + KNO_3$

(d) $NH_3 + AgCl \longrightarrow Ag(NH_3)_2Cl$

(e) $Sb_2S_3 + Na_2S \longrightarrow NaSbS_2$

Ans.

(a) $H_2S + CuCl_2 \longrightarrow 2HCl + CuS$

(b) $Hg_2Cl_2 + 2NH_3 \longrightarrow HgNH_2Cl + Hg + NH_4Cl$

(c) $Pb(NO_3)_2 + 2KI \longrightarrow PbI_2 + 2KNO_3$

(d) $2NH_3 + AgCl \longrightarrow Ag(NH_3)_2Cl$

(e) $Sb_2S_3 + Na_2S \longrightarrow 2NaSbS_2$

7.15. Why is the catalyst not merely placed on both sides of the arrow, since it comes out of the reaction with the same composition as it started with?

Ans.   That would imply a certain mole ratio to the other reactants and products, which is not correct.

PREDICTING THE PRODUCTS OF A REACTION

7.16. In the list of reactivities of metals, Table 7-1, are all alkali metals more reactive than all alkaline earth metals, or are all elements of both groups of metals more active than any other metals?

Ans.   Both groups of metals are more active than any other metals. Actually, some alkaline earth metals are more active than some alkali metals, and vice versa.

7.17. (*a*) What type of reaction requires knowledge of solubility properties of compounds? (*b*) What type requires knowledge of reactivities of elements?

Ans.   (*a*) Double-substitution reaction. (*b*) Substitution reaction.

7.18. Complete and balance the following equations:

(a) $C + O_2$ (limited amount) $\longrightarrow$

(b) $C + O_2$ (excess amount) $\longrightarrow$

Ans.

(a) $2C + O_2$ (limited amount) $\longrightarrow 2CO$

(b) $C + O_2$ (excess amount) $\longrightarrow CO_2$

If sufficient O_2 is available, CO_2 is the product.

7.19. What type of chemical reaction is represented by each of the following? Complete and balance the equation for each.

(a) $Cl_2 + NaI \longrightarrow$

(b) $Cl_2 + Na \longrightarrow$

(c) $Na + ZnCl_2 \longrightarrow$

(d) $ZnCl_2 + AgC_2H_3O_2 \longrightarrow$

(e) $C_3H_8 + O_2$ (excess) $\longrightarrow$

Ans. (*a*) Substitution

$$Cl_2 + 2\,NaI \longrightarrow 2\,NaCl + I_2$$

(*b*) Combination

$$Cl_2 + 2\,Na \longrightarrow 2\,NaCl$$

(*c*) Substitution

$$2\,Na + ZnCl_2 \longrightarrow Zn + 2\,NaCl$$

(*d*) Double substitution

$$ZnCl_2 + 2\,AgC_2H_3O_2 \longrightarrow Zn(C_2H_3O_2)_2 + 2\,AgCl$$

(*e*) Combustion

$$C_3H_8 + 5\,O_2 \longrightarrow 3\,CO_2 + 4\,H_2O$$

7.20. What type of chemical reaction is represented by each of the following? Complete and balance the equation for each.

(*a*) $\qquad\qquad\qquad\qquad Cl_2 + FeCl_2 \longrightarrow$

(*b*) $\qquad\qquad\qquad\qquad CO + O_2 \longrightarrow$

(*c*) $\qquad\qquad\qquad\qquad CaCO_3 \xrightarrow{\text{heat}}$

(*d*) $\qquad\qquad\qquad\qquad ZnCl_2 + Cl_2 \longrightarrow$

(*e*) $\qquad\qquad C_3H_8 + O_2 \text{ (limited quantity)} \longrightarrow$

Ans. (*a*) Combination

$$Cl_2 + 2\,FeCl_2 \longrightarrow 2\,FeCl_3$$

(*b*) Combination (or combustion)

$$2\,CO + O_2 \longrightarrow 2\,CO_2$$

(*c*) Decomposition

$$CaCO_3 \xrightarrow{\text{heat}} CO_2 + CaO$$

(*d*) No reaction

$$ZnCl_2 + Cl_2 \longrightarrow NR$$

(*e*) Combustion

$$2\,C_3H_8 + 7\,O_2 \longrightarrow 6\,CO + 8\,H_2O$$

7.21. Which of the following is soluble in water—$CuCl$ or $CuCl_2$?

Ans. $CuCl_2$ is soluble; $CuCl$ is one of the four chlorides which are listed as insoluble (Table 7-2).

7.22. Complete and balance the following equation. If no reaction occurs, indicate that fact by writing "NR."

$$HCl + Na_3PO_4 \longrightarrow$$

Ans.

$$3\,HCl + Na_3PO_4 \longrightarrow H_3PO_4 + 3\,NaCl$$

The phosphoric acid produced is weak, that is, mostly covalent, and the formation of the H_3PO_4 is the driving force for this reaction. (HCl is one of the seven strong acids listed in Table 7-3.)

7.23. Is F_2 soluble in water?

Ans. No, it reacts with water, liberating oxygen:

$$2\,F_2 + 2\,H_2O \longrightarrow 4\,HF + O_2$$

Supplementary Problems

7.24. What is the difference between dissolving and reacting?

7.25. How can you tell that the following are combination reactions rather than replacement or double-replacement reactions?

$$FeCl_2 + Cl_2 \longrightarrow$$
$$CO + O_2 \longrightarrow$$
$$HgCl_2 + Hg \longrightarrow$$
$$CaO + CO_2 \longrightarrow$$

7.26. State why the equation of Problem 7.2 is unusual.

Ans.   It is a substitution reaction in which a nonmetal replaces a metal. Carbon, at high temperatures, can replace relatively inactive metals from their oxides.

7.27. Complete and balance each of the following. If no reaction occurs, write "NR."

  (a) $SO_3 + H_2O \longrightarrow$ (m) $BaSO_4 + NaCl \longrightarrow$

  (b) $CaO + H_2O \longrightarrow$ (n) $NaBr(l) \xrightarrow{\text{electricity}}$

  (c) $Ca + O_2 \longrightarrow$ (o) $BaCl_2 + H_2SO_4 \longrightarrow$

  (d) $Li + S \longrightarrow$ (p) $AgNO_3 + HCl \longrightarrow$

  (e) $CaO + CO_2 \longrightarrow$ (q) $KNO_3 + HCl \longrightarrow$

  (f) $Al + Cl_2 \longrightarrow$ (r) $C + O_2 \text{ (excess)} \longrightarrow$

  (g) $Cu + ZnCl_2 \longrightarrow$ (s) $C + O_2 \text{ (limited)} \longrightarrow$

  (h) $Cl_2 + KI \longrightarrow$ (t) $C_3H_6 + O_2 \text{ (excess)} \longrightarrow$

  (i) $KOH + HCl \longrightarrow$ (u) $C_3H_6 + O_2 \text{ (limited)} \longrightarrow$

  (j) $HNO_3 + ZnO \longrightarrow$ (v) $C_3H_6O + O_2 \text{ (excess)} \longrightarrow$

  (k) $NaOH + H_2SO_3 \longrightarrow$ (w) $C_3H_6O + O_2 \text{ (limited)} \longrightarrow$

  (l) $Ba(OH)_2 + HClO_3 \longrightarrow$

Ans.   (a) $SO_3 + H_2O \longrightarrow H_2SO_4$

        (b) $CaO + H_2O \longrightarrow Ca(OH)_2$

        (c) $2Ca + O_2 \longrightarrow 2CaO$

        (d) $2Li + S \longrightarrow Li_2S$

        (e) $CaO + CO_2 \longrightarrow CaCO_3$

        (f) $2Al + 3Cl_2 \longrightarrow 2AlCl_3$

        (g) $Cu + ZnCl_2 \longrightarrow NR$

        (h) $Cl_2 + 2KI \longrightarrow 2KCl + I_2$

        (i) $KOH + HCl \longrightarrow KCl + H_2O$

        (j) $2HNO_3 + ZnO \longrightarrow Zn(NO_3)_2 + H_2O$

        (k) $NaOH + H_2SO_3 \longrightarrow NaHSO_3 + H_2O$

             or $2NaOH + H_2SO_3 \longrightarrow Na_2SO_3 + 2H_2O$

        (l) $Ba(OH)_2 + 2HClO_3 \longrightarrow Ba(ClO_3)_2 + 2H_2O$

        (m) $BaSO_4 + NaCl \longrightarrow NR$

        (n) $2NaBr(l) \xrightarrow{\text{electricity}} 2Na + Br_2$

        (o) $BaCl_2 + H_2SO_4 \longrightarrow BaSO_4 + 2HCl$

        (p) $AgNO_3 + HCl \longrightarrow AgCl + HNO_3$

(q) $KNO_3 + HCl \longrightarrow NR$

(r) $C + O_2$ (excess) $\longrightarrow CO_2$

(s) $2C + O_2$ (limited) $\longrightarrow 2CO$

(t) $2C_3H_6 + 9O_2$ (excess) $\longrightarrow 6CO_2 + 6H_2O$

(u) $C_3H_6 + 3O_2$ (limited) $\longrightarrow 3CO + 3H_2O$

(v) $C_3H_6O + 4O_2$ (excess) $\longrightarrow 3CO_2 + 3H_2O$

(w) $2C_3H_6O + 5O_2$ (limited) $\longrightarrow 6CO + 6H_2O$

Chapter 8

Stoichiometry

8.1 MOLE-TO-MOLE CALCULATIONS

In chemical work, it is important to be able to calculate how much raw material is needed to prepare a certain quantity of products. It is also useful to know if a certain reaction method can prepare more product from a given quantity of material than another reaction method. Analyzing materials means finding out how much of each element is present. To do the measurements, we often convert parts of the material to compounds that are easy to separate, and we then measure those compounds. All these measurements involve chemical *stoichiometry*, the science of measuring how much of one thing can be produced from certain amounts of others.

From the practical viewpoint of a student, this chapter is extremely important. The calculations introduced here are also used in the chapters on gas laws, thermochemistry, thermodynamics, solution chemistry, electrochemistry, equilibrium, kinetics, and other topics.

In Chap. 7, the balanced chemical equation was introduced. The equation expresses the ratios of numbers of formula units of each chemical involved in the reaction. Thus, for the reaction of nitrogen with hydrogen to produce ammonia:

$$N_2 + 3H_2 \longrightarrow 2NH_3$$

the equation states that the chemicals react in the ratio of one molecule of nitrogen (N_2) with three molecules of hydrogen (H_2) to produce two molecules of ammonia (NH_3). Thus, if 4 molecules of nitrogen react, they will react with 12 molecules of hydrogen, and 8 molecules of ammonia will be produced.

The balanced chemical equation may also be used to express the ratios of *moles* of reactants and products involved. Thus, for the reaction whose equation is given above, 1 mol of N_2 reacts with 3 mol of H_2 to produce 2 mol of NH_3. It is also true that 4 mol of nitrogen can react with 12 mol of hydrogen to produce 8 mol of ammonia, and so on.

EXAMPLE 8.1. How many moles of ammonia can be prepared from the reaction of 0.250 mol of H_2 plus sufficient N_2?

The first step in any stoichiometry problem is to write the balanced chemical equation:

$$N_2 + 3H_2 \longrightarrow 2NH_3$$

Then the coefficients in the balanced chemical equation can be used as factors in the factor-label method to convert from moles of one chemical to moles of any other in the equation:

$$0.250 \, \text{mol H}_2 \left(\frac{2 \, \text{mol NH}_3}{3 \, \text{mol H}_2} \right) = 0.167 \, \text{mol NH}_3$$

In the example given in the text above, with 4 mol of nitrogen, it was not necessary to use the factor-label method; the numbers were easy enough to work with. However, when the numbers get even slightly complicated, it is useful to use the factor-label method. Note that any of the following factors could be used for this equation, but we used the one above because it is the one that changes moles of hydrogen to moles of ammonia.

$$\frac{2 \, \text{mol NH}_3}{3 \, \text{mol H}_2} \qquad \frac{2 \, \text{mol NH}_3}{\text{mol N}_2} \qquad \frac{3 \, \text{mol H}_2}{\text{mol N}_2}$$

$$\frac{3 \, \text{mol H}_2}{2 \, \text{mol NH}_3} \qquad \frac{1 \, \text{mol N}_2}{3 \, \text{mol H}_2} \qquad \frac{1 \, \text{mol N}_2}{2 \, \text{mol NH}_3}$$

130

EXAMPLE 8.2. How many moles of nitrogen does it take to react with 0.750 mol of hydrogen?

   According to the equation in Example 8.1, it takes

$$0.750 \, \text{mol H}_2 \left(\frac{1 \, \text{mol N}_2}{3 \, \text{mol H}_2} \right) = 0.250 \, \text{mol N}_2$$

EXAMPLE 8.3. How many moles of NO is produced by the reaction at high temperature of 0.500 mol of N_2 with excess O_2?

   The balanced equation is

$$N_2 + O_2 \longrightarrow 2 \, NO$$

$$0.500 \, \text{mol N}_2 \left(\frac{2 \, \text{mol NO}}{\text{mol N}_2} \right) = 1.00 \, \text{mol NO}$$

   A simple figure linking the quantities, with the factor label as a bridge, is shown in Fig. 8-1.

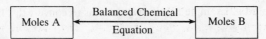

Fig. 8-1 The conversion of moles of one reagent to moles of another, using a ratio of the coefficients of the balanced chemical equation as a factor label

   In all the problems given above, a sufficient or excess quantity of a second reactant was stated in the problem. If nothing is stated about the quantity of a second (or third, etc.) reactant, it may be assumed to be present in sufficient quantity to allow the reaction to take place. Otherwise, no calculation can be done.

EXAMPLE 8.4. How many moles of NO are produced by the reaction of 0.500 mol N_2 (with O_2 at high temperature)?

   This is the same problem as given in Example 8.3. If we do not assume that there is a source of oxygen present, we cannot assume that any product will be produced. As long as we have enough oxygen, we can base the calculation on the quantity of nitrogen stated.

8.2 CALCULATIONS INVOLVING MASS

   The balanced equation expresses quantities in moles, but it is seldom possible to measure out quantities in moles directly. If the quantities given or required are expressed in other units, it is necessary to convert them to moles before using the factors of the balanced chemical equation. Conversion of mass to moles and vice versa was considered in Sec. 4.5. Here we will use that knowledge first to calculate the number of moles of reactant or product, and then use that value to calculate the number of moles of other reactant or product.

EXAMPLE 8.5. How many moles of Na_2SO_4 can be produced by reaction of 100.0 g of NaOH with sufficient H_2SO_4?

   Again, the first step is to write the balanced chemical equation:

$$2 \, NaOH + H_2SO_4 \longrightarrow Na_2SO_4 + 2 \, H_2O$$

Since the *mole ratio* is given by the equation, we must convert the 100.0 g of NaOH to moles:

$$100.0 \, \text{g NaOH} \left(\frac{1 \, \text{mol NaOH}}{40.0 \, \text{g NaOH}} \right) = 2.50 \, \text{mol NaOH}$$

Now we can solve the problem just as we did those above:

$$2.50 \, \text{mol NaOH} \left(\frac{1 \, \text{mol Na}_2\text{SO}_4}{2 \, \text{mol NaOH}} \right) = 1.25 \, \text{mol Na}_2\text{SO}_4$$

Figure 8-2 illustrates the additional step required for this calculation.

Fig. 8-2   Conversion of mass of a reactant to moles of a product

It is also possible to calculate the mass of product from the number of moles of product.

EXAMPLE 8.6.   How many grams of Na_2SO_4 can be produced from 100.0 g NaOH with sufficient H_2SO_4?

The first steps were presented in Example 8.5. It only remains to convert the 1.25 mol Na_2SO_4 to grams (see Fig. 8-3):

$$1.25 \, \text{mol Na}_2\text{SO}_4 \left(\frac{142 \, \text{g Na}_2\text{SO}_4}{\text{mol Na}_2\text{SO}_4} \right) = 178 \, \text{g Na}_2\text{SO}_4$$

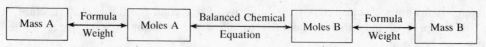

Fig. 8-3   Conversion of mass of a reactant to mass of a product

Not only mass, but any measurable quantity that can be converted to moles may be treated in this manner to determine the quantity of product or reactant involved in a reaction from the quantity of any other reactant or product. In later chapters, the volumes of gases and the volumes of solutions of known concentrations will be used to determine the numbers of moles of a reactant or product. We can illustrate the process with the following problem:

EXAMPLE 8.7.   How many grams of Na_3PO_4 can be produced by the reaction of 2.50×10^{23} molecules of H_3PO_4 with excess NaOH?

$$H_3PO_4 + 3\,NaOH \longrightarrow Na_3PO_4 + 3\,H_2O$$

Since the equation states the mole ratio, we first convert the number of molecules of H_3PO_4 to moles:

$$2.50 \times 10^{23} \, \text{molecules H}_3\text{PO}_4 \left(\frac{1 \, \text{mol H}_3\text{PO}_4}{6.02 \times 10^{23} \, \text{molecules H}_3\text{PO}_4} \right) = 0.415 \, \text{mol H}_3\text{PO}_4$$

Next we convert that number of moles of H_3PO_4 to moles and then to grams of Na_3PO_4:

$$0.415 \, \text{mol H}_3\text{PO}_4 \left(\frac{1 \, \text{mol Na}_3\text{PO}_4}{1 \, \text{mol H}_3\text{PO}_4} \right) = 0.415 \, \text{mol Na}_3\text{PO}_4$$

$$0.415 \, \text{mol Na}_3\text{PO}_4 \left(\frac{164 \, \text{g Na}_3\text{PO}_4}{\text{mol Na}_3\text{PO}_4} \right) = 68.1 \, \text{g Na}_3\text{PO}_4$$

Figure 8-4 illustrates the process.

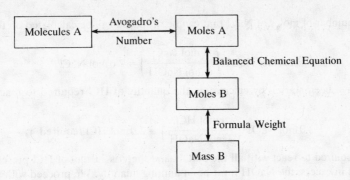

Fig. 8-4   Conversion of number of molecules of a reactant to mass of a product

8.3   LIMITING QUANTITIES

In Secs. 8.1 and 8.2, there was always sufficient (or excess) of all reactants except the one whose quantity was given. The quantity of only one reactant or product was stated in the problem. In this section, the quantities of more than one reactant will be stated. This type of problem is called a *limiting quantities problem*.

How much NH_3 can be prepared from 2.0 mol of H_2 and 0.0 mol of N_2? The first step, as usual, is to write the balanced chemical equation:

$$N_2 + 3H_2 \longrightarrow 2NH_3$$

It should be obvious that with no N_2, there can be no NH_3 produced by this reaction. This problem is not the one which is likely to appear on examinations.

How much sulfur dioxide is produced by the reaction of 1.00 g S and all the oxygen in the atmosphere of the earth? (If you strike a match outside, do you really have to worry about not having enough oxygen to burn all the sulfur in the match head?) This problem has the quantity of each of two reactants stated, but it is obvious that the sulfur will be used up before all the oxygen. It is also obvious that not all the oxygen will react! (Otherwise, we are all in trouble.) The problem is solved just like the problems in Sec. 8.2.

To solve a limiting quantities problem in which the reactant in excess is not obvious, you should:

1. Calculate the number of moles of one reactant required to react with all the other reactant present.
2. Compare the number of moles of that one reactant which is present and the number of moles required (calculated in step 1). This comparison will tell you which reactant is present in excess, and which one is in limiting quantity.
3. Calculate the quantity of reaction (reactants used up and products produced) on the basis of the quantity of reactant in limiting quantity. Perform this calculation just as was done in Sec. 8.1 or 8.2.

EXAMPLE 8.8.   How many moles of NaCl can be produced by the reaction of 2.0 mol NaOH and 3.0 mol HCl?

First, write the balanced chemical equation:

$$NaOH + HCl \longrightarrow NaCl + H_2O$$

Step 1: Determine the number of moles of NaOH required to react completely with 3.0 mol of HCl:

$$3.0 \, \text{mol HCl} \left(\frac{1 \, \text{mol NaOH}}{1 \, \text{mol HCl}} \right) = 3.0 \, \text{mol NaOH required}$$

Step 2: Since there is 2.0 mol NaOH present but the HCl present would require 3.0 mol NaOH, there is not enough NaOH to react completely with the HCl. The NaOH is present in limiting quantity. Put it another way, after 2.0 mol NaOH reacts with 2.0 mol HCl, there will be 1.0 mol of HCl and 0.0 mol of NaOH left. After the 2.0 mol of each has reacted, the rest of the reaction is like the first example in this section: There can be no further reaction. Once the reactant in limiting quantity is used up, the reactant in excess cannot react any more.

Step 3: Now the number of moles of NaCl that can be produced can be calculated on the basis of the 2.0 mol NaOH present:

$$2.0 \, \text{mol NaOH} \left(\frac{1 \, \text{mol NaCl}}{1 \, \text{mol NaOH}} \right) = 2.0 \, \text{mol NaCl}$$

The problem could have been started by calculating the quantity of HCl required to react completely with the NaOH present:

$$2.0 \, \text{mol NaOH} \left(\frac{1 \, \text{mol HCl}}{1 \, \text{mol NaOH}} \right) = 2.0 \, \text{mol HCl required}$$

Since 2.0 mol HCl is required to react with all the NaOH and there is 3.0 mol of HCl present, the HCl is present in excess. If the HCl is in excess, the NaOH must be in limiting quantity. We proceed with the calculation just as above. The quantity of either reactant can be used to determine how much of the other is required. The same result will be obtained no matter which is used. It should be emphasized that it is not necessary to do both calculations. See how much HCl is required to react with the NaOH present or see how much NaOH is required to react with the HCl present; do not do both.

EXAMPLE 8.9. How many moles of PbI_2 can be prepared by the reaction of 0.235 mol of $Pb(NO_3)_2$ and 0.423 mol NaI?

The balanced equation:

$$Pb(NO_3)_2 + 2 \, NaI \longrightarrow PbI_2 + 2 \, NaNO_3$$

Let us see how many moles of NaI is required to react with all the $Pb(NO_3)_2$ present:

$$0.235 \, \text{mol Pb(NO}_3)_2 \left(\frac{2 \, \text{mol NaI}}{\text{mol Pb(NO}_3)_2} \right) = 0.470 \, \text{mol NaI required}$$

Since there is more NaI required (0.470 mol) than present (0.423 mol), the NaI is in limiting quantity.

$$0.423 \, \text{mol NaI} \left(\frac{1 \, \text{mol PbI}_2}{2 \, \text{mol NaI}} \right) = 0.212 \, \text{mol PbI}_2$$

Note especially that the number of moles of NaI exceeds the number of moles of $Pb(NO_3)_2$ present, but that these numbers are not what must be compared. Compare the number of moles of one reactant present with the number of moles of that same reactant required! The ratio of moles of NaI to $Pb(NO_3)_2$ in the equation is 2:1, but in the sample that ratio is less than 2:1; therefore, the NaI is in limiting quantity.

EXAMPLE 8.10. How many grams of $Ca(ClO_4)_2$ can be prepared by treatment of 12.5 g CaO with 75.0 g $HClO_4$?

The balanced equation:

$$CaO + 2 \, HClO_4 \longrightarrow Ca(ClO_4)_2 + H_2O$$

This problem gives the quantities of the two reactants in grams; we must first change them to moles:

$$12.5 \, \text{g CaO} \left(\frac{1 \, \text{mol CaO}}{56.0 \, \text{g CaO}} \right) = 0.223 \, \text{mol CaO}$$

$$75.0 \, \text{g HClO}_4 \left(\frac{1 \, \text{mol HClO}_4}{100 \, \text{g HClO}_4} \right) = 0.750 \, \text{mol HClO}_4$$

Now the problem can be done as in the last example:

$$0.750 \, \text{mol HClO}_4 \left(\frac{1 \, \text{mol CaO}}{2 \, \text{mol HClO}_4} \right) = 0.375 \, \text{mol CaO required}$$

Since there is 0.375 mol CaO required and 0.223 mol CaO present, the CaO is in limiting quantity.

$$0.223 \, \text{mol CaO} \left(\frac{1 \, \text{mol Ca(ClO}_4)_2}{\text{mol CaO}} \right) = 0.223 \, \text{mol Ca(ClO}_4)_2$$

$$0.223 \, \text{mol Ca(ClO}_4)_2 \left(\frac{239 \, \text{g Ca(ClO}_4)_2}{\text{mol Ca(ClO}_4)_2} \right) = 53.3 \, \text{g Ca(ClO}_4)_2 \, \text{produced}$$

If the quantities of both reactants are in exactly the correct ratio for the balanced chemical equation, then *either* reactant may be used to calculate the quantity of product produced. (If on a quiz or examination it is obvious that they are in the correct ratio, you should state that they are so that your instructor will understand that you recognize the problem to be a limiting quantities problem.)

EXAMPLE 8.11.   How many moles of lithium nitride, Li_3N, can be prepared by the reaction of 12.0 mol Li and 2.00 mol N_2?

$$6\,Li + N_2 \longrightarrow 2\,Li_3N$$

Since the ratio of moles of lithium present to moles of nitrogen present is 6:1, just as is required for the balanced equation, either reactant may be used.

$$12.0\,\text{mol Li}\left(\frac{2\,\text{mol Li}_3\text{N}}{6\,\text{mol Li}}\right) = 4.00\,\text{mol Li}_3\text{N}$$

or

$$2.00\,\text{mol N}_2\left(\frac{2\,\text{mol Li}_3\text{N}}{\text{mol N}_2}\right) = 4.00\,\text{mol Li}_3\text{N}$$

(The first sentence of this answer should be stated on an examination.)

EXAMPLE 8.12.   How many grams of excess reactant will remain after the reaction of 12.5 g CaO and 75.0 g $HClO_4$?

This problem is started just as was Example 8.10. The limiting quantity is CaO, and the quantity of $HClO_4$ which reacts is therefore

$$0.223\,\text{mol CaO}\left(\frac{2\,\text{mol HClO}_4}{\text{mol CaO}}\right) = 0.446\,\text{mol HClO}_4\ \text{reacting}$$

Since there is 0.750 mol $HClO_4$ present and 0.446 mol reacts, there is

$$0.750\,\text{mol} - 0.446\,\text{mol} = 0.304\,\text{mol remaining unreacted}$$

$$0.304\,\text{mol HClO}_4\left(\frac{100\,\text{g HClO}_4}{\text{mol HClO}_4}\right) = 30.4\,\text{g HClO}_4\ \text{unreacted}$$

Solved Problems

MOLE-TO-MOLE CALCULATIONS

8.1.   Can the balanced chemical equation dictate to a chemist how much reactant to place in a reaction vessel?

Ans.   The chemist can put in as little as is weighable or as much as the vessel will hold. For example, the fact that a reactant has a coefficient of 2 in the balanced chemical equation does not mean that the chemist must put two moles into the reaction vessel. The chemist might decide to add the reactants in the ratio of the balanced chemical equation, but even that is not required. And even in that case, the numbers of moles of each reactant might be twice the respective coefficients or one-tenth those values, etc. The equation merely states the *reacting ratio*.

8.2.   How many moles of Na_2HPO_4 can be prepared from 3.0 mol NaOH and sufficient H_3PO_4?

Ans.

$$2\,NaOH + H_3PO_4 \longrightarrow Na_2HPO_4 + 2\,H_2O$$

$$3.0\,\text{mol NaOH}\left(\frac{1\,\text{mol Na}_2\text{HPO}_4}{2\,\text{mol NaOH}}\right) = 1.5\,\text{mol Na}_2\text{HPO}_4$$

8.3. How many moles of H_2O will react with 1.25 mol PCl_3 to form HCl and H_3PO_3?

Ans.
$$3\,H_2O + PCl_3 \longrightarrow 3\,HCl + H_3PO_3$$
$$1.25\text{ mol }PCl_3\left(\frac{3\text{ mol }H_2O}{\text{mol }PCl_3}\right) = 3.75\text{ mol }H_2O$$

8.4. How many factor labels can be used corresponding to each of the following balanced equations?

(*a*)
$$2\,Na + Cl_2 \longrightarrow 2\,NaCl$$

(*b*)
$$Ba(OH)_2 + 2\,HBr \longrightarrow BaBr_2 + 2\,H_2O$$

Ans.   (*a*)   Six:

$$\frac{2\text{ mol Na}}{\text{mol }Cl_2} \qquad \frac{2\text{ mol Na}}{2\text{ mol NaCl}} \qquad \frac{1\text{ mol }Cl_2}{2\text{ mol Na}}$$

$$\frac{1\text{ mol }Cl_2}{2\text{ mol NaCl}} \qquad \frac{2\text{ mol NaCl}}{2\text{ mol Na}} \qquad \frac{2\text{ mol NaCl}}{\text{mol }Cl_2}$$

(*b*)   Twelve: Each of the four compounds as numerators with the three others as denominators—$4 \times 3 = 12$.

8.5. Which of the factors of Problem 8.4(*a*) would be used to convert (*a*) the number of moles of Cl_2 to number of moles of NaCl? (*b*) Na to Cl_2? (*c*) Cl_2 to Na?

Ans.   (*a*) $\dfrac{2\text{ mol NaCl}}{\text{mol }Cl_2}$ (*b*) $\dfrac{1\text{ mol }Cl_2}{2\text{ mol Na}}$ (*c*) $\dfrac{2\text{ mol Na}}{\text{mol }Cl_2}$

8.6. Balance the following equation. Calculate the number of moles of CO_2 that can be prepared by the reaction of 7.50 mol of $Ca(HCO_3)_2$.

$$Ca(HCO_3)_2 + HCl \longrightarrow CaCl_2 + CO_2 + H_2O$$

Ans.
$$Ca(HCO_3)_2 + 2\,HCl \longrightarrow CaCl_2 + 2\,CO_2 + 2\,H_2O$$
$$7.50\text{ mol }Ca(HCO_3)_2\left(\frac{2\text{ mol }CO_2}{\text{mol }Ca(HCO_3)_2}\right) = 15.0\text{ mol }CO_2$$

8.7. (*a*) How many moles of $BaCl_2$ can be prepared by the reaction of 2.50 mol HCl with excess $Ba(OH)_2$? (*b*) How many moles of NaCl can be prepared by the reaction of 2.50 mol HCl with excess NaOH?

Ans.

(*a*)
$$Ba(OH)_2 + 2\,HCl \longrightarrow BaCl_2 + 2\,H_2O$$
$$2.50\text{ mol }HCl\left(\frac{1\text{ mol }BaCl_2}{2\text{ mol }HCl}\right) = 1.25\text{ mol }BaCl_2$$

(*b*)
$$NaOH + HCl \longrightarrow NaCl + H_2O$$
$$2.50\text{ mol }HCl\left(\frac{1\text{ mol }NaCl}{\text{mol }HCl}\right) = 2.50\text{ mol }NaCl$$

8.8. How many moles of H_2O are prepared along with 5.0 mol Na_3PO_4 in a reaction of NaOH and H_3PO_4?

Ans.
$$H_3PO_4 + 3\,NaOH \longrightarrow Na_3PO_4 + 3\,H_2O$$
$$5.0\text{ mol }Na_3PO_4\left(\frac{3\text{ mol }H_2O}{\text{mol }Na_3PO_4}\right) = 15\text{ mol }H_2O$$

8.9. How many moles of CuO does it take to convert 3.0 mol CH_4 to H_2O, CO_2, and Cu, according to the equation

$$4\,CuO + CH_4 \xrightarrow{\text{heat}} 4\,Cu + 2\,H_2O + CO_2$$

Ans.

$$3.0\,\text{mol CH}_4\left(\frac{4\,\text{mol CuO}}{\text{mol CH}_4}\right) = 12\,\text{mol CuO}$$

8.10. Consider the complicated-looking equation

$$KMnO_4 + 5\,FeCl_2 + 8\,HCl \longrightarrow MnCl_2 + 5\,FeCl_3 + 4\,H_2O + KCl$$

How many moles of $FeCl_3$ will be produced by the reaction of 1.00 mol of HCl?

Ans. No matter how complicated the equation, the reacting ratio is still given by the coefficients. The coefficients of interest are 8 for HCl and 5 for $FeCl_3$.

$$1.00\,\text{mol HCl}\left(\frac{5\,\text{mol FeCl}_3}{8\,\text{mol HCl}}\right) = 0.625\,\text{mol FeCl}_3$$

Note: The hard part of this problem is balancing the equation, which will be presented in Chap. 13. Since the balanced equation was given in the statement of the problem, the problem is as easy to solve as the previous ones.

CALCULATIONS INVOLVING MASS

8.11. Which earlier sections must be understood before mass-to-mass conversions can be studied profitably?

Ans. Section 2.4, factor-label method; Sec. 4.4, calculation of formula weights; Sec. 4.5, changing moles to grams and vice versa; Sec. 4.5, Avogadro's number; and/or Sec. 7.2, balancing chemical equations.

8.12. Figure 8-3 is a combination of which two earlier figures?

Ans. Figures 8-1 and 4-2.

8.13. In a stoichiometry problem, (*a*) if the mass of a reactant is given, what conversions (if any) should be made? (*b*) If a number of molecules is given, what conversions (if any) should be made? (*c*) If a number of moles is given, what conversions (if any) should be made?

Ans. (*a*) The mass should be converted to moles. (*b*) The number of molecules should be converted to moles. (*c*) No conversion need be done; the quantity is given in moles.

8.14. Sulfurous acid reacts with sodium hydroxide to produce sodium sulfite and water. (*a*) Write a balanced chemical equation for the reaction. (*b*) Determine the number of moles of sulfurous acid in 50.0 g sulfurous acid. (*c*) How many moles of sodium sulfite will be produced by the reaction of this number of moles of sulfurous acid? (*d*) How many grams of sodium sulfite will be produced? (*e*) How many moles of sodium hydroxide will it take to react with this quantity of sulfurous acid? (*f*) How many grams of sodium hydroxide will be used up?

Ans.

(*a*) $$H_2SO_3 + 2\,NaOH \longrightarrow Na_2SO_3 + 2\,H_2O$$

(*b*) $$50.0\,\text{g H}_2SO_3\left(\frac{1\,\text{mol H}_2SO_3}{82.0\,\text{g H}_2SO_3}\right) = 0.610\,\text{mol H}_2SO_3$$

(*c*) $$0.610\,\text{mol H}_2SO_3\left(\frac{1\,\text{mol Na}_2SO_3}{\text{mol H}_2SO_3}\right) = 0.610\,\text{mol Na}_2SO_3$$

(d)
$$0.610 \text{ mol Na}_2\text{SO}_3\left(\frac{126 \text{ g Na}_2\text{SO}_3}{\text{mol Na}_2\text{SO}_3}\right) = 76.9 \text{ g Na}_2\text{SO}_3$$

(e)
$$0.610 \text{ mol H}_2\text{SO}_3\left(\frac{2 \text{ mol NaOH}}{\text{mol Na}_2\text{SO}_3}\right) = 1.22 \text{ mol NaOH}$$

(f)
$$1.22 \text{ mol NaOH}\left(\frac{40.0 \text{ g NaOH}}{\text{mol NaOH}}\right) = 48.8 \text{ g NaOH}$$

8.15. In a certain experiment, 100 g of ethane, C_2H_6, is burned. (a) Write a balanced chemical equation for the combustion of ethane to produce CO_2 and water. (b) Determine the number of moles of ethane in 100 g of ethane. (c) Determine the number of moles of CO_2 that would be produced by the combustion of that number of moles of ethane. (d) Determine the mass of CO_2 that can be produced by the combustion of 100 g of ethane.

Ans.

(a)
$$2 C_2H_6 + 7 O_2 \longrightarrow 4 CO_2 + 6 H_2O$$

(b)
$$100 \text{ g C}_2\text{H}_6\left(\frac{1 \text{ mol C}_2\text{H}_6}{30.0 \text{ g C}_2\text{H}_6}\right) = 3.33 \text{ mol C}_2\text{H}_6$$

(c)
$$3.33 \text{ mol C}_2\text{H}_6\left(\frac{4 \text{ mol CO}_2}{2 \text{ mol C}_2\text{H}_6}\right) = 6.66 \text{ mol CO}_2$$

(d)
$$6.66 \text{ mol CO}_2\left(\frac{44.0 \text{ g CO}_2}{\text{mol CO}_2}\right) = 293 \text{ g CO}_2$$

8.16. (a) Write the balanced chemical equation for the reaction of sodium with chlorine. (b) How many moles of Cl_2 are there in 10.0 g chlorine? (c) How many moles of NaCl will that number of moles of chlorine produce? (d) What mass of NaCl is that number of moles of NaCl?

Ans.

(a)
$$2 \text{Na} + Cl_2 \longrightarrow 2 \text{NaCl}$$

(b)
$$10.0 \text{ g Cl}_2\left(\frac{1 \text{ mol Cl}_2}{70.90 \text{ g Cl}_2}\right) = 0.141 \text{ mol Cl}_2$$

(c)
$$0.141 \text{ mol Cl}_2\left(\frac{2 \text{ mol NaCl}}{\text{mol Cl}_2}\right) = 0.282 \text{ mol NaCl}$$

(d)
$$0.282 \text{ mol NaCl}\left(\frac{58.45 \text{ g NaCl}}{\text{mol NaCl}}\right) = 16.5 \text{ g NaCl}$$

8.17. How many formula units of sodium hydroxide, along with H_2SO_4, does it take to make 5.00×10^{22} formula units of Na_2SO_4?

Ans.

$$2 \text{NaOH} + H_2SO_4 \longrightarrow Na_2SO_4 + 2 H_2O$$

$$5.00 \times 10^{22} \text{ units Na}_2\text{SO}_4\left(\frac{1 \text{ mol Na}_2\text{SO}_4}{6.02 \times 10^{23} \text{ units}}\right) = 0.0831 \text{ mol Na}_2\text{SO}_4$$

$$0.0831 \text{ mol Na}_2\text{SO}_4\left(\frac{2 \text{ mol NaOH}}{\text{mol Na}_2\text{SO}_4}\right) = 0.166 \text{ mol NaOH}$$

$$0.166 \text{ mol NaOH}\left(\frac{6.02 \times 10^{23} \text{ units}}{\text{mol NaOH}}\right) = 1.00 \times 10^{23} \text{ units NaOH}$$

Since the balanced chemical equation also relates the numbers of formula units of reactants and products, the problem can be solved by converting directly with the factor label from the balanced equation:

$$5.00 \times 10^{22} \text{ units Na}_2\text{SO}_4 \left(\frac{2 \text{ units NaOH}}{\text{unit Na}_2\text{SO}_4} \right) = 1.00 \times 10^{23} \text{ units NaOH}$$

8.18. Draw a figure like Fig. 8-3 for Problem 8.17.

   Ans.   See Fig. 8-5.

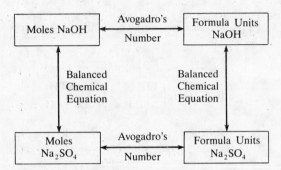

Fig. 8-5   Conversion of formula units of a reactant to formula units of a product

8.19. To change from mass of one substance to mass of another substance in the same balanced chemical equation, we use factors such as

$$\frac{x \text{ g A}}{\text{mol A}} \quad \text{and} \quad \frac{y \text{ mol A}}{\text{mol B}}$$

That is, we use a factor to change the units (mol to g) for one substance (A) or we use a factor to express the ratio of substances (B to A) in terms of the same unit (mol). If we follow the diagrams in the figures in this chapter, do we ever use one factor to make both a conversion for substance and a conversion of units simultaneously?

   Ans.   No. (It is possible to change both, but that factor then represents more than one step in the process.)

8.20. How many grams of barium hydroxide will be used up in the reaction with hydrogen chloride (hydrochloric acid) to produce 45.00 g of barium chloride plus some water?

   Ans.

$$\text{Ba(OH)}_2 + 2\,\text{HCl} \longrightarrow \text{BaCl}_2 + 2\,\text{H}_2\text{O}$$

$$45.00 \text{ g BaCl}_2 \left(\frac{1 \text{ mol BaCl}_2}{208.2 \text{ g BaCl}_2} \right) \left(\frac{1 \text{ mol Ba(OH)}_2}{\text{mol BaCl}_2} \right) \left(\frac{171.3 \text{ g Ba(OH)}_2}{\text{mol Ba(OH)}_2} \right)$$

$$= 37.02 \text{ g Ba(OH)}_2$$

8.21. (*a*) What other reactant may be treated with phosphoric acid to produce 3.00 mol of potassium phosphate (plus some water)? (*b*) How many moles of phosphoric acid will it take? (*c*) How many moles of the other reactant is required? (*d*) How many grams?

Ans.

(*a*) $H_3PO_4 + 3\,KOH \longrightarrow K_3PO_4 + 3\,H_2O$

KOH may be used.

(*b*) $3.00\,\text{mol K}_3PO_4\left(\dfrac{1\,\text{mol H}_3PO_4}{\text{mol K}_3PO_4}\right) = 3.00\,\text{mol H}_3PO_4$

(*c*) $3.00\,\text{mol K}_3PO_4\left(\dfrac{3\,\text{mol KOH}}{\text{mol K}_3PO_4}\right) = 9.00\,\text{mol KOH}$

(*d*) $9.00\,\text{mol KOH}\left(\dfrac{56.1\,\text{g KOH}}{\text{mol KOH}}\right) = 505\,\text{g KOH}$

8.22. How many grams of Hg_2Cl_2 can be prepared using 10.0 mL of mercury (density 13.6 g/mL)?

Ans.

$$2\,Hg + Cl_2 \longrightarrow Hg_2Cl_2$$

$$10.0\,\text{mL Hg}\left(\dfrac{13.6\,\text{g Hg}}{\text{mL Hg}}\right)\left(\dfrac{1\,\text{mol Hg}}{200.6\,\text{g Hg}}\right) = 0.678\,\text{mol Hg}$$

$$0.678\,\text{mol Hg}\left(\dfrac{1\,\text{mol Hg}_2Cl_2}{2\,\text{mol Hg}}\right) = 0.339\,\text{mol Hg}_2Cl_2$$

$$0.339\,\text{mol Hg}_2Cl_2\left(\dfrac{471\,\text{g Hg}_2Cl_2}{\text{mol Hg}_2Cl_2}\right) = 160\,\text{g Hg}_2Cl_2$$

8.23. Determine the number of grams of chloric acid that will just react with 20.0 g of calcium carbonate to produce carbon dioxide, water, and calcium chlorate.

Ans.

$$CaCO_3 + 2\,HClO_3 \longrightarrow Ca(ClO_3)_2 + H_2O + CO_2$$

$$20.0\,\text{g CaCO}_3\left(\dfrac{1\,\text{mol CaCO}_3}{100\,\text{g CaCO}_3}\right)\left(\dfrac{2\,\text{mol HClO}_3}{\text{mol CaCO}_3}\right)\left(\dfrac{84.47\,\text{g HClO}_3}{\text{mol HClO}_3}\right)$$

$$= 33.8\,\text{g HClO}_3$$

8.24. Nitrogen trichloride reacts with water to produce ammonia (NH_3) and hypochlorous acid ($HClO$). Calculate the number of grams of ammonia that can be produced from 20.0 g of nitrogen trichloride.

Ans.

$$NCl_3 + 3\,H_2O \longrightarrow NH_3 + 3\,HClO$$

$$20.0\,\text{g NCl}_3\left(\dfrac{1\,\text{mol NCl}_3}{120.4\,\text{g NCl}_3}\right)\left(\dfrac{1\,\text{mol NH}_3}{\text{mol NCl}_3}\right)\left(\dfrac{17.0\,\text{g NH}_3}{\text{mol NH}_3}\right) = 2.82\,\text{g NH}_3$$

8.25. Sulfur trioxide reacts (violently) with water to produce sulfuric acid. Calculate the number of grams of sulfur trioxide necessary to produce 1.00 kg of sulfuric acid.

Ans.

$$SO_3 + H_2O \longrightarrow H_2SO_4$$

$$1\,000\,\text{g H}_2SO_4\left(\dfrac{1\,\text{mol H}_2SO_4}{98.08\,\text{g H}_2SO_4}\right)\left(\dfrac{1\,\text{mol SO}_3}{\text{mol H}_2SO_4}\right)\left(\dfrac{80.06\,\text{g SO}_3}{\text{mol SO}_3}\right) = 816\,\text{g SO}_3$$

8.26. How many grams of methyl alcohol, CH_3OH, can be obtained in an industrial process from 10.0×10^6 g (10.0 metric tons) of CO plus hydrogen gas? To calculate the answer, (*a*) write a balanced chemical equation for the process. (*b*) Calculate the number of moles of CO in 10.0×10^6 g CO. (*c*) Calculate the number of moles of CH_3OH obtainable from that number of moles of CO. (*d*) Calculate the number of grams of CH_3OH obtainable.

Ans.

(*a*) $$CO + 2H_2 \xrightarrow{\text{special conditions}} CH_3OH$$

(*b*) $$10.0 \times 10^6 \text{ g CO} \left(\frac{1 \text{ mol CO}}{28.0 \text{ g CO}} \right) = 3.57 \times 10^5 \text{ mol CO}$$

(*c*) $$3.57 \times 10^5 \text{ mol CO} \left(\frac{1 \text{ mol CH}_3\text{OH}}{\text{mol CO}} \right) = 3.57 \times 10^5 \text{ mol CH}_3\text{OH}$$

(*d*) $$3.57 \times 10^5 \text{ mol CH}_3\text{OH} \left(\frac{32.0 \text{ g CH}_3\text{OH}}{\text{mol CH}_3\text{OH}} \right) = 11.4 \times 10^6 \text{ g CH}_3\text{OH}$$

8.27. Determine the number of grams of lithium oxide necessary to prepare 35.0 g of lithium hydroxide by addition of excess water.

Ans.

$$Li_2O + H_2O \longrightarrow 2 LiOH$$

$$35.0 \text{ g LiOH} \left(\frac{1 \text{ mol LiOH}}{23.9 \text{ g LiOH}} \right) \left(\frac{1 \text{ mol Li}_2\text{O}}{2 \text{ mol LiOH}} \right) \left(\frac{29.9 \text{ g Li}_2\text{O}}{\text{mol Li}_2\text{O}} \right) = 21.9 \text{ g Li}_2\text{O}$$

8.28. Determine the number of grams of barium hydroxide it would take to neutralize (just react completely, with none left over) 80.0 g of phosphoric acid.

Ans.

$$3 Ba(OH)_2 + 2 H_3PO_4 \longrightarrow Ba_3(PO_4)_2 + 6 H_2O$$

$$80.0 \text{ g H}_3\text{PO}_4 \left(\frac{1 \text{ mol H}_3\text{PO}_4}{98.0 \text{ g H}_3\text{PO}_4} \right) \left(\frac{3 \text{ mol Ba(OH)}_2}{2 \text{ mol H}_3\text{PO}_4} \right) \left(\frac{171 \text{ g Ba(OH)}_2}{\text{mol Ba(OH)}_2} \right)$$
$$= 209 \text{ g Ba(OH)}_2$$

8.29. Calculate the number of moles of Al_2O_3 needed to prepare 5 000 g of Al metal in the Hall process:

$$Al_2O_3 + 3C \xrightarrow{\text{electricity}} 2 Al + 3 CO$$

Ans.

$$5 000 \text{ g Al} \left(\frac{1 \text{ mol Al}}{27.0 \text{ g Al}} \right) = 185 \text{ mol Al}$$

$$185 \text{ mol Al} \left(\frac{1 \text{ mol Al}_2\text{O}_3}{2 \text{ mol Al}} \right) = 92.5 \text{ mol Al}_2\text{O}_3$$

8.30. Calculate the number of moles of NaOH required to remove the SO_2 from 10.0 metric tons $(10.0 \times 10^6$ g) of atmosphere if the SO_2 is 0.10% by mass. (Na_2SO_3 and water are the products.)

Ans.

$$2 NaOH + SO_2 \longrightarrow Na_2SO_3 + H_2O$$

$$10.0 \times 10^6 \text{ g atmosphere} \left(\frac{0.10 \text{ g SO}_2}{100 \text{ g atmosphere}} \right) = 10.0 \times 10^3 \text{ g SO}_2$$

$$10.0 \times 10^3 \text{ g SO}_2 \left(\frac{1 \text{ mol SO}_2}{64.0 \text{ g SO}_2} \right) = 156 \text{ mol SO}_2$$

$$156 \text{ mol SO}_2 \left(\frac{2 \text{ mol NaOH}}{\text{mol SO}_2} \right) = 312 \text{ mol NaOH}$$

8.31. In chemistry recitation, a student hears (incorrectly) the instructor say "hydrogen chloride reacts with NaOH" when the instructor has actually said "hydrogen fluoride reacts with NaOH." The instructor then asked how much product is formed. The student answers the question correctly. Which section, 8.1 or 8.2, were they discussing? Explain.

Ans.   They were discussing Sec. 8.1. Since the student got the answer correct despite hearing the wrong name, they must have been discussing the number of moles of reactants and products. The numbers of moles of HF and HCl would be the same in the reaction, but since they have different formula weights, their masses would be different.

8.32. How many grams of NaCl can be produced from 10.0 g chlorine?

Ans.   This problem is the same as Problem 8.16. Problem 8.16 was stated in steps, and this problem is not, but you must do the same steps whether or not they are explicitly stated.

8.33. Consider the equation

$$KMnO_4 + 5\,FeCl_2 + 8\,HCl \longrightarrow MnCl_2 + 5\,FeCl_3 + 4\,H_2O + KCl$$

How many grams of $FeCl_3$ will be produced by the reaction of 105 g of HCl?

Ans.   The reacting ratio is given by the coefficients. The coefficients of interest are 8 for HCl and 5 for $FeCl_3$.

$$105\,\text{g HCl}\left(\frac{1\,\text{mol HCl}}{36.5\,\text{g HCl}}\right) = 2.88\,\text{mol HCl}$$

$$2.88\,\text{mol HCl}\left(\frac{5\,\text{mol FeCl}_3}{8\,\text{mol HCl}}\right) = 1.80\,\text{mol FeCl}_3\left(\frac{162\,\text{g FeCl}_3}{\text{mol FeCl}_3}\right) = 292\,\text{g FeCl}_3$$

8.34. What mass of $ZnCl_2$ can be prepared by treating 10.0 g HCl in water solution with excess zinc metal?

Ans.

$$Zn + 2\,HCl \longrightarrow ZnCl_2 + H_2$$

$$10.0\,\text{g HCl}\left(\frac{1\,\text{mol HCl}}{36.5\,\text{g HCl}}\right) = 0.274\,\text{mol HCl}$$

$$0.274\,\text{mol HCl}\left(\frac{1\,\text{mol ZnCl}_2}{2\,\text{mol HCl}}\right) = 0.137\,\text{mol ZnCl}_2$$

$$0.137\,\text{mol ZnCl}_2\left(\frac{136\,\text{g ZnCl}_2}{\text{mol ZnCl}_2}\right) = 18.6\,\text{g ZnCl}_2$$

8.35. How much AgCl can be prepared with 20.0 g $BaCl_2$ and excess $AgNO_3$?

Ans.

$$BaCl_2 + 2\,AgNO_3 \longrightarrow Ba(NO_3)_2 + 2\,AgCl$$

$$20.0\,\text{g BaCl}_2\left(\frac{1\,\text{mol BaCl}_2}{208\,\text{g BaCl}_2}\right) = 0.0962\,\text{mol BaCl}_2$$

$$0.0962\,\text{mol BaCl}_2\left(\frac{2\,\text{mol AgCl}}{\text{mol BaCl}_2}\right) = 0.192\,\text{mol AgCl}$$

$$0.192\,\text{mol AgCl}\left(\frac{143\,\text{g AgCl}}{\text{mol AgCl}}\right) = 27.5\,\text{g AgCl}$$

8.36. If, under certain conditions, 1 mol of CO_2 occupies 25.0 L, how many liters of CO_2 can be prepared from 150.0 g of $CaCO_3$ plus excess HCl?

Ans.

$$CaCO_3 + 2\,HCl \longrightarrow CaCl_2 + H_2O + CO_2$$

$$150.0\,\text{g } CaCO_3\left(\frac{1\,\text{mol } CaCO_3}{100\,\text{g } CaCO_3}\right) = 1.50\,\text{mol } CaCO_3$$

$$1.50\,\text{mol } CaCO_3\left(\frac{1\,\text{mol } CO_2}{\text{mol } CaCO_3}\right) = 1.50\,\text{mol } CO_2$$

$$1.50\,\text{mol } CO_2\left(\frac{25.0\,\text{L } CO_2}{\text{mol } CO_2}\right) = 37.5\,\text{L } CO_2$$

8.37. How much $KClO_3$ must be decomposed thermally to produce 1.50 g O_2?

Ans.

$$2\,KClO_3 \longrightarrow 2\,KCl + 3\,O_2$$

$$1.50\,\text{g } O_2\left(\frac{1\,\text{mol } O_2}{32.0\,\text{g } O_2}\right) = 0.0469\,\text{mol } O_2$$

$$0.0469\,\text{mol } O_2\left(\frac{2\,\text{mol } KClO_3}{3\,\text{mol } O_2}\right) = 0.0313\,\text{mol } KClO_3 \text{ decomposed}$$

$$0.0313\,\text{mol } KClO_3\left(\frac{122.6\,\text{g } KClO_3}{\text{mol } KClO_3}\right) = 3.84\,\text{g } KClO_3 \text{ decomposed}$$

LIMITING QUANTITIES

8.38. How many sandwiches, each containing 1 slice of cheese and 2 slices of bread, can you make with 30 slices of bread and 20 slices of cheese?

Ans. With 30 slices of bread, the most sandwiches you can make is 15. The bread is the limiting quantity.

8.39. How much sulfur dioxide is produced by the reaction of 1.00 g S and all the oxygen in the atmosphere of the earth?

Ans. In this problem, it is obvious that the oxygen in the entire earth's atmosphere is in excess, so that no preliminary calculation need be done.

$$S + O_2 \longrightarrow SO_2$$

$$1.00\,\text{g } S\left(\frac{1\,\text{mol } S}{32.06\,\text{g } S}\right) = 0.0312\,\text{mol } S$$

$$0.0312\,\text{mol } S\left(\frac{1\,\text{mol } SO_2}{\text{mol } S}\right) = 0.0312\,\text{mol } SO_2\left(\frac{64.1\,\text{g } SO_2}{\text{mol } SO_2}\right) = 2.00\,\text{g } SO_2$$

8.40. How can you recognize a limiting quantities problem?

Ans. The quantities of two reactants (or products) are given in the problem. They might be stated in any terms—moles, mass, etc.—but they must be given for the problem to be a limiting quantities problem.

8.41. (a) How many moles of Cl_2 will react with 7.0 mol Na to produce NaCl? (b) If 5.0 mol Cl_2 is treated with 7.0 mol Na, how much Cl_2 will react? (c) What is the limiting quantity in this problem?

Ans. (a)
$$2\,Na + Cl_2 \longrightarrow 2\,NaCl$$

$$7.0\,\text{mol Na}\left(\frac{1\,\text{mol } Cl_2}{2\,\text{mol Na}}\right) = 3.5\,\text{mol } Cl_2 \text{ reacts}$$

(b) We calculated in part (a) that 3.5 mol of Cl_2 is needed to react. As long as we have at least 3.5 mol present, 3.5 mol Cl_2 will react. (c) Since we have more Cl_2 than that, the Na is in limiting quantity.

8.42. If you were treating two chemicals, one cheap and one expensive, to produce a product, which one would you use in excess, if one of them had to be in excess?

Ans. Economically, it would be advisable to use the cheap one in excess, since more product could be obtained per dollar by using up all the expensive reactant.

8.43. (a) The price of pistachio nuts is $5.00 per pound. If a grocer has 13 pounds for sale, and a buyer has $24.00 to buy nuts with, what is the maximum number of pounds that can be sold? (b) Consider the reaction

$$KMnO_4 + 5\,FeCl_2 + 8\,HCl \longrightarrow MnCl_2 + 5\,FeCl_3 + 4\,H_2O + KCl$$

If 24.0 mol of $FeCl_2$ and 13.0 mol of $KMnO_4$ are mixed with excess HCl, how many moles of $MnCl_2$ can be formed?

Ans. (a) With $24, the buyer can buy

$$24\,\text{dollars}\left(\frac{1\,\text{pound}}{5\,\text{dollars}}\right) = 4.8\,\text{pounds}$$

Since the seller has more nuts than that, the money is in limiting quantity, and controls the amount of the sale.

(b) With 24.0 mol $FeCl_2$,

$$24.0\,\text{mol } FeCl_2\left(\frac{1\,\text{mol } KMnO_4}{5\,\text{mol } FeCl_2}\right) = 4.80\,\text{mol } KMnO_4 \text{ required}$$

Since the number of moles of $KMnO_4$ present (13.0 mol) exceeds that number, the limiting quantity is the number of moles of $FeCl_2$.

$$24.0\,\text{mol } FeCl_2\left(\frac{1\,\text{mol } MnCl_2}{5\,\text{mol } FeCl_2}\right) = 4.80\,\text{mol } MnCl_2$$

8.44. In each of the following cases, determine which reactant is present in excess, and tell how many moles in excess it is.

	Equation	Moles Present
(a)	$2\,Na + Cl_2 \longrightarrow 2\,NaCl$	7.9 mol Na, 3.0 mol Cl_2
(b)	$P_4O_{10} + 6\,H_2O \longrightarrow 4\,H_3PO_4$	0.50 mol P_4O_{10}, 3.0 mol H_2O
(c)	$HNO_3 + NaOH \longrightarrow NaNO_3 + H_2O$	2.0 mol acid, 2.1 mol base
(d)	$Ca(HCO_3)_2 + 2\,HCl \longrightarrow CaCl_2 + 2\,CO_2 + 2\,H_2O$	3.5 mol HCl, 1 mol $Ca(HCO_3)_2$
(e)	$H_3PO_4 + 3\,NaOH \longrightarrow Na_3PO_4 + 3\,H_2O$	0.13 mol acid, 1.05 mol NaOH

Ans. (a) $3.0\,\text{mol } Cl_2\left(\frac{2\,\text{mol Na}}{\text{mol } Cl_2}\right) = 6.0\,\text{mol Na required}$

CHAP. 8] STOICHIOMETRY 145

There is 1.9 mol more Na present than is required.

(b) $0.50\,\text{mol P}_4\text{O}_{10}\left(\dfrac{6\,\text{mol H}_2\text{O}}{\text{mol P}_4\text{O}_{10}}\right) = 3.0\,\text{mol H}_2\text{O required}$

Neither reagent is in excess; there is just enough H_2O to react with all the P_4O_{10}.

(c) $2.1\,\text{mol NaOH}\left(\dfrac{1\,\text{mol HNO}_3}{\text{mol NaOH}}\right) = 2.1\,\text{mol HNO}_3 \text{ required}$

There is not enough HNO_3 present; NaOH is in excess.

$$2.0\,\text{mol HNO}_3\left(\dfrac{1\,\text{mol NaOH}}{\text{mol HNO}_3}\right) = 2.0\,\text{mol NaOH required}$$

There is 0.1 mol NaOH in excess.

(d) $1.0\,\text{mol Ca(HCO}_3)_2\left(\dfrac{2\,\text{mol HCl}}{\text{mol Ca(HCO}_3)_2}\right) = 2.0\,\text{mol HCl required}$

There is 1.5 mol HCl in excess.

(e) $0.13\,\text{mol H}_3\text{PO}_4\left(\dfrac{3\,\text{mol NaOH}}{\text{mol H}_3\text{PO}_4}\right) = 0.39\,\text{mol NaOH required}$

There is $1.05 - 0.39 = 0.66$ mol NaOH in excess.

8.45. For the following reaction,

$$2\,\text{NaOH} + \text{H}_2\text{SO}_4 \longrightarrow \text{Na}_2\text{SO}_4 + 2\,\text{H}_2\text{O}$$

(a) How many moles of NaOH would react with $0.750\,\text{mol H}_2\text{SO}_4$? How many moles of Na_2SO_4 would be produced?
(b) If $0.750\,\text{mol}$ of H_2SO_4 and $2.00\,\text{mol}$ NaOH were mixed, how much NaOH would react?
(c) If $73.5\,\text{g H}_2\text{SO}_4$ and $80.0\,\text{g}$ NaOH were mixed, how many grams of Na_2SO_4 would be produced?

Ans. (a)

$$0.750\,\text{mol H}_2\text{SO}_4\left(\dfrac{2\,\text{mol NaOH}}{\text{mol H}_2\text{SO}_4}\right) = 1.50\,\text{mol NaOH}$$

$$0.750\,\text{mol H}_2\text{SO}_4\left(\dfrac{1\,\text{mol Na}_2\text{SO}_4}{\text{mol H}_2\text{SO}_4}\right) = 0.750\,\text{mol Na}_2\text{SO}_4$$

(b) 1.50 mol NaOH, as calculated in part (a).
(c) This is really the same problem as part (b), except that it is stated in grams, because $73.5\,\text{g}$ H_2SO_4 is $0.750\,\text{mol}$ and $80.0\,\text{g}$ NaOH is $2.00\,\text{mol}$ NaOH.

$$0.750\,\text{mol Na}_2\text{SO}_4\left(\dfrac{142\,\text{g Na}_2\text{SO}_4}{\text{mol Na}_2\text{SO}_4}\right) = 106\,\text{g Na}_2\text{SO}_4$$

8.46. For the reaction

$$3\,\text{HCl} + \text{Na}_3\text{PO}_4 \longrightarrow \text{H}_3\text{PO}_4 + 3\,\text{NaCl}$$

a chemist added $3.9\,\text{mol}$ of HCl and a certain quantity of Na_3PO_4 to a reaction vessel; $1.1\,\text{mol}$ H_3PO_4 was produced. Which one of the reactants was in excess, assuming that one of them was.

Ans. $$3.9\,\text{mol HCl}\left(\dfrac{1\,\text{mol H}_3\text{PO}_4}{3\,\text{mol HCl}}\right) = 1.3\,\text{mol H}_3\text{PO}_4$$

would be produced by reaction of all the HCl. Since the actual quantity of H_3PO_4 produced is $1.1\,\text{mol}$, not all the HCl was used up, and the Na_3PO_4 must be the limiting quantity. The HCl was in excess.

8.47. How much CO_2 can be produced by the combustion of $1.00\,kg$ of octane, C_8H_{18}, in $4.00\,kg$ of oxygen?

Ans.

$$2\,C_8H_{18} + 25\,O_2 \longrightarrow 16\,CO_2 + 18\,H_2O$$

$$1.00\,kg\,C_8H_{18}\left(\frac{1\,000\,g}{kg}\right)\left(\frac{1\,mol\,C_8H_{18}}{114\,g\,C_8H_{18}}\right) = 8.77\,mol\,C_8H_{18}\ present$$

$$4.00\,kg\,O_2\left(\frac{1\,000\,g}{kg}\right)\left(\frac{1\,mol\,O_2}{32.0\,g\,O_2}\right) = 125\,mol\,O_2\ present$$

$$8.77\,mol\,C_8H_{18}\left(\frac{25\,mol\,O_2}{2\,mol\,C_8H_{18}}\right) = 110\,mol\,O_2\ required$$

Since $110\,mol$ of O_2 is required and $125\,mol$ O_2 is present, the C_8H_{18} is in limiting quantity.

$$8.77\,mol\,C_8H_{18}\left(\frac{16\,mol\,CO_2}{2\,mol\,C_8H_{18}}\right) = 70.2\,mol\,CO_2\ produced$$

$$70.2\,mol\,CO_2\left(\frac{44.0\,g\,CO_2}{mol\,CO_2}\right) = 3\,090\,g = 3.09\,kg\,CO_2\ produced$$

8.48. How many grams of CO can be made by burning $100\,g$ C_8H_{18} in $100\,g$ O_2?

Ans.

$$2\,C_8H_{18} + 17\,O_2 \longrightarrow 16\,CO + 18\,H_2O$$

$$100\,g\,C_8H_{18}\left(\frac{1\,mol\,C_8H_{18}}{114\,g\,C_8H_{18}}\right) = 0.877\,mol\,C_8H_{18}$$

$$100\,g\,O_2\left(\frac{1\,mol\,O_2}{32.0\,g\,O_2}\right) = 3.125\,mol\,O_2$$

$$3.125\,mol\,O_2\left(\frac{2\,mol\,C_8H_{18}}{17\,mol\,O_2}\right) = 0.368\,mol\,C_8H_{18}\ required$$

Since there is more C_8H_{18} present than is required, the O_2 is in limiting quantity.

$$3.125\,mol\,O_2\left(\frac{16\,mol\,CO}{17\,mol\,O_2}\right) = 2.94\,mol\,CO\ produced$$

$$2.94\,mol\,CO\left(\frac{28.0\,g\,CO}{mol\,CO}\right) = 82.3\,g\,CO\ produced$$

Supplementary Problems

8.49. Read each of the other supplementary problems and state which ones are limiting quantities problems.

8.50. Calculate the number of grams of methyl alcohol, CH_3OH, obtainable in an industrial process from $10.0 \times 10^6\,g$ (10.0 metric tons) of CO plus hydrogen gas.

Ans. See Problem 8.26.

8.51. What percentage of $5.00\,g$ $KClO_3$ must be decomposed thermally to produce $1.50\,g$ O_2?

Ans. The statement of the problem implies that not all the $KClO_3$ is decomposed; it must be in excess for the purposes of producing the $1.50\,g$ O_2. Hence we can base the solution on the quantity of O_2.

In Problem 8.37 we found that 3.84 g of $KClO_3$ decomposed. The percent $KClO_3$ decomposed is then

$$\left(\frac{3.84\,g}{5.00\,g}\right) \times 100\% = 76.8\%$$

8.52. The percent yield is 100 times the amount of a product actually prepared during a reaction divided by the amount theoretically possible to be prepared according to the balanced chemical equation. (Some reactions are slow, and sometimes not enough time is allowed for their completion; some reactions are accompanied by side reactions which consume a portion of the reactants; some reactions never get to completion.) If 3.00 g $C_6H_{11}Br$ is prepared by treating 3.00 g C_6H_{12} with excess Br_2, what is the percent yield? The equation is

$$C_6H_{12} + Br_2 \longrightarrow C_6H_{11}Br + HBr$$

Ans.

$$3.00\,g\ C_6H_{12}\left(\frac{1\,mol\ C_6H_{12}}{84.2\,g\ C_6H_{12}}\right) = 0.0356\,mol\ C_6H_{12}$$

$$0.0356\,mol\ C_6H_{12}\left(\frac{1\,mol\ C_6H_{11}Br}{mol\ C_6H_{12}}\right) = 0.0356\,mol\ C_6H_{11}Br$$

$$0.0356\,mol\ C_6H_{11}Br\left(\frac{163\,g\ C_6H_{11}Br}{mol\ C_6H_{11}Br}\right) = 5.80\,g\ C_6H_{11}Br$$

$$\left(\frac{3.00\,g\ C_6H_{11}Br\ obtained}{5.80\,g\ C_6H_{11}Br\ possible}\right) \times 100 = 51.7\%\ yield$$

8.53. If 3.00 g $C_6H_{11}Br$ is prepared by treating 3.00 g C_6H_{12} with 5.00 g Br_2, what is the percent yield? The equation is

$$C_6H_{12} + Br_2 \longrightarrow C_6H_{11}Br + HBr$$

Ans.

$$3.00\,g\ C_6H_{12}\left(\frac{1\,mol\ C_6H_{12}}{84.2\,g\ C_6H_{12}}\right) = 0.0356\,mol\ C_6H_{12}$$

$$5.00\,g\ Br_2\left(\frac{1\,mol\ Br_2}{160\,g\ Br_2}\right) = 0.0312\,mol\ Br_2$$

Since the reactants react in a 1:1 mole ratio, the Br_2 is in limiting quantity:

$$0.0312\,mol\ Br_2\left(\frac{1\,mol\ C_6H_{11}Br}{mol\ Br_2}\right) = 0.0312\,mol\ C_6H_{11}Br$$

$$0.0312\,mol\ C_6H_{11}Br\left(\frac{163\,g\ C_6H_{11}Br}{mol\ C_6H_{11}Br}\right) = 5.09\,g\ C_6H_{11}Br$$

$$\left(\frac{3.00\,g\ C_6H_{11}Br\ obtained}{5.09\,g\ C_6H_{11}Br\ possible}\right) \times 100 = 58.9\%\ yield$$

8.54. Make up your own stoichiometry problem using the equation

$$BaCl_2 + 2\,AgNO_3 \longrightarrow Ba(NO_3)_2 + 2\,AgCl$$

8.55. In a certain experiment, 100 g of ethane, C_2H_6, is burned. (*a*) Write a balanced chemical equation for the combustion of ethane to produce CO_2 and water. (*b*) Determine the number of moles of ethane in 100 g of ethane. (*c*) Determine the number of moles of CO_2 that would be produced by the combustion of that number of moles of ethane. (*d*) Determine the mass of CO_2 that can be produced by the combustion of 100 g of ethane.

Ans.

(*a*) $$2\,C_2H_6 + 7\,O_2 \longrightarrow 4\,CO_2 + 6\,H_2O$$

(*b*) $$100\,g\,C_2H_6 \left(\frac{1\,mol\,C_2H_6}{30.1\,g\,C_2H_6} \right) = 3.32\,mol\,C_2H_6$$

(*c*) $$3.32\,mol\,C_2H_6 \left(\frac{4\,mol\,CO_2}{2\,mol\,C_2H_6} \right) = 6.64\,mol\,CO_2$$

(*d*) $$6.64\,mol\,CO_2 \left(\frac{44.0\,g\,CO_2}{mol\,CO_2} \right) = 292\,g\,CO_2$$

8.56. Determine the number of grams of SO_3 produced by treating excess SO_2 with $500\,g$ of oxygen.

Ans.

$$2\,SO_2 + O_2 \longrightarrow 2\,SO_3$$

$$500\,g\,O_2 \left(\frac{1\,mol\,O_2}{32.0\,g\,O_2} \right) = 15.6\,mol\,O_2$$

$$15.6\,mol\,O_2 \left(\frac{2\,mol\,SO_3}{mol\,O_2} \right) = 31.2\,mol\,SO_3$$

$$31.2\,mol\,SO_3 \left(\frac{80.0\,g\,SO_3}{mol\,SO_3} \right) = 2\,500\,g\,SO_3 = 2.50\,kg\,SO_3$$

8.57. List the steps necessary to do the following problems. (Then, for additional practice, solve them.)

(*a*) How many grams of $NaClO_3$ can be prepared by treating $10.0\,g$ of $HClO_3$ with excess $NaOH$?
(*b*) How many moles of $(NH_4)_3PO_4$ fertilizer can be made from 2.0×10^{25} molecules of ammonia?
(*c*) How many grams of $(NH_4)_3PO_4$ can be made by the process of part (*b*)?
(*d*) How many grams of $Ca(HCO_3)_2$ can be made by treating $40.0\,g\,Ca(OH)_2$ with $20.0\,g\,CO_2$?

Ans. (*a*) Change the grams of $HClO_3$ to moles $HClO_3$ with the formula weight of $HClO_3$.
Change the moles of $HClO_3$ to moles $NaClO_3$, using the coefficients of the balanced chemical equation.
Change the moles of $NaClO_3$ to grams with the formula weight of $NaClO_3$.
(*b*) Change the molecules of NH_3 to moles of NH_3 with Avogadro's number.
Change the moles of NH_3 to moles of $(NH_4)_3PO_4$ with the balanced chemical equation.
(*c*) Add the following step to those of part (*b*):
Change the moles of $(NH_4)_3PO_4$ to grams with the formula weight.
(*d*) Find the number of moles of each reactant with the formula weights.
Find the number of moles of CO_2 needed to react with the number of moles of $Ca(OH)_2$ present, using the balanced chemical equation.
Compare the number of moles of CO_2 needed and the number present; the smaller number is used to complete the problem.

8.58. How many grams of Na_2CO_3 will be produced by thermal decomposition of $50.0\,g\,NaHCO_3$?

Ans.

$$2\,NaHCO_3 \xrightarrow{\text{heat}} Na_2CO_3 + CO_2 + H_2O$$

$$50.0\,g\,NaHCO_3 \left(\frac{1\,mol\,NaHCO_3}{84.0\,g\,NaHCO_3} \right) \left(\frac{1\,mol\,Na_2CO_3}{2\,mol\,NaHCO_3} \right) \left(\frac{106\,g\,Na_2CO_3}{mol\,Na_2CO_3} \right)$$

$$= 31.5\,g\,Na_2CO_3$$

8.59. How many grams of NaCl can be prepared by treating 97.5 g $BaCl_2$ with 111 g Na_2SO_4?

Ans.

$$BaCl_2 + Na_2SO_4 \longrightarrow BaSO_4 + 2\,NaCl$$

$$97.5\,\text{g BaCl}_2\left(\frac{1\,\text{mol BaCl}_2}{208\,\text{g BaCl}_2}\right) = 0.469\,\text{mol BaCl}_2$$

$$111\,\text{g Na}_2\text{SO}_4\left(\frac{1\,\text{mol Na}_2\text{SO}_4}{142\text{g Na}_2\text{SO}_4}\right) = 0.782\,\text{mol Na}_2\text{SO}_4$$

Since they react in a $1:1$ mole ratio, the $BaCl_2$ is obviously in limiting quantity.

$$0.469\,\text{mol BaCl}_2\left(\frac{2\,\text{mol NaCl}}{\text{mol BaCl}_2}\right)\left(\frac{58.5\,\text{g NaCl}}{\text{mol NaCl}}\right) = 54.9\,\text{g NaCl}$$

8.60. How many moles of H_2 can be prepared by treating 2.00 g Na with 3.00 g H_2O? CAUTION: This reaction is very energetic, and can even cause an explosion.

Ans.

$$2\,\text{Na} + 2\,\text{H}_2\text{O} \longrightarrow 2\,\text{NaOH} + \text{H}_2$$

$$2.00\,\text{g Na}\left(\frac{1\,\text{mol Na}}{23.0\,\text{g Na}}\right) = 0.0870\,\text{mol Na}$$

$$3.00\,\text{g H}_2\text{O}\left(\frac{1\,\text{mol H}_2\text{O}}{18.0\,\text{g H}_2\text{O}}\right) = 0.167\,\text{mol H}_2\text{O}$$

Since they react in a $1:1$ mole ratio, the Na is obviously in limiting quantity.

$$0.0870\,\text{mol Na}\left(\frac{1\,\text{mol H}_2}{2\,\text{mol Na}}\right) = 0.0435\,\text{mol H}_2$$

8.61. If 50.0 g Zn is treated with 150 g $AgNO_3$, how many grams of Ag metal can be prepared?

Ans.

$$2\,\text{AgNO}_3 + \text{Zn} \longrightarrow 2\,\text{Ag} + \text{Zn(NO}_3)_2$$

$$50.0\,\text{g Zn}\left(\frac{1\,\text{mol Zn}}{65.4\,\text{g Zn}}\right) = 0.765\,\text{mol Zn present}$$

$$150\,\text{g AgNO}_3\left(\frac{1\,\text{mol AgNO}_3}{170\,\text{g AgNO}_3}\right) = 0.882\,\text{mol AgNO}_3 \text{ present}$$

$$0.882\,\text{mol AgNO}_3\left(\frac{1\,\text{mol Zn}}{2\,\text{mol AgNO}_3}\right) = 0.441\,\text{mol Zn required}$$

Since more moles of Zn is present than is required to react with all the $AgNO_3$, the Zn is present in excess. We base the calculation on the $AgNO_3$ present.

$$0.882\,\text{mol AgNO}_3\left(\frac{1\,\text{mol Ag}}{\text{mol AgNO}_3}\right)\left(\frac{108\,\text{g Ag}}{\text{mol Ag}}\right) = 95.3\,\text{g Ag}$$

8.62. What, if any, is the difference between the following three exam questions?
How much NaCl is produced by treating 5.00 g NaOH with excess HCl?
How much NaCl is produced by treating 5.00 g NaOH with sufficient HCl?
How much NaCl is produced by treating 5.00 g NaOH with HCl?

Ans.   There is no difference.

8.63. Some pairs of substances undergo different reactions if one or the other is in excess. For example,

$$\text{with excess } O_2: \quad C + O_2 \longrightarrow CO_2$$
$$\text{with excess } C: \quad 2C + O_2 \longrightarrow 2CO$$

Does this fact make the limiting quantities calculations in the text untrue for such pairs?

Ans.   No. The principles are true for *each* possible reaction.

8.64. A manufacturer produces ammonium phosphate for fertilizer. Treatment of 63.1 g of product with excess NaOH produces 20.0 g NH_3. What percentage of the product is pure $(NH_4)_3PO_4$? (Assume that any impurity contains no ammonium compound.)

Ans. $$3\,NaOH + (NH_4)_3PO_4 \longrightarrow 3\,NH_3 + 3\,H_2O + Na_3PO_4$$

$$20.0\,g\,NH_3\left(\frac{1\,mol\,NH_3}{17.0\,g\,NH_3}\right)\left(\frac{1\,mol\,(NH_4)_3PO_4}{3\,mol\,NH_3}\right)\left(\frac{149\,g\,(NH_4)_3PO_4}{mol\,(NH_4)_3PO_4}\right) = 58.4\,g\,(NH_4)_3PO_4$$

$$\left(\frac{58.4\,g\,(NH_4)_3PO_4}{63.1\,g\,product}\right) \times 100\% = 92.6\%\ pure$$

8.65. (*a*) When 10.0 g NaOH reacts with 20.0 g HCl, how much NaCl is produced? (*b*) When 20.0 g NaOH reacts with 10.0 g HCl, how much NaCl is produced?

Ans.

$$HCl + NaOH \longrightarrow NaCl + H_2O$$

(*a*) $$10.0\,g\,NaOH\left(\frac{1\,mol\,NaOH}{40.0\,g\,NaOH}\right) = 0.250\,mol\,NaOH$$

$$20.0\,g\,HCl\left(\frac{1\,mol\,HCl}{36.5\,g\,HCl}\right) = 0.548\,mol\,HCl$$

Since the reactants react in a 1 : 1 ratio, the NaOH is the limiting quantity:

$$0.250\,mol\,NaOH\left(\frac{1\,mol\,NaCl}{mol\,NaOH}\right)\left(\frac{58.5\,g\,NaCl}{mol\,NaCl}\right) = 14.6\,g\,NaCl$$

(*b*) $$20.0\,g\,NaOH\left(\frac{1\,mol\,NaOH}{40.0\,g\,NaOH}\right) = 0.500\,mol\,NaOH$$

$$10.0\,g\,HCl\left(\frac{1\,mol\,HCl}{36.5\,g\,HCl}\right) = 0.274\,mol\,HCl$$

Since the reactants react in a 1 : 1 ratio, the HCl is the limiting quantity:

$$0.274\,mol\,HCl\left(\frac{1\,mol\,NaCl}{mol\,HCl}\right)\left(\frac{58.5\,g\,NaCl}{mol\,NaCl}\right) = 16.0\,g\,NaCl$$

8.66. Consider the reaction

$$Ba(OH)_2 + 2\,HCl \longrightarrow BaCl_2 + 2\,H_2O$$

If exactly 20.0 g of $Ba(OH)_2$ reacts according to this equation, do you know how much $Ba(OH)_2$ was added to the HCl? Do you know how much HCl was added to the $Ba(OH)_2$? Explain.

Ans.   You cannot tell. Either reactant might have been in excess. The information given allows calculations of how much *reacted* and how much of the products was produced, but not how much was added in the first place.

8.67. Redo Problem 8.21 without bothering to solve for intermediate answers.

8.68. (a) How many sandwiches of each type can you make, each containing 1 or 2 slices of cheese and 2 slices of bread, with 30 slices of bread and 20 slices of cheese, assuming that all the cheese and all the bread are used up? (b) How much CO and how much CO_2 can you make with 20.0 g of carbon and 45.0 g of oxygen? Assume that both reactants are used up.

Ans. (a) With 30 slices of bread, the most sandwiches you can make is 15. Let x = number of one-slice sandwiches. Then $15 - x$ = number of two-slice sandwiches.

$$x + 2(15 - x) = 20$$
$$x = 10 = \text{number of one-slice sandwiches}$$
$$15 - x = 5 = \text{number of two-slice sandwiches}$$

Alternatively, if you use 1 slice of cheese for each sandwich, you will have 5 slices of cheese left over. You then could make five 2-slice sandwiches and have ten 1-slice sandwiches left.

(b)
$$20.0 \, \text{g C} \left(\frac{1 \, \text{mol C}}{12.0 \, \text{g C}} \right) = 1.67 \, \text{mol C}$$

$$45.0 \, \text{g O} \left(\frac{1 \, \text{mol O}}{16.0 \, \text{g O}} \right) = 2.81 \, \text{mol O}$$

With 1.67 mol C, the total of CO and CO_2 is 1.67 mol. Let x = number of moles of CO. Then $1.67 - x$ = number of moles of CO_2.

$$x + 2(1.67 - x) = 2.81$$
$$x = 0.53 \, \text{mol CO}$$
and
$$1.67 - x = 1.14 \, \text{mol CO}_2$$

Alternatively, if you make all CO, you will have $2.81 - 1.67 = 1.14$ mol O left over. Use that 1.14 mol O to make 1.14 mol CO_2, leaving 0.53 mol CO.

$$2 \, CO + O_2 \longrightarrow 2 \, CO_2$$

8.69. Determine the number of grams of SO_3 produced by treating excess SO_2 with 500 g of oxygen.

Ans. 2 500 g.

8.70. A sample of a hydrocarbon (a compound of carbon and hydrogen only) is burned, and 2.20 g CO_2 and 0.900 g H_2O are produced. What is the empirical formula of the hydrocarbon?

Ans. The empirical formula can be determined from the ratio of moles of carbon atoms to moles of hydrogen atoms. From the masses of products, we can get the numbers of moles of products, from which we can get the numbers of moles of C and H. The mole ratio of these two elements is the same in the products as it is in the original compound.

$$2.20 \, \text{g CO}_2 \left(\frac{1 \, \text{mol CO}_2}{44.0 \, \text{g CO}_2} \right) = 0.0500 \, \text{mol CO}_2$$

$$0.900 \, \text{g H}_2\text{O} \left(\frac{1 \, \text{mol H}_2\text{O}}{18.0 \, \text{g H}_2\text{O}} \right) = 0.0500 \, \text{mol H}_2\text{O}$$

0.0500 mol CO_2 contains 0.0500 mol C

0.0500 mol H_2O contains 0.100 mol H

The mole ratio is 1:2, and the empirical formula is CH_2.

8.71. How many moles of each substance is left in solution after 0.100 mol of NaOH is added to 0.200 mol of $HC_2H_3O_2$?

Ans. The balanced equation is

$$NaOH + HC_2H_3O_2 \longrightarrow NaC_2H_3O_2 + H_2O$$

The limiting quantity is NaOH, and so 0.100 mol NaOH reacts with 0.100 mol $HC_2H_3O_2$ to produce 0.100 mol $NaC_2H_3O_2$ + 0.100 mol H_2O. There is also 0.100 mol excess $HC_2H_3O_2$ left in the solution.

8.72.   Combine Figs. 4-5 and 8-5 to show all the conversions learned so far.

   Ans.   The figure is shown in Fig. 8-6.

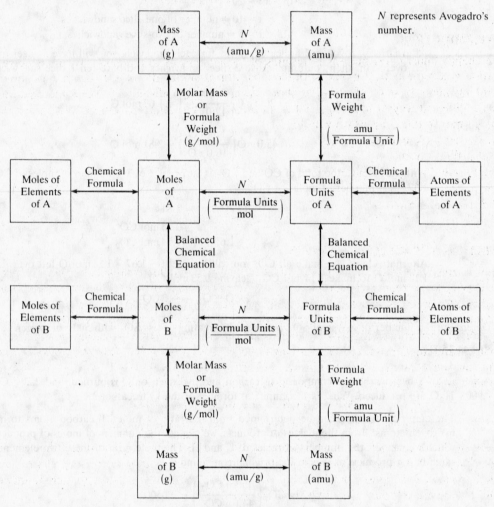

Fig. 8-6   Conversions.

Chapter 9

Net Ionic Equations

9.1 INTRODUCTION

The chemist must know literally thousands of facts. One way to remember such a wide variety of information is to systematize it. The periodic table, for example, allows us to learn data about whole groups of elements instead of learning about each element individually. Net ionic equations give chemists a different way of learning a lot of information with relatively little effort, rather than one piece at a time. In this chapter we will learn

1. What a net ionic equation means
2. What it does not mean
3. How to write net ionic equations
4. How to use net ionic equations
5. What they cannot tell us

9.2 WRITING NET IONIC EQUATIONS

When a substance made up of ions is dissolved in water, the dissolved ions behave independently. That is, they undergo their own characteristic reactions regardless of what other ions may be present. For example, barium ions in solution, Ba^{2+}, always react with sulfate ions in solution, SO_4^{2-}, to form an insoluble ionic compound, $BaSO_4(s)$, no matter what other ions are present in the barium solution. If a solution of barium chloride, $BaCl_2$, and a solution of sodium sulfate, Na_2SO_4, are mixed, a white solid, barium sulfate, is produced. The solid can be separated from the solution by filtration, and the resulting solution contains sodium chloride, just as it would if solid NaCl were added to water. In other words, when the two solutions are mixed, the following reaction occurs:

$$BaCl_2 + Na_2SO_4 \longrightarrow BaSO_4(s) + 2\,NaCl$$

or

$$Ba^{2+} + 2\,Cl^- + 2\,Na^+ + SO_4^{2-} \longrightarrow BaSO_4(s) + 2\,Na^+ + 2\,Cl^-$$

Written in the latter manner, the equation shows that in effect the sodium ions and the chloride ions have not changed. They began as ions in solution and wound up as those same ions in solution. They are called *spectator ions*. Since they have not reacted, it is really not necessary to include them in the equation. If they are left out, a *net ionic equation* results:

$$Ba^{2+} + SO_4^{2-} \longrightarrow BaSO_4(s)$$

This equation may be interpreted to mean that *any* soluble barium salt will react with *any* soluble sulfate to produce (insoluble) barium sulfate.

EXAMPLE 9.1. Write four equations that the preceding net ionic equation can represent.

The following equations represent four of many possible equations:

$$Ba(NO_3)_2 + Na_2SO_4 \longrightarrow BaSO_4(s) + 2\,NaNO_3$$
$$BaBr_2 + K_2SO_4 \longrightarrow BaSO_4(s) + 2\,KBr$$
$$BaCl_2 + FeSO_4 \longrightarrow BaSO_4(s) + FeCl_2$$
$$BaI_2 + ZnSO_4 \longrightarrow BaSO_4(s) + ZnI_2$$

Obviously, it is easier to remember the net ionic equation than the many possible overall equations that it represents. (See Problem 9.15.)

Net ionic equations may be written whenever reactions occur in solution in which some of the ions originally present are removed from solution or when ions not originally present are formed. Usually, ions are removed from solution by one or more of the following processes:

1. Formation of an insoluble ionic compound (see Table 7-2)
2. Formation of molecules containing only covalent bonds
3. Formation of new ionic species
4. Formation of a gas

Examples of these processes include:

1. $AgNO_3 + NaCl \longrightarrow AgCl(s) + NaNO_3$

   $Ag^+ + Cl^- \longrightarrow AgCl(s)$

2. $HCl + NaOH \longrightarrow H_2O + NaCl$

   $H^+ + OH^- \longrightarrow H_2O$

3. $Zn + CuSO_4 \longrightarrow ZnSO_4 + Cu$

   $Zn + Cu^{2+} \longrightarrow Zn^{2+} + Cu$

4. $Na_2CO_3 + 2\,HCl \longrightarrow CO_2 + H_2O + 2\,NaCl$

   $CO_3{}^{2-} + 2\,H^+ \longrightarrow CO_2 + H_2O$

The question arises how the student can tell whether a compound is ionic or covalent. The following generalizations will be of some help in deciding:

1. Binary compounds of two nonmetals are covalently bonded. However, strong acids in water form ions completely.
2. Binary compounds of a metal and nonmetal are usually ionic.
3. Ternary compounds are usually ionic, at least in part, except if they contain no metal atoms or ammonium ion.

EXAMPLE 9.2. Predict which of the following will contain ionic bonds: (a) NaBr, (b) CO, (c) CaO, (d) NH_4Cl, (e) H_2SO_4, and (f) HCl.

(a) NaBr, (c) CaO, and (d) NH_4Cl contain ionic bonds. NH_4Cl also has covalent bonds within the ammonium ion. (e) H_2SO_4 and (f) HCl would form ions if allowed to react with water.

When do we write compounds as separate ions, and when do we write them as complete compounds? Ions can act independently in solution, and so we write ionic compounds as separate ions only when they are soluble. We write compounds together when they are not ionic or when they are not soluble.

EXAMPLE 9.3. Write each of the following compounds as it should be written in an ionic equation. (a) NaCl, (b) AgCl, (c) CO_2, and (d) $NaHCO_3$.

(a) $Na^+ + Cl^-$, (b) AgCl (insoluble), (c) CO_2 (not ionic), and (d) $Na^+ + HCO_3{}^-$.

The insoluble compound is written as one compound even though it is ionic. The covalent compound is written together because it is not ionic.

EXAMPLE 9.4. Each of the following reactions produces 56 kJ per mole of water produced. Is this just a coincidence? If not, explain why the same value is obtained each time.

$$NaOH + HCl \longrightarrow NaCl + H_2O$$

$$KOH + HNO_3 \longrightarrow KNO_3 + H_2O$$

$$RbOH + HBr \longrightarrow RbBr + H_2O$$

$$LiOH + HI \longrightarrow LiI + H_2O$$

The same quantity of heat is generated per mole of water formed in each reaction because it is really the same reaction in each case:

$$OH^- + H^+ \longrightarrow H_2O$$

It does not matter whether it is an Na^+ ion in solution that undergoes no reaction or a K^+ ion in solution that undergoes no reaction. As long as the spectator ions undergo no reaction, they do not contribute anything to the heat of the reaction.

EXAMPLE 9.5. Write a net ionic equation for the reaction of $Ba(OH)_2$ with HCl in water.

The overall equation is

$$Ba(OH)_2 + 2\,HCl \longrightarrow BaCl_2 + 2\,H_2O$$

In ionic form:

$$Ba^{2+} + 2\,OH^- + 2\,H^+ + 2\,Cl^- \longrightarrow Ba^{2+} + 2\,Cl^- + 2\,H_2O$$

Leaving out the spectator ions yields:

$$2\,OH^- + 2\,H^+ \longrightarrow 2\,H_2O$$

Dividing each side by 2 yields the net ionic equation:

$$OH^- + H^+ \longrightarrow H_2O$$

The net ionic equation is the same as that in Example 9.4.

Writing net ionic equations does not imply that any solution can contain only positive ions or only negative ions. For example, the net ionic equation

$$Ba^{2+} + SO_4^{2-} \longrightarrow BaSO_4(s)$$

does not imply that there is any solution containing Ba^{2+} ions with no negative ions, or any solution containing SO_4^{2-} ions with no positive ions. It merely implies that whatever negative ion is present with the barium ion and whatever positive ion is present with the sulfate ion, these unspecified ions do not make any difference to the reaction that will occur.

Net ionic equations must always have the same net charge on each side of the equation. (The same number of each type of spectator ion must be omitted from both sides of the equation.) For example, the equation

$$Cu + Ag^+ \longrightarrow Cu^{2+} + Ag \qquad \text{(unbalanced)}$$

has the same number of each type of atom on its two sides, but it is still not balanced. (One cannot add just one nitrate ion to the left side of an equation and two to the right.) The net charge must also be balanced:

$$Cu + 2\,Ag^+ \longrightarrow Cu^{2+} + 2\,Ag$$

There is a net 2+ charge on each side of the balanced equation.

9.3 CALCULATIONS BASED ON NET IONIC EQUATIONS

The net ionic equation, like all balanced chemical equations, gives the ratio of moles of each substance to moles of each of the others. It does not immediately yield information about the mass of the entire salt, however. (One cannot weigh out only Ba^{2+} ions.) Therefore, when masses of reactants are required, the specific compound used must be included in the calculation. The use of net ionic equations in stoichiometric calculations will be more important after study of molarity (Chap. 10).

EXAMPLE 9.6. (*a*) How many moles of silver ion is required to make 100 g AgCl? (*b*) What mass of silver nitrate is required to prepare 100 g AgCl?

$$Ag^+ + Cl^- \longrightarrow AgCl$$

(*a*) The formula weight of AgCl, the sum of the atomic weights of Ag and Cl, is $108 + 35 = 143$ amu. Therefore, there is 143 g/mol of AgCl. In 100 g of AgCl, which is to be prepared, there is

$$100 \text{ g AgCl}\left(\frac{1 \text{ mol AgCl}}{143 \text{ g AgCl}}\right) = 0.699 \text{ mol AgCl}$$

Hence, from the balanced chemical equation, 0.699 mol of Ag^+ is required:

$$0.699 \text{ mol AgCl}\left(\frac{1 \text{ mol Ag}^+}{1 \text{ mol AgCl}}\right) = 0.699 \text{ mol Ag}^+$$

(*b*) The 0.699 mol of Ag^+ may be furnished from 0.699 mol of $AgNO_3$. Then

$$0.699 \text{ mol AgNO}_3\left(\frac{170 \text{ g AgNO}_3}{\text{mol AgNO}_3}\right) = 119 \text{ g AgNO}_3$$

Hence, when treated with enough chloride ion, 119 g of $AgNO_3$ will produce 100 g of AgCl.

EXAMPLE 9.7. What is the maximum mass of $BaSO_4$ that can be produced when a solution containing 10.0 g of Na_2SO_4 is added to another solution containing an excess of Ba^{2+}?

$$Ba^{2+} + SO_4^{2-} \longrightarrow BaSO_4$$

$$10.0 \text{ g Na}_2\text{SO}_4\left(\frac{1 \text{ mol Na}_2\text{SO}_4}{142 \text{ g Na}_2\text{SO}_4}\right) = 0.0704 \text{ mol Na}_2\text{SO}_4$$

$$0.0704 \text{ mol Na}_2\text{SO}_4\left(\frac{1 \text{ mol SO}_4^{2-}}{1 \text{ mol Na}_2\text{SO}_4}\right) = 0.0704 \text{ mol SO}_4^{2-}$$

$$0.0704 \text{ mol SO}_4^{2-}\left(\frac{1 \text{ mol BaSO}_4}{1 \text{ mol SO}_4^{2-}}\right) = 0.0704 \text{ mol BaSO}_4$$

$$0.0704 \text{ mol BaSO}_4\left(\frac{233 \text{ g BaSO}_4}{\text{mol BaSO}_4}\right) = 16.4 \text{ g BaSO}_4$$

Solved Problems

WRITING NET IONIC EQUATIONS

9.1. Calcium and oxygen form a binary ionic compound. Write its formula.

Ans.   CaO.

9.2. How many types of ions are generally found in any ionic compound?

Ans.   Two—one type of cation and one type of anion. (Some alums have three types of ions.)

9.3. Write the formulas for the ions that are present in each of the following compounds: (*a*) NaCl, (*b*) $BaCl_2$, (*c*) $NaNO_3$, (*d*) $Ba(ClO)_2$, (*e*) $NaClO_2$, (*f*) $(NH_4)_3PO_4$, (*g*) $Co(ClO_3)_2$, (*h*) $BaSO_4(s)$, and (*i*) $NaHCO_3$.

Ans.   (*a*) Na^+ and Cl^- (*f*) NH_4^+ and PO_4^{3-}
        (*b*) Ba^{2+} and Cl^- (*g*) Co^{2+} and ClO_3^-
        (*c*) Na^+ and NO_3^- (*h*) Ba^{2+} and SO_4^{2-} (even though it is a solid)
        (*d*) Ba^{2+} and ClO^- (*i*) Na^+ and HCO_3^-
        (*e*) Na^+ and ClO_2^-

9.4. Which of the following compounds are ionic? Which are soluble? Which would be written as separate ions in an ionic equation as written for the first equation in this section? Write the species as they would appear in an ionic equation. (a) $NaC_2H_3O_2$, (b) $BaSO_4$, (c) $PbCl_2$, (d) Na_2SO_4, (e) $FeCl_3$, and (f) CH_3OH.

Ans.

		Ionic	Soluble	Written Separately	Formulas in Ionic Equation
(a)	$NaC_2H_3O_2$	Yes	Yes	Yes	$Na^+ + C_2H_3O_2^-$
(b)	$BaSO_4$	Yes	No	No	$BaSO_4$
(c)	$PbCl_2$	Yes	No	No	$PbCl_2$
(d)	Na_2SO_4	Yes	Yes	Yes	$2\,Na^+ + SO_4^{2-}$
(e)	$FeCl_3$	Yes	Yes	Yes	$Fe^{3+} + 3\,Cl^-$
(f)	CH_3OH	No	Yes	No	CH_3OH

9.5. Given that $BaCO_3$ is insoluble in water and that NH_3 and H_2O are covalent compounds, write net ionic equations for the following processes:

(a) $$NH_4Cl + KOH \longrightarrow NH_3 + H_2O + KCl$$

(b) $$BaCl_2 + (NH_4)_2CO_3 \longrightarrow BaCO_3 + 2\,NH_4Cl$$

Ans.

(a) $$NH_4^+ + OH^- \longrightarrow NH_3 + H_2O$$

(b) $$Ba^{2+} + CO_3^{2-} \longrightarrow BaCO_3$$

9.6. Write a net ionic equation for the equation in each of the following parts:

(a) $$NaI + AgNO_3 \longrightarrow AgI(s) + NaNO_3$$

(b) $$Li_2SO_4 + BaCl_2 \longrightarrow BaSO_4 + 2\,LiCl$$

Ans.

(a) $$I^- + Ag^+ \longrightarrow AgI$$

(b) $$SO_4^{2-} + Ba^{2+} \longrightarrow BaSO_4$$

9.7. Write a net ionic equation for the equation in each of the following parts:

(a) $$HCl + NaHCO_3 \longrightarrow NaCl + CO_2 + H_2O$$

(b) $$NaOH + NH_4Cl \longrightarrow NH_3 + H_2O + NaCl$$

(c) $$HC_2H_3O_2 + NaOH \longrightarrow NaC_2H_3O_2 + H_2O$$

(d) $$Ba(OH)_2 + 2\,HCl \longrightarrow BaCl_2 + 2\,H_2O$$

Ans.

(a) $$H^+ + HCO_3^- \longrightarrow CO_2 + H_2O \qquad \text{(HCl is strong)}$$

(b) $$OH^- + NH_4^+ \longrightarrow NH_3 + H_2O$$

(c) $$HC_2H_3O_2 + OH^- \longrightarrow C_2H_3O_2^- + H_2O \qquad \text{(HC}_2\text{H}_3\text{O}_2 \text{ is weak)}$$

(d) $$OH^- + H^+ \longrightarrow H_2O$$

9.8. Write a net ionic equation for the following overall equation:

$$Fe + 2\,FeCl_3 \longrightarrow 3\,FeCl_2$$

Ans.

$$Fe + 2\,Fe^{3+} \longrightarrow 3\,Fe^{2+}$$

Iron ions appear on both sides of this equation, but they are not spectator ions since they are not identical. One is a 3+ ion and the other is a 2+ ion. The neutral atom is different from both of these.

9.9.   Write a net ionic equation for the following overall equation:

$$AlCl_3 + 4\,NaOH \longrightarrow NaAl(OH)_4(aq) + 3\,NaCl$$

Ans.

$$Al^{3+} + 4\,OH^- \longrightarrow Al(OH)_4^-$$

9.10. Write a net ionic equation for each of the following overall equations:

(*a*) $\qquad\qquad\qquad H_3PO_4 + 2\,NaOH \longrightarrow Na_2HPO_4 + 2\,H_2O$

(*b*) $\qquad\qquad\qquad CaCO_3(s) + CO_2 + H_2O \longrightarrow Ca(HCO_3)_2$

(*c*) $\qquad\qquad\qquad NaHCO_3 + NaOH \longrightarrow Na_2CO_3 + H_2O$

Ans.

(*a*) $\qquad\qquad\qquad H_3PO_4 + 2\,OH^- \longrightarrow HPO_4^{2-} + 2\,H_2O$

(*b*) $\qquad\qquad\qquad CaCO_3 + CO_2 + H_2O \longrightarrow Ca^{2+} + 2\,HCO_3^-$

(*c*) $\qquad\qquad\qquad HCO_3^- + OH^- \longrightarrow CO_3^{2-} + H_2O$

9.11. Write a net ionic equation for each of the following overall equations:

(*a*) $\qquad\qquad\qquad 2\,AgNO_3 + H_2S \longrightarrow Ag_2S(s) + 2\,HNO_3$

(*b*) $\qquad\qquad\qquad Zn + 2\,HCl \longrightarrow ZnCl_2 + H_2$

(*c*) $\qquad\qquad\qquad Zn + CuCl_2 \longrightarrow ZnCl_2 + Cu$

(*d*) $\qquad\qquad\qquad Zn + HgCl_2 \longrightarrow ZnCl_2 + Hg$

(*e*) $\qquad\qquad\qquad Zn + 2\,AgNO_3 \longrightarrow Zn(NO_3)_2 + 2\,Ag$

Ans.

(*a*) $\qquad\qquad\qquad 2\,Ag^+ + H_2S \longrightarrow Ag_2S + 2\,H^+$

(*b*) $\qquad\qquad\qquad Zn + 2\,H^+ \longrightarrow Zn^{2+} + H_2$

(*c*) $\qquad\qquad\qquad Zn + Cu^{2+} \longrightarrow Zn^{2+} + Cu$

(*d*) $\qquad\qquad\qquad Zn + Hg^{2+} \longrightarrow Zn^{2+} + Hg$

(*e*) $\qquad\qquad\qquad Zn + 2\,Ag^+ \longrightarrow Zn^{2+} + 2\,Ag$

(Note the overall charge balance.)

9.12. Write a net ionic equation for each of the following overall equations:

(*a*) $\qquad\qquad\qquad HClO_3 + NaOH \longrightarrow NaClO_3 + H_2O$

(*b*) $\qquad\qquad\qquad HCl + NaOH \longrightarrow NaCl + H_2O$

(*c*) $\qquad\qquad\qquad HClO_4 + NaOH \longrightarrow NaClO_4 + H_2O$

(*d*) $\qquad\qquad\qquad HBr + NaOH \longrightarrow NaBr + H_2O$

(*e*) $\qquad\qquad\qquad HClO_3 + KOH \longrightarrow KClO_3 + H_2O$

Ans.   In each case, the net ionic equation is

$$H^+ + OH^- \longrightarrow H_2O$$

9.13. Write a net ionic equation for each of the following overall equations:

(*a*) $\qquad\qquad\qquad HClO_3 + LiOH \longrightarrow LiClO_3 + H_2O$

(*b*) $\qquad\qquad\qquad HClO_3 + RbOH \longrightarrow RbClO_3 + H_2O$

Ans.   In each case, the net ionic equation is

$$H^+ + OH^- \longrightarrow H_2O$$

9.14. Write one or more complete equations for each of the following net ionic equations:

(a) $$NH_3 + H^+ \longrightarrow NH_4^+$$

(b) $$Co^{2+} + S^{2-} \longrightarrow CoS$$

(c) $$CO_2 + 2OH^- \longrightarrow CO_3^{2-} + H_2O$$

Ans.

(a) $$NH_3 + HCl \longrightarrow NH_4Cl$$

or $$NH_3 + HNO_3 \longrightarrow NH_4NO_3$$

or ammonia plus any other strong acid to yield the ammonium salt.

(b) $$CoCl_2 + K_2S \longrightarrow CoS + 2KCl$$

or $$Co(ClO_3)_2 + BaS \longrightarrow CoS + Ba(ClO_3)_2$$

or $$CoSO_4 + (NH_4)_2S \longrightarrow CoS + (NH_4)_2SO_4$$

or any soluble cobalt(II) salt with any soluble sulfide.

(c) $$CO_2 + 2NaOH \longrightarrow Na_2CO_3 + H_2O$$

or CO_2 plus any other soluble hydroxide to give a soluble carbonate, but not

$$CO_2 + Ba(OH)_2 \longrightarrow BaCO_3 + H_2O$$

because $BaCO_3$ is insoluble.

9.15. Write 12 more equations that can be represented by the net ionic equation of Example 9.1. Use all the same compounds that are used on the left in the equation in Example 9.1.

Ans.

$$BaBr_2 + Na_2SO_4 \longrightarrow BaSO_4(s) + 2NaBr$$

$$BaCl_2 + K_2SO_4 \longrightarrow BaSO_4(s) + 2KCl$$

$$BaI_2 + FeSO_4 \longrightarrow BaSO_4(s) + FeI_2$$

$$Ba(NO_3)_2 + ZnSO_4 \longrightarrow BaSO_4(s) + Zn(NO_3)_2$$

$$BaCl_2 + Na_2SO_4 \longrightarrow BaSO_4(s) + 2NaCl$$

$$BaI_2 + K_2SO_4 \longrightarrow BaSO_4(s) + 2KI$$

$$Ba(NO_3)_2 + FeSO_4 \longrightarrow BaSO_4(s) + Fe(NO_3)_2$$

$$BaBr_2 + ZnSO_4 \longrightarrow BaSO_4(s) + ZnBr_2$$

$$BaI_2 + Na_2SO_4 \longrightarrow BaSO_4(s) + 2NaI$$

$$Ba(NO_3)_2 + K_2SO_4 \longrightarrow BaSO_4(s) + 2KNO_3$$

$$BaBr_2 + FeSO_4 \longrightarrow BaSO_4(s) + FeBr_2$$

$$BaCl_2 + ZnSO_4 \longrightarrow BaSO_4(s) + ZnCl_2$$

9.16. A student used $HC_2H_3O_2$ to prepare a test solution that was supposed to contain acetate ions. Criticize this choice.

Ans.   Acetic acid is weak, and relatively little acetate ion is present. The student should have used an ionic acetate—sodium acetate or ammonium acetate, for example.

9.17. What chemical would you use to prepare a solution to be used for a test requiring the presence of Ba^{2+} ions?

Ans. $BaCl_2$, $Ba(ClO_3)_2$, or any other soluble barium salt.

9.18. What chemical would you use to prepare a solution to be used for a test requiring the presence of Cl^- ions?

Ans.   HCl(aq), $BaCl_2$, NaCl, or any other soluble, ionic chloride.

9.19. What chemical would you use to prepare a solution to be used for a test requiring the presence of SO_4^{2-} ions?

Ans. $FeSO_4$, Na_2SO_4, or any other soluble sulfate.

CALCULATIONS BASED ON NET IONIC EQUATIONS

9.20. What mass of each of the following silver salts would be required to react completely with a solution containing 3.55 g of chloride ion to form (insoluble) silver chloride? (*a*) $AgNO_3$, (*b*) Ag_2SO_4, and (*c*) $AgC_2H_3O_2$.

Ans.

$$3.55\,g\,Cl^-\left(\frac{1\,mol\,Cl^-}{35.5\,g\,Cl^-}\right)\left(\frac{1\,mol\,Ag^+}{mol\,Cl^-}\right)=0.100\,mol\,Ag^+$$

(*a*) $$0.100\,mol\,Ag^+\left(\frac{1\,mol\,AgNO_3}{mol\,Ag^+}\right)\left(\frac{169.9\,g\,AgNO_3}{mol\,AgNO_3}\right)=17.0\,g\,AgNO_3$$

(*b*) $$0.100\,mol\,Ag^+\left(\frac{1\,mol\,Ag_2SO_4}{2\,mol\,Ag^+}\right)\left(\frac{311.8\,g\,Ag_2SO_4}{mol\,Ag_2SO_4}\right)=15.6\,g\,Ag_2SO_4$$

(*c*) $$0.100\,mol\,Ag^+\left(\frac{1\,mol\,AgC_2H_3O_2}{mol\,Ag^+}\right)\left(\frac{166.9\,g\,AgC_2H_3O_2}{mol\,AgC_2H_3O_2}\right)=16.7\,g\,AgC_2H_3O_2$$

Supplementary Problems

9.21. Write a net ionic equation for the following overall equation:

$$4\,Au+16\,KCN+6\,H_2O+3\,O_2\longrightarrow 4\,KAu(CN)_4(aq)+12\,KOH$$

Ans.

$$4\,Au+16\,CN^-+6\,H_2O+3\,O_2\longrightarrow 4\,Au(CN)_4^-+12\,OH^-$$

9.22. Write a net ionic equation for each of the following equations:

(*a*) $CdS(s)+I_2\longrightarrow CdI_2(aq)+S$

(*b*) $2\,HI+2\,HNO_2\longrightarrow 2\,NO+2\,H_2O+I_2$

(*c*) $4\,KOH+4\,KMnO_4(aq)\longrightarrow 4\,K_2MnO_4(aq)+O_2+2\,H_2O$

Ans.

(*a*) $CdS(s)+I_2\longrightarrow Cd^{2+}+2\,I^-+S$

(*b*) $2\,H^++2\,I^-+2\,HNO_2\longrightarrow 2\,NO+2\,H_2O+I_2$ (HNO$_2$ is weak)

(*c*) $4\,OH^-+4\,MnO_4^-\longrightarrow 4\,MnO_4^{2-}+O_2+2\,H_2O$

9.23. Write 12 more equations represented by the net ionic equation given in Example 9.4, using only the reactants used in that example.

Ans.

$$KOH + HCl \longrightarrow KCl + H_2O$$
$$RbOH + HNO_3 \longrightarrow RbNO_3 + H_2O$$
$$LiOH + HBr \longrightarrow LiBr + H_2O$$
$$NaOH + HI \longrightarrow NaI + H_2O$$
$$RbOH + HCl \longrightarrow RbCl + H_2O$$
$$LiOH + HNO_3 \longrightarrow LiNO_3 + H_2O$$
$$NaOH + HBr \longrightarrow NaBr + H_2O$$
$$KOH + HI \longrightarrow KI + H_2O$$
$$LiOH + HCl \longrightarrow LiCl + H_2O$$
$$NaOH + HNO_3 \longrightarrow NaNO_3 + H_2O$$
$$KOH + HBr \longrightarrow KBr + H_2O$$
$$RbOH + HI \longrightarrow RbI + H_2O$$

9.24. Would the following reaction yield 56 kJ of heat per mole of water formed, as the reactions in Example 9.4 do? Explain.

$$HC_2H_3O_2 + NaOH \longrightarrow NaC_2H_3O_2 + H_2O$$

Ans.   No. Since $HC_2H_3O_2$ is a weak acid, there is a different net ionic equation, and a different amount of heat:

$$HC_2H_3O_2 + OH^- \longrightarrow C_2H_3O_2^- + H_2O$$

9.25. Per mole of water formed, how much heat is generated by the reaction of Example 9.5?

Ans.   56 kJ per mole. It is the same reaction as that of Example 9.4.

9.26. Would 56 kJ per mole of water formed be generated by the following reaction? Compare your answer with that of the last problem.

$$Ba(OH)_2(s) + 2 HCl \longrightarrow BaCl_2 + 2 H_2O$$

Ans.   No. It is not represented by the same net ionic equation. Some heat is involved in dissolving the solid $Ba(OH)_2$.

9.27. Write a net ionic equation for each of the following overall equations:

(*a*) $$Cu(OH)_2(s) + 2 HClO_3 \longrightarrow Cu(ClO_3)_2 + 2 H_2O$$
(*b*) $$CuCl_2 + H_2S \longrightarrow CuS(s) + 2 HCl$$
(*c*) $$ZnS(s) + 2 HCl \longrightarrow H_2S + ZnCl_2$$

Ans.

(*a*) $$Cu(OH)_2(s) + 2 H^+ \longrightarrow Cu^{2+} + 2 H_2O$$
(*b*) $$Cu^{2+} + H_2S \longrightarrow CuS + 2 H^+$$
(*c*) $$ZnS + 2 H^+ \longrightarrow H_2S + Zn^{2+}$$

9.28. What is the difference between reactions involving 5.00 g Cl^- and 5.00 g NaCl?

Ans.   The former might be part of any ionic chloride, but more important, the former has a greater mass of chlorine, because 5.00 g NaCl has less than 5.00 g Cl^-.

9.29. How can you weigh out just the chloride ion of Problem 9.20?

Ans. You cannot. The chloride ion does not exist alone. Writing it as Cl^- indicates that the positive ion is not important in the reaction, not that it is not present.

9.30. Balance the following net ionic equations:

(*a*)
$$Ag^+ + Fe \longrightarrow Fe^{2+} + Ag$$
(*b*)
$$Fe^{3+} + I^- \longrightarrow Fe^{2+} + I_2$$
(*c*)
$$Cu^{2+} + I^- \longrightarrow CuI + I_2$$
(*d*)
$$Zn + Cr^{3+} \longrightarrow Cr^{2+} + Zn^{2+}$$

Ans. In each part, the net charge as well as the number of each type of atom must balance.

(*a*)
$$2\,Ag^+ + Fe \longrightarrow Fe^{2+} + 2\,Ag$$
(*b*)
$$2\,Fe^{3+} + 2\,I^- \longrightarrow 2\,Fe^{2+} + I_2$$
(*c*)
$$2\,Cu^{2+} + 4\,I^- \longrightarrow 2\,CuI + I_2$$
(*d*)
$$Zn + 2\,Cr^{3+} \longrightarrow 2\,Cr^{2+} + Zn^{2+}$$

9.31. Try to write a complete equation corresponding to the unbalanced and the balanced net ionic equations of Problem 9.30. What do you find?

Ans. You cannot write a complete equation for an unbalanced net ionic equation. [In part (*a*), for example, you might have one nitrate ion on the left and two on the right.] The complete equation for the balanced net ionic equation might be

$$2\,AgNO_3 + Fe \longrightarrow Fe(NO_3)_2 + 2\,Ag$$

Chapter 10

Molarity

10.1 INTRODUCTION

Many quantitative chemical reactions are carried out in solution because volumes are easier to measure than masses. If you dissolve a certain mass of a chemical in a measured volume of solution, you can take measured fractions of the total volume of the solution and know how much chemical it contains.

Solute is defined as the material that is dissolved in a *solvent*. For example, if you dissolve sugar in water, the sugar is the solute and the water is the solvent. In the discussions of this chapter, the solvent will be water; that is, we will discuss *aqueous* solutions only.

Perhaps the most useful measure of concentration is *molarity*. Molarity is defined as the number of moles of solute per liter of solution:

$$\text{molarity} = \frac{\text{number of moles of solute}}{\text{liter of solution}}$$

Note particularly that it is liters of *solution* that is used, not liters of *solvent*. Chemists often abbreviate the definition to merely "moles per liter," but the shortening does not change the way the unit is actually defined. The unit of molarity is *molar*, symbolized M. Special care must be taken with abbreviations in this chapter. M stands for molar; mol for mole(s). (Some authors use M to stand for molarity as well as molar. They may use italics for one and not the other. Check your text and follow its convention.)

EXAMPLE 10.1. What is the molarity of the solution produced by dissolving 4.0 mol of solute in enough water to make 2.0 L of solution?

$$\text{molarity} = \frac{4.0 \, \text{mol}}{2.0 \, \text{L}} = 2.0 \, M$$

The answer is stated out loud "The molarity is 2.0 molar," or more commonly, "The solution is 2.0 molar." Note the difference between the word molarity (the quantity) and molar (the unit of molarity).

10.2 MOLARITY CALCULATIONS

Molarity, being a ratio, can be used as a factor. Anywhere M (for molar) is used, it can be replaced by mol/L. The reciprocal of molarity can also be used.

EXAMPLE 10.2. How many moles of solute are contained in 2.0 L of 3.0 M solution?

As usual, the quantity given is put down first and multiplied by a ratio—in this case, the molarity. It is easy to visualize the solution, pictured in Fig. 10-1.

$$2.0 \, \text{L} \left(\frac{3.0 \, \text{mol}}{\text{L}} \right) = 6.0 \, \text{mol}$$

EXAMPLE 10.3. What volume of 3.0 M solution contains 6.0 mol of solute?

$$6.0 \, \text{mol} \left(\frac{1 \, \text{L}}{3.0 \, \text{mol}} \right) = 2.0 \, \text{L}$$

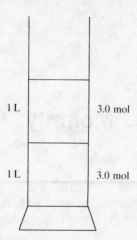

Fig. 10-1   2.0 L of 3.0 M solution

EXAMPLE 10.4.   What is the molarity of a solution prepared by dissolving 117 g NaCl in enough water to make 500 mL of solution?

Molarity is defined in moles per liter. We convert each of the quantities given to those used in the definition:

$$117 \text{ g NaCl}\left(\frac{1 \text{ mol NaCl}}{58.5 \text{ g NaCl}}\right) = 2.00 \text{ mol NaCl}$$

$$500 \text{ mL}\left(\frac{1 \text{ L}}{1\,000 \text{ mL}}\right) = 0.500 \text{ L}$$

$$\text{molarity} = \frac{2.00 \text{ mol}}{0.500 \text{ L}} = 4.00 \text{ M}$$

EXAMPLE 10.5.   What is the concentration of a solution containing 6.35 mmol in 1.00 mL?

$$\frac{6.35 \text{ mmol}\left(\dfrac{1 \text{ mol}}{1\,000 \text{ mmol}}\right)}{1.00 \text{ mL}\left(\dfrac{1 \text{ L}}{1\,000 \text{ mL}}\right)} = \frac{6.35 \text{ mol}}{L} = 6.35 \text{ M} \qquad C? = \frac{m}{V}$$

The numeric value is the same in mmol/mL as in mol/L. Thus, molarity can be defined as the number of millimoles of solute per milliliter of solution.

EXAMPLE 10.6.   How much water was added to the solute in Example 10.1?

We have no way of knowing how much water was added. We know the final volume of the solution, but not the volume of the solvent.

When water is added to a solution, the volume increases but the number of moles of the solutes does not change. The molarity of every solute in the solution therefore decreases.

EXAMPLE 10.7.   What is the final concentration of 2.0 L of 3.0 M solution if enough water is added to dilute the solution to 5.0 L?

The concentration is the number of moles of solute per liter of solution, equal to the number of moles of solute divided by the total volume in liters. The original number of moles of solute does not change:

$$\text{number of moles} = 2.0 \text{ L}\left(\frac{3.0 \text{ mol}}{L}\right) = 6.0 \text{ mol}$$

The final volume is 5.0 L, as stated in the problem.

$$\text{molarity} = \frac{6.0\,\text{mol}}{5.0\,\text{L}} = 1.2\,M$$

EXAMPLE 10.8. What is the final concentration of 2.0 L of 3.0 M solution if 5.0 L of water is added to dilute the solution?

Note the difference in the wording of this example and the last. Here the final volume is 7.0 L. (When you mix solutions, unless they have identical compositions, the final volume might not be exactly equal to the sum of the individual volumes. When only dilute aqueous solutions and water are involved, the volumes are very nearly additive, however.)

$$\text{molarity} \cong \frac{6.0\,\text{mol}}{7.0\,\text{L}} = 0.86\,M$$

EXAMPLE 10.9. (a) A car was driven 30 miles per hour for 1.0 hour and then 40 miles per hour for 1.0 hour. What was the average speed over the whole trip? (b) If 1.0 L of 3.0 M NaCl solution is added to 1.0 L of 4.0 M NaCl solution, what is the final molarity?

(a) The average speed is equal to the total distance divided by the total time. The total time is 2.0 hours. The total distance is

$$\text{distance} = 1.0\,\text{hour}\left(\frac{30\,\text{miles}}{\text{hour}}\right) + 1.0\,\text{hour}\left(\frac{40\,\text{miles}}{\text{hour}}\right) = 70\,\text{miles}$$

$$\text{average speed} = \frac{70\,\text{miles}}{2.0\,\text{hours}} = \frac{35\,\text{miles}}{\text{hour}}$$

Note that we cannot merely add the speeds here to get the speed for the entire trip.

(b) The final concentration is the total number of moles of NaCl divided by the total volume. The total volume is about 2.0 L. The total number of moles is

$$1.0\,\text{L}\left(\frac{3.0\,\text{mol}}{\text{L}}\right) + 1.0\,\text{L}\left(\frac{4.0\,\text{mol}}{\text{L}}\right) = 7.0\,\text{mol}$$

The final concentration is $(7.0\,\text{mol})/(2.0\,\text{L}) = 3.5\,M$. Note that we cannot merely add the concentrations to get the final concentration.

EXAMPLE 10.10. Calculate the final concentration if 2.0 L of 3.0 M NaCl and 4.0 L of 1.5 M NaCl are mixed.

$$\text{final volume} = 6.0\,\text{L}$$

$$\text{final number of moles} = 2.0\,\text{L}\left(\frac{3.0\,\text{mol}}{\text{L}}\right) + 4.0\,\text{L}\left(\frac{1.5\,\text{mol}}{\text{L}}\right) = 12.0\,\text{mol}$$

$$\text{molarity} = \frac{12.0\,\text{mol}}{6.0\,\text{L}} = 2.0\,M$$

Note that the answer is reasonable; the final concentration is between the concentrations of the two original solutions.

EXAMPLE 10.11. Calculate the final concentration if 2.0 L of 3.0 M NaCl, 4.0 L of 1.5 M NaCl, and 4.0 L of water are mixed.

The final volume is about 10.0 L. The final number of moles of NaCl is 12.0 mol, the same as in Example 10.10, since there was no NaCl in the 4.0 L of water. Hence, the final concentration is

$$\text{molarity} = \frac{12.0\,\text{mol}}{10.0\,\text{L}} = 1.20\,M$$

Note that the concentration is lower than it was in Example 10.10 despite the presence of the same number of moles of NaCl, since there is a greater volume.

10.3   MOLARITIES OF IONS

The concentration of ionic solutes can be described in terms of the concentrations of the individual ions in solution. The molarity of Ba^{2+} in 2.0 M $BaCl_2$ is 2.0 M. The molarity of Cl^- in the same solution is 4.0 M, because there are twice as many Cl^- ions as Ba^{2+} ions per unit volume. When different salts containing one ion in common are dissolved in the same solution, the concentration of the individual ions becomes more important.

EXAMPLE 10.12.   (*a*) What are the molarities of the ions in a solution prepared by dissolving 3.0 mol of $AlCl_3$ in enough water to make 1.0 L of solution? (*b*) What are the molarities of the ions in a 3.0 M $AlCl_3$ solution?

(*a*)

$$\frac{3.0 \text{ mol } Al^{3+}}{1.0 \text{ L}} = 3.0 \text{ } M \text{ } Al^{3+}$$

$$\frac{3.0 \text{ mol } AlCl_3}{1.0 \text{ L}} \left(\frac{3 \text{ mol } Cl^-}{\text{mol } AlCl_3} \right) = \frac{9.0 \text{ mol } Cl^-}{\text{L}} = 9.0 \text{ } M \text{ } Cl^-$$

(*b*)   In this wording, the "3.0 M $AlCl_3$" means a solution such as described in part (*a*). There are *no molecules* of $AlCl_3$ in the solution, and so we can use this wording to describe the process in part (*a*). The answer is the same.

EXAMPLE 10.13.   Calculate the concentration of all the ions in solution if 1.0 mol HCl and 2.0 mol NaCl are dissolved in sufficient water to make 6.0 L of a single solution.

These two compounds do not react with each other.
1.0 mol HCl consists of 1.0 mol H^+ and   1.0 mol Cl^-.
2.0 mol NaCl consists of   2.0 mol Na^+ and 2.0 mol Cl^-.
The solution contains   1.0 mol H^+, 2.0 mol Na^+, and 3.0 mol Cl^-.
The concentrations are

$$\frac{1.0 \text{ mol } H^+}{6.0 \text{ L}} = 0.17 M \text{ } H^+ \qquad \frac{3.0 \text{ mol } Cl^-}{6.0 \text{ L}} = 0.50 \text{ } M \text{ } Cl^-$$

$$\frac{2.0 \text{ mol } Na^+}{6.0 \text{ L}} = 0.33 \text{ } M \text{ } Na^+$$

EXAMPLE 10.14.   Calculate the concentrations of all the ions in solution if 3.0 mol KNO_3 and 2.6 mol $Ba(NO_3)_2$ are dissolved in sufficient water to make 2.0 L of a single solution.

These two compounds do not react with each other.
3.0 mol KNO_3 consists of 3.0 mol K^+ and   3.0 mol NO_3^-.
2.6 mol $Ba(NO_3)_2$ consists of   2.6 mol Ba^{2+} and 5.2 mol NO_3^-.
The solution contains   3.0 mol K^+, 2.6 mol Ba^{2+}, and 8.2 mol NO_3^-.
The concentrations are

$$\frac{3.0 \text{ mol } K^+}{2.0 \text{ L}} = 1.5 \text{ } M \text{ } K^+ \qquad \frac{8.2 \text{ mol } NO_3^-}{2.0 \text{ L}} = 4.1 \text{ } M \text{ } NO_3^-$$

$$\frac{2.6 \text{ mol } Ba^{2+}}{2.0 \text{ L}} = 1.3 \text{ } M \text{ } Ba^{2+}$$

EXAMPLE 10.15.   Calculate the final concentration of each ion in solution if 2.0 L of 3.0 M $Mg(NO_3)_2$ is added to 3.0 L of 2.5 M NaCl.

The compounds do not react. The solution contains $6.0 \, \text{mol} \, Mg^{2+}$, $12 \, \text{mol} \, NO_3^-$, $7.5 \, \text{mol} \, Na^+$, and $7.5 \, \text{mol}$ Cl^-. The concentrations are

$$\frac{6.0 \, \text{mol} \, Mg^{2+}}{5.0 \, \text{L}} = 1.2 \, M \, Mg^{2+} \qquad \frac{12 \, \text{mol} \, NO_3^-}{5.0 \, \text{L}} = 2.4 \, M \, NO_3^-$$

$$\frac{7.5 \, \text{mol} \, Na^+}{5.0 \, \text{L}} = 1.5 \, M \, Na^+ \qquad \frac{7.5 \, \text{mol} \, Cl^-}{5.0 \, \text{L}} = 1.5 \, M \, Cl^-$$

10.4   REACTIONS IN SOLUTION

Thus far, we have added only compounds that did not react chemically with each other. If the compounds react, the ions added will be reduced in concentration and the products, if soluble, will be increased in concentration. If a solid is produced, its concentration will not be calculable. If water is produced, it does not affect the concentrations of water or any other reactant or product. The concentration of water in dilute aqueous solutions is so large that a little more added by a reaction is immaterial. The water produced does not materially affect the volume of the solution either, because the ions from which it was made also had volume. The concentration of any spectator ions (Chap. 9) is not affected by the reaction and may be calculated as if no chemical reaction had occurred.

EXAMPLE 10.16.   Calculate the concentration of all ions in solution after $2.0 \, \text{L}$ of $1.5 \, M$ HCl is added to $4.0 \, \text{L}$ of $0.50 \, M$ NaOH.

Before reaction there are

$$2.0 \, \text{L} \left(\frac{1.5 \, \text{mol}}{\text{L}} \right) = 3.0 \, \text{mol} \, HCl \qquad \text{consisting of } 3.0 \, \text{mol} \, H^+ \text{ and } 3.0 \, \text{mol} \, Cl^-$$

$$4.0 \, \text{L} \left(\frac{0.50 \, \text{mol}}{\text{L}} \right) = 2.0 \, \text{mol} \, NaOH \qquad \text{consisting of } 2.0 \, \text{mol} \, OH^- \text{ and } 2.0 \, \text{mol} \, Na^+$$

The H^+ reacts with the OH^- according to the net ionic equation

$$H^+ + OH^- \longrightarrow H_2O$$

The limiting quantity (Sec. 8.3) is OH^-, so $2.0 \, \text{mol} \, OH^-$ reacts with $2.0 \, \text{mol} \, H^+$ to give $2.0 \, \text{mol} \, H_2O$, leaving $1.0 \, \text{mol} \, H^+$ in solution. The Na^+ and Cl^- are spectator ions; they do not react. The final volume is $6.0 \, \text{L}$, and the final concentrations are

$$\frac{1.0 \, \text{mol} \, H^+}{6.0 \, \text{L}} = 0.17 M \, H^+$$

$$\frac{2.0 \, \text{mol} \, Na^+}{6.0 \, \text{L}} = 0.33 \, M \, Na^+$$

$$\frac{3.0 \, \text{mol} \, Cl^-}{6.0 \, \text{L}} = 0.50 \, M \, Cl^-$$

Alternatively, one could have calculated that after the reaction there was $2.0 \, M$ NaCl and $1.0 \, M$ HCl in the $6.0 \, \text{L}$ of solution (Example 8.8). The ion concentrations then could have been calculated as in Example 10.13, and the results just shown would have been obtained.

EXAMPLE 10.17.   Calculate the final concentration of all ions in solution after $2.00 \, \text{L}$ of $1.30 \, M$ $Ba(OH)_2$ is treated with $3.00 \, \text{L}$ of $2.00 \, M$ HCl.

$$(2.00 \, \text{L})(1.30 \, \text{mol/L}) = 2.60 \, \text{mol} \, Ba(OH)_2 \qquad \text{consisting of } 2.60 \, \text{mol} \, Ba^{2+} \text{ and } 5.20 \, \text{mol} \, OH^-$$

$$(3.00 \, \text{L})(2.00 \, \text{mol/L}) = 6.00 \, \text{mol} \, HCl \qquad \text{consisting of } 6.00 \, \text{mol} \, Cl^- \text{ and } 6.00 \, \text{mol} \, H^+$$

The H^+ reacts with the OH^- according to the net ionic equation

$$H^+ + OH^- \longrightarrow H_2O$$

The limiting quantity is OH^-, so 5.20 mol OH^- reacts with 5.20 mol H^+ to give 5.20 mol H_2O, leaving 0.80 mol H^+ in solution. The final volume is 5.00 L, and the final concentrations are

$$\frac{0.80 \text{ mol } H^+}{5.00 \text{ L}} = 0.16 \, M \, H^+$$

$$\frac{2.60 \text{ mol } Ba^{2+}}{5.00 \text{ L}} = 0.520 \, M \, Ba^{2+}$$

$$\frac{6.00 \text{ mol } Cl^-}{5.00 \text{ L}} = 1.20 \, M \, Cl^-$$

10.5 TITRATION

More often than adding two solutions of known concentration and volume, we treat one solution of known concentration and volume with another solution until the mole ratio is exactly what is required by the balanced chemical equation. Then from the known volumes of both reactants, the concentration of the second reactant can be calculated. In this manner, we can determine the concentrations of solutions without weighing out the solute, which is often difficult or even impossible. (For example, some reactants absorb water from the air so strongly that you do not know how much reactant and how much water you are weighing.) The procedure used is called *titration*. In a typical titration experiment, 25.00 mL of 2.000 M HCl is pipetted into a conical flask. A *pipet* (Fig. 10-2) is a piece of glassware that is calibrated to deliver an exact volume of liquid. A solution of NaOH of unknown concentration is placed in a *buret* (Fig. 10-2), and some is allowed to drain out the bottom to ensure that the portion below the stopcock is filled. The buret volume is read before any NaOH is added from it to the HCl (say 3.30 mL) and again after the NaOH is added (say 45.32 mL). The volume of added NaOH is merely the difference in readings (45.32 mL − 3.30 mL = 42.02 mL). The concentration of NaOH may now be calculated because the exact number of moles of NaOH has been added to exactly react with the HCl.

$$25.00 \text{ mL HCl} \left(\frac{2.000 \text{ mmol HCl}}{\text{mL HCl}} \right) = 50.00 \text{ mmol HCl}$$

The number of millimoles of NaOH is exactly the same, since the addition was stopped when the reaction was just complete. Therefore, there is 50.00 mmol NaOH in the volume of NaOH added,

$$\frac{50.00 \text{ mmol NaOH}}{42.02 \text{ mL}} = 1.190 \, M \, \text{NaOH}$$

Two questions should immediately arise. (*a*) How can you know exactly when to stop the titration so that the number of moles of NaOH is equal to the number of moles of HCl? (*b*) What is the use of determining the concentration of a solution of NaOH when the NaOH has now been used up reacting with the HCl? (*a*) An *indicator* is used to tell us when to stop the titration. Typically an indicator is a compound that is one color (or colorless) in an acidic solution and a second color in a basic solution. Thus, we add NaOH slowly, drop by drop toward the end, until a permanent color change takes place. At that point, the *end point* has been reached, and the titration is complete. The HCl has just been used up. (*b*) The purpose of doing a titration is to determine the concentration of a solution. If the concentration of a liter of NaOH is to be determined, a small portion of it is used in the titration. The rest has the same concentration, and although the part used in the titration is no longer useful, the concentration of the bulk of the solution is now known.

EXAMPLE 10.18. What is the concentration of a $Ba(OH)_2$ solution if it takes 43.50 mL to neutralize 25.00 mL of 1.453 M HCl?

The number of moles of HCl is easily calculated.

$$0.02500 \text{ L} \left(\frac{1.453 \text{ mol}}{\text{L}} \right) = 0.03632 \text{ mol HCl}$$

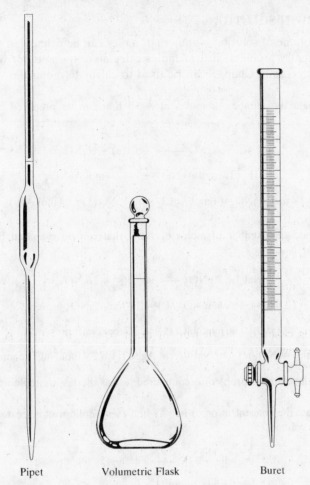

Pipet   Volumetric Flask   Buret

Fig. 10-2   Volumetric glassware
The pipet is designed to deliver an exact volume of liquid; the volumetric flask is designed to hold an
exact volume of liquid; and the buret is designed to deliver precisely measurable volumes of liquid.
(Not drawn to scale.)

The balanced chemical equation shows that the ratio of moles of HCl to $Ba(OH)_2$ is $2:1$.

$$2\,HCl + Ba(OH)_2 \longrightarrow BaCl_2 + 2\,H_2O$$

$$0.03632\,mol\ HCl\left(\frac{1\,mol\ Ba(OH)_2}{2\,mol\ HCl}\right) = 0.01816\,mol\ Ba(OH)_2$$

The molarity is given by

$$\frac{0.01816\,mol\ Ba(OH)_2}{0.04350\,L} = 0.4175\,M\ Ba(OH)_2$$

Note that you cannot calculate this concentration with the equation

$$M_1V_1 = M_2V_2 \qquad \text{(limited applications)}$$

where M is molarity and V is volume. This equation is given in some texts for simple dilution problems. The
equation reduces to $\text{moles}_1 = \text{moles}_2$, which is not true in cases in which there is not a $1:1$ ratio in the balanced
chemical equation.

10.6   STOICHIOMETRY IN SOLUTION

With molarity and volume of solution, numbers of moles can be calculated. The numbers of moles may be used in stoichiometry problems just as moles calculated in any other way are used. Also, the number of moles calculated as in Chap. 8 can be used to calculate molarities or volumes of solution.

EXAMPLE 10.19. Calculate the number of moles of AgCl that can be prepared by mixing 2.0 L of 1.2 M $AgNO_3$ with excess NaCl.

$$NaCl + AgNO_3 \longrightarrow AgCl(s) + NaNO_3$$

$$(2.0\,L)(1.2\,mol\ AgNO_3/L) = 2.4\,mol\ AgNO_3$$

$$(2.4\,mol\ AgNO_3)(1\,mol\ AgCl/mol\ AgNO_3) = 2.4\,mol\ AgCl$$

EXAMPLE 10.20. Calculate the number of moles of AgCl that can be prepared by mixing 2.0 L of 1.2 M $AgNO_3$ with 3.0 L of 0.90 M NaCl.

$$NaCl + AgNO_3 \longrightarrow AgCl(s) + NaNO_3$$

$$(2.0\,L)(1.2\,mol\ AgNO_3/L) = 2.4\,mol\ AgNO_3\ present$$

$$(3.0\,L)(0.90\,mol\ NaCl/L) = 2.7\,mol\ NaCl\ present$$

$$(2.4\,mol\ AgNO_3)(1\,mol\ NaCl/mol\ AgNO_3) = 2.4\,mol\ NaCl\ required$$

$AgNO_3$ is in limiting quantity, and the problem is completed just as the last example was done.

EXAMPLE 10.21. Calculate the concentration of Fe^{2+} when excess solid iron is treated with 50.0 mL of 6.00 M HCl. Assume no change in volume.

$$Fe + 2H^+ \longrightarrow Fe^{2+} + H_2$$

$$50.0\,mL\left(\frac{6.00\,mmol\ H^+}{mL}\right) = 300\,mmol\ H^+$$

$$300\,mmol\ H^+\left(\frac{1\,mmol\ Fe^{2+}}{2\,mmol\ H^+}\right) = 150\,mmol\ Fe^{2+}$$

$$\frac{150\,mmol\ Fe^{2+}}{50.0\,mL} = 3.00\ M\ Fe^{2+}$$

Solved Problems

INTRODUCTION

10.1. Which, if either, has more sugar in it: (a) a half cup of tea with one lump of sugar or (b) a whole cup of tea with two lumps of sugar?

   Ans.   Two lumps is more than one; the whole cup has more sugar. Note the difference between *amount* of sugar and *concentration* of sugar.

10.2. Which, if either, of the following tastes sweeter: (a) a half cup of tea with one lump of sugar or (b) a whole cup of tea with two lumps of sugar?

Ans. They both taste equally sweet, since their concentrations are equal.

MOLARITY CALCULATIONS

10.3. Calculate the molarity of each of the following solutions: (*a*) 3.0 mol solute in 2.0 L of solution, (*b*) 4.0 mol solute in 800 mL of solution, (*c*) 0.30 mol solute in 0.12 L of solution.

Ans.

(*a*)
$$\frac{3.0\,\text{mol}}{2.0\,\text{L}} = 1.5\,M$$

(*b*)
$$\frac{4.0\,\text{mol}}{0.800\,\text{L}} = 5.0\,M$$

(Note: moles per liter, not moles per milliliter)

(*c*)
$$\frac{0.30\,\text{mol}}{0.12\,\text{L}} = 2.5\,M$$

10.4. A 100-mL solution contains 24.0 mmol of solute. What is its molarity?

Ans.

$$\frac{24.0\,\text{mmol}}{100\,\text{mL}} = 0.240\,M$$

Molarity can be calculated by dividing millimoles by milliliters.

10.5. Calculate the number of moles of solute in each of the following solutions: (*a*) 3.0 L of 1.5 *M* solution, (*b*) 2.22 L of 0.750 *M* solution, (*c*) 250 mL of 1.50 *M* solution, and (*d*) 25.0 mL of 2.00 *M* solution.

Ans.

(*a*)
$$3.0\,\text{L}\left(\frac{1.5\,\text{mol}}{\text{L}}\right) = 4.5\,\text{mol}$$

(*b*)
$$2.22\,\text{L}\left(\frac{0.750\,\text{mol}}{\text{L}}\right) = 1.66\,\text{mol}$$

(*c*)
$$0.250\,\text{L}\left(\frac{1.50\,\text{mol}}{\text{L}}\right) = 0.375\,\text{mol}$$

(*d*)
$$0.0250\,\text{L}\left(\frac{2.00\,\text{mol}}{\text{L}}\right) = 0.0500\,\text{mol}$$

10.6. How can you make 2.0 L of 3.0 *M* sugar solution?

Ans. The solution will contain 6.0 mol sugar (see Example 10.2). Thus, place 6.0 mol sugar in a liter or so of water, mix until dissolved, dilute the resulting solution to 2.0 L, and mix thoroughly.

10.7. Calculate the volume of 1.5 *M* solution required to contain 3.0 mol of solute.

Ans.

$$3.0\,\text{mol}\left(\frac{1\,\text{L}}{1.5\,\text{mol}}\right) = 2.0\,\text{L}$$

10.8. Calculate the number of milliliters of 2.50 *M* solution required to contain 0.330 mol of solute.

Ans.

$$0.330\,\text{mol}\left(\frac{1\,\text{L}}{2.50\,\text{mol}}\right) = 0.132\,\text{L} = 132\,\text{mL}$$

10.9. Calculate the volume of 0.750 M solution required to contain 1.30 mol of solute.

Ans.

$$1.30 \, \text{mol} \left(\frac{1 \, \text{L}}{0.750 \, \text{mol}} \right) = 1.73 \, \text{L}$$

10.10. What is the concentration of a solution prepared by diluting 2.0 L of 4.0 M solution to 9.0 L with water?

Ans.　The number of moles of solute is not changed by addition of the water. The number of moles in the original solution is

$$2.0 \, \text{L} \left(\frac{4.0 \, \text{mol}}{\text{L}} \right) = 8.0 \, \text{mol}$$

That 8.0 mol is now dissolved in 9.0 L, and its concentration is

$$\frac{8.0 \, \text{mol}}{9.0 \, \text{L}} = 0.89 \, \dot{M}$$

10.11. What is the concentration of a solution prepared by diluting 1.50 L of 3.00 M solution to 4.50 L with solvent?

Ans.　The number of moles of solute is not changed by addition of the solvent. The number of moles in the original solution is

$$1.50 \, \text{L} \left(\frac{3.00 \, \text{mol}}{\text{L}} \right) = 4.50 \, \text{mol}$$

That 4.50 mol is now dissolved in 4.50 L (compare Problem 10.12), and its concentration is

$$\frac{4.50 \, \text{mol}}{4.50 \, \text{L}} = 1.00 \, M$$

10.12. What is the concentration of a solution prepared by diluting 1.50 L of 3.00 M solution with 4.50 L of solvent?

Ans.　The number of moles of solute is not changed by addition of the solvent. The number of moles in the original solution is

$$1.50 \, \text{L} \left(\frac{3.00 \, \text{mol}}{\text{L}} \right) = 4.50 \, \text{mol}$$

What is the final volume of the solution? If we add 4.50 L of solvent, it will be about 6.00 L. Compare this wording with that of Problem 10.11. The 4.50 mol is now dissolved in 6.00 L, and its concentration is

$$\frac{4.50 \, \text{mol}}{6.00 \, \text{L}} = 0.750 \, M$$

10.13. What is the concentration of a solution prepared by diluting 250 mL of 3.00 M solution to 600 mL?

Ans.　The number of moles of solute is not changed by addition of the solvent. The number of moles in the original solution is

$$(0.250 \, \text{L})(3.00 \, \text{mol/L}) = 0.750 \, \text{mol}$$

That 0.750 mol is now dissolved in 0.600 L, and its concentration is

$$\frac{0.750 \, \text{mol}}{0.600 \, \text{L}} = 1.25 \, M$$

Alternatively, the number of millimoles is given by

$$(250 \, \text{mL})(3.00 \, \text{mmol/mL}) = 750 \, \text{mmol}$$

The concentration is

$$\frac{750\,\text{mmol}}{600\,\text{mL}} = 1.25\,M$$

The use of millimoles and milliliters saves conversions of the milliliters to liters.

10.14. What concentration of salt is obtained by mixing 200 mL of 3.0 M salt solution with 300 mL of 2.0 M salt solution?

Ans. The final concentration is the total number of moles divided by the total number of liters. The volume is 0.200 L + 0.300 L = 0.500 L. The total number of moles is given by

$$(0.200\,\text{L})(3.0\,\text{mol/L}) + (0.300\,\text{L})(2.0\,\text{mol/L}) = 1.20\,\text{mol}$$

The concentration is 1.20 mol/0.500 L = 2.40 M.

10.15. What concentration of salt is obtained by mixing 200 mL of 3.0 M salt solution with 300 mL of 2.0 M salt solution and diluting with water to 1.00 L?

Ans. The final concentration is the total number of moles divided by the total number of liters. The volume is 1.00 L. Since there is no solute in the water, the total number of moles is given by

$$(0.200\,\text{L})(3.0\,\text{mol/L}) + (0.300\,\text{L})(2.0\,\text{mol/L}) = 1.20\,\text{mol}$$

The concentration is 1.20 mol/1.00 L = 1.20 M. The concentration is lower than that in Problem 10.14 despite the same number of moles of solute, because of the greater volume.

10.16. Calculate the concentration of sugar in a solution prepared by mixing 2.0 L of 3.0 M sugar with 2.5 L of 4.0 M salt.

Ans. The sugar concentration is given by dividing the number of moles of sugar by the total volume. The moles of salt makes no difference in this problem because the problem does not ask about salt concentration and the salt does not react. This is simply a dilution problem for the sugar.

$$(2.0\,\text{L})(3.0\,\text{mol/L}) = 6.0\,\text{mol sugar}$$

$$\frac{6.0\,\text{mol sugar}}{4.5\,\text{L}} = 1.3\,M\ \text{sugar}$$

10.17. How many moles of solute are present in 325 mL of 0.894 M solution?

Ans.

$$(0.325\,\text{L})(0.894\,\text{mol/L}) = 0.291\,\text{mol}$$

10.18. What is the concentration of a solution prepared by dissolving 25.0 g NaCl in sufficient water to make 540 mL of solution?

Ans. Molarity is in moles per liter. We must change the grams of NaCl to moles and the milliliters of solution to liters.

$$25.0\,\text{g NaCl}\left(\frac{1\,\text{mol NaCl}}{58.5\,\text{g NaCl}}\right) = 0.427\,\text{mol}$$

$$540\,\text{mL}\left(\frac{1\,\text{L}}{1\,000\,\text{mL}}\right) = 0.540\,\text{L}$$

$$(0.427\,\text{mol})/(0.540\,\text{L}) = 0.791\,M$$

10.19. What volume of 2.00 M NaCl solution contains 85.0 g NaCl?

Ans.

$$85.0\,\text{g NaCl}\left(\frac{1\,\text{mol NaCl}}{58.5\,\text{g NaCl}}\right) = 1.45\,\text{mol NaCl}$$

$$1.45\,\text{mol NaCl}\left(\frac{1\,\text{L solution}}{2.00\,\text{mol NaCl}}\right) = 0.725\,\text{L} = 725\,\text{mL}$$

10.20. How many grams of NaCl are present in 548 mL of 0.944 M NaCl?

Ans.

$$(0.548\,\text{L})(0.944\,\text{mol/L}) = 0.517\,\text{mol}$$

$$(0.517\,\text{mol})(58.5\,\text{g NaCl/mol}) = 30.2\,\text{g NaCl}$$

10.21. How many milligrams of NaOH are present in 25.0 mL of 1.63 M NaOH?

Ans.

$$25.0\,\text{mL}\left(\frac{1.63\,\text{mmol}}{\text{mL}}\right)\left(\frac{40.0\,\text{mg}}{\text{mmol}}\right) = 1\,630\,\text{mg}$$

MOLARITIES OF IONS

10.22. Calculate the molarity of each ion in solution in (*a*) 2.0 M NaCl, (*b*) 2.0 M BaCl$_2$, (*c*) 2.0 M AlCl$_3$, (*d*) 2.0 M Cr(ClO$_3$)$_3$, and (*e*) 0.20 M Fe$_2$(SO$_4$)$_3$.

Ans. (*a*) 2.0 M NaCl is 2.0 M Na$^+$ and 2.0 M Cl$^-$. (*b*) 2.0 M BaCl$_2$ is 2.0 M Ba^{2+} and 4.0 M Cl$^-$. (There is twice the concentration of Cl$^-$ as Ba^{2+}.) (*c*) 2.0 M AlCl$_3$ is 2.0 M Al^{3+} and 6.0 M Cl$^-$. (*d*) 2.0 M Cr(ClO$_3$)$_3$ is 2.0 M Cr^{3+} and 6.0 M ClO$_3^-$. (*e*) 0.20 M Fe$_2$(SO$_4$)$_3$ is 0.40 M Fe^{3+} and 0.60 M SO$_4^{2-}$.

10.23. What is the chloride ion concentration in a solution prepared by dissolving 33.0 g BaCl$_2$ in sufficient water to make 250 mL of solution?

Ans.

$$33.0\,\text{g BaCl}_2\left(\frac{1\,\text{mol BaCl}_2}{208\,\text{g BaCl}_2}\right) = 0.159\,\text{mol BaCl}_2$$

$$0.159\,\text{mol BaCl}_2\left(\frac{2\,\text{mol Cl}^-}{\text{mol BaCl}_2}\right) = 0.318\,\text{mol Cl}^-$$

$$\frac{0.318\,\text{mol Cl}^-}{0.250\,\text{L}} = 1.27\,M\ \text{Cl}^-$$

10.24. In Example 10.15, the total concentration of positive ions does not equal the total concentration of negative ions. What is true about these concentrations?

Ans. The total concentration of positive charge is equal to the total concentration of negative charge. That is, in this case, twice the concentration of Mg^{2+} (since it is 2+) plus the concentration of Na$^+$ is equal to the sum of the concentrations of Cl$^-$ and NO$_3^-$:

$$2(1.2\,M) + 1.5\,M = 2.4\,M + 1.5\,M$$

10.25. Calculate the concentration of each ion in solution if 2.0 L of 3.0 M NaCl is mixed with 1.0 L of 1.5 M BaCl$_2$.

Ans. The 6.0 mol NaCl consists of 6.0 mol Na$^+$ and 6.0 mol Cl$^-$.
The 1.5 mol BaCl$_2$ consists of   3.0 mol Cl$^-$ and 1.5 mol Ba^{2+}.
The total number of moles of Cl$^-$ is   9.0 mol.
The concentrations are

$$\frac{6.0\,\text{mol Na}^+}{3.0\,\text{L}} = 2.0\,M\ \text{Na}^+ \qquad\qquad \frac{1.5\,\text{mol Ba}^{2+}}{3.0\,\text{L}} = 0.50\,M\ \text{Ba}^{2+}$$

$$\frac{9.0\,\text{mol Cl}^-}{3.0\,\text{L}} = 3.0\,M\ \text{Cl}^-$$

Note particularly that the numbers of moles of chloride ion from the two solutions are added, but that the moles of sodium ion and the moles of barium ion are not added.

10.26. Calculate the concentration of each ion in solution if 2.0 L of 3.0 M $NaClO_2$ is mixed with 1.0 L of 1.5 M $Ba(ClO)_2$.

    Ans.　The 6.0 mol $NaClO_2$ consists of 6.0 mol Na^+ and 6.0 mol ClO_2^-. The 1.5 mol $Ba(ClO)_2$ consists of 1.5 mol Ba^{2+} and 3.0 mol ClO^-. There are four different ions present. The concentrations are

$$\frac{6.0\,\text{mol Na}^+}{3.0\,\text{L}} = 2.0\,M\ \text{Na}^+ \qquad \frac{6.0\,\text{mol ClO}_2^-}{3.0\,\text{L}} = 2.0\,M\ \text{ClO}_2^-$$

$$\frac{1.5\,\text{mol Ba}^{2+}}{3.0\,\text{L}} = 0.50\,M\ \text{Ba}^{2+} \qquad \frac{3.0\,\text{mol ClO}^-}{3.0\,\text{L}} = 1.0\,M\ \text{ClO}^-$$

10.27. Calculate the concentration of each ion in solution if 2.0 L of 3.0 M NaCl and 2.0 L of 0.50 M $CuCl_2$ are diluted with water to 5.0 L.

    Ans.　There are 6.0 mol Na^+, 1.0 mol Cu^{2+}, and $(6.0 + 2.0)$ mol Cl^- present. The concentrations are

$$\frac{6.0\,\text{mol Na}^+}{5.0\,\text{L}} = 1.2\,M\ \text{Na}^+$$

$$\frac{1.0\,\text{mol Cu}^{2+}}{5.0\,\text{L}} = 0.20\,M\ \text{Cu}^{2+}$$

$$\frac{8.0\,\text{mol Cl}^-}{5.0\,\text{L}} = 1.6\,M\ \text{Cl}^-$$

REACTIONS IN SOLUTION

10.28. Calculate the concentration of each ion in solution when 2.00 L of 0.561 M HCl and 1.50 L of 2.31 M NaOH are mixed. Assume that the volume of the final solution is the sum of the two initial volumes.

    Ans.

$$(2.00\,\text{L})(0.561\,\text{mol/L}) = 1.12\,\text{mol HCl}$$

$$(1.50\,\text{L})(2.31\,\text{mol/L}) = 3.46\,\text{mol NaOH}$$

$$\text{NaOH} + \text{HCl} \longrightarrow \text{NaCl} + \text{H}_2\text{O}$$

We have 1.12 mol Cl^-, 3.46 mol Na^+, 0 mol H^+, since it is all used up, and $3.46 - 1.12 = 2.34$ mol OH^- in excess. The concentrations are

$$\frac{1.12\,\text{mol Cl}^-}{3.50\,\text{L}} = 0.320\,M\ \text{Cl}^-$$

$$\frac{3.46\,\text{mol Na}^+}{3.50\,\text{L}} = 0.989\,M\ \text{Na}^+$$

$$\frac{2.34\,\text{mol OH}^-}{3.50\,\text{L}} = 0.669\,M\ \text{OH}^-$$

TITRATION

10.29. A 25.00-mL sample of 1.000 M HCl is titrated with 43.06 mL of NaOH. What is the concentration of the base?

    Ans.

$$(25.00\,\text{mL HCl})(1.000\,\text{mmol/mL}) = 25.00\,\text{mmol HCl}$$

Since the reagents react in a 1:1 ratio, there is 25.00 mmol NaOH in the 43.06 mL of base.

$$\frac{25.00\,\text{mmol NaOH}}{43.06\,\text{mL}} = 0.5806\,M\ \text{NaOH}$$

10.30. A 25.00-mL sample of 1.000 M H_2SO_4 is titrated with 43.06 mL of NaOH until both hydrogen atoms of each molecule of the acid are just neutralized. What is the concentration of the base?

Ans.

$$(25.00 \text{ mL } H_2SO_4)(1.000 \text{ mmol/mL}) = 25.00 \text{ mmol } H_2SO_4$$

$$H_2SO_4 + 2 NaOH \longrightarrow Na_2SO_4 + 2 H_2O$$

$$25.00 \text{ mmol } H_2SO_4 \left(\frac{2 \text{ mmol NaOH}}{\text{mmol } H_2SO_4} \right) = 50.00 \text{ mmol NaOH}$$

$$\frac{50.00 \text{ mmol NaOH}}{43.06 \text{ mL}} = 1.161 \ M \text{ NaOH}$$

10.31. (*a*) A solid acid containing one H atom per molecule is titrated with 1.000 M NaOH. If 47.44 mL of base is used in the titration, how many moles of base are present? (*b*) How many moles of acid? (*c*) If the mass of the acid was 5.789 g, what is the molecular weight of the acid?

Ans.

(*a*) $$(47.44 \text{ mL})(1.000 \text{ mmol/mL}) = 47.44 \text{ mmol base}$$

(*b*) Since the acid has one hydrogen atom per molecule, it will react in a 1:1 ratio with the base. The equation might be written as

$$HX + NaOH \longrightarrow NaX + H_2O$$

where X stands for the anion of the acid, whatever it might be (just as x is often used for an unknown in algebra). The quantity of acid is therefore 47.44 mmol, or 0.04744 mol.

(*c*) $$\text{molecular weight} = \frac{5.789 \text{ g}}{0.04744 \text{ mol}} = 122.0 \text{ g/mol}$$

STOICHIOMETRY IN SOLUTION

10.32. Calculate the number of grams of $BaSO_4$ that can be prepared by treating 3.00 L of 0.253 M $BaCl_2$ with excess Na_2SO_4.

Ans.

$$BaCl_2 + Na_2SO_4 \longrightarrow BaSO_4 + 2 NaCl$$

$$(3.00 \text{ L})(0.253 \text{ mol } BaCl_2/\text{L}) = 0.759 \text{ mol } BaCl_2$$

$$0.759 \text{ mol } BaCl_2 \left(\frac{1 \text{ mol } BaSO_4}{\text{mol } BaCl_2} \right) \left(\frac{233 \text{ g } BaSO_4}{\text{mol } BaSO_4} \right) = 177 \text{ g } BaSO_4$$

10.33. What mass of $H_2C_2O_4$ can react with 500 mL of 2.50 M $KMnO_4$ according to the following equation?

$$5 H_2C_2O_4 + 2 KMnO_4 + 3 H_2SO_4 \longrightarrow 2 MnSO_4 + K_2SO_4 + 10 CO_2 + 8 H_2O$$

Ans.

$$(0.500 \text{ L})(2.50 \text{ mol/L}) = 1.25 \text{ mol } KMnO_4$$

$$1.25 \text{ mol } KMnO_4 \left(\frac{5 \text{ mol } H_2C_2O_4}{2 \text{ mol } KMnO_4} \right) \left(\frac{90.0 \text{ g } H_2C_2O_4}{\text{mol } H_2C_2O_4} \right) = 281 \text{ g } H_2C_2O_4$$

10.34. When 200 mL of 2.21 M AgNO$_3$ is added to 350 mL of 0.912 M CuCl$_2$, how many grams of AgCl will be produced?

Ans.

$$0.200 \, \text{L} \left(\frac{2.21 \, \text{mol AgNO}_3}{\text{L}} \right) \left(\frac{1 \, \text{mol Ag}^+}{\text{mol AgNO}_3} \right) = 0.442 \, \text{mol Ag}^+ \, \text{present}$$

$$0.350 \, \text{L} \left(\frac{0.912 \, \text{mol CuCl}_2}{\text{L}} \right) \left(\frac{2 \, \text{mol Cl}^-}{\text{mol CuCl}_2} \right) = 0.638 \, \text{mol Cl}^- \, \text{present}$$

$$\text{Ag}^+ + \text{Cl}^- \longrightarrow \text{AgCl}$$

The chloride is in excess. The 0.442 mol Ag$^+$ will produce 0.442 mol AgCl.

$$(0.442 \, \text{mol AgCl})(143 \, \text{g AgCl/mol AgCl}) = 63.2 \, \text{g AgCl}$$

10.35. What concentration of NaCl will be produced when 2.00 L of 1.11 M HCl and 500 mL of 4.44 M NaOH are mixed? Assume that the volume of the final solution is the sum of the two initial volumes.

Ans.

$$(2.00 \, \text{L})(1.11 \, \text{mol/L}) = 2.22 \, \text{mol HCl}$$

$$(0.500 \, \text{L})(4.44 \, \text{mol/L}) = 2.22 \, \text{mol NaOH}$$

$$\text{NaOH} + \text{HCl} \longrightarrow \text{NaOH} + \text{H}_2\text{O}$$

2.22 mol NaCl will be produced, in 2.50 L:

$$\frac{2.22 \, \text{mol}}{2.50 \, \text{L}} = 0.888 \, M$$

Supplementary Problems

10.36. What is the percent by mass of NaCl in 1.82 M NaCl solution? Assume that the solution has a density of 1.07 g/mL.

Ans.   Percent by mass is the number of grams of NaCl in 100 g solution.

$$\frac{1.82 \, \text{mol NaCl}}{\text{L}} \left(\frac{58.5 \, \text{g NaCl}}{\text{mol NaCl}} \right) \left(\frac{1 \, \text{L}}{1\,070 \, \text{g solution}} \right) = \frac{0.0995 \, \text{g NaCl}}{\text{g solution}}$$

For 100 g of solution, multiply the numerator and denominator by 100:

$$\frac{0.0995 \, \text{g NaCl}}{\text{g solution}} = \frac{9.95 \, \text{g NaCl}}{100 \, \text{g solution}} = 9.95\% \, \text{NaCl}$$

10.37. Describe in detail how you would prepare 250.0 mL of 2.000 M NaCl solution.

Ans.   First, figure out how much NaCl you need:

$$(0.2500 \, \text{L})(2.000 \, \text{mol/L}) = 0.5000 \, \text{mol}$$

Since the laboratory balance does not weigh out in moles, convert this quantity to grams:

$$(0.5000 \, \text{mol NaCl})(58.45 \, \text{g/mol}) = 29.22 \, \text{g NaCl}$$

Weigh out 29.22 g of NaCl and dissolve it in a portion of water in a 250-mL volumetric flask (Fig. 10-2). After the salt has dissolved, dilute the solution with water until the volume reaches the calibration mark on the flask (250.0 mL). Mix the solution thoroughly by inverting and shaking the stoppered flask several times.

10.38. What is the concentration of all ions in solution after 2.0 L of 3.0 M NaOH is added to 4.0 L of 2.0 M HCl.

Ans.   Present before the reaction are

$$6.0 \, \text{mol NaOH} \qquad \text{consisting of } 6.0 \, \text{mol Na}^+ \text{ and } 6.0 \, \text{mol OH}^-$$

$$8.0 \, \text{mol HCl(aq)} \qquad \text{consisting of } 8.0 \, \text{mol H}^+ \text{ and } 8.0 \, \text{mol Cl}^-$$

The H^+ and OH^- react according to the net ionic equation

$$H^+ + OH^- \longrightarrow H_2O$$

leaving 0 mol OH^- and 2.0 mol H^+ in the solution. Since the Na^+ and Cl^- are spectator ions, the number of moles of these ions does not change. The final concentrations then are

$$\frac{6.0 \, \text{mol Na}^+}{6.0 \, \text{L}} = 1.0 \, M \, \text{Na}^+$$

$$\frac{8.0 \, \text{mol Cl}^-}{6.0 \, \text{L}} = 1.3 \, M \, \text{Cl}^-$$

$$\frac{2.0 \, \text{mol H}^+}{6.0 \, \text{L}} = 0.33 \, M \, \text{H}^+$$

10.39. What is the concentration of all ions in solution after 2.00 L of 0.300 M $Ba(OH)_2$ is added to 4.00 L of 0.200 M HCl.

Ans.   Present before the reaction are

$$0.600 \, \text{mol Ba(OH)}_2 \qquad \text{consisting of } 0.600 \, \text{mol of Ba}^{2+} \text{ and } 1.20 \, \text{mol OH}^-$$

$$0.800 \, \text{mol HCl(aq)} \qquad \text{consisting of } 0.800 \, \text{mol H}^+ \text{ and } 0.800 \, \text{mol Cl}^-$$

The H^+ and OH^- react according to the net ionic equation

$$H^+ + OH^- \longrightarrow H_2O$$

leaving 0.40 mol OH^- and 0 mol H^+ in the solution. Since the Ba^{2+} and Cl^- are spectator ions, the number of moles of these ions does not change. The final concentrations then are

$$\frac{0.600 \, \text{mol Ba}^{2+}}{6.00 \, \text{L}} = 0.100 \, M \, \text{Ba}^{2+}$$

$$\frac{0.800 \, \text{mol Cl}^-}{6.00 \, \text{L}} = 0.133 \, M \, \text{Cl}^-$$

$$\frac{0.40 \, \text{mol OH}^-}{6.00 \, \text{L}} = 0.067 \, M \, \text{OH}^-$$

10.40. Calculate the acetate ion concentration in a solution prepared by dissolving 0.500 mol sodium acetate and 0.250 mol acetic acid in sufficient water to make 2.00 L.

Ans.   The acetic acid is weak (Sec. 6.4), and provides essentially no acetate ions to the solution. The acetate ion is provided only by the sodium acetate, and its concentration therefore is (0.500 mol)/(2.00 L) = 0.250 M.

10.41. Add boxes to Fig. 8-6 to show the conversions involving volumes of solutions of solutes A and B.

Ans.   See Fig. 10-3.

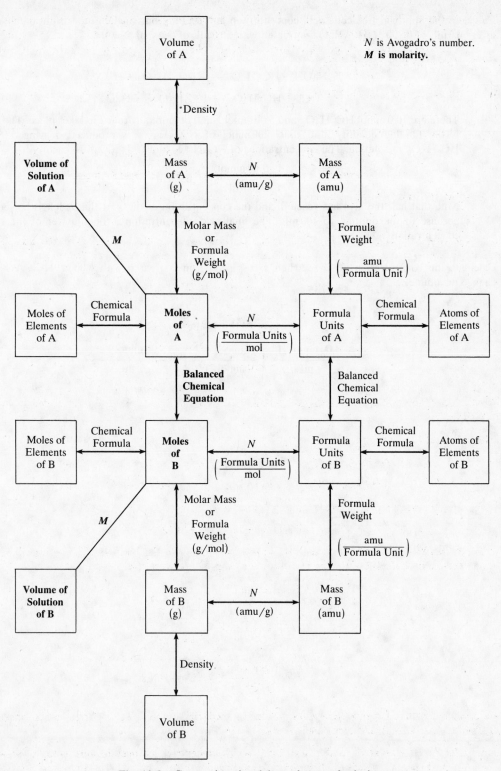

Fig. 10-3 Conversions involving volumes of solutions

10.42. Calculate the approximate acetic acid concentration and acetate ion concentration in the final solution after 50.0 mL of 0.200 M $HC_2H_3O_2$ is treated with 50.0 mL of 0.100 M NaOH.

Ans.

$$NaOH + HC_2H_3O_2 \longrightarrow NaC_2H_3O_2 + H_2O$$

or

$$OH^- + HC_2H_3O_2 \longrightarrow C_2H_3O_2^- + H_2O$$

There are 10.0 mmol $HC_2H_3O_2$ and 5.00 mmol NaOH before any reaction takes place. The NaOH is the limiting quantity, and thus 5.00 mmol of $NaC_2H_3O_2$ is produced, leaving 5.00 mmol $HC_2H_3O_2$ in solution. The concentrations of $HC_2H_3O_2$ and $C_2H_3O_2^-$ are each

$$\frac{5.00 \, \text{mmol}}{100 \, \text{mL}} = 0.0500 \, M$$

Note that half the acid has reacted, and the concentration of the part that is left is halved again because of the dilution to 100 mL. The final acid concentration is one-quarter of the original concentration.

10.43. Calculate the concentration of Fe^{2+} when excess solid iron is treated with 6.00 M HCl. Assume no change in volume.

Ans.

$$Fe + 2H^+ \longrightarrow Fe^{2+} + H_2$$

$$\frac{6.00 \, \text{mmol} \, H^+}{\text{mL}} \left(\frac{1 \, \text{mmol} \, Fe^{2+}}{2 \, \text{mmol} \, H^+} \right) = \frac{3.00 \, \text{mmol} \, Fe^{2+}}{\text{mL}}$$

$$= 3.00 \, M \, Fe^{2+}$$

Chapter 11

Gases

11.1 INTRODUCTION

Long before the science of chemistry was established, materials were described as existing in one of three physical states. There are rigid, *solid* objects, having a definite volume and a fixed shape; there are nonrigid *liquids*, having no fixed shape other than that of their containers but having definite volumes; and there are *gases*, which have neither fixed shape nor fixed volume. (Indeed, ancient scholars considered that everything in the universe was made up of four "elements," including earth, which represented solid objects; water, which represented liquids; air, which represented gases; and fire, which represented heat or energy. It is in this outdated connotation that poets speak of "the fury of the elements.")

The techniques used for handling various materials depend on their physical states as well as their chemical properties. While it is comparatively easy to handle liquids and solids, it is not as convenient to measure out a quantity of a gas. Fortunately, except under rather extreme conditions, all gases have similar physical properties, and the chemical identity of the substance does not influence those properties. For example, all gases expand when they are heated in a nonrigid container and contract when they are cooled or subjected to increased pressure. They readily diffuse through other gases. Any quantity of gas will occupy the entire volume of its container, regardless of the size of the container.

11.2 PRESSURE OF GASES

Pressure is defined as force per unit area. All *fluids* (liquids and gases) exert pressure in all directions. For example, in an inflated balloon, the gas inside pushes against the interior walls of the balloon with such force that the walls stretch. The pressure of the gas is merely the force per square centimeter exerted on the interior surface of the balloon. The pressure *of* a gas is equal to the pressure *on* the gas. For example, if the atmosphere presses on a piston against a gas with a pressure of 15 pounds per square inch, then the pressure of the gas must also be 15 pounds per square inch. A way of measuring the pressure of the atmosphere is by means of a *barometer* (Fig. 11-1). The *standard atmosphere* (abbreviated atm) is defined as the pressure that will support a column of mercury to a vertical height of 760 mm at a temperature of 0 °C. It is convenient to express the measured gas pressure in terms of the vertical height of a mercury column that the gas is capable of supporting. Thus, if the gas supports a column of mercury to a height of only 76 mm, the gas is exerting a pressure of 0.10 atm:

$$76 \, \text{mm} \left(\frac{1 \, \text{atm}}{760 \, \text{mm}} \right) = 0.10 \, \text{atm}$$

EXAMPLE 11.1. What is the pressure in atmospheres of a gas that supports a column of mercury to a height of 1 090 mm?

$$\text{pressure} = 1 \, 090 \, \text{mm} \left(\frac{1 \, \text{atm}}{760 \, \text{mm}} \right) = 1.43 \, \text{atm}$$

Note that the dimension 1 atmosphere (1 atm) is not the same as *atmospheric pressure*. The atmospheric pressure—the pressure of the atmosphere—varies widely from day to day and from place

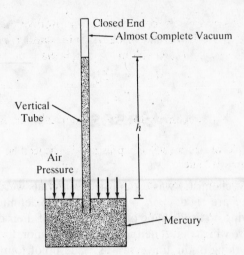

Fig. 11-1 Simple barometer
Air pressure on the surface of the open dish is balanced by the extra pressure caused by the weight of the mercury in the closed tube above the mercury level in the dish. The greater the air pressure, the higher the mercury stands in the vertical tube.

to place, whereas the dimension 1 atm has a fixed value by definition.

The unit *torr* is currently used to indicate the pressure necessary to support mercury to a vertical height of 1 mm. Thus, 1 atm is by definition equal to 760 torr.

11.3 BOYLE'S LAW

Robert Boyle (1627–1691) studied the effect of changing the pressure of a gas on its volume at constant temperature. He measured the volume of a given quantity of gas at a given pressure, changed its pressure, and measured the volume again. He obtained data similar to the data shown in Table 11-1. After repeating the process many times with several different gases, he concluded that

At constant temperature, the volume of a given sample of a gas is inversely proportional to its pressure.

This statement is known as *Boyle's law.*

Table 11-1 Typical Set of Data Showing Boyle's Law

Pressure, P (atm)	Volume, V (L)
4.0	1.0
2.0	2.0
1.0	4.0
0.50	8.0

The term *inversely proportional* in the statement of Boyle's law means that as the pressure increases, the volume becomes smaller by the same factor. That is, if the pressure is doubled, the volume is halved; if the pressure goes up to 8 times its previous value, the volume goes down to $\frac{1}{8}$ times its previous value. This relationship can be represented mathematically by any of the following:

$$P \propto 1/V \qquad P = k/V \qquad PV = k$$

where P represents the pressure, V represents the volume, and k is a constant.

EXAMPLE 11.2. What is the value for the constant k for the sample of gas for which data are given in Table 11-1?

For each case, multiplying the observed pressure by the volume gives the value $4.0\,L \cdot atm$. Therefore, the constant k is $4.0\,L \cdot atm$.

If for a given sample of a gas at a given temperature, the product PV is a constant, changing the pressure from some initial value P_1 to a new value P_2 will cause a corresponding change in the volume from the original volume V_1 to a new volume V_2 such that

$$P_1V_1 = k = P_2V_2$$

or

$$P_1V_1 = P_2V_2$$

This last equation means that it is unnecessary to solve numerically for k.

EXAMPLE 11.3. A 7.00-L sample of gas at 800 torr pressure is changed at constant temperature until its pressure is 1 000 torr. What is its new volume?

A useful first step in doing this type of problem is to tabulate clearly all the information given in terms of initial and final conditions:

$$P_1 = 800 \, \text{torr} \qquad P_2 = 1\,000 \, \text{torr}$$
$$V_1 = 7.00 \, \text{L} \qquad V_2 = ?$$

Let the new, unknown volume be represented by V_2. Since the temperature is constant, Boyle's law is used:

$$P_1V_1 = P_2V_2$$

Solving for V_2 by dividing each side by P_2 yields

$$\frac{P_1V_1}{P_2} = V_2$$

Substituting the values for P_1, V_1, and P_2 given in the statement of the problem:

$$V_2 = \frac{(800 \, \text{torr})(7.00 \, \text{L})}{(1\,000 \, \text{torr})} = 5.60 \, \text{L}$$

The final volume is 5.60 L. The answer is seen to be reasonable by noting that since the pressure increases, the volume must decrease.

Note that the units of the constant k are determined by the units used to express the volume and the pressure. Consequently, the units of pressure and of volume must be explicitly stated.

EXAMPLE 11.4. To what pressure must a sample of gas be subjected at constant temperature in order to compress it from 500 mL to 300 mL if its original pressure is 1.71 atm?

$$V_1 = 500 \, \text{mL} \qquad V_2 = 300 \, \text{mL}$$
$$P_1 = 1.71 \, \text{atm} \qquad P_2 = ?$$
$$P_1V_1 = P_2V_2$$
$$P_2 = \frac{P_1V_1}{V_2} = \frac{(1.71 \, \text{atm})(500 \, \text{mL})}{(300 \, \text{mL})} = 2.85 \, \text{atm}$$

The pressure necessarily is increased to make the volume smaller.

11.4 GRAPHIC REPRESENTATION OF DATA

Often in scientific work it is useful to report data in the form of a graph in order to enable immediate visualization of general trends and relationships. Another advantage of plotting data in the form of a graph is to be able to estimate values for points between and beyond the experimental points. For example, in Fig. 11-2 the data of Table 11-1 are plotted on a graph using P as a vertical axis (y axis) and V as the horizontal axis (x axis). Note that in plotting a graph based on experimental data, the numerical scales of the axes should be chosen so that they can be read to the same number of significant figures as was used in reporting the measurements. It can be seen that as the magnitude of the pressure decreases, the magnitude of the volume increases. It is possible to obtain values of the

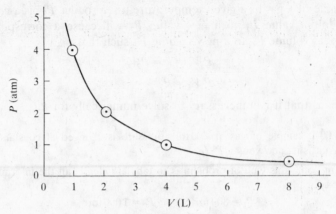

Fig. 11-2   Plot of PV data for the gas in Table 11-1

volume at intermediate values of the pressure merely by reading from points on the curve (interpolating).

EXAMPLE 11.5.   For the sample of gas described in Fig. 11-2, what pressure is required to make the volume 3.0 L?

It is apparent from the graph that a pressure of 1.3 atm is required.

If the graph of one variable against another should happen to be a straight line, the relationship between the variables can be expressed by a simple algebraic equation. If the data fall on a straight line that goes through the origin (the 0, 0 point), then the two variables are *directly proportional*. As one goes up, the other goes up by the same factor. For example, as one doubles, the other doubles. When one finds such a direct proportionality, for example the distance traveled at constant velocity by an automobile and the time of travel, one can immediately write a mathematical equation relating the two variables.

$$\text{distance} = \text{speed} \times \text{time}$$

In this case, the longer the time spent traveling at constant speed, the greater the distance traveled.

Often it is worthwhile to plot data in several ways until a straight-line graph is obtained. Consider the plot of volume and pressure shown in Fig. 11-2, which is definitely not a straight line. What will the pressure of the sample be at a volume of 10 L? It is somewhat difficult to estimate past the data points (to extrapolate) in this case, because the curve is not a straight line, and it is difficult to know how much it will bend. However, these data can be replotted in the form of a straight line by using the reciprocal of one of the variables. The straight line can be extended past the experimental points (extrapolated) rather easily, and the desired number can be estimated. For example, if the data of Table 11-1 are retabulated as in Table 11-2, and an additional column is added with the reciprocal of the pressure $1/P$, it is possible to plot $1/P$ as the vertical axis against V on the horizontal axis, and a straight line through the origin is obtained (Fig. 11-3). Thus, it is proper to say that the quantity $1/P$ is directly proportional to V. When the *reciprocal* of a quantity is *directly* proportional to a second quantity, the first quantity itself is *inversely* proportional to the second.

Table 11-2   Reciprocal of Pressure Data

P (atm)	$1/P$ (1 / atm)	V (L)
4.0	0.25	1.0
2.0	0.50	2.0
1.0	1.0	4.0
0.50	2.0	8.0

If $1/P$ is directly proportional to V, that is, $(1/P)k = V$, then P is inversely proportional to V, that is, $k = PV$. The straight line found by plotting $1/P$ versus V can easily be extended to the point where $V = 10\,\text{L}$. The $1/P$ point is 2.5/atm; therefore,

$$P = \frac{1}{2.5/\text{atm}} = 0.40\,\text{atm}$$

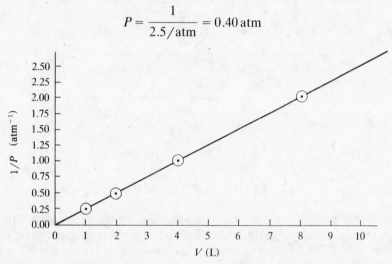

Fig. 11-3   Plot of $1/P$ versus V for the gas in Table 11-2 to show proportionality

11.5   CHARLES' LAW

If a given quantity of gas is heated at constant pressure in a container that has a movable wall, such as a piston (Fig. 11-4), the volume of the gas will increase. If a given quantity of gas is heated in a container that has a fixed volume (Fig. 11-5), its pressure will increase. Conversely, cooling a gas at constant pressure causes a decrease in its volume, while cooling it at constant volume causes a decrease in its pressure.

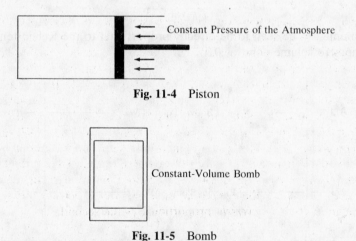

Fig. 11-4   Piston

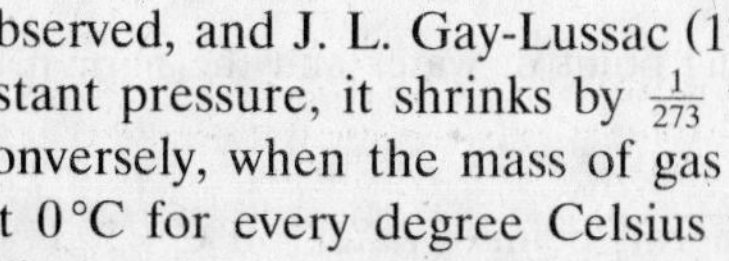

Fig. 11-5   Bomb

J. A. C. Charles (1746–1823) observed, and J. L. Gay-Lussac (1778–1850) confirmed, that when a given mass of gas is cooled at constant pressure, it shrinks by $\frac{1}{273}$ times its volume at $0\,°\text{C}$ for every degree Celsius that it is cooled. Conversely, when the mass of gas is heated at constant pressure, it expands by $\frac{1}{273}$ times its volume at $0\,°\text{C}$ for every degree Celsius that it is heated. The changes in volume with temperature of two different-sized samples of a gas are shown in Fig. 11-6.

The chemical identity of the gas has no influence on the volume changes as long as the gas does not liquefy in the range of temperatures studied. It is seen in Fig. 11-6 that for each sample, the

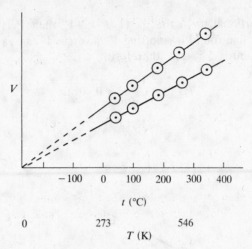

Fig. 11-6 Two samples of gas

volume of the gas changes linearly with temperature. A straight line can be drawn through all the points. If it were assumed that the gas does not liquefy at very low temperatures, each sample would have zero volume at $-273\,°C$. Of course, any real gas could never have zero volume. (Gases are expected to liquefy before this very cold temperature is reached.) Nevertheless, $-273\,°C$ is the temperature at which a sample of gas would theoretically have zero volume. This type of graph can be plotted for any mass of any gas, and in every case the observed volume will change with temperature at such a rate that at $-273\,°C$ the volume would be 0. Therefore, the temperature $-273\,°C$ can be regarded as the *absolute zero* of temperature. Since there cannot be less than zero volume, there can be no colder temperature than $-273\,°C$. The temperature scale that has been devised using this fact is called the *Kelvin*, or *absolute*, temperature scale (Sec. 2.7). A comparison of the Kelvin scale and the Celsius scale is shown in Fig. 11-7. It is seen that any temperature in degrees Celsius (t, °C) may be converted to kelvins (T, K) by adding 273°. It is customary to use capital T to represent Kelvin temperatures and small t to represent Celsius temperatures.

$$T = t + 273°$$

In Fig. 11-6, the volume can be seen to be directly proportional to the Kelvin temperature, since the lines go through the zero volume point at $0\,K$.

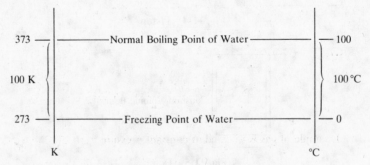

Fig. 11-7 Comparison of Celsius and Kelvin temperature scales

EXAMPLE 11.6.   What is the freezing point of water and the normal boiling point of water on the absolute scale?

$$\text{freezing point of water} = 0\,°C + 273° = 273\,K$$

$$\text{normal boiling point of water} = 100\,°C + 273° = 373\,K$$

The fact that the volume of a gas varies linearly with temperature is combined with the concept of absolute temperature to give a statement of *Charles' law*:

At constant pressure, the volume of a given sample of gas is directly proportional to its absolute temperature.

Expressed mathematically,

$$V = kT \qquad \text{(constant pressure)}$$

This equation can be rearranged to give

$$V/T = k$$

Since V/T is a constant, this ratio for a given sample of gas at one volume and temperature is equal to the same ratio at any other volumes and temperatures. See Table 11-3. That is, for a given sample at constant pressure,

$$\frac{V_1}{T_1} = \frac{V_2}{T_2} \qquad \text{(constant pressure)}$$

One can see from the data of Table 11-3 that *absolute* temperatures must be used.

Table 11-3   Volumes and Temperatures of a Given Sample of Gas

Volume (V) (L)	Temperature (T) (K)	Temperature (t) (°C)	V/T (L/K)
0.50	125°	−148°	4.0×10^{-3}
1.0	250°	− 23°	4.0×10^{-3}
1.5	375°	102°	4.0×10^{-3}
2.0	500°	227°	4.0×10^{-3}

EXAMPLE 11.7.   A 4.50-L sample of gas is warmed at constant pressure from 300 K to 350 K. What will its final volume be?

	1	2
V	4.50 L	V_2
T	300 K	350 K

$$\frac{V_1}{T_1} = \frac{V_2}{T_2}$$

$$V_2 = \frac{V_1 T_2}{T_1} = \frac{(4.50\,\text{L})(350\,\text{K})}{(300\,\text{K})} = 5.25\,\text{L}$$

EXAMPLE 11.8.   A 4.50-L sample of gas is warmed at constant pressure from 27 °C to 77 °C. What will its final volume be?

This example is a restatement of Example 11.7. The conditions are precisely the same; the only difference is that the temperatures are expressed in degrees Celsius and they must first be converted to kelvins.

$$27\,°\text{C} = 27\,°\text{C} + 273° = 300\,\text{K}$$
$$77\,°\text{C} = 77\,°\text{C} + 273° = 350\,\text{K}$$

The example is solved as shown above. Note that V is directly proportional to T, but not to t; for example, when V doubles, T doubles, but t does not double.

11.6   COMBINED GAS LAW

Suppose it is desired to calculate the final volume V_2 of a gas originally at volume V_1 when its temperature is changed from T_1 to T_2 at the same time its pressure is changed from P_1 to P_2? One might consider the two effects separately, for example that first the pressure is changed at constant temperature T_1 and calculate a new volume V_{new} using Boyle's law. Then, using Charles' law, one would calculate how the new volume V_{new} would change to V_2 when the temperature is changed from T_1 to T_2 at the constant pressure P_2 (Fig. 11-8).

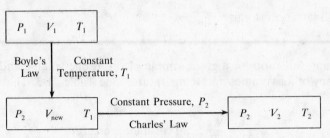

Fig. 11-8   Change in gas volume with pressure then temperature

It would be equally correct to consider that first the temperature of the gas was changed from T_1 to T_2 at the constant pressure P_1, for which a new volume V_{new} could be calculated using Charles' law. Then, assuming that the temperature is held constant at T_2, calculate how the volume would change as the pressure is changed from P_1 to P_2 (Fig. 11-9).

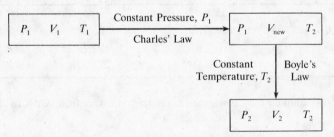

Fig. 11-9   Change in gas volume with temperature then pressure

However, the fact that the volume V of a given mass of gas is inversely proportional to its pressure P and directly proportional to its absolute temperature T can be combined mathematically to give the single equation:

$$V = k\left(\frac{T}{P}\right)$$

where k is the proportionality constant. Rearranging the variables gives the following equation:

$$\frac{PV}{T} = k$$

That is, for a given mass of gas, the ratio PV/T remains a constant, and therefore

$$\frac{P_1 V_1}{T_1} = \frac{P_2 V_2}{T_2}$$

This expression is a mathematical statement of the *combined* (or general) gas law. In words, the volume of a given sample of gas is inversely proportional to its pressure and directly proportional to its absolute temperature.

Note that if the temperature is constant, $T_1 = T_2$, then the expression reduces to that for Boyle's law, $P_1 V_1 = P_2 V_2$. Alternatively, if the pressure is constant, $P_1 = P_2$, the expression is equivalent to

Charles' law,

$$\frac{V_1}{T_1} = \frac{V_2}{T_2}$$

EXAMPLE 11.9. A sample of gas is pumped from a 12.0-L vessel at 27 °C and 760 torr pressure to a 3.5-L vessel at 52 °C. What is its final pressure?

	1	2
P	760 torr	P_2
V	12.0 L	3.5 L
T	27 °C = 300 K	52 °C = 325 K

$$\frac{P_1 V_1}{T_1} = \frac{P_2 V_2}{T_2}$$

$$P_2 = \frac{P_1 V_1}{T_1} \frac{T_2}{V_2} = \frac{(760 \text{ torr})(12.0 \text{ L})(325 \text{ K})}{(300 \text{ K})(3.5 \text{ L})} = 2.8 \times 10^3 \text{ torr}$$

Standard Conditions

According to the combined gas law, the volume of a given mass of gas can have any value, depending on its temperature and pressure. To compare the quantities of gas present in two different samples, it is useful to adopt a set of standard conditions of temperature and pressure. By universal agreement, the standard temperature is chosen as 273 K (0 °C) and the standard pressure is chosen as exactly 1 atm (760 torr). Together, these conditions are referred to as *standard conditions* or as *standard temperature and pressure* (STP). While there is nothing special about STP, some authors and some instructors find it convenient to use this short notation for this particular temperature and pressure.

EXAMPLE 11.10. A sample of gas occupies a volume of 14.0 L at 27 °C and 2.00 atm pressure. What is the volume of this sample at STP?

$$P_1 = 2.00 \text{ atm} \qquad\qquad P_2 = 1.00 \text{ atm}$$
$$V_1 = 14.0 \text{ L} \qquad\qquad V_2 = ?$$
$$T_1 = 27 \,°\text{C} + 273\,° = 300 \text{ K} \qquad T_2 = 273 \text{ K}$$

$$\frac{P_1 V_1}{T_1} = \frac{P_2 V_2}{T_2}$$

$$V_2 = \frac{P_1 V_1}{T_1} \frac{T_2}{P_2} = \frac{(2.00 \text{ atm})(14.0 \text{ L})(273 \text{ K})}{(300 \text{ K})(1.00 \text{ atm})} = 25.5 \text{ L}$$

11.7 IDEAL GAS LAW

All the gas laws described so far worked only for a given sample of gas. If a gas is produced during a chemical reaction or some of the gas under study escapes during processing, these gas laws do not apply. The *ideal gas law* works (at least approximately) for any sample of gas. Consider a given sample of gas, for which

$$\frac{PV}{T} = k \qquad \text{(a constant)}$$

If we increase the number of moles of gas at constant pressure and temperature, the volume must also increase. Thus, we can conclude that the constant k can be regarded as a product of two constants, one of which represents the number of moles of gas. We then get

$$\frac{PV}{T} = nR \qquad \text{or} \qquad PV = nRT$$

where P, V, and T have their usual meanings, n is the number of moles of gas *molecules*, and R is a new constant that is valid for any sample of any gas. This equation is known as the *ideal gas law*. You should remember the following value of R; other values of R in other units will be introduced later.

$$R = 0.0821 \, \text{L} \cdot \text{atm/mol} \cdot \text{K}$$

(Note that there is a 0 after the decimal point.)

In the simplest ideal gas law problems, values for three of the four variables are given, and you are asked to calculate the value of the fourth. As usual with the gas laws, the temperature must be given as an absolute temperature, in kelvins. The units of P and V are most conveniently given in atmospheres and liters because the units of R with the value given above are in terms of these units. If other units are given for pressure or volume, convert them to atmospheres and liters, respectively.

EXAMPLE 11.11.   How many moles of O_2 is present in a 2.47-L sample at 45 °C and 1.10 atm?

$$T = 45\,°\text{C} + 273 = 318\,\text{K}$$

$$n = \frac{PV}{RT} = \frac{(1.10\,\text{atm})(2.47\,\text{L})}{(0.0821\,\text{L} \cdot \text{atm/mol} \cdot \text{K})(318\,\text{K})} = 0.104\,\text{mol}\,O_2$$

Note that T must be in kelvins.

EXAMPLE 11.12.   How many moles of O_2 is present in a 500-mL sample at 45 °C and 795 torr?

Since R is defined in terms of liters and atmospheres, the pressure and volume are first converted to those units. Temperature as usual is given in kelvins. First, change the pressure to atmospheres and the volume to liters:

$$795\,\text{torr}\left(\frac{1\,\text{atm}}{760\,\text{torr}}\right) = 1.05\,\text{atm} \qquad 500\,\text{mL}\left(\frac{1\,\text{L}}{1\,000\,\text{mL}}\right) = 0.500\,\text{L}$$

$$n = \frac{PV}{RT} = \frac{(1.05\,\text{atm})(0.500\,\text{L})}{(0.0821\,\text{L} \cdot \text{atm/mol} \cdot \text{K})(318\,\text{K})} = 0.0201\,\text{mol}$$

Alternately, we can do the entire calculation, including the conversions, with one equation:

$$n = \frac{PV}{RT} = \frac{(795\,\text{torr})(1.00\,\text{atm/760\,torr})(500\,\text{mL})(1\,\text{L/1}\,000\,\text{mL})}{(0.0821\,\text{L} \cdot \text{atm/mol} \cdot \text{K})(318\,\text{K})} = 0.0200\,\text{mol}$$

EXAMPLE 11.13.   At what temperature will 0.100 mol of CO_2 occupy 2.00 L at 1.31 atm?

$$T = \frac{PV}{nR} = \frac{(1.31\,\text{atm})(2.00\,\text{L})}{(0.0821\,\text{L} \cdot \text{atm/mol} \cdot \text{K})(0.100\,\text{mol})} = 319\,\text{K}$$

EXAMPLE 11.14.   What use was made of the information about the chemical identity of the gas in Examples 11.12 and 11.13?

None. The ideal gas law works no matter what gas is being used.

EXAMPLE 11.15.   What volume will 2.00 g of O_2 occupy at STP?

The value of n is not given explicitly in the problem, but the mass is given, with which we can calculate the number of moles:

$$2.00\,\text{g}\,O_2\left(\frac{1\,\text{mol}\,O_2}{32.0\,\text{g}\,O_2}\right) = 0.0625\,\text{mol}\,O_2$$

$$V = \frac{nRT}{P} = \frac{(0.0625\,\text{mol})(0.0821\,\text{L} \cdot \text{atm/mol} \cdot \text{K})(273\,\text{K})}{(1.00\,\text{atm})} = 1.40\,\text{L}$$

The identity of the gas is important here to determine the number of moles.

EXAMPLE 11.16. How many moles of O atoms were present in Example 11.11?

$$0.104\,\text{mol}\,O_2\left(\frac{2\,\text{mol}\,O}{\text{mol}\,O_2}\right) = 0.208\,\text{mol}\,O\ \text{atoms}$$

Note that n in the ideal gas equation in the example refers to moles of O_2 molecules.

EXAMPLE 11.17. What quantity in moles of He atoms is present in 2.47 L at 45 °C and 1.10 atm?

$$n = \frac{PV}{RT} = \frac{(1.10\,\text{atm})(2.47\,\text{L})}{(0.0821\,\text{L}\cdot\text{atm}/\text{mol}\cdot\text{K})(318\,\text{K})} = 0.104\,\text{mol He}$$

The He molecules are individual He atoms, and thus there is 0.104 mol He atoms present.

As soon as P, V, and T data are recognized in a problem, you can calculate n. In a complicated problem, if you do not see how to do the whole problem, first calculate n and then see what you can do with it.

EXAMPLE 11.18. If 2.06 g of a gas occupies 3.33 L at 17 °C and 700 torr, what is the molecular weight of the gas?

If you do not see at first how to solve this problem to completion, at least you can recognize that P, V, and T data are given. First calculate the number of moles of gas present:

$$700\,\text{torr}\left(\frac{1\,\text{atm}}{760\,\text{torr}}\right) = 0.921\,\text{atm}$$

$$17\,°\text{C} + 273 = 290\,\text{K}$$

$$n = \frac{PV}{RT} = \frac{(0.921\,\text{atm})(3.33\,\text{L})}{(0.0821\,\text{L}\cdot\text{atm}/\text{mol}\cdot\text{K})(290\,\text{K})} = 0.129\,\text{mol}$$

We now know the mass of the gas and the number of moles. That is enough to calculate the molecular weight:

$$\frac{2.06\,\text{g}}{0.129\,\text{mol}} = 16.0\,\text{g/mol}$$

The gas has a molecular weight of 16.0 amu.

EXAMPLE 11.19. What volume would be occupied by the oxygen liberated by heating 0.200 g of $KClO_3$ until it completely decomposes to KCl and oxygen? The gas is collected at STP.

Although the temperature and pressure of the gas are given, the number of moles of gas is not. Can we get it somewhere? The chemical reaction of $KClO_3$ yields the oxygen, and the rules of stoichiometry (Chap. 8) may be used to calculate the number of moles of gas. Note that the number of moles of $KClO_3$ is *not* used in the ideal gas law equation.

$$2\,KClO_3 \xrightarrow{\text{heat}} 2\,KCl + 3\,O_2$$

$$0.200\,\text{g}\,KClO_3\left(\frac{1\,\text{mol}\,KClO_3}{122\,\text{g}\,KClO_3}\right)\left(\frac{3\,\text{mol}\,O_2}{2\,\text{mol}\,KClO_3}\right) = 0.00246\,\text{mol}\,O_2$$

Now that we know the number of moles, the pressure, and the temperature of the O_2, we can calculate its volume:

$$V = \frac{nRT}{P} = \frac{(0.00246\,\text{mol})(0.0821\,\text{L}\cdot\text{atm}/\text{mol}\cdot\text{K})(273\,\text{K})}{(1.00\,\text{atm})} = 0.0551\,\text{L}$$

11.8 DALTON'S LAW OF PARTIAL PRESSURES

When two or more gases are mixed, they each occupy the entire volume of the container. They each have the same temperature as the other(s). However, each gas exerts its own pressure, independent of the other gases. Moreover, according to Dalton's law of partial pressures, their pressures must add up to the total pressure of the gas mixture.

EXAMPLE 11.20. (*a*) If I try to put a 1.00-L sample of O_2 at 300 K and 1.00 atm plus a 1.00-L sample of N_2 at 300 K and 1.00 atm into a rigid 1.00-L container at 300 K, will they fit? (*b*) If so, what will be their total volume and total pressure?

(*a*) The gases will fit; gases expand or contract to fill their containers. (*b*) The total volume is the volume of the container—1.00 L. The temperature is 300 K, given in the problem. The total pressure is the sum of the two partial pressures. *Partial pressure* is the pressure of each gas (as if the other were not present). The oxygen pressure is 1.00 atm. The oxygen has been moved from a 1.00-L container at 300 K to another 1.00-L container at 300 K, and so its pressure does not change. The nitrogen pressure is 1.00 atm for the same reason. The total pressure is 1.00 atm + 1.00 atm = 2.00 atm.

EXAMPLE 11.21. A 1.00-L sample of O_2 at 300 K and 1.00 atm plus a 0.500-L sample of N_2 at 300 K and 1.00 atm are put into a rigid 1.00-L container at 300 K. What will be their total volume, temperature, and total pressure?

The total volume is the volume of the container—1.00 L. The temperature is 300 K, given in the problem. The total pressure is the sum of the two partial pressures. The oxygen pressure is 1.00 atm. (See Example 11.20.) The nitrogen pressure is 0.500 atm, since it was moved from 0.500 L at 1.00 atm to 1.00 L at the same temperature (Boyle's law). The total pressure is

$$1.00 \text{ atm} + 0.500 \text{ atm} = 1.50 \text{ atm}$$

EXAMPLE 11.22. If the N_2 of the last example were added to the O_2 in the container originally containing the O_2, how would the problem be affected?

It would not change; the final volume would still be 1.00 L.

EXAMPLE 11.23. If the O_2 of Example 11.21 were added to the N_2 in the container originally containing the N_2, how would the problem change?

The pressure would be doubled, because the final volume would be 0.500 L.

The ideal gas law applies to each individual gas in a gas mixture as well as to the gas mixture as a whole. Thus, in a mixture of nitrogen and oxygen, one can apply the ideal gas law to the oxygen, to the nitrogen, and to the mixture as a whole.

$$PV = nRT$$

The V and the T, as well as the R, refer to each gas and the total mixture. If we want the number of moles of O_2, we can use the pressure of O_2. If we want the number of moles of N_2, we can use the pressure of N_2. If we want the total number of moles, we use the total pressure.

EXAMPLE 11.24. Calculate the number of moles of O_2 in Example 11.21, both before and after mixing.

Before mixing:

$$n = \frac{PV}{RT} = \frac{(1.00 \text{ atm})(1.00 \text{ L})}{(0.0821 \text{ L} \cdot \text{atm/mol} \cdot \text{K})(300 \text{ K})} = 0.0406 \text{ mol}$$

After mixing:

$$n_{O_2} = \frac{P_{O_2}V}{RT} = \frac{(1.00 \text{ atm})(1.00 \text{ L})}{(0.0821 \text{ L} \cdot \text{atm/mol} \cdot \text{K})(300 \text{ K})} = 0.0406 \text{ mol}$$

Water Vapor

At 25 °C, water is ordinarily a liquid. However, even at 25 °C, water evaporates. In a closed container at 25 °C, water evaporates enough to get a 27 torr water vapor pressure in its container. The pressure of the gaseous water is called its *vapor pressure* at that temperature. At different temperatures, it evaporates to different extents to give different vapor pressures. As long as there is liquid water present, however, the vapor pressure above pure water depends on the temperature alone. Only the nature of the liquid and the temperature affect the vapor pressure; the volume of the container does not affect the final pressure.

The water vapor mixes with any other gas(es) present, and the mixture is governed by Dalton's law of partial pressures, just like any other gas mixture.

EXAMPLE 11.25. O_2 is collected in a bottle over water at 25°C at 1.00 atm barometric pressure. (a) What gas(es) is (are) in the bottle? (b) What is (are) the pressure(s)?

(a) O_2 and water vapor are both in the bottle. (b) The total pressure is the barometric pressure, 760 torr. The water vapor pressure is 27 torr, given above for the gas phase above liquid water at 25°C. The pressure of the O_2 must be

$$760 \text{ torr} - 27 \text{ torr} = 733 \text{ torr}$$

EXAMPLE 11.26. How many moles of oxygen is contained in a 1.00-L vessel over water at 25°C and a barometric pressure of 1.00 atm?

The barometric pressure is the total pressure of the gas mixture. The pressure of O_2 is 760 torr − 27 torr = 733 torr. Since we want to know about moles of O_2, we need to use the pressure of O_2 in the ideal gas law:

$$n_{O_2} = P_{O_2} V / RT$$

$$= \frac{((733/760) \text{ atm})(1.00 \text{ L})}{(0.0821 \text{ L} \cdot \text{atm}/\text{mol} \cdot \text{K})(298 \text{ K})} = 0.0394 \text{ mol}$$

Solved Problems

INTRODUCTION

11.1. What is the difference between gas and gasoline?

Ans. Gas is a state of matter. Gasoline is a liquid, used mainly for fuel, with a nickname "gas." Do not confuse the two. This chapter is about the gas phase, not about liquid gasoline.

11.2. What is a fluid?

Ans. A gas or a liquid.

PRESSURE OF GASES

11.3. Change 750 mm Hg to (a) torr and (b) atm.

Ans. (a) 750 torr and (b) $750 \text{ torr}\left(\dfrac{1 \text{ atm}}{760 \text{ torr}}\right) = 0.987 \text{ atm}.$

11.4. Change (a) 950 torr to atm, (b) 1.350 atm to torr, (c) 755 mm Hg to torr, and (d) 0.970 atm to mm Hg.

Ans.

$$(a) \qquad 950 \text{ torr}\left(\frac{1 \text{ atm}}{760 \text{ torr}}\right) = 1.25 \text{ atm}$$

$$(b) \qquad 1.350 \text{ atm}\left(\frac{760.0 \text{ torr}}{1 \text{ atm}}\right) = 1\,026 \text{ torr}$$

$$(c) \qquad 755 \text{ mm Hg}\left(\frac{1 \text{ torr}}{1 \text{ mm Hg}}\right) = 755 \text{ torr}$$

$$(d) \qquad 0.970 \text{ atm}\left(\frac{760 \text{ mm Hg}}{\text{atm}}\right) = 737 \text{ mm Hg}$$

11.5. How many pounds force does the atmosphere exert on the side of a metal can that measures 5.0 in. by 7.0 in.? (1 atm = 14.7 lb/in.²)

Ans.

$$\text{area} = 5.0 \text{ in.} \times 7.0 \text{ in.} = 35 \text{ in.}^2$$

$$35 \text{ in.}^2 \left(\frac{14.7 \text{ lb}}{\text{in.}^2} \right) = 510 \text{ lb}$$

BOYLE'S LAW

11.6. What law may be stated qualitatively, "When you squeeze a gas, it gets smaller."?

Ans. Boyle's law.

11.7. Calculate the product of the pressure and volume for each point in Table 11-1. What can you conclude?

Ans. $PV = 4.0 \text{ L} \cdot \text{atm}$ in each case. You can conclude that PV is a constant for this sample of gas, and that Boyle's law is obeyed.

11.8. If 4.00 L of gas at 1.04 atm is changed to 745 torr at constant temperature, what is its final volume?

Ans.

$$1.04 \text{ atm} \left(\frac{760 \text{ torr}}{\text{atm}} \right) = 790 \text{ torr}$$

$$P_1 V_1 = P_2 V_2$$

$$V_2 = \frac{P_1 V_1}{P_2} = \frac{(790 \text{ torr})(4.00 \text{ L})}{(745 \text{ torr})} = 4.24 \text{ L}$$

11.9. If 2.00 L of gas at 790 torr is changed to 1 565 mL at constant temperature, what is its final pressure?

Ans.

$$P_2 = \frac{P_1 V_1}{V_2} = \frac{(790 \text{ torr})(2\,000 \text{ mL})}{(1\,565 \text{ mL})} = 1\,010 \text{ torr}$$

11.10. A sample of gas occupies 3.44 L. What will be its new volume if its pressure is doubled at constant temperature?

Ans. According to Boyle's law, doubling the pressure will cut the volume in half; the new volume will be 1.72 L. A second method allows us to use unknown variables for the pressure:

	1	2
P	P_1	$P_2 = 2P_1$
V	3.44 L	V_2

$$V_2 = \frac{P_1 V_1}{P_2} = \frac{P_1 V_1}{2 P_1} = \frac{V_1}{2} = \frac{(3.44 \text{ L})}{2} = 1.72 \text{ L}$$

11.11. A 1.00-L sample of gas at 18 °C and 1.00 atm pressure is changed to 2.00 atm pressure at 18 °C. What law may be used to determine its final volume?

Ans. Boyle's law may be used because the temperature is unchanged. Alternately, the combined gas law may be used, with the Kelvin equivalent of 18 °C used for both T_1 and T_2.

GRAPHIC REPRESENTATION OF DATA

11.12. On a graph, what criteria represent direct proportionality?

Ans.   (*a*) The plot is a straight line; (*b*) the plot passes through the origin.

11.13. What would be the volume of the gas described in Table 11-1 at 1.5 atm pressure? Answer first by calculating with Boyle's law, second by reading from Fig. 11-2, and third by reading from Fig. 11-3. Which determination is easiest (assuming that the graphs have already been drawn)?

Ans.   The volume of the gas is 2.7 L. The second method involves merely reading a point from a graph. To use Fig. 11-3, you first have to calculate the reciprocal of the pressure. None of the methods is difficult, however.

11.14. Plot the following data.

V (L)	*P* (atm)
0.500	4.00
1.00	2.00
2.00	1.00
4.00	0.500

Replot the data, using the volume and the reciprocal of the pressure. Do these values fall on a straight line? Are volume and the reciprocal of pressure directly proportional? Are volume and pressure directly proportional?

Ans.   The points of the second plot fall on a straight line through the origin, and so the volume and reciprocal of pressure are directly proportional to each other, making the volume inversely proportional to the pressure.

CHARLES' LAW

11.15. Plot the following data:

V (L)	*t* (°C)
4.00	100
3.46	50
2.93	0
2.39	−50

Do the values fall on a straight line? Are the volume and Celsius temperatures directly proportional? Replot the volumes versus the Kelvin temperatures. Are the volume and Kelvin temperature directly proportional?

Ans.   The points of the first plot fall on a straight line, but that line does not go through the origin (the point at 0 L and 0 °C), and so these quantities are *not* directly proportional. When volume is replotted versus Kelvin temperature, the resulting straight line goes through the origin, and thus volume and *absolute temperature* are directly proportional.

11.16. If 4.00 L of gas at 33 °C is changed to 66 °C at constant pressure, what is its final volume?

Ans.   Absolute temperatures must be used:

$$33\,°C + 273° = 306\,K$$
$$66\,°C + 273° = 339\,K$$

According to Charles' law:

$$\frac{V_1}{T_1} = \frac{V_2}{T_2}$$

$$V_2 = \frac{V_1 T_2}{T_1} = \frac{(4.00\,\text{L})(339\,\text{K})}{(306\,\text{K})} = 4.43\,\text{L}$$

The volume is *not* doubled by doubling the Celsius temperature.

11.17. If 2.00 L of gas at 0 °C is changed to 1575 mL at constant pressure, what is its final temperature?

  Ans. Using the reciprocal of the equation usually used for Charles' law:

$$\frac{T_2}{V_2} = \frac{T_1}{V_1}$$

and solving for T_2:

$$T_2 = \frac{T_1 V_2}{V_1} = \frac{(273\,\text{K})(1.575\,\text{L})}{(2.00\,\text{L})} = 215\,\text{K}$$

The temperature must have been lowered to reduce the volume.

COMBINED GAS LAW

11.18. Calculate the missing value for each set of data in the following table:

	P_1	V_1	T_1	P_2	V_2	T_2
(a)	1.00 atm	1.00 L	273 K	2.00 atm	2.00 L	——
(b)	7.00 atm	——	300 K	3.30 atm	6.00 L	400 K
(c)	1.00 atm	3.65 L	30 °C	——	5.43 L	30 °C
(d)	——	2.91 L	45 °C	780 torr	2.22 L	77 °C
(e)	2.00 atm	——	28 °C	2.00 atm	750 mL	53 °C
(f)	721 torr	200 mL	——	1.21 atm	0.850 L	100 °C

  Ans. Each problem is solved by rearranging the equation

$$\frac{P_1 V_1}{T_1} = \frac{P_2 V_2}{T_2}$$

All temperatures must be in kelvins.

(a) $$T_2 = \frac{T_1 P_2 V_2}{P_1 V_1} = \frac{(273\,\text{K})(2.00\,\text{atm})(2.00\,\text{L})}{(1.00\,\text{atm})(1.00\,\text{L})} = 1\,090\,\text{K}$$

(b) $$V_1 = \frac{P_2 V_2 T_1}{T_2 P_1} = \frac{(3.30\,\text{atm})(6.00\,\text{L})(300\,\text{K})}{(7.00\,\text{atm})(400\,\text{K})} = 2.12\,\text{L}$$

(c) $$P_2 = \frac{T_2 P_1 V_1}{T_1 V_2} = \frac{(303\,\text{K})(1.00\,\text{atm})(3.65\,\text{L})}{(303\,\text{K})(5.43\,\text{L})} = 0.672\,\text{atm}$$

Since $T_1 = T_2$, we could have used $P_2 = P_1 V_1 / V_2$ and arrived at the same conclusion.

(d) $$P_1 = \frac{P_2 V_2 T_1}{T_2 V_1} = \frac{(780\,\text{torr})(2.22\,\text{L})(318\,\text{K})}{(350\,\text{K})(2.91\,\text{L})} = 541\,\text{torr}$$

(e) $$V_1 = \frac{P_2 V_2 T_1}{T_2 P_1} = \frac{(2.00\,\text{atm})(750\,\text{mL})(301\,\text{K})}{(2.00\,\text{atm})(326\,\text{K})} = 692\,\text{mL}$$

Since P did not change, Charles' law in the form

$$V_1 = V_2 T_1 / T_2$$

could have been used.

(f) $T_1 = \dfrac{P_1 V_1 T_2}{P_2 V_2} = \dfrac{(721\,\text{torr})(0.200\,\text{L})(373\,\text{K})}{(1.21\,\text{atm})(760\,\text{torr}/\text{atm})(0.850\,\text{L})} = 68.8\,\text{K}$

The units of P and V each must be the same in state 1 and state 2. Since each of them is given in different units, one of each must be changed.

11.19. A 4.00-L sample of gas has its pressure doubled while its absolute temperature is increased by 25%. What is its new volume?

Ans.   Doubling P would halve the volume to 2.00 L. Then increasing T by 25% (multiplying it by 1.25) would increase the volume by a factor of 1.25 to 2.50 L. That is,

$$V_2 = \left(\frac{P_1 V_1}{T_1}\right)\left(\frac{T_2}{P_2}\right) = \left(\frac{P_1}{P_2}\right)\left(\frac{T_2}{T_1}\right)(V_1)$$

$$= \left(\frac{1}{2}\right)\left(\frac{1.25}{1}\right)(4.00\,\text{L}) = 2.50\,\text{L}$$

IDEAL GAS LAW

11.20. How can you recognize an ideal gas law problem?

Ans.   Ideal gas law problems involve moles. If the number of moles of gas is given or asked for, or if a quantity that involves moles is given or asked for, the problem is most likely an ideal gas law problem. Thus, any problem involving masses of gas (which can be converted to moles of gas) or molecular weights (grams per mole) or numbers of individual molecules (which can be converted to moles) and so forth is an ideal gas law problem. Problems that involve an unchanging mass of gas are most likely not ideal gas law problems.

11.21. Calculate R, the gas law constant, in units of $L \cdot torr/mol \cdot K$ and in units of $mL \cdot atm/mol \cdot K$.

Ans.

$$R = \frac{0.0821\,\text{L} \cdot \text{atm}}{\text{mol} \cdot \text{K}}\left(\frac{760\,\text{torr}}{\text{atm}}\right) = \frac{62.4\,\text{L} \cdot \text{torr}}{\text{mol} \cdot \text{K}}$$

$$R = \frac{0.0821\,\text{L} \cdot \text{atm}}{\text{mol} \cdot \text{K}}\left(\frac{1\,000\,\text{mL}}{\text{L}}\right) = \frac{82.1\,\text{mL} \cdot \text{atm}}{\text{mol} \cdot \text{K}}$$

11.22. Calculate the volume of 3.00 mol of a gas at 20 °C and 770 torr.

Ans.

$$V = \frac{nRT}{P} = \frac{(3.00\,\text{mol})(0.0821\,\text{L} \cdot \text{atm}/\text{mol} \cdot \text{K})(293\,\text{K})}{(770\,\text{torr})(1\,\text{atm}/760\,\text{torr})} = 71.2\,\text{L}$$

11.23. Calculate the pressure of 2.50 mol of a gas that occupies 47.0 L at 47 °C.

Ans.

$$P = \frac{nRT}{V} = \frac{(2.50\,\text{mol})(0.0821\,\text{L} \cdot \text{atm}/\text{mol} \cdot \text{K})(320\,\text{K})}{47.0\,\text{L}} = 1.40\,\text{atm}$$

11.24. Calculate the absolute temperature of 0.250 mol of a gas that occupies 10.0 L at 1.05 atm.

Ans.

$$T = \frac{PV}{nR} = \frac{(1.05\,\text{atm})(10.0\,\text{L})}{(0.250\,\text{mol})(0.0821\,\text{L} \cdot \text{atm}/\text{mol} \cdot \text{K})} = 512\,\text{K}$$

11.25. Calculate the volume of 3.00 mol of H_2O at 1.00 atm pressure and a temperature of 25 °C.

Ans. H_2O is not a gas under these conditions, and so the equation $PV = nRT$ does not apply. (The ideal gas law can be used for water vapor—for example, water over 100 °C at 1 atm or water at lower temperatures mixed with air.) At 1 atm pressure and 25 °C, water is a liquid with a density about 1.00 g/mL.

$$3.00 \, \text{mol} \left(\frac{18.0 \, \text{g}}{\text{mol}} \right) \left(\frac{1 \, \text{mL}}{1.00 \, \text{g}} \right) = 54.0 \, \text{mL}$$

11.26. Calculate the number of moles of a gas that occupies 4.00 L at 278 K and 1.09 atm.

Ans.

$$n = \frac{PV}{RT} = \frac{(1.09 \, \text{atm})(4.00 \, \text{L})}{(0.0821 \, \text{L} \cdot \text{atm/mol} \cdot \text{K})(278 \, \text{K})} = 0.191 \, \text{mol}$$

11.27. Calculate the number of moles of gas present in Problem 11.18(*a*).

Ans.   Since the number of moles of gas does not change during the process of changing from state 1 to state 2, either set of data can be used to calculate the number of moles of gas. Since all three values are given for state 1 it is easiest to use that state:

$$n = \frac{PV}{RT} = \frac{(1.00 \, \text{atm})(1.00 \, \text{L})}{(0.0821 \, \text{L} \cdot \text{atm/mol} \cdot \text{K})(273 \, \text{K})} = 0.0446 \, \text{mol}$$

11.28. Calculate the value of R if 1.00 mol of gas occupies 22.4 L at STP.

Ans.   STP means 1.00 atm and 273 K (0 °C). Thus,

$$R = \frac{PV}{nT} = \frac{(1.00 \, \text{atm})(22.4 \, \text{L})}{(1.00 \, \text{mol})(273 \, \text{K})} = 0.0821 \, \text{L} \cdot \text{atm/mol} \cdot \text{K}$$

11.29. (*a*) Compare qualitatively the volumes at STP of 2.1 mol N_2 and 2.1 mol H_2. (*b*) Compare the volumes at STP of 2.1 g N_2 and 2.1 g H_2.

Ans.   (*a*) Since the pressures, temperature, and numbers of moles are the same, the volumes also must be the same. (*b*) The number of moles of nitrogen is lower than the number of moles of hydrogen because its molecular weight is greater. Since the number of moles of nitrogen is lower, so is its volume.

11.30. Calculate the volume of 3.00 g of helium at 27 °C and 1.00 atm.

Ans.

$$3.00 \, \text{g He} \left(\frac{1 \, \text{mol He}}{4.00 \, \text{g He}} \right) = 0.750 \, \text{mol He}$$

$$V = \frac{nRT}{P} = \frac{(0.750 \, \text{mol})(0.0821 \, \text{L} \cdot \text{atm/mol} \cdot \text{K})(300 \, \text{K})}{1.00 \, \text{atm}} = 18.5 \, \text{L}$$

11.31. Repeat the last problem using hydrogen gas instead of helium. Explain why the occurrence of hydrogen gas in diatomic molecules is so important.

Ans.

$$n = 1.50 \, \text{mol} \qquad V = 37.0 \, \text{L}$$

The gas laws work with moles of molecules, not atoms. It is necessary to know that hydrogen gas occurs in diatomic molecules so that the proper number of moles of gas may be calculated from the mass of the gas.

DALTON'S LAW OF PARTIAL PRESSURES

11.32. What is the pressure of H_2 if 1.00 mol of H_2 and 2.00 mol of He are placed in a 10.0-L vessel at 27 °C?

  Ans.   The presence of the He makes no difference. The pressure of H_2 is calculated with the ideal gas law, using the number of moles of H_2.

$$P = \frac{nRT}{V} = \frac{(1.00\,\text{mol})(0.0821\,\text{L} \cdot \text{atm}/\text{mol} \cdot \text{K})(300\,\text{K})}{10.0\,\text{L}} = 2.46\,\text{atm}$$

11.33. What is the total pressure of a gas mixture containing H_2 at 0.25 atm, N_2 at 0.55 atm, and NO at 0.30 atm?

  Ans.

$$P_{\text{total}} = P_{H_2} + P_{N_2} + P_{NO} = 1.10\,\text{atm}$$

11.34. What is the total pressure of a gas mixture containing He at 0.25 atm, Ne at 0.55 atm, and Ar at 0.30 atm?

  Ans.

$$P_{\text{total}} = P_{He} + P_{Ne} + P_{Ar} = 0.25\,\text{atm} + 0.55\,\text{atm} + 0.30\,\text{atm} = 1.10\,\text{atm}$$

11.35. What is the difference between Problems 11.33 and 11.34?

  Ans.   In Problem 11.33, molecules are involved. In Problem 11.34, uncombined atoms are involved. In both cases, the same laws apply, and it is best to regard an uncombined atom of He, for example, as a molecule.

11.36. The total pressure of a 200-mL sample of oxygen collected over water at 25 °C is 1.030 atm. (*a*) How many moles of gas is present? (*b*) How many moles of oxygen is present? (*c*) How many moles of water vapor is present?

  Ans.

  (*a*)
$$n = \frac{PV}{RT} = \frac{(1.030\,\text{atm})(0.200\,\text{L})}{(0.0821\,\text{L} \cdot \text{atm}/\text{mol} \cdot \text{K})(298\,\text{K})} = 8.42 \times 10^{-3}\,\text{mol total}$$

  (*b*) The gas is composed of O_2 and water vapor. The pressure of the water vapor at 25 °C, given in the last part of Sec. 11.8 or in tables in your text, is 27 torr. In atmospheres:

$$27\,\text{torr}\left(\frac{1\,\text{atm}}{760\,\text{torr}}\right) = 0.036\,\text{atm}$$

  The pressure of the O_2 is therefore

$$1.030\,\text{atm} - 0.036\,\text{atm} = 0.994\,\text{atm}$$

$$n_{O_2} = \frac{PV}{RT} = \frac{(0.994\,\text{atm})(0.200\,\text{L})}{(0.0821\,\text{L} \cdot \text{atm}/\text{mol} \cdot \text{K})(298\,\text{K})} = 8.13 \times 10^{-3}\,\text{mol}\,O_2$$

  (*c*)
$$n_{H_2O} = \frac{PV}{RT} = \frac{(0.036\,\text{atm})(0.200\,\text{L})}{(0.0821\,\text{L} \cdot \text{atm}/\text{mol} \cdot \text{K})(298\,\text{K})} = 2.9 \times 10^{-4}\,\text{mol}\,H_2O$$

  Check:

$$8.13 \times 10^{-3}\,\text{mol} + 0.29 \times 10^{-3}\,\text{mol} = 8.42 \times 10^{-3}\,\text{mol}$$

11.37. In Dalton's law problems, what is the difference in the behavior of water vapor mixed with air and helium mixed with air?

  Ans.   The pressure of water vapor, if it is in contact with liquid water, is governed by the temperature only. More water can evaporate or some water vapor can condense if the pressure is not equal to the tabulated vapor pressure at the given temperature. Helium is a gas under most conditions, and is not capable of adjusting its pressure in the same way.

11.38. A mixture of gases contains He, Ne, and Ar. The pressure of the He is 0.500 atm. The volume of the Ne is 6.00 L. The temperature of the Ar is 27 °C. What value can be calculated from these data? Explain.

Ans. The temperature of the Ar is the temperature of all the gases, since they are all mixed together. The volume of the Ne is the volume of each of the gases and the total volume too. Therefore, the number of moles of He can be calculated because its temperature, pressure, and volume are known:

$$n = \frac{PV}{RT} = \frac{(0.500\,\text{atm})(6.00\,\text{L})}{(0.0821\,\text{L}\cdot\text{atm}/\text{mol}\cdot\text{K})(300\,\text{K})} = 0.122\,\text{mol He}$$

Supplementary Problems

11.39. Explain why gas law problems are not given with data to four or five significant figures.

Ans. The laws are only approximate, and having better data would not necessarily yield more accurate answers.

11.40. Under what conditions of temperature and pressure do the gas laws work best?

Ans. Under low-pressure and high-temperature conditions (far from the possibility of change to liquid state).

11.41. If 0.259 g of a gas occupies 3.33 L at 17 °C and 700 torr, what is the identity of the gas?

Ans. If you do not see at first how to solve this problem to completion, at least you can recognize that P, V, and T data are given. First calculate the number of moles of gas present:

$$700\,\text{torr}\left(\frac{1\,\text{atm}}{760\,\text{torr}}\right) = 0.921\,\text{atm}$$

$$17\,°\text{C} = (17 + 273)\,\text{K} = 290\,\text{K}$$

$$n = \frac{PV}{RT} = \frac{(0.921\,\text{atm})(3.33\,\text{L})}{(0.0821\,\text{L}\cdot\text{atm}/\text{mol}\cdot\text{K})(290\,\text{K})} = 0.129\,\text{mol}$$

We now know the mass of the gas and the number of moles. That is enough to calculate the molecular weight:

$$\frac{0.259\,\text{g}}{0.129\,\text{mol}} = 2.01\,\text{g}/\text{mol}$$

The gas has a molecular weight of 2.01 amu. It could only be hydrogen, H_2, because no other gas has a molecular weight that low.

11.42. Which temperature scale must be used in (*a*) Charles' law problems? (*b*) ideal gas law problems? (*c*) combined gas law problems? (*d*) Boyle's law problems?

Ans. (*a*) through (*c*) Kelvin. (*d*) No temperature is used in the calculation for Boyle's law problems since the temperature must be constant.

11.43. Calculate the mass of $KClO_3$ required to decompose to provide 3.00 L of O_2 at 24 °C and 1.02 atm.

Ans. The volume, temperature, and pressure given allow us to calculate the number of moles of *oxygen*.

$$n_{O_2} = \frac{PV}{RT} = \frac{(1.02\,\text{atm})(3.00\,\text{L})}{(0.0821\,\text{L}\cdot\text{atm}/\text{mol}\cdot\text{K})(297\,\text{K})} = 0.125\,\text{mol}$$

The number of moles of $KClO_3$ may be calculated from the number of moles of O_2 by means of the balanced chemical equation, and that value is then converted to mass.

$$2\,KClO_3 \longrightarrow 2\,KCl + 3\,O_2$$

$$0.125\,mol\,O_2\left(\frac{2\,mol\,KClO_3}{3\,mol\,O_2}\right)\left(\frac{122\,g\,KClO_3}{mol\,KClO_3}\right) = 10.2\,g\,KClO_3$$

11.44. Write an equation for the combined gas law using temperature in Celsius. Explain why the Kelvin scale is more convenient.

Ans.

$$\frac{P_1V_1}{t_1 + 273} = \frac{P_2V_2}{t_2 + 273}$$

The law is simpler with the Kelvin temperatures.

11.49. The total pressure of a mixture of gases is 1.20 atm. The mixture contains 0.10 mol of N_2 and 0.20 mol of O_2. What is the partial pressure of O_2?

Ans.

$$\frac{n_{N_2}}{n_{O_2}} = \frac{0.10\,mol\,N_2}{0.20\,mol\,O_2} = \frac{P_{N_2}V/RT}{P_{O_2}V/RT} = \frac{P_{N_2}}{P_{O_2}} = 0.50$$

$$P_{N_2} = 0.50P_{O_2}$$

$$P_{N_2} + P_{O_2} = 1.20\,atm$$

$$0.50P_{O_2} + P_{O_2} = 1.20\,atm$$

$$P_{O_2} = \frac{1.20\,atm}{1.50} = 0.80\,atm$$

11.46. Two samples of gas at equal pressures and temperatures are held in containers of equal volume. What can be stated about the comparative number of molecules in each gas sample?

Ans. Since the volumes, temperatures, and pressures are the same, the numbers of moles of the two gases are the same. Therefore, there are equal numbers of molecules of the two gases.

11.47. P is inversely proportional to V. Write three mathematical expressions that relate this fact.

Ans.

$$P \propto 1/V \qquad P = k/V \qquad PV = k$$

11.48. Calculate the molecular weight of a gas if 6.00 g occupies 4.06 L at 1.05 atm and 22 °C.

Ans. From the pressure, volume, and temperature data, we can calculate the number of moles of gas present. From the number of moles and the mass, we can calculate the molecular weight.

$$n = \frac{PV}{RT} = \frac{(1.05\,atm)(4.06\,L)}{(0.0821\,L \cdot atm/mol \cdot K)(295\,K)} = 0.176\,mol$$

$$molecular\,weight = (6.00\,g)/(0.176\,mol) = 34.1\,g/mol$$

11.49. What volume of a CO_2 and H_2O mixture at 1.00 atm and 400 K can be prepared by the thermal decomposition of 2.00 g $NaHCO_3$?

Ans.

$$2\,NaHCO_3 \xrightarrow{\text{heat}} Na_2CO_3 + CO_2 + H_2O$$

$$2.00\,g\,NaHCO_3\left(\frac{1\,mol\,NaHCO_3}{84.0\,g\,NaHCO_3}\right) = 0.0238\,mol\,NaHCO_3$$

$$0.0238\,mol\,NaHCO_3\left(\frac{1\,mol\,CO_2}{2\,mol\,NaHCO_3}\right) = 0.0119\,mol\,CO_2$$

The same number of moles of gaseous water (at 400 K) is obtained. The total number of moles of gas is 0.0238 mol. The volume is given by

$$V = \frac{nRT}{P} = \frac{(0.0238\,\text{mol})(0.0821\,\text{L} \cdot \text{atm}/\text{mol} \cdot \text{K})(400\,\text{K})}{1.00\,\text{atm}} = 0.782\,\text{L}$$

Note: At 400 K and 1.00 atm, water is completely in the gas phase. (The vapor pressure of water at 400 K is greater than 1.00 atm.)

11.50. What volume of O_2 at STP can be prepared by the thermal decomposition of 2.00 g of HgO?

Ans.

$$2\,\text{HgO} \xrightarrow{\text{heat}} 2\,\text{Hg} + O_2$$

$$2.00\,\text{g HgO}\left(\frac{1\,\text{mol HgO}}{216\,\text{g HgO}}\right) = 0.00926\,\text{mol HgO}$$

$$0.00926\,\text{mol HgO}\left(\frac{1\,\text{mol } O_2}{2\,\text{mol HgO}}\right) = 0.00463\,\text{mol } O_2$$

$$V = \frac{nRT}{P} = \frac{(0.00463\,\text{mol})(0.0821\,\text{L} \cdot \text{atm}/\text{mol} \cdot \text{K})(273\,\text{K})}{1.00\,\text{atm}} = 0.104\,\text{L}$$

11.51. What volume will 33.0 g of CH_4 occupy at 27°C and 0.950 atm?

Ans.

$$33.0\,\text{g } CH_4\left(\frac{1\,\text{mol } CH_4}{16.0\,\text{g } CH_4}\right) = 2.06\,\text{mol } CH_4$$

$$V = \frac{nRT}{P} = \frac{(2.06\,\text{mol})(0.0821\,\text{L} \cdot \text{atm}/\text{mol} \cdot \text{K})(300\,\text{K})}{0.950\,\text{atm}} = 53.4\,\text{L}$$

11.52. Show that the volumes of individual gases involved in a chemical reaction (before they are mixed or after they are separated), all measured at the same temperature and pressure, are in proportion to their numbers of moles. Use the following reaction as an example:

$$2\,\text{NO} + O_2 \longrightarrow 2\,\text{NO}_2$$

Ans.

$$\frac{n_{O_2}}{n_{NO}} = \frac{PV_{O_2}/RT}{PV_{NO}/RT} = \frac{V_{O_2}}{V_{NO}}$$

If gas *mixtures* were involved, their volumes would necessarily be the same and their *pressures* would be in the ratios of their numbers of moles.

11.53. What is the ratio of volume of NO_2 produced to O_2 used up, both at the same temperature and pressure, for the reaction in the last problem?

Ans. The ratio is 2 : 1.

11.54. (*a*) Calculate the initial volume of 5.00 L of gas whose pressure has been increased from 1.00 atm to 2.00 atm. (*b*) Calculate the final volume of 5.00 L of gas whose pressure has been increased from 1.00 atm to 2.00 atm.

Ans. Read the problem carefully. What must be assumed in part (*a*)? in part (*b*)? Since no mention was made of temperature, we must assume that the temperature is constant in both parts, or we cannot do the problem. In part (*a*), the initial volume was asked for, so we must assume that the 5.00 L is the final volume.

$$V_1 = \frac{V_2 P_2}{P_1} = \frac{(5.00\,\text{L})(2.00\,\text{atm})}{(1.00\,\text{atm})} = 10.0\,\text{L}$$

(b) The final volume was asked for, so the 5.00 L must be the initial volume.

$$V_2 = \frac{V_1 P_1}{P_2} = \frac{(5.00\,\text{L})(1.00\,\text{atm})}{(2.00\,\text{atm})} = 2.50\,\text{L}$$

(In each case, doubling the pressure reduced the volume to half its initial volume.)

11.55. Calculate the final volume of 5.00 L of gas at constant temperature whose pressure is doubled.

Ans.   The final volume is 2.50 L. According to Boyle's law, doubling the pressure causes V to be halved.

11.56. In a certain experiment, when 3.000 g of $KClO_3$ was heated, some O_2 was driven off. After the experiment, 2.890 g of solid was left. (Not all the $KClO_3$ decomposed.) (a) Write a balanced chemical equation for the reaction. (b) What compounds make up the solid? (c) What causes the loss of mass of the sample? (d) Calculate the volume of oxygen produced at STP.

Ans.

(a) $2\,KClO_3 \longrightarrow 2\,KCl + 3\,O_2$

(b) The solid is KCl and unreacted $KClO_3$.
(c) The loss of mass is caused solely by the escape of O_2.

(d) $0.110\,\text{g}\,O_2 \left(\dfrac{1\,\text{mol}\,O_2}{32.0\,\text{g}\,O_2} \right) = 0.00344\,\text{mol}\,O_2$

$$V = \frac{nRT}{P} = \frac{(0.00344\,\text{mol})(0.0821\,\text{L}\cdot\text{atm}/\text{mol}\cdot\text{K})(273\,\text{K})}{1.00\,\text{atm}} = 0.0771\,\text{L}$$
$$= 77.1\,\text{mL}$$

11.57. Replot the data of Table 11-1 using P and $1/V$. Is the result a straight line? Explain.

Ans.   The result is still a straight line. $PV = k$ so $P = k(1/V)$ and also $V = k(1/P)$.

11.58. Calculate the ratios of volume to Celsius temperature for the data in Table 11-3. Are the ratios the same for all the temperatures?

11.59. What is the final temperature or pressure in each of the following parts? (a) The pressure of a sample of gas at STP is raised 3 atm. (b) The pressure of a sample of gas at STP is raised to 3 atm. (c) The temperature of a sample of gas at STP is raised 30 °C. (d) The temperature of a sample of gas at STP is raised to 30 °C. (e) The Kelvin temperature of a sample of gas at STP is raised 300 °C. (f) The Kelvin temperature of a sample of gas at STP is raised to 300 °C.

11.60. For a mixture of two gases—gas 1 and gas 2—put in equal signs or plus signs in the appropriate places in the equations below:

$$P_{\text{total}} = P_1 \quad P_2$$
$$V_{\text{total}} = V_1 \quad V_2$$
$$T_{\text{total}} = T_1 \quad T_2$$
$$n_{\text{total}} = n_1 \quad n_2$$

Ans.

$$P_{\text{total}} = P_1 + P_2$$
$$V_{\text{total}} = V_1 = V_2$$
$$T_{\text{total}} = T_1 = T_2$$
$$n_{\text{total}} = n_1 + n_2$$

11.61. Draw a figure to represent all the conversions learned so far, including volumes of gases. Start with Fig. 10-3.

Ans.   See Fig. 11-10.

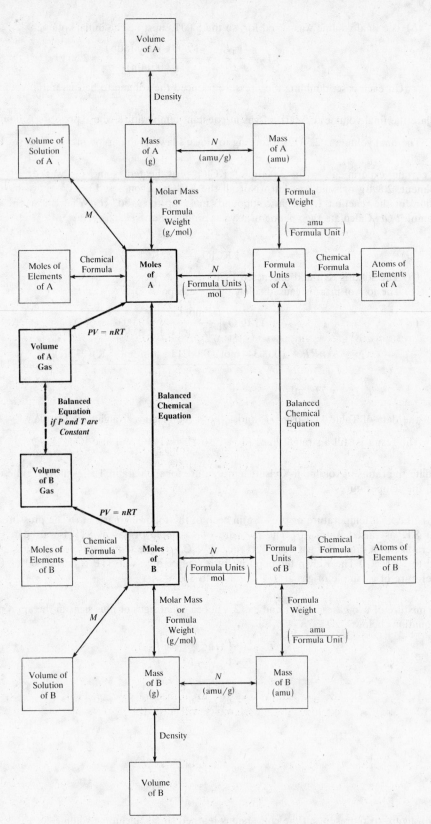

Fig. 11-10 Conversions including gases

Chapter 12

Kinetic Molecular Theory

12.1 INTRODUCTION

In Chap. 11 the laws governing the behavior of gases were presented. The fact that gases exert pressure was stated, but no reasons why gases should exhibit such behavior were given. The *kinetic molecular theory* explains all the gas laws that we have studied and some additional ones also. It describes gases in terms of the behavior of the molecules that make them up.

12.2 POSTULATES OF THE KINETIC MOLECULAR THEORY

All gases, under ordinary conditions of temperature and pressure, are made of molecules (including one-atom molecules such as are present in samples of the noble gases). That is, ionic substances do not form gases under conditions prevalent on earth. The molecules of a gas act according to the following postulates:
1. Molecules are in constant random motion.
2. The molecules exhibit negligible intermolecular attractions or repulsions except when they collide.
3. Molecular collisions are elastic, which means that although the molecules transfer energy from one to another, as a whole they do not lose kinetic energy when they collide with each other or with the walls of their container.
4. The molecules occupy a negligible fraction of the volume occupied by the gas as a whole.
5. The *average kinetic energy* of the gas molecules is directly proportional to the absolute temperature of the gas.

$$\overline{\text{KE}} = \tfrac{3}{2}kT = \tfrac{1}{2}\overline{mv^2}$$

The overbar means "average." The k in the proportionality constant is called the *Boltzmann constant*. It is equal to R, the gas constant, divided by Avogadro's number. Note that k is the same for all gases.

Postulate 1 means that the molecules move in any direction whatever until they collide with another molecule or a wall, whereupon they bounce off and move in another direction until their next collision. Postulate 2 means that the molecules move in a straight line at constant speed between collisions. Postulate 3 means that there is no friction in molecular collisions. The molecules have the same total kinetic energy after the collision as before. Postulate 4 concerns the volume of the molecules themselves versus the volume of the container they occupy. The individual particles do not occupy the entire container. If the molecules of gas had zero volumes and zero intermolecular attractions and repulsions, the gas would obey the ideal gas law exactly. Postulate 5 means that if two gases are at the same temperature, their molecules will have the same average kinetic energies.

12.3 GAS PRESSURE, BOYLE'S LAW, AND CHARLES' LAW

Kinetic molecular theory explains why gases exert pressure. The constant bombardment of the walls of the vessel by the gas molecules, like the hitting of a target by machine gun bullets, causes a constant force to be applied to the wall. The force applied, divided by the area of the wall, is the pressure of the gas.

Boyle's law may be explained using the kinetic molecular theory by considering the box illustrated in Fig. 12-1. If a sample of gas is placed in the left half of the box shown in the figure, it will exert a certain pressure. If the volume is doubled by extending the right wall to include the entire box shown in the figure, the pressure should fall to half its original value. Why should that happen? In an oversimplified picture, the molecules bouncing back and forth between the right and left walls now have twice as far to travel, and thus hit each wall only half as often in a given time. Therefore, the pressure is only half what it was. How about the molecules that are traveling up and down or in and out? There are as many such molecules as there were before and they hit the walls as often, but they are now striking an area twice as large, and so the pressure is half what it was originally. Thus, doubling the volume halves the pressure. This can be shown to be true no matter what the shape of the container.

Fig. 12-1   Explanation of Boyle's law

Charles' law governs the volume of a gas at constant pressure when its temperature is changed. When the absolute temperature of a gas is multiplied by 4, for example, the average kinetic energy of its molecules is also multiplied by 4 (postulate 5). The kinetic energy of any particle is given by $KE = \frac{1}{2}mv^2$, where m is the mass of particle and v is its velocity. When the kinetic energy is multiplied by 4, what happens to the velocity? It is doubled.

$$KE_1 = \tfrac{1}{2}mv_1^{\,2}$$
$$KE_2 = \tfrac{1}{2}mv_2^{\,2} = 4KE_1 = \tfrac{1}{2}m(2v_1)^2$$

The velocity v_2 is equal to $2v_1$. On average, in a sample of gas the molecules are going twice as fast at the higher temperature. They therefore hit the walls (1) twice as often per unit time and (2) twice as hard each time they hit, for a combined effect of 4 times the pressure (in a given volume). If we want a constant pressure, we have to expand the volume to 4 times what it was before, and we see that multiplying the absolute temperature by 4 must be accompanied by a fourfold increase in volume if the pressure is to be kept constant.

12.4   GRAHAM'S LAW

An experimental law not yet discussed is *Graham's law*, which states that the rate of effusion or diffusion of a gas is inversely proportional to the square root of its molecular weight. *Effusion* is the passage of a gas through small holes in its container, such as the pores in a porous cup. The deflation of a helium-filled party balloon over several days results from the helium atoms (molecules) effusing through the tiny pores of the balloon wall. *Diffusion* is the passage of a gas through another gas. For example, if a bottle of ammonia water is spilled in one corner of a room, soon the odor of ammonia is apparent throughout the room. The ammonia molecules have diffused through the air molecules. Consider two gases with molecular weights MW_1 and MW_2. The ratio of their rates of effusion or diffusion is given by

$$\frac{r_1}{r_2} = \sqrt{\frac{MW_2}{MW_1}}$$

That is, the heavier a molecule of the gas, the slower it effuses or diffuses.

Graham's law may be explained in terms of the kinetic molecular theory as follows: Since the two gases are at the same temperature, their average kinetic energies are the same:

$$\overline{KE_1} = \overline{KE_2} = \tfrac{1}{2}m_1\overline{v_1^{\,2}} = \tfrac{1}{2}m_2\overline{v_2^{\,2}}$$

Multiplying the last of these equations by 2 yields

$$m_1\overline{v_1}^2 = m_2\overline{v_2}^2$$

Simplifying:

$$\frac{m_1}{m_2} = \frac{\overline{v_2}^2}{\overline{v_1}^2}$$

Since the masses of the molecules are proportional to their molecular weights and the average velocity of the molecules is a measure of the rate of effusion or diffusion, all we have to do to this equation to get Graham's law is to take its square root. (The square root of $\overline{v}^2$ is not quite equal to the average velocity, but is a quantity called the "root mean square velocity." See Problem 12.18.)

Solved Problems

POSTULATES OF THE KINETIC MOLECULAR THEORY

12.1. (a) Calculate the volume at 100 °C of 18.0 g of liquid water, assuming the density to be 1.00 g/mL. (b) Calculate the volume of 18.0 g of water vapor at 100 °C and 1.00 atm pressure using the ideal gas law. (c) Assuming that the volume of the liquid is the total volume of the molecules themselves, calculate the percentage of the gas volume occupied by molecules.

Ans.

(a)
$$18.0 \, g\left(\frac{1.00 \, mL}{1.00 \, g}\right) = 18.0 \, mL = 0.0180 \, L$$

(b)
$$18.0 \, g\left(\frac{1 \, mol}{18.0 \, g}\right) = 1.00 \, mol$$

$$V = \frac{nRT}{P} = \frac{(1.00 \, mol)(0.0821 \, L \cdot atm/mol \cdot K)(373 \, K)}{1.00 \, atm} = 30.6 \, L$$

(c) The percentage occupied by the molecules is

$$\left(\frac{0.0180 \, L}{30.6 \, L}\right) \times 100\% = 0.0588\%$$

12.2. If two different gases are at the same temperature, which of the following must also be equal? (a) Their pressures, (b) their average molecular velocities, or (c) the average kinetic energies of their molecules.

Ans. (c) Their average kinetic energies must be equal since the temperatures are equal.

12.3. Does the kinetic molecular theory state that all the molecules of a given sample of gas have the same velocity since they are all at one temperature?

Ans. No. The kinetic molecular theory states that the *average* kinetic energy is related to the temperature, not the velocity or kinetic energy of any one molecule. The velocity of each individual molecule changes as it strikes other molecules or the walls.

12.4. Calculate the temperature at which oxygen molecules have the same "average" velocity as hydrogen molecules have 273 K.

Ans. Let

$$\overline{v} = \overline{v_{O_2}} = \overline{v_{H_2}}$$

$$\frac{T_{O_2}}{T_{H_2}} = \frac{\overline{KE_{O_2}}}{\overline{KE_{H_2}}} = \frac{m_{O_2}\overline{v^2}}{m_{H_2}\overline{v^2}} = \frac{m_{O_2}}{m_{H_2}} = \frac{32.00}{2.016} = 15.87$$

$$T_{O_2} = 15.87 T_{H_2} = 15.87(273 \, K) = 4330 \, K$$

12.5. Calculate the value for k, the Boltzmann constant, using the following value for R:
$$R = 8.31 \text{ J/mol} \cdot \text{K}$$

Ans.

$$k = \frac{R}{N} = \frac{8.31 \text{ J/mol} \cdot \text{K}}{6.02 \times 10^{23} \text{ molecules/mol}} = \frac{1.38 \times 10^{-23} \text{ J}}{\text{molecule} \cdot \text{K}}$$

12.6. If the molecules of a gas are compressed so that their average distance of separation gets smaller, what should happen to the forces between them? To their ideal behavior?

Ans. Their average distance gets smaller, and so the intermolecular forces increase. As the intermolecular forces increase and the volume of the molecules themselves become a more significant fraction of the total gas volume, their behavior becomes further from ideal.

GAS PRESSURE, BOYLE'S LAW, AND CHARLES' LAW

12.7. Suppose that we double the length of each side of a rectangular box containing a gas. (*a*) What would happen to the volume? (*b*) What would happen to the pressure? (*c*) Explain the effect on the pressure on the basis of the kinetic molecular theory.

Ans. (*a*) The volume will increase by a factor of $(2)^3 = 8$. (*b*) The pressure will fall to one-eighth its original value. (*c*) In each direction, the molecules would hit the wall only half as often and the force on each wall would drop to half of what it was originally. Each wall has 4 times the area, and so the pressure will be reduced to one-fourth its original value because of this effect. The total reduction in pressure is $\frac{1}{2} \times \frac{1}{4} = \frac{1}{8}$, in agreement with Boyle's law.

GRAHAM'S LAW

12.8. (*a*) If the velocity of a single gas molecule doubles, what happens to its kinetic energy? (*b*) If the average velocity of the molecules of a gas doubles, what happens to the temperature of the gas?

Ans.

(*a*) $v_2 = 2v_1$

$$\text{KE}_2 = \tfrac{1}{2}mv_2^2 = \tfrac{1}{2}m(2v_1)^2 = 4\left(\tfrac{1}{2}mv_1^2\right) = 4\text{KE}_1$$

The kinetic energy is increased by a factor of 4.
(*b*) The absolute temperature is increased by a factor of 4.

12.9. Would it be possible to separate isotopes using the principle of Graham's law? Explain what factors would be important.

Ans. Since the molecules containing different isotopes have different masses, it is possible to separate them on the basis of their different "average" molecular velocities. One would have to have gaseous molecules in which the element being separated into its isotopes is the only element present in more than one isotope. For example, if uranium is being separated, a gaseous compound of uranium is needed. The compound could be made with chlorine perhaps, but chlorine exists naturally in two isotopes of its own, and many different masses of molecules could be prepared with the naturally occurring mixture. Fluorine exists 100% as ^{19}F, and its gaseous compound with uranium, UF_6, will have two different masses, corresponding to $^{235}UF_6$ and $^{238}UF_6$. Uranium has been separated into its isotopes by repeated effusion of UF_6 through towers of porous dividers. Each process enriches the individual isotopes a little, and many repetitions are required to get relatively pure isotopes.

12.10. List the different molecular masses possible in UCl_3 with ^{238}U and ^{235}U as well as ^{35}Cl and ^{37}Cl.

Ans.

$^{238}U(^{35}Cl)_3$	343 amu
$^{238}U(^{35}Cl)_2(^{37}Cl)$	345
$^{238}U(^{35}Cl)(^{37}Cl)_2$	347
$^{238}U(^{37}Cl)_3$	349
$^{235}U(^{35}Cl)_3$	340
$^{235}U(^{35}Cl)_2(^{37}Cl)$	342
$^{235}U(^{35}Cl)(^{37}Cl)_2$	344
$^{235}U(^{37}Cl)_3$	346

12.11. Calculate the ratio of rates of effusion of $^{238}UF_6$ and $^{235}UF_6$.

Ans.   The molecular masses are 352 amu and 349 amu. The relative rates of effusion are

$$\sqrt{\frac{352}{349}} = 1.004$$

The molecules containing the lighter isotope will travel on average 1.004 times as fast as those containing the heavier one.

12.12. What possible complications would there be in trying to separate hydrogen into ^1H and ^2H by gaseous diffusion?

Ans.   Hydrogen occurs as diatomic molecules, and it would be easy to separate $^1H_2$, $^1H^2H$, and $^2H_2$, but not the individual atoms. But there would be relatively little $^2H_2$ since in abundance the heavy isotope accounts for only 0.015% of naturally occurring hydrogen atoms.

Supplementary Problems

12.13. (*a*) Is the ratio

$$\frac{\text{total volume of gas molecules}}{\text{volume of gas}}$$

smaller for a given sample of gas at constant pressure at 300 K or at 400 K? (*b*) Would the gas exhibit more ideal behavior at 300 K or at 400 K?

Ans.   (*a*) The ratio is smaller at 400 K. The volume of the molecules themselves would not change appreciably between the two temperatures, but the gas volume changes according to Charles' law. (*b*) Since the gas volume is larger at 400 K, the gas molecules are farther apart at that temperature, and exhibit lower intermolecular forces. The gas would therefore be more ideal at the higher temperature.

12.14. (*a*) Is the ratio

$$\frac{\text{total volume of gas molecules}}{\text{volume of gas}}$$

smaller for a given sample of gas at constant temperature at 1.00 atm or at 2.00 atm? (*b*) Would the gas exhibit more ideal behavior at 1.00 atm or at 2.00 atm?

Ans.   (*a*) The ratio is smaller at 1.00 atm. The volume of the molecules themselves would not change appreciably between the two pressures, but the gas volume changes according to Boyle's law. (*b*) Since the gas volume is larger at 1.00 atm, the gas molecules are farther apart at that temperature, and exhibit lower intermolecular forces. The gas would therefore be more ideal at the lower pressure.

12.15. (a) Calculate the average kinetic energy of H_2 molecules at STP. (b) Does the pressure matter? (c) Does the identity of the gas matter?

    Ans.

      (a)
$$\overline{KE} = \tfrac{3}{2}kT$$
$$= 1.5(1.38 \times 10^{-23}\,\text{J/molecule} \cdot \text{K})(273\,\text{K})$$
$$= 5.65 \times 10^{-21}\,\text{J}$$

    (b) and (c) The pressure and the identity of the gas do not matter.

12.16. (a) Calculate the "average" velocity of H_2 molecules at STP. (b) Does the pressure matter? (c) Does the identity of the gas matter?

    Ans.

      (a)
$$\tfrac{1}{2}m\overline{v^2} = 5.65 \times 10^{-21}\,\text{J} \quad \text{(from the prior problem)}$$
$$m_{H_2} = 2.016\,\text{amu}\left(\frac{1\,\text{g}}{6.02 \times 10^{23}\,\text{amu}}\right)\left(\frac{1\,\text{kg}}{1\,000\,\text{g}}\right) = 3.35 \times 10^{-27}\,\text{kg}$$
$$\overline{v^2} = 3.37 \times 10^6\,(\text{m/s})^2$$
$$v_{\text{rms}} = 1.84 \times 10^3\,\text{m/s} = 1.84\,\text{km/s} \quad \text{(more than 1 mile/s)}$$

    The square root of the average of the square of the velocity, v_{rms}, is not the average velocity, but a quantity called the "root mean square velocity." (b) The pressure of the gas does not matter. (c) The identity of the gas is important, because the mass of the molecule is included in the calculation. (Contrast this conclusion with that of the previous problem.)

12.17. Explain why helium atoms are classified as molecules in this chapter.

    Ans.   The gas laws work for unbonded atoms as well as for multiatom molecules, and so it is convenient to classify the unbonded atoms as molecules. If these atoms were not classified as molecules, it would be harder to state the postulates of the kinetic molecular theory. For example, postulate 1 would have to be stated: "Molecules or unbonded atoms are in constant random motion."

12.18. (a) Calculate the square of each of the following numbers: 1, 2, 3, 4, and 5. (b) Calculate the average of the numbers. (c) Calculate the average of the squares. (d) Is the square root of the average of the squares equal to the average of the numbers? (e) Explain why quotation marks are used with "average" velocity in the text for the velocity of molecules with average kinetic energies.

    Ans.   (a)

Number	Square
1	1
2	4
3	9
4	16
5	25

    (b) and (c)   Averages: 3   11

    (d) The square root of the average of the squares (11) is not equal to the average of the numbers (3).

    (e) The velocity, equal to the square root of the quotient of the average kinetic energy divided by the molecular mass, is not really the average velocity. That is, the square root of $\overline{v^2}$ does not give $\overline{v}$, but the root mean square velocity.

12.19. Contrast the motions of the molecules of a gas at rest to those in a hurricane wind.

    Ans.   At rest, as many molecules are traveling on average in any given direction as in the opposite direction, and at the same average speeds. In a hurricane, the molecules on average are traveling somewhat faster in the direction of the wind than in the opposite direction.

12.20. Hydrogen gas and helium gas are at the same temperature. What is the ratio of the "average" velocities of their molecules?

   Ans. Since the temperatures are the same, so are the average kinetic energies of their molecules. From Graham's law,

$$\frac{v_1}{v_2} = \sqrt{\frac{MW_2}{MW_1}} = \sqrt{\frac{4.0}{2.0}} = 1.4$$

   The hydrogen molecules are moving, on average, 1.4 times as fast as the helium molecules.

Chapter 13

Oxidation Numbers

13.1 INTRODUCTION

We learned to write formulas of ionic compounds in Chaps. 5 and 6. We balanced the charges to determine the number of each ion to use in the formula. We could not do the same thing for atoms of elements in covalent compounds, because in these compounds the atoms do not have charges. In order to overcome this difficulty, we define *oxidation numbers*, also called *oxidation states*.

In Sec. 13.2 we will learn to determine oxidation numbers from the formulas of compounds and ions. We will learn how to assign oxidation numbers from electron dot diagrams and more quickly from a short set of rules. We use these oxidation numbers for naming the compounds or ions (Chap. 6 and Sec. 13.4) and to balance equations for oxidation-reduction reactions (Sec. 13.5). In Sec. 13.3 we will learn to predict oxidation numbers for the elements from their positions in the periodic table in order to be able to predict formulas for their compounds and ions.

13.2 ASSIGNING OXIDATION NUMBERS

In Sec. 5.4, electron dot diagrams were introduced. The electrons shared between atoms were counted as "belonging" to both atoms. We thus counted more valence electrons than we actually had. For oxidation numbers, however, we can count each electron only once. If electrons are shared, we arbitrarily assign "control" of them to the more electronegative atom. For atoms of the same element, each atom is assigned half of the shared electrons. The oxidation number is then defined as the number of valence electrons in the free atom minus the number "controlled" by the atom in the compound. If we actually transfer the electrons from one atom to another, the oxidation number equals the resulting charge. If we share the electrons, the oxidation number does not equal the charge; there may be no charge. In this case, "control" is not meant literally, but is just a term to describe the counting procedure. For example, the electron dot diagram of CO_2 may be written

$$:\ddot{O}::C::\ddot{O}:$$

Since O is to the right of C in the second period of the periodic table, O is more electronegative, and we assign "control" of all eight shared electrons to the two O atoms. (It does not really have complete control of the electrons; if it did, the compound would be ionic.) Thus, the oxidation number of each atom is calculated as follows:

	C	Each O atom
number of valence electrons in free atom	4	6
− number of valence electrons "controlled"	0	8
oxidation number	+4	−2

Like the charge on an ion, *each* atom is assigned an oxidation number. Do not say that oxygen has an oxidation number of −4 because the two oxygen atoms together control four more electrons in the compound than they would in the free atoms. *Each* oxygen atom has an oxidation number of −2.

Although the assignment of control of electrons is somewhat arbitrary, the total number of electrons is accurately counted, which leads to a main principle of oxidation numbers:

The total of the oxidation numbers of all the atoms (not just all the elements) is equal to the net charge on the molecule or ion.

For example, the total of the three oxidation numbers in CO_2 is $4 + 2(-2) = 0$, and the charge on CO_2 is 0.

One principal source of student errors is confusion between the charge and oxidation number. Do not confuse them. To help keep them distinct, use Roman numerals to represent positive oxidation numbers. (The Romans did not have negative numbers.) In this book, charges have the numeral first followed by the sign; oxidation numbers have the sign first followed by the numeral. You might want to write oxidation numbers encircled and under the symbol for the element. Individual covalently bonded atoms do not have easily calculated charges, but they do have oxidation numbers. For example, the oxidation number of each element and the charge on the ion for CO_3^{2-} and for Cl^- are shown below.

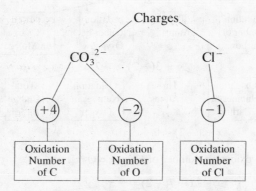

It is too time-consuming to calculate oxidation numbers by drawing electron dot diagrams each time. We can speed up the process by learning the following simple rules:
1. The sum of all the oxidation numbers in a species is equal to the charge on the species.
2. The oxidation number of uncombined elements is equal to 0.
3. The oxidation number of every monatomic ion is equal to its charge.
4. In its compounds, the oxidation number of every alkali metal and alkaline earth metal is equal to its group number.
5. The oxidation number of hydrogen in compounds is $+1$ except when the hydrogen is combined with active metals; then it is -1.
6. The oxidation number of oxygen in its compounds is -2 except in peroxides (where it is -1), superoxides (where it is $-\frac{1}{2}$), or in OF_2 and O_2F_2 (where it is positive). The peroxides and superoxides generally occur only with other elements in their maximum oxidation states. You will be able to recognize peroxides or superoxides by the presence of pairs of oxygen atoms and by the fact that if the compounds were normal oxides, the other element present would have too high an oxidation number (Sec. 13.3).

Na_2O_2	sodium peroxide	oxidation number of sodium = 1
BaO_2	barium peroxide	oxidation number of barium = 2
PbO_2	lead(IV) oxide	oxidation number of lead = 4 (permitted)
KO_2	potassium superoxide	oxidation number of potassium = 1
H_2O_2	hydrogen peroxide	oxidation number of hydrogen = 1

7. The oxidation number of every halogen atom in its compounds is -1 except for a chlorine, bromine, or iodine atom combined with oxygen or a halogen atom higher in the periodic table. For example, the chlorine atoms in each of the following compounds have oxidation numbers of -1:

$$CCl_4 \quad HCl \quad PbCl_2 \quad SCl_2 \quad NaCl$$

The chlorine atom in each of the following has an oxidation number different from -1:

$$ClO_2 \quad ClO_2^- \quad ClF \quad \text{(F has an oxidation number} = -1)$$

With these rules, we can quickly and easily calculate the oxidation numbers of an element most of the time from the formulas of its compounds.

EXAMPLE 13.1.   What are the oxidation numbers of chlorine and iodine in ICl_3?

Cl has an oxidation state of -1 (rule 7). I has an oxidation number of $+3$ (rule 1). I is not -1 because it is combined with a halogen higher in the periodic table.

EXAMPLE 13.2.   Calculate the oxidation number of C in (a) CO_2, (b) $CO_3{}^{2-}$, and (c) CO.

Let x = oxidation number of C in each case:

(a)

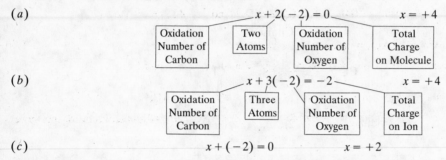

$$x + 2(-2) = 0 \qquad x = +4$$

| Oxidation Number of Carbon | Two Atoms | Oxidation Number of Oxygen | Total Charge on Molecule |

(b)

$$x + 3(-2) = -2 \qquad x = +4$$

| Oxidation Number of Carbon | Three Atoms | Oxidation Number of Oxygen | Total Charge on Ion |

(c)

$$x + (-2) = 0 \qquad x = +2$$

EXAMPLE 13.3.   Calculate the oxidation number of Cr in $Cr_2O_7{}^{2-}$.

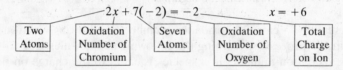

$$2x + 7(-2) = -2 \qquad x = +6$$

| Two Atoms | Oxidation Number of Chromium | Seven Atoms | Oxidation Number of Oxygen | Total Charge on Ion |

Oxidation numbers are most often, but not always, integers. If there are nonintegral oxidation numbers, there must be multiple atoms so that the number of electrons is an integer.

EXAMPLE 13.4.   Calculate the oxidation number of N in NaN_3, sodium azide.

Na has an oxidation number of $+1$. Therefore, the total charge on the three nitrogen atoms is -1 and the average charge, which is equal to the oxidation number, is $-\frac{1}{3}$. (Three nitrogen atoms, times $-\frac{1}{3}$ each, have a total of oxidation numbers equal to an integer, -1.)

13.3   PERIODIC RELATIONSHIPS OF OXIDATION NUMBERS

Oxidation numbers are very useful in correlating and systematizing a lot of inorganic chemistry. For example, the metals in very high oxidation states behave like nonmetals. They form oxyanions like $CrO_4{}^{2-}$, but do not form highly charged monatomic ions, for example. A few simple rules allow the prediction of the formulas of covalent compounds, just as predictions were made for ionic compounds in Chap. 5 using the charges on the ions. We can learn over 200 possible oxidation numbers with relative ease by learning the following rules. We will learn other oxidation numbers as we progress.
1.   All elements when uncombined have oxidation numbers equal to 0. (Some also have oxidation numbers equal to 0 in some of their compounds, by the way.)
2.   The maximum oxidation number of any atom in any of its compounds is equal to its periodic group number. There are three groups that have atoms in excess of the group number, and thus are exceptions to this rule. The coinage metals have the following maximum oxidation numbers: Cu, $+2$; Ag, $+2$; and Au, $+3$. Some of the noble gases (group 0) have positive oxidation numbers. Some lanthanoid and actinoid element oxidation numbers exceed $+3$, their nominal group number.
3.   The minimum oxidation number of any nonmetallic atom is equal to its group number minus 8. The minimum oxidation number of any metallic atom is 0.

EXAMPLE 13.5.   Give three possible oxidation numbers for sulfur.

S is in periodic group VIA, and so its maximum oxidation number is +6 and its minimum oxidation number is $6 - 8 = -2$. It also has an oxidation number of 0 when it is a free element.

EXAMPLE 13.6.   Give the possible oxidation numbers for sodium.

Na can have an oxidation number of 0 when it is a free element and +1 in all its compounds. (See rule 4, Sec. 13.2.)

EXAMPLE 13.7.   What is the maximum oxidation number of (*a*) Mn, (*b*) Os, (*c*) Mg, and (*d*) N?

(*a*) +7 (group VIIB), (*b*) +8 (group VIII), (*c*) +2 (group IIA), and (*d*) +5 (group VA).

EXAMPLE 13.8.   Can titanium (Ti) exist in an oxidation state +5?

No. Its maximum oxidation state is +4 since it is in group IV in the periodic table.

EXAMPLE 13.9.   What is the minimum oxidation state of (*a*) P, (*b*) Cl, and (*c*) Mg?

(*a*) −3 (group number − 8 = −3), (*b*) −1 (group number − 8 = −1), and (*c*) 0 (metal atoms do not have negative oxidation states).

EXAMPLE 13.10.   Name one possible binary compound of (*a*) S and O and (*b*) C and F.

The more electronegative element will take the negative oxidation state. (*a*) The maximum oxidation state of sulfur is +6; the most common negative oxidation number of oxygen is −2. Therefore, it takes three oxygen atoms to balance one sulfur atom, and the formula is SO_3. (*b*) The maximum oxidation state of carbon is +4; the only oxidation number of fluorine in its compounds is −1. Therefore, it takes four fluorine atoms to balance one carbon atom, and the formula is CF_4.

EXAMPLE 13.11.   Explain why phosphorus pentoxide cannot be PO_5.

If phosphorus pentoxide were PO_5, phosphorus would have an oxidation number of +10, which exceeds its group number. The maximum oxidation number that phosphorus can have is +5 (from group VA), and so the formula is P_2O_5. (The compound cannot be a peroxide; it was named "oxide.")

The rules above gave maximum and minimum oxidation numbers, but those might not be the only oxidation numbers or even the most important oxidation numbers for an element. Elements of the last six groups of the periodic table for example may have several oxidation numbers in their compounds, most of which vary from each other in steps of 2. For example, the major oxidation states of chlorine in its compounds are −1, +1, +3, +5, and +7. The transition metals have oxidation numbers that may vary from each other in steps of 1. The inner transition elements mostly form oxidation states of +3, but the first part of the actinoid series acts more like transition elements and the elements have

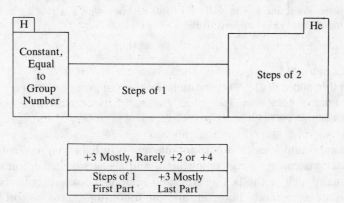

Fig. 13-1   Possible oxidation numbers

maximum oxidation numbers that increase from $+4$ for Th to $+6$ for U. These generalizations are not absolute rules, but allow students to make educated guesses about possible compound formation without exhaustive memorization. These possibilities are illustrated in Fig. 13-1.

EXAMPLE 13.12. Determine the formula of two oxides of sulfur.

The oxygen must exist in a -2 oxidation state, because it is more electronegative than is sulfur. Therefore, sulfur must exist in two different positive oxidation states in the two compounds. Its maximum oxidation state is $+6$, corresponding to its position in periodic group VIA. It also has an oxidation state of $+4$, 2 less than its maximum (see Fig. 13-1). The formulas therefore are SO_2 and SO_3.

13.4   OXIDATION NUMBERS IN INORGANIC NOMENCLATURE

In Chap. 6 we placed Roman numerals at the ends of names of metals to distinguish the charge on monatomic cations. This nomenclature system is called the *Stock system*. It is really the oxidation number that is in parentheses. For monatomic ions, the oxidation number is equal to the charge. For other cations, again the oxidation number is used in the name. For example, Hg_2^{2+} is named mercury(I) ion. Its charge is $2+$; the oxidation number of each atom is $+1$. Oxidation numbers are also used for other cations, such as oxovanadium(IV) ion, VO^{2+}. The prefix *oxo-* stands for oxygen. Oxidation numbers can be used with nonmetal-nonmetal compounds, as in sulfur(VI) oxide for SO_3, but the older system using prefixes (Table 6-2) is still used more often.

EXAMPLE 13.13. Name the following according to the Stock system: (*a*) $FeCl_2$, (*b*) UO_2SO_4, and (*c*) P_4O_{10}.

(*a*) Iron(II) chloride, (*b*) dioxouranium(VI) sulfate, and (*c*) phosphorus(V) oxide.

13.5   BALANCING OXIDATION-REDUCTION EQUATIONS

In every reaction in which the oxidation number of an element in one reactant (or more than one) goes up, an element in some reactant (or more than one) must go down in oxidation number. An increase in oxidation number is called an *oxidation*. A decrease in oxidation number is called a *reduction*. The term *redox* (the first letters of reduction and oxidation) is often used as a synonym for oxidation-reduction. The total change in oxidation number (change in each atom times number of atoms) must be the same in the oxidation as in the reduction, because the number of electrons transferred from one species must be the same as the number transferred to the other. The species that causes another to be reduced is called the *reducing agent*; in the process, it is oxidized. The species that causes the oxidation is called the *oxidizing agent*; in the process, it is reduced.

EXAMPLE 13.14. (*a*) In drying dishes, a dish towel could be termed a drying agent, and the dish a wetting agent. What happens to the towel and to the dish? (*b*) In its reaction with $FeCl_2$, Cl_2 is the oxidizing agent and the $FeCl_2$ is the reducing agent. What happens to the Cl_2 and to the Fe^{2+}?

(*a*) The towel, the drying agent, gets wet. The dish, the wetting agent, gets dry.

(*b*) $2\,FeCl_2 + Cl_2 \longrightarrow 2\,FeCl_3$

Cl_2, the oxidizing agent, is reduced to Cl^-. The oxidation number goes from 0 to -1. Fe^{2+}, the reducing agent, is oxidized. Its oxidation number goes from $+2$ to $+3$. Just as the water must go somewhere in part (*a*), the electrons must go somewhere in part (*b*).

One of the most important uses of oxidation numbers is in balancing redox (oxidation-reduction) equations. These equations can get very complicated, and a systematic method of balancing them is essential. There are many such methods, however, and each textbook seems to use its own. There are many similarities among the methods, however, and the following discussion will help no matter what method your instructor and your textbook use.

It is of critical importance in this section to keep in mind the difference between oxidation number and charge. You balance charge one way and changes in oxidation number another way.

There are two essentially different methods to balance redox reactions—the oxidation number change method and the ion-electron method. The first of these is perhaps easier, and the second is somewhat more useful, especially for electrochemical reactions (Chap. 14).

Oxidation Number Change

The total of the oxidation numbers gained in a reaction must equal the total of the oxidation numbers lost, since the numbers of electrons gained and lost must be equal. Therefore, we can balance the species in which the elements that are oxidized and reduced appear by the changes in oxidation numbers. We use the numbers of atoms of each of these elements which will give us equal numbers of electrons gained and lost. If necessary, first balance the number of atoms of the element oxidized and/or the number of atoms of the element reduced. Finally, we balance the rest of the species by inspection, as we did in Chap. 7.

EXAMPLE 13.15.   Balance the following equation:

$$? \, HCl + ? \, HNO_3 + ? \, FeCl_2 \longrightarrow ? \, FeCl_3 + ? \, NO + ? \, H_2O$$

By inspecting the oxidation states of all the elements, we find that the Fe goes from $+2$ to $+3$ and the N goes from $+5$ to $+2$.

$$
\begin{array}{c}
(+2 \rightarrow +3) = +1 \\
? \, HCl + ? \, HNO_3 + ? \, \overline{FeCl_2} \longrightarrow ? \, FeCl_3 + ? \, NO + ? \, H_2O \\
(+5 \rightarrow +2) = -3
\end{array}
$$

We see that Fe and N are the only elements undergoing change in oxidation number. To balance the oxidation numbers lost and gained, we need three $FeCl_2$ and three $FeCl_3$ for each N atom reduced:

$$? \, HCl + 1 \, HNO_3 + 3 \, FeCl_2 \longrightarrow 3 \, FeCl_3 + 1 \, NO + ? \, H_2O$$

It is now easy to balance the HCl by balancing the Cl atoms and to balance the H_2O by balancing the O atoms:

$$3 \, HCl + 1 \, HNO_3 + 3 \, FeCl_2 \longrightarrow 3 \, FeCl_3 + 1 \, NO + 2 \, H_2O$$

or

$$3 \, HCl + HNO_3 + 3 \, FeCl_2 \longrightarrow 3 \, FeCl_3 + NO + 2 \, H_2O$$

There are now four H atoms on each side, and the equation is balanced.

EXAMPLE 13.16.   Balance the following equation:

$$HCl + KMnO_4 + H_2C_2O_4 \longrightarrow CO_2 + MnCl_2 + KCl + H_2O$$

Here, the Mn is reduced and the C is oxidized. Before attempting to balance the oxidization numbers gained and lost, we first have to balance the number of carbon atoms. Before the carbon atoms are balanced, we cannot say that there is one carbon (as in the CO_2) or two (as in the $H_2C_2O_4$).

$$HCl + KMnO_4 + H_2C_2O_4 \longrightarrow 2 \, CO_2 + MnCl_2 + KCl + H_2O$$

Now proceed as before:

$$
\begin{array}{c}
(+7 \rightarrow +2) = -5 \\
HCl + KMnO_4 + H_2\underline{C_2}O_4 \longrightarrow 2 \, \overline{CO_2} + MnCl_2 + KCl + H_2O \\
2(+3 \rightarrow +4) = +2
\end{array}
$$

$$HCl + 2 \, KMnO_4 + 5 \, H_2C_2O_4 \longrightarrow 10 \, CO_2 + 2 \, MnCl_2 + KCl + H_2O$$

Balance the H_2O from the number of O atoms, the KCl by the number of K atoms, and finally the HCl by the number of Cl or H atoms. Check.

$$6 \, HCl + 2 \, KMnO_4 + 5 \, H_2C_2O_4 \longrightarrow 10 \, CO_2 + 2 \, MnCl_2 + 2 \, KCl + 8 \, H_2O$$

It is easier to balance redox equations in net ionic form than in overall form.

EXAMPLE 13.17.   Balance:

$$H^+ + MnO_4^- + H_2C_2O_4 \longrightarrow CO_2 + Mn^{2+} + H_2O$$

First balance the C atoms. Then balance the elements changing oxidation state. Then balance the rest of the atoms by inspection.

$$6H^+ + 2MnO_4^- + 5H_2C_2O_4 \longrightarrow 10CO_2 + 2Mn^{2+} + 8H_2O$$

The net charge (4+) and the numbers of atoms of each element are all equal on the two sides.

Ion-Electron, Half-Reaction Method

In the ion-electron method of balancing redox equations, an equation for the oxidation half-reaction and one for the reduction half-reaction are written and balanced separately. Only when each of these is complete and balanced are the two combined into one complete equation for the reaction as a whole. It is worthwhile to balance the half-reactions separately since the two half-reactions can be carried out in separate vessels if they are suitably connected electrically. (See Chap. 14.) In general, net ionic equations are used in this process; certainly some ions are required in each half-reaction. In the equations for the two half-reactions, electrons appear explicitly; in the equation for the complete reaction—the combination of the two half-reactions—no electrons are included.

During the balancing of redox equations by the ion-electron method, species may be added to one side of the equation or the other. These species are present in the solution, and their inclusion in the equation indicates that they also react. Water is a prime example. When it reacts, the chemist is not likely to notice that there is less water left. In general, all of the formulas for the atoms that change oxidation number will be given to you in equations to balance. Sometimes, the formula of a compound or ion of oxygen will be omitted, but rarely will the formula for a compound or ion of some other element not be provided. If a formula is not given, see if you can figure out what that formula might be by considering the possible oxidation states (Sec. 13.3).

EXAMPLE 13.18.   Guess a formula for the oxygen-containing product in the following redox reaction:

$$H_2O_2 + I^- \longrightarrow I_2 +$$

Since the iodine is oxidized (from -1 to 0), the oxygen must be reduced. It starts out in the -1 oxidation state; it is reduced to the -2 oxidation state. Water (or OH^-) is the probable product.

There are many methods of balancing redox equations by the half-reaction method. One such method is presented here. You should do steps 1 through 5 for one half-reaction and then those same steps for the other half-reaction before proceeding to the rest of the steps.
1. Identify the element(s) oxidized and the element(s) reduced. Write a separate half-reaction for each of these.
2. Balance these elements.
3. Balance the change in oxidation number by adding electrons to the side with the higher total of oxidation numbers. That is, add electrons on the left for a reduction half-reaction and on the right for an oxidation half-reaction. One way to remember on which side to add the electrons is the following mnemonic:

Loss of Electrons is Oxidation (LEO).
Gain of Electrons is Reduction (GER). "LEO the Lion says 'GER'."

4. In acid solution, balance the net charge with hydrogen ions, H^+.
5. Balance the hydrogen and oxygen atoms with water.
6. Multiply every item in one or both equations by small integers, if necessary, so that the number of electrons is the same in each. The same small integer is used throughout each half-reaction, and is different from that used in the other half-reaction. Then add the two half-reactions.

7. Cancel all species that appear on both sides of the equation. All the electrons must cancel out in this step, and often some hydrogen ions and water molecules also cancel.

8. Check to see that atoms of all the elements are balanced and that the net charge is the same on both sides of the equation.

Note that all the added atoms in steps 4 and 5 have the same oxidation number as the atoms already in the equation. The atoms changing oxidation number have already been balanced in steps 1 and 2.

EXAMPLE 13.19.   Balance the following equation by the ion-electron, half-reaction method:

$$Fe^{2+} + H^+ + NO_3^- \longrightarrow Fe^{3+} + NO + H_2O$$

Step 1: $Fe^{2+} \longrightarrow Fe^{3+}$ $\qquad\qquad\qquad\qquad\qquad\qquad\qquad\qquad\qquad NO_3^- \longrightarrow NO$

Step 2: The Fe and N are already balanced.

Step 3: $Fe^{2+} \longrightarrow Fe^{3+} + e^-$ $\qquad\qquad\qquad\qquad\qquad\qquad\quad 3e^- + NO_3^- \longrightarrow NO$

Step 4: $Fe^{2+} \longrightarrow Fe^{3+} + e^-$ $\qquad\qquad\qquad\qquad\qquad\quad 4H^+ + 3e^- + NO_3^- \longrightarrow NO$

Step 5: $Fe^{2+} \longrightarrow Fe^{3+} + e^-$ $\qquad\qquad\qquad\quad 4H^+ + 3e^- + NO_3^- \longrightarrow NO + 2H_2O$

Step 6: Multiplying by 3:

$$3Fe^{2+} \longrightarrow 3Fe^{3+} + 3e^-$$

Adding:

$$3Fe^{2+} + 4H^+ + 3e^- + NO_3^- \longrightarrow NO + 2H_2O + 3Fe^{3+} + 3e^-$$

Step 7: $3Fe^{2+} + 4H^+ + NO_3^- \longrightarrow NO + 2H_2O + 3Fe^{3+}$

Step 8: The equation is balanced. There are three Fe atoms, four H atoms, one N atom, and three O atoms on each side. The net charge on each side is 9+.

EXAMPLE 13.20.   Complete and balance the following equation in acid solution:

$$Cr_2O_7^{2-} + I_2 \longrightarrow IO_3^- + Cr^{3+}$$

Step 1: $Cr_2O_7^{2-} \longrightarrow Cr^{3+}$

$$I_2 \longrightarrow IO_3^-$$

Step 2: $Cr_2O_7^{2-} \longrightarrow 2Cr^{3+}$

$$I_2 \longrightarrow 2IO_3^-$$

Step 3: 2 atoms are reduced 3 units each and 2 atoms are oxidized 5 units each

$$6e^- + Cr_2O_7^{2-} \longrightarrow 2Cr^{3+}$$

$$I_2 \longrightarrow 2IO_3^- + 10e^-$$

Step 4: $14H^+ + 6e^- + Cr_2O_7^{2-} \longrightarrow 2Cr^{3+}$

$$I_2 \longrightarrow 2IO_3^- + 10e^- + 12H^+$$

Step 5: $14H^+ + 6e^- + Cr_2O_7^{2-} \longrightarrow 2Cr^{3+} + 7H_2O$

$$6H_2O + I_2 \longrightarrow 2IO_3^- + 10e^- + 12H^+$$

Step 6: $5[14\,H^+ + 6\,e^- + Cr_2O_7^{2-} \longrightarrow 2\,Cr^{3+} + 7\,H_2O]$

$$3[6\,H_2O + I_2 \longrightarrow 2\,IO_3^- + 10\,e^- + 12\,H^+]$$

$$70\,H^+ + 30\,e^- + 5\,Cr_2O_7^{2-} \longrightarrow 10\,Cr^{3+} + 35\,H_2O$$

$$18\,H_2O + 3\,I_2 \longrightarrow 6\,IO_3^- + 30\,e^- + 36\,H^+$$

Step 7: $18\,H_2O + 3\,I_2 + 70\,H^+ + 30\,e^- + 5\,Cr_2O_7^{2-} \longrightarrow 10\,Cr^{3+} + 35\,H_2O + 6\,IO_3^- + 30\,e^- + 36\,H^+$

$$3\,I_2 + 34\,H^+ \qquad\quad + 5\,Cr_2O_7^{2-} \longrightarrow 10\,Cr^{3+} + 17\,H_2O + 6\,IO_3^-$$

Step 8:   The equation is balanced. There are 6 I atoms, 34 H atoms, 10 Cr atoms, and 35 O atoms on each side, as well as a net 24+ charge on each side.

If a reaction is carried out in basic solution, the same process may be followed. After all the other steps have been completed, any H^+ can be neutralized by adding OH^- ions to each side, creating water and excess OH^- ions. The water created may be combined with any water already on that side or may cancel any water on the other side.

EXAMPLE 13.21.   Complete and balance the following equation in basic solution:

$$Fe(OH)_2 + CrO_4^{2-} \longrightarrow Fe(OH)_3 + Cr(OH)_3$$

Step 1: $Fe(OH)_2 \longrightarrow Fe(OH)_3$

$$CrO_4^{2-} \longrightarrow Cr(OH)_3$$

Step 2:   Already done.

Step 3: $Fe(OH)_2 \longrightarrow Fe(OH)_3 + e^-$

$$3\,e^- + CrO_4^{2-} \longrightarrow Cr(OH)_3$$

Step 4: $Fe(OH)_2 \longrightarrow Fe(OH)_3 + e^- + H^+$

$$5\,H^+ + 3\,e^- + CrO_4^{2-} \longrightarrow Cr(OH)_3$$

Step 5: $H_2O + Fe(OH)_2 \longrightarrow Fe(OH)_3 + e^- + H^+$

$$5\,H^+ + 3\,e^- + CrO_4^{2-} \longrightarrow Cr(OH)_3 + H_2O$$

Step 6: $3\,H_2O + 3\,Fe(OH)_2 \longrightarrow 3\,Fe(OH)_3 + 3\,e^- + 3\,H^+$

$$5\,H^+ + 3\,e^- + CrO_4^{2-} \longrightarrow Cr(OH)_3 + H_2O$$

$$5\,H^+ + 3\,e^- + CrO_4^{2-} + 3\,H_2O + 3\,Fe(OH)_2 \longrightarrow 3\,Fe(OH)_3 + 3\,e^- + 3\,H^+ + Cr(OH)_3 + H_2O$$

$$2\,H^+ + CrO_4^{2-} + 2\,H_2O + 3\,Fe(OH)_2 \longrightarrow 3\,Fe(OH)_3 + Cr(OH)_3$$

To eliminate the H^+, which cannot exist in basic solution, add $2\,OH^-$ to each side, forming $2\,H_2O$ on the left:

$$2\,H_2O + CrO_4^{2-} + 2\,H_2O + 3\,Fe(OH)_2 \longrightarrow 3\,Fe(OH)_3 + Cr(OH)_3 + 2\,OH^-$$

Finally,

$$4\,H_2O + CrO_4^{2-} + 3\,Fe(OH)_2 \longrightarrow 3\,Fe(OH)_3 + Cr(OH)_3 + 2\,OH^-$$

Solved Problems

INTRODUCTION

13.1.   (a) What is the formula of a compound of two ions: X^{2+} and Y^-? (b) What is the formula of a compound of two elements: W with an oxidation state of $+2$ and Z with an oxidation state of -1?

Ans. We treat the oxidation states in part (*b*) just like the charges in part (*a*). In this manner, we can predict formulas for covalent and ionic compounds. (*a*) XY_2 and (*b*) WZ_2.

ASSIGNING OXIDATION NUMBERS

13.2. Draw an electron dot diagram for H_2O_2. Assign an oxidation number to oxygen on this basis. Compare this number with that assigned by rule 6 (Sec. 13.2).

Ans.

$$H:\ddot{O}:\ddot{O}:H \qquad \begin{array}{lr} \text{free atom} & 6 \\ -\text{controlled} & 7 \\ \hline \text{oxidation number} & -1 \end{array}$$

The electrons shared between the oxygen atoms are counted one for each atom. Peroxide oxygen is assigned an oxidation state of -1 by rule 6 also.

13.3. What is the sum of the oxidation numbers of all the *atoms* in the following compounds or ions? (*a*) ClO_3^-, (*b*) VO_2^+, (*c*) PO_4^{3-}, (*d*) $Cr_2O_7^{2-}$, (*e*) $NaCl$, and (*f*) CCl_4.

Ans. The sum equals the charge on the species in each case: (*a*) -1, (*b*) $+1$, (*c*) -3, (*d*) -2, (*e*) 0, and (*f*) 0.

13.4. Show that rules 2 and 3 (Sec. 13.2) are corollaries of rule 1.

Ans. Rule 2: Uncombined elements have zero charges, and so the oxidation numbers must add up to 0. Since all the atoms are the same, all the oxidation numbers must be the same—0. Rule 3: Monatomic ions have the oxidation numbers of all the atoms add up to the charge on the ion. Since there is only one atom (monatomic), the oxidation number of that atom must add up to the charge on the ion; that is, it is equal to the charge on the ion.

13.5. What is the oxidation number of chlorine in each of the following? (*a*) ClO_2, (*b*) ClO_2^-, and (*c*) ClF.

Ans. (*a*) $+4$, (*b*) $+3$, and (*c*) $+1$.

13.6. Determine the oxidation number for the underlined element: (*a*) $\underline{P}OCl_3$, (*b*) $H\underline{N}O_3$, (*c*) $Na_2\underline{S}O_3$, (*d*) $\underline{P}Cl_3$, and (*e*) $\underline{N}_2O_5$.

Ans. (*a*) $+5$, (*b*) $+5$, (*c*) $+4$, (*d*) $+3$, and (*e*) $+5$.

13.7. Determine the oxidation number for the underlined element: (*a*) $\underline{Cl}O_3^-$, (*b*) $\underline{P}O_4^{3-}$, (*c*) $\underline{C}O_3^{2-}$, and (*d*) $\underline{V}O^{2+}$.

Ans. (*a*) $+5$, (*b*) $+5$, (*c*) $+4$, and (*d*) $+4$.

13.8. Determine the oxidation number of oxygen in: (*a*) H_2O_2, (*b*) KO_2, and (*c*) OF_2.

Ans. (*a*) -1, (*b*) $-\frac{1}{2}$, and (*c*) $+2$. For part (*b*), $+1 + 2x = 0$, therefore $x = -\frac{1}{2}$.

13.9. Determine the oxidation numbers for the underlined elements: $(\underline{V}O_2)_2\underline{S}O_4$.

Ans. We recognize the sulfate ion, SO_4^{2-}. Each VO_2 ion therefore must have a $1+$ charge. The oxidation numbers are $+6$ for S and $+5$ for V.

13.10. What oxyacid of nitrogen can be prepared by adding water to N_2O_5? Hint: Both compounds have nitrogen in the same oxidation state.

Ans. HNO_3.

13.11. What is the oxidation number of Si in $Si_6O_{18}{}^{12-}$?

   Ans.

$$6x + 18(-2) = -12$$
$$x = +4$$

PERIODIC RELATIONSHIPS OF OXIDATION NUMBERS

13.12. Predict the formulas of two compounds of each of the following pairs of elements: (*a*) C and O, (*b*) Cl and O, (*c*) P and F, (*d*) P and S, (*e*) S and F, and (*f*) I and F.

   Ans.   The first element in each part is given in its highest oxidation state and in an oxidation state 2 less than the highest. The second element is in its minimum oxidation state. (*a*) CO_2 and CO, (*b*) Cl_2O_7 and Cl_2O_5, (*c*) PF_5 and PF_3, (*d*) P_2S_5 and P_2S_3, (*e*) SF_6 and SF_4, and (*f*) IF_7 and IF_5.

13.13. Write the formulas for two monatomic ions for each of the following metals: (*a*) Pb, (*b*) Tl, (*c*) Sn, and (*d*) Cu.

   Ans.   (*a*) Pb^{4+} and Pb^{2+}. (The maximum oxidation state of a group IV element and the state 2 less than the maximum.) (*b*) Tl^{3+} and Tl^+. (The maximum oxidation state of a group III element and the state 2 less than the maximum.) (*c*) Sn^{4+} and Sn^{2+}. (The maximum oxidation state of a group IV element and the state 2 less than the maximum.) (*d*) Cu^+ and Cu^{2+}. (The maximum oxidation state for the coinage metals is greater than the group number.)

13.14. Predict the formulas of three fluorides of sulfur.

   Ans. SF_6, SF_4, and SF_2. The oxidation states of sulfur in these compounds correspond to the maximum oxidation state for a group VI element and to states 2 and 4 lower. (See Fig. 13-1.)

OXIDATION NUMBERS IN INORGANIC NOMENCLATURE

13.15. Name NO_2 and N_2O_4 using the Stock system. Explain why the older system using prefixes is still useful.

   Ans.   Both compounds have nitrogen in the +4 oxidation state, so if we call NO_2 nitrogen(IV) oxide, what do we call N_2O_4? We actually use the older system for N_2O_4—nitrogen tetroxide (or dinitrogen tetroxide).

BALANCING OXIDATION-REDUCTION EQUATIONS

13.16. Why is it possible to add H^+ and/or H_2O to an equation for a reaction carried out in aqueous acid solution when none seems to be appearing or disappearing?

   Ans.   The H_2O and H^+ are present in excess in the solution. Therefore, they will react or be produced without the change being noticed much.

13.17. Identify (*a*) the oxidizing agent, (*b*) the reducing agent, (*c*) the element oxidized, and (*d*) the element reduced in the following reaction:

$$8\,H^+ + MnO_4{}^- + 5\,Fe^{2+} \longrightarrow Mn^{2+} + 5\,Fe^{3+} + 4\,H_2O$$

   Ans.   (*a*) $MnO_4{}^-$, (*b*) Fe^{2+}, (*c*) Fe, and (*d*) Mn. An element in the reducing agent is oxidized; an element in the oxidizing agent is reduced.

13.18. Balance the equation for the reduction of HNO_3 to NH_4NO_3 by Zn by the oxidation change method. Add other compounds as needed.

   Ans.

$$HNO_3 + Zn \longrightarrow NH_4NO_3 + Zn(NO_3)_2$$

The zinc goes up two oxidation numbers, and some of the nitrogen atoms are reduced from $+5$ to -3:

$$\overbrace{HNO_3 + Zn}^{(0 \to +2) = +2} \longrightarrow NH_4NO_3 + Zn(NO_3)_2$$
$$\underbrace{}_{(+5 \to -3) = -8}$$

It takes four Zn atoms per N atom reduced:

$$HNO_3 + 4\,Zn \longrightarrow 1\,NH_4NO_3 + 4\,Zn(NO_3)_2$$

There are other N atoms that were not reduced, but still are present in the nitrate ions. Additional HNO_3 is needed to provide for these.

$$10\,HNO_3 + 4\,Zn \longrightarrow NH_4NO_3 + 4\,Zn(NO_3)_2$$

Water is also produced, needed to balance both the H and O atoms:

$$10\,HNO_3 + 4\,Zn \longrightarrow NH_4NO_3 + 4\,Zn(NO_3)_2 + 3\,H_2O$$

The equation is now balanced, having 10 H atoms, 10 N atoms, 30 O atoms, and 4 Zn atoms on each side.

13.19. How many electrons are involved in a reaction of one atom with a change of oxidation number from $+3$ to -3?

   Ans. $3 - (-3) = 6$ 6 electrons are involved.

13.20. How many electrons are involved in a reaction in which one atom is reduced from an oxidation state $+5$ to an oxidation state -3?

   Ans. $5 - (-3) = 8$ 8 electrons are involved.

Supplementary Problems

13.21. Explain why we were able to use the charges on the cations in names in Chap. 6 instead of the required oxidation numbers.

   Ans.   For monatomic ions, the charge is equal to the oxidation number.

13.22. Complete and balance the following redox equations:

(a) $$MnO_4^- + Co^{2+} \longrightarrow Co^{3+} + Mn^{2+}$$
(b) $$Cr^{2+} + H_2O_2 \longrightarrow Cr^{3+} + H_2O$$
(c) $$Br_2 + OH^- \longrightarrow Br^- + BrO_3^-$$
(d) $$P_4 + OH^- \longrightarrow PH_3 + HPO_3^{2-}$$
(e) $$H_2C_2O_4 + Cr_2O_7^{2-} \longrightarrow Cr^{3+} + CO_2$$
(f) $$Fe(OH)_2 + H_2O_2 \longrightarrow Fe(OH)_3$$

Ans.

(a) $$8\,H^+ + MnO_4^- + 5\,Co^{2+} \longrightarrow 5\,Co^{3+} + Mn^{2+} + 4\,H_2O$$
(b) $$2\,H^+ + 2\,Cr^{2+} + H_2O_2 \longrightarrow 2\,Cr^{3+} + 2\,H_2O$$
(c) $$3\,Br_2 + 6\,OH^- \longrightarrow 5\,Br^- + BrO_3^- + 3\,H_2O$$
(d) $$2\,H_2O + P_4 + 4\,OH^- \longrightarrow 2\,PH_3 + 2\,HPO_3^{2-}$$
(e) $$8\,H^+ + 3\,H_2C_2O_4 + Cr_2O_7^{2-} \longrightarrow 2\,Cr^{3+} + 6\,CO_2 + 7\,H_2O$$
(f) $$2\,Fe(OH)_2 + H_2O_2 \longrightarrow 2\,Fe(OH)_3$$

13.23. Which of the equations in Problem 13.22 represent reactions in basic solution? How can you tell?

Ans. Reactions (c), (d), and (f) occur in basic solution. The presence of OH^- ions shows immediately that the solution is basic. (The presence of NH_3 also indicates basic solution; in acid solution, this base would react to form NH_4^+.) The other reactions are not in base; H^+ is present. Also, if they were in base, the metal ions would react to form insoluble hydroxides and their ionic formulas would not be given.

13.24. Balance the following equation by the oxidation number change method:

$$NH_3 + O_2 \longrightarrow NO + H_2O$$

Ans. The O_2 is reduced to the -2 oxidation state; both products contain the reduction product.

$$\overset{(-3 \,\rightarrow\, +2) \,=\, +5}{\overbrace{NH_3 + \underbrace{O_2 \longrightarrow NO + H_2O}_{2(0 \,\rightarrow\, -2) \,=\, -4}}}$$

We know that five O_2 molecules are required, but we still do not know from this reduction how many of the oxygen atoms go to NO or H_2O.

$$4\,NH_3 + 5\,O_2 \longrightarrow 4\,NO + H_2O$$
$$4\,NH_3 + 5\,O_2 \longrightarrow 4\,NO + 6\,H_2O$$

13.25. Complete and balance the following equations:

(a)
$$Zn + H^+ + NO_3^- \longrightarrow NH_4^+ + Zn^{2+}$$

(b)
$$Zn + HNO_3 \longrightarrow NH_4NO_3 + Zn(NO_3)_2$$

(c) How are these equations related? Which is easier to balance?

Ans.

(a) $\qquad\qquad\qquad\qquad\qquad\qquad\qquad\qquad\qquad\qquad\qquad Zn \longrightarrow Zn^{2+} + 2\,e^-$

$$NO_3^- \longrightarrow NH_4^+$$
$$8\,e^- + NO_3^- \longrightarrow NH_4^+$$
$$10\,H^+ + 8\,e^- + NO_3^- \longrightarrow NH_4^+$$
$$10\,H^+ + 8\,e^- + NO_3^- \longrightarrow NH_4^+ + 3\,H_2O$$

$$4\,Zn \longrightarrow 4\,Zn^{2+} + 8\,e^-$$

$$10\,H^+ + 4\,Zn + NO_3^- \longrightarrow NH_4^+ + 3\,H_2O + 4\,Zn^{2+}$$

(b) Either add $9\,NO_3^-$ to each side of the equation in part (a):

$$10\,HNO_3 + 4\,Zn \longrightarrow NH_4NO_3 + 3\,H_2O + 4\,Zn(NO_3)_2$$

or
$$2\,HNO_3 \longrightarrow NH_4NO_3$$
$$8\,e^- + 2\,HNO_3 \longrightarrow NH_4NO_3$$
$$8\,H^+ + 8\,e^- + 2\,HNO_3 \longrightarrow NH_4NO_3$$
$$8\,H^+ + 8\,e^- + 2\,HNO_3 \longrightarrow NH_4NO_3 + 3\,H_2O$$

$$Zn \longrightarrow Zn(NO_3)_2 + 2\,e^-$$
$$2\,HNO_3 + Zn \longrightarrow Zn(NO_3)_2 + 2\,e^-$$
$$2\,HNO_3 + Zn \longrightarrow Zn(NO_3)_2 + 2\,e^- + 2\,H^+$$
$$8\,HNO_3 + 4\,Zn \longrightarrow 4\,Zn(NO_3)_2 + 8\,e^- + 8\,H^+$$

$$10\,HNO_3 + 4\,Zn \longrightarrow 4\,Zn(NO_3)_2 + NH_4NO_3 + 3\,H_2O$$

(c) Part (a) is the net ionic equation for part (b). It is easier to balance part (a). To balance part (b), either amend the balanced net ionic equation of part (a), or do part (b) starting from scratch. (There must always be at least one type of ion represented in a balanced half-reaction equation.)

13.26. Complete and balance the following equations:

(a) $$Br^- + BrO_3^- \longrightarrow Br_2 + H_2O$$

(b) $$H_2O_2 + Sn^{2+} \longrightarrow Sn^{4+} + H_2O$$

(c) $$Zn + OH^- \longrightarrow Zn(OH)_4^{2-} + H_2$$

(d) $$Ce^{4+} + SO_3^{2-} \longrightarrow Ce^{3+} + SO_4^{2-}$$

(e) $$Cu_2O + H^+ \longrightarrow Cu + Cu^{2+} + H_2O$$

(f) $$H_2SO_4(conc) + Zn \longrightarrow H_2S + Zn^{2+}$$

(g) $$HNO_3 + Sn(NO_3)_2 \longrightarrow Sn(NO_3)_4 + NO$$

(h) $$H^+ + NO_3^- + Pb^{2+} \longrightarrow NO + Pb^{4+}$$

(i) $$Ag^+ + S_2O_3^{2-} \longrightarrow S_4O_6^{2-} + Ag$$

(j) $$SO_4^{2-} + I^- + H^+ \longrightarrow SO_2 + I_2 + H_2O$$

(k) $$Cr^{3+} + MnO_4^- \longrightarrow Cr_2O_7^{2-} + Mn^{2+}$$

Ans.

(a) $$6H^+ + 5Br^- + BrO_3^- \longrightarrow 3Br_2 + 3H_2O$$

(b)
$$\overset{+2}{H_2O_2 + \underset{\underset{2(-1\,\to\,-2)\,=\,-2}{\rule{0pt}{0pt}}}{Sn^{2+}} \longrightarrow Sn^{4+} + 2H_2O}$$

$$2H^+ + H_2O_2 + Sn^{2+} \longrightarrow Sn^{4+} + 2H_2O$$

(c)
$$4OH^- + Zn \longrightarrow Zn(OH)_4^{2-} + 2e^-$$
$$2e^- + 2H_2O \longrightarrow 2OH^- + H_2$$
$$\rule{6cm}{0.4pt}$$
$$2H_2O + 2OH^- + Zn \longrightarrow Zn(OH)_4^{2-} + H_2$$

(d)
$$e^- + Ce^{4+} \longrightarrow Ce^{3+}$$
$$H_2O + SO_3^{2-} \longrightarrow SO_4^{2-} + 2e^- + 2H^+$$
$$\rule{6cm}{0.4pt}$$
$$2Ce^{4+} + H_2O + SO_3^{2-} \longrightarrow SO_4^{2-} + 2H^+ + 2Ce^{3+}$$

(e) $$Cu_2O + 2H^+ \longrightarrow Cu + Cu^{2+} + H_2O$$

Simple Cu(I) compounds are not stable in acid solution. Cu(I) is stable in solid compounds like Cu_2O, but when that reacts with an acid, the Cu(I) disproportionates—reacts with itself—to produce a lower and a higher oxidation state.

(f)
$$8e^- + 8H^+ + H_2SO_4 \longrightarrow H_2S + 4H_2O$$
$$Zn \longrightarrow Zn^{2+} + 2e^-$$
$$\rule{6cm}{0.4pt}$$
$$4Zn + 8H^+ + H_2SO_4 \longrightarrow H_2S + 4H_2O + 4Zn^{2+}$$

(g)
$$\overset{(+2\,\to\,+4)\,=\,+2}{\underset{(+5\,\to\,+2)\,=\,-3}{HNO_3 + Sn(NO_3)_2 \longrightarrow Sn(NO_3)_4 + NO}}$$

$$\underset{\substack{\text{(for the extra}\\ \text{nitrate ions)}}}{6HNO_3} + 2HNO_3 + 3Sn(NO_3)_2 \longrightarrow 3Sn(NO_3)_4 + 2NO + 4H_2O$$

$$8HNO_3 + 3Sn(NO_3)_2 \longrightarrow 3Sn(NO_3)_4 + 2NO + 4H_2O$$

(h) $$8\,H^+ + 2\,NO_3^- + 3\,Pb^{2+} \longrightarrow 2\,NO + 3\,Pb^{4+} + 4\,H_2O$$

(i) $$2\,Ag^+ + 2\,S_2O_3^{2-} \longrightarrow S_4O_6^{2-} + 2\,Ag$$

(j) $$SO_4^{2-} + 2\,I^- + 4\,H^+ \longrightarrow SO_2 + I_2 + 2\,H_2O$$

(k) $$10\,Cr^{3+} + 6\,MnO_4^- + 11\,H_2O \longrightarrow 5\,Cr_2O_7^{2-} + 6\,Mn^{2+} + 22\,H^+$$

13.27. What is the oxidation number of sulfur in $S_2O_8^{2-}$?

Ans. If you calculate the oxidation number assuming that the oxygen atoms are normal oxide ions, you get an answer $+7$, which is greater than the maximum oxidation number for sulfur. That must mean that one of the pairs of oxygen atoms is a peroxide, and thus the sulfur must be in its highest oxidation number, $+6$. The ion is peroxydisulfate:

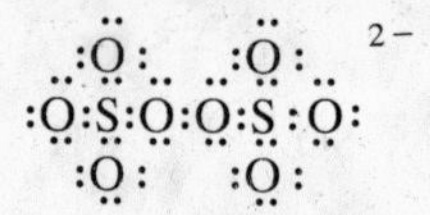

$$
\begin{array}{c}
\ddot{\text{:O:}} \quad \ddot{\text{:O:}} \\
\text{:O:S:O:O:S:O:} \\
\ddot{\text{:O:}} \quad \ddot{\text{:O:}}
\end{array}
\right)^{2-}
$$

13.28. Consider the following part of an equation:

$$NO_2 + MnO_4^- \longrightarrow Mn^{2+} +$$

(a) If one half-reaction is a reduction, what must the other half-reaction be? (b) To what oxidation state can the nitrogen be changed? (c) Complete and balance the equation.

Ans. (a) Since one half-reaction is a reduction, the other half-reaction must be an oxidation. (b) The maximum oxidation state for nitrogen is $+5$, because nitrogen is in periodic group V. Since it starts out in oxidation number $+4$, it must be oxidized to $+5$.

(c) $$NO_2 + MnO_4^- \longrightarrow Mn^{2+} + NO_3^-$$

$$5\,NO_2 + MnO_4^- \longrightarrow Mn^{2+} + 5\,NO_3^-$$

$$H_2O + 5\,NO_2 + MnO_4^- \longrightarrow Mn^{2+} + 5\,NO_3^- + 2\,H^+$$

13.29. Which of the following reactions (indicated by unbalanced equations) occur in acid solution and which occur in basic solution?

(a) $$HNO_3 + Fe^{2+} \longrightarrow NO + Fe^{3+}$$

(b) $$CrO_4^{2-} + Fe(OH)_2 \longrightarrow Cr(OH)_3 + Fe(OH)_3$$

(c) $$N_2H_4 + I^- \longrightarrow NH_4^+ + I_2$$

Ans. (a) Acid solution (HNO_3 would not be present in base). (b) Basic solution (the hydroxides would not be present in acid). (c) Acid solution (NH_3 rather than NH_4^+ would be present in base).

13.30. Complete and balance the following equations:

(a) $$Cl^- + MnO_2 + H^+ \longrightarrow Mn^{2+} + H_2O + Cl_2$$

(b) $$ClO_3^- \longrightarrow Cl^- + ClO_4^-$$

(c) $$Pb + PbO_2 + SO_4^{2-} \longrightarrow PbSO_4 + H_2O$$

(d) $$I_3^- + Fe(CN)_6^{4-} \longrightarrow Fe(CN)_6^{3-} + I^-$$

(e) $$V^{2+} + H_3AsO_4 \longrightarrow HAsO_2 + VO^{2+}$$

(f) $$VO^{2+} + AsO_4^{3-} \longrightarrow AsO_2^- + VO_2^+$$

(g) $$Hg_2^{2+} + CN^- \longrightarrow C_2N_2 + Hg$$

Ans.

(a) $$2\,Cl^- + MnO_2 + 4\,H^+ \longrightarrow Mn^{2+} + 2\,H_2O + Cl_2$$

(b) $$4\,ClO_3^- \longrightarrow Cl^- + 3\,ClO_4^-$$

(c) $$4\,H^+ + Pb + PbO_2 + 2\,SO_4{}^{2-} \longrightarrow 2\,PbSO_4 + 2\,H_2O$$

(d) $$I_3{}^- + 2\,Fe(CN)_6{}^{4-} \longrightarrow 2\,Fe(CN)_6{}^{3-} + 3\,I^-$$

(e) $$V^{2+} + H_3AsO_4 \longrightarrow HAsO_2 + VO^{2+} + H_2O$$

(f) $$2\,VO^{2+} + AsO_4{}^{3-} \longrightarrow AsO_2{}^- + 2\,VO_2{}^+$$

(g) $$Hg_2{}^{2+} + 2\,CN^- \longrightarrow C_2N_2 + 2\,Hg$$

13.31. Complete and balance the following equation:

$$Sb_2S_3 + HNO_3 \longrightarrow H_3SbO_4 + SO_2 + NO$$

Ans.

$$3\,Sb_2S_3 + 22\,HNO_3 \longrightarrow 6\,H_3SbO_4 + 9\,SO_2 + 22\,NO + 2\,H_2O$$

13.32. The oxidizing ability of H_2SO_4 depends on its concentration. Which element is reduced by reaction of Zn on H_2SO_4 in each of the following reactions?

$$Zn + H_2SO_4(\text{dilute}) \longrightarrow ZnSO_4 + H_2$$

$$4\,Zn + 5\,H_2SO_4(\text{concentrated}) \longrightarrow H_2S + 4\,H_2O + 4\,ZnSO_4$$

Ans. In the first reaction, hydrogen is reduced. In the second reaction sulfur is reduced.

13.33. What is the maximum oxidation state of fluorine in any compound?

Ans. The only oxidation state of fluorine in a compound is -1; it is the most electronegative element. (It always has "control" of any shared electrons, except in the element, F_2.)

13.34. What is the more likely formula for bismuth in the $+5$ oxidation state—Bi^{5+} or $BiO_3{}^-$?

Ans. $BiO_3{}^-$. There are no $5+$ monatomic ions.

Chapter 14

Electrochemistry

14.1 INTRODUCTION

The interaction of electricity with matter was introduced in Chap. 7, where the electrical decomposition of a melted salt was used to prepare active elements from their compounds. An illustration of an electrolysis was shown in Fig. 5-2. Chemical reactions occur at the two electrodes. The electrode at which oxidation occurs is called the *anode*; the one at which reduction takes place is called the *cathode*. There are two different types of interaction of electricity and matter, as follows:

Electrolysis: Electric current causes chemical reaction.
Galvanic cell action: Chemical reaction causes electric current, as in a battery.
Two distinctly different types of problems are associated with these types of interactions.

14.2 ELECTRICAL UNITS

The units important for the discussion of electrochemistry in this chapter are presented in Table 14-1. The passage of electrons through a wire or the passage of ions through a solution constitutes an electric *current*. The basic unit of electric charge is the *coulomb*, C. The unit of electric current is the *ampere*, A. The passage of $1\,C/s$ is a current of $1\,A$.

$$1\,A = 1\,C/s$$

Table 14-1   Electrical Quantities and Units

Quantity	Symbol	Unit	Symbol
Charge	Q	Coulomb	C
Current	I	Amphere	A
Potential	ε	Volt	V
Energy	E	Joule	J
Power	P	Watt	W
Resistance	R	Ohm	Ω

One mole of electrons has a total charge of $96\,500\,C$. The charge on one mole of electrons is called a *Faraday*, F.

$$1\,F = 96\,500\,C = 1\,mol\,e^-$$

EXAMPLE 14.1.   Calculate the charge in coulombs on one electron.

$$1\,e^-\left(\frac{1\,mol\,e^-}{6.02 \times 10^{23}\,e^-}\right)\left(\frac{96\,500\,C}{mol\,e^-}\right) = 1.60 \times 10^{-19}\,C$$

The charge on one electron is equal to the charge on a mole of electrons divided by Avogadro's number.

Electricity passes through a *circuit* under the influence of a *potential* or *voltage*, the driving force of the movement of charge.

228

14.3   ELECTROLYSIS

The quantity of chemical change associated with a given quantity of charge can be stated in terms of Faraday's law, but even easier is to use the factor $(96\,500\,C/mol\ e^-)$ and the fact that the number of coulombs per second is equal to the number of amperes, along with the factors used earlier.

EXAMPLE 14.2. How many grams of silver metal can be produced from $AgNO_3$ by the passage of $4\,010\,C$ of charge?

The net ionic equation for the reaction is

$$Ag^+ + e^- \longrightarrow Ag$$

$$4\,010\,C\left(\frac{1\,mol\ e^-}{96\,500\,C}\right)\left(\frac{1\,mol\ Ag}{mol\ e^-}\right)\left(\frac{108\,g\ Ag}{mol\ Ag}\right) = 4.49\,g\ Ag$$

$$\underset{\substack{\text{from the}\\ \text{definition}\\ \text{of a faraday}}}{}$$

EXAMPLE 14.3. How many grams of copper can be deposited from $CuSO_4$ solution by the passage of $3.00\,A$ for $1\,230\,s$?

The net ionic equation for the reaction is

$$Cu^{2+} + 2\,e^- \longrightarrow Cu$$

$$1\,230\,s\left(\frac{3.00\,C}{s}\right)\left(\frac{1\,mol\ e^-}{96\,500\,C}\right)\left(\frac{1\,mol\ Cu}{2\,mol\ e^-}\right)\left(\frac{63.5\,g\ Cu}{mol\ Cu}\right) = 1.21\,g\ Cu$$

$$\underset{\substack{\text{from the}\\ \text{current}}}{}$$

EXAMPLE 14.4. How many hours time is required to produce $30.0\,g$ of gold by passage of a 4.00-A current through a solution of a gold(III) compound?

Each mole of gold(III) requires $3\,mol$ electrons to be reduced to gold metal.

$$30.0\,g\ Au\left(\frac{1\,mol\ Au}{197\,g\ Au}\right)\left(\frac{3\,mol\ e^-}{mol\ Au}\right)\left(\frac{96\,500\,C}{mol\ e^-}\right)\left(\frac{1\,s}{4.00\,C}\right)\left(\frac{1\,hr}{3\,600\,s}\right) = 3.06\,hr$$

Note that there are *not* 60 seconds per hour.

The following are the requirements for electrolysis:
1. Ions. (There must be charged particles to carry current. It might not be the ions that react, however.)
2. Liquid, either a pure liquid or a solution, so that the ions can migrate.
3. Source of potential (in a galvanic cell, the chemical reaction is the source of potential, but not in an electrolysis cell).
4. Mobile ions, complete circuit (including wires to carry electrons and a salt bridge to carry the ions), and electrodes (at which the current changes from the flow of electrons to the movement of ions or vice versa).

If you electrolyze a solution containing a compound of a very active metal and/or a very active nonmetal, the water (or other solvent) might be electrolyzed instead of the ion. For example, if you electrolyze molten sodium chloride, you get the free elements:

$$2\,NaCl \xrightarrow{\text{electricity}} 2\,Na + Cl_2$$

However, if you electrolyze a dilute aqueous solution of NaCl, the water is decomposed. (The NaCl is necessary to conduct the current, but neither the Na^+ nor the Cl^- reacts at the electrodes.)

$$2\,H_2O \xrightarrow[\text{NaCl(dilute)}]{\text{electricity}} 2\,H_2 + O_2$$

If you electrolyze a concentrated solution of NaCl instead, H_2 is produced at the cathode and Cl_2 is produced at the anode:

$$2e^- + 2H_2O \xrightarrow{\text{electricity}} H_2 + 2OH^-$$

$$2Cl^- \xrightarrow{\text{electricity}} Cl_2 + 2e^-$$

It is obvious that the reaction conditions are very important to the products.

14.4 GALVANIC CELLS

When you place a piece of zinc metal into a solution of $CuSO_4$, you expect a chemical reaction because the more active zinc displaces the less active copper from its compound (Sec. 7.3). We learned in Chap. 13 that this is an oxidation-reduction reaction, involving transfer of electrons from the zinc to the copper.

$$Zn \longrightarrow Zn^{2+} + 2e^-$$

$$Cu^{2+} + 2e^- \longrightarrow Cu$$

It is possible to carry out these two half-reactions in different places if we connect them suitably. We must deliver the electrons from the Zn to the Cu^{2+} and we must have a complete circuit. The apparatus is shown in Fig. 14-1. A galvanic cell with this particular combination of reactants is called a *Daniell cell*. The pieces of zinc and copper serve as electrodes, at which chemical reaction takes place. It is at the electrodes that the electron current is changed into an ion current or vice versa. The salt bridge is necessary to complete the circuit. If it were not there, the buildup of charge in each beaker (positive in the left, negative in the right) would stop the reaction extremely quickly (less than 1 s). The same chemical reactions are taking place in this apparatus as would take place if we dipped the zinc metal in the $CuSO_4$ solution, but the zinc half-reaction is taking place in the left beaker and the copper half-reaction is taking place in the right beaker. Electrons flow from left to right in the wire, and they could be made to do electrical work, such as lighting a small bulb. To keep the beakers from acquiring a charge, cations flow through the salt bridge toward the right and anions flow to the left. The salt bridge is filled with a solution of an unreacting salt, such as KNO_3.

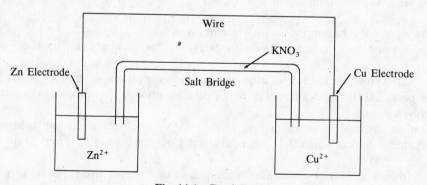

Fig. 14-1   Daniell cell

One such combination of anode and cathode is called a *cell*. A combination of cells is a *battery*.

In Table 7-1 the relative tendencies of certain elements to react were listed qualitatively. We can give a quantitative measure of relative tendency to react, called *standard reduction potential*, as shown in Table 14-2. In this table, the standard half-cell potential for each half-reaction, as a reduction, is tabulated in order with the highest potential first. If we turn these half-reactions around, we change the signs of the potentials and we get oxidation potentials. We thus have half-reactions including both elementary metals and elementary nonmetals in the same table, as well as many half-reactions that do not involve uncombined elements at all. For reactions not involving a solid conductor of electricity, such as a piece of metal, an inert electrode such as platinum is used.

Table 14-2 Standard Reduction Potentials at 25 °C

	$\varepsilon°$ (V)
$F_2 + 2e^- \longrightarrow 2F^-$	2.87
$Co^{3+} + e^- \longrightarrow Co^{2+}$	1.82
$MnO_4^- + 8H^+ + 5e^- \longrightarrow Mn^{2+} + 4H_2O$	1.51
$Cl_2 + 2e^- \longrightarrow 2Cl^-$	1.36
$Br_2 + 2e^- \longrightarrow 2Br^-$	1.09
$Ag^+ + e^- \longrightarrow Ag$	0.80
$Fe^{3+} + e^- \longrightarrow Fe^{2+}$	0.77
$I_2 + 2e^- \longrightarrow 2I^-$	0.54
$Cu^{2+} + 2e^- \longrightarrow Cu$	0.34
$Sn^{4+} + 2e^- \longrightarrow Sn^{2+}$	0.13
$2H^+ + 2e^- \longrightarrow H_2$	0.0000
$Pb^{2+} + 2e^- \longrightarrow Pb$	−0.13
$PbSO_4(s) + 2e^- \longrightarrow Pb + SO_4^{2-}$	−0.31
$Fe^{2+} + 2e^- \longrightarrow Fe$	−0.44
$Zn^{2+} + 2e^- \longrightarrow Zn$	−0.76
$Al^{3+} + 3e^- \longrightarrow Al$	−1.66
$Mg^{2+} + 2e^- \longrightarrow Mg$	−2.37
$Na^+ + e^- \longrightarrow Na$	−2.71
$Ba^{2+} + 2e^- \longrightarrow Ba$	−2.73
$Li^+ + e^- \longrightarrow Li$	−3.05

We can combine the half-cell potentials for any two half-reactions in the table to get a complete cell potential. The chemical reaction may proceed spontaneously if the complete cell potential is positive. Otherwise, the opposite reaction may proceed spontaneously. We combine half-cells by adding the chemical reactions and by adding the corresponding half-cell potentials. We must first get the correct chemical reactions and corresponding half-cell potentials for the half-reactions, as follows:
1. If you reverse the direction of the chemical reaction, change the sign of the potential. You can ensure getting a positive cell potential by reversing the lower half-reaction from Table 14.2.
2. If you multiply the coefficients of the chemical equation by any number, leave the half-cell potential *unchanged*! ε is intensive; it does not depend on the number of moles of chemical involved.

EXAMPLE 14.5.   Calculate the standard cell potential of the Daniell cell, shown in Fig. 14-1.

$$Cu^{2+} + 2e^- \longrightarrow Cu \qquad \varepsilon° = 0.34\,V$$
$$Zn^{2+} + 2e^- \longrightarrow Zn \qquad \varepsilon° = -0.76\,V$$

To get the proper chemical equation, reverse the zinc half-reaction:

$$Zn \longrightarrow Zn^{2+} + 2e^- \qquad \varepsilon° = +0.76\,V \qquad (\text{Note change of sign.})$$

Then add the copper half-cell reduction to the zinc half-cell oxidation and add the half-cell potentials:

$$Cu^{2+} + Zn \longrightarrow Zn^{2+} + Cu \qquad \varepsilon° = +1.10\,V$$

EXAMPLE 14.6.   Calculate the standard cell potential of the Ag/Ag^+, Cu/Cu^{2+} cell.

$$Ag^+ + e^- \longrightarrow Ag \qquad \varepsilon° = 0.80\,V$$
$$Cu^{2+} + 2e^- \longrightarrow Cu \qquad \varepsilon° = 0.34\,V$$

To get the proper chemical equation, reverse the copper half-reaction:

$$Cu \longrightarrow Cu^{2+} + 2e^- \qquad \varepsilon^\circ = -0.34\,V$$

Then double silver half-cell to get 2 mol e^- for the reduction.

$$2\,Ag^+ + 2\,e^- \longrightarrow 2\,Ag \qquad \varepsilon^\circ = 0.80\,V \qquad (\text{Note } \varepsilon^\circ \text{ unchanged.})$$

Add the resulting equation to the copper half-cell oxidation equation and add the corresponding half-cell potentials:

$$2\,Ag^+ + Cu \longrightarrow Cu^{2+} + 2\,Ag \qquad \varepsilon^\circ = +0.46\,V$$

14.5   THE NERNST EQUATION

The superscript $^\circ$ used in the last section means *standard* potential, corresponding to the potential when all the reagents involved are in their *standard states*. (This is not STP.) Standard state means
1.   1 M for solutes.
2.   1 atm for gases.
3.   pure substances for liquids and solids.

If the concentration or pressure differs from these conditions, the potential will vary from the standard potential. The actual potential will be denoted ε (without the $^\circ$), and its value is given by the Nernst equation:

$$\varepsilon = \varepsilon^\circ - \frac{0.0592}{n} \log Q$$

In this equation, ε is the actual potential and ε° is the standard potential, n is the number of electrons involved, and Q is a ratio of concentration terms. Q is equal to the ratio of concentrations of *products* to concentrations of *reactants*, each raised to the power corresponding to the coefficient in the balanced chemical equation. Pure solids and liquids and the solvent water are not included in Q; their effective concentrations are assumed to be 1. Gas pressures in atmospheres are used instead of concentrations. For a general reaction

$$a\,A + b\,B \longrightarrow c\,C + d\,D + n\,e^-$$

$$Q = \frac{[C]^c[D]^d}{[A]^a[B]^b}$$

The square brackets refer to the concentration of the species inside. Note that the concentrations are multiplied or divided, not added or subtracted, and the concentrations of the products are in the numerator. The coefficients in the balanced chemical equation have become exponents in this mathematical equation.

EXAMPLE 14.7.   Write the Nernst equation for the Cu/Ag cell of Example 14.6.

$$\varepsilon = \varepsilon^\circ - \frac{0.0592}{n} \log Q$$

There are two electrons involved in each half-reaction, even though none appears in the overall equation. Thus,

$$\varepsilon = 0.46 - \frac{0.0592}{2} \log \frac{[Cu^{2+}]}{[Ag^+]^2}$$

Metallic Ag and Cu have effective concentrations of 1.

EXAMPLE 14.8.   Calculate ε for the Ag/Cu cell containing 0.100 M Ag^+ and 4.00 M Cu^{2+}.

Using the equation from the last example,

$$\varepsilon = 0.46 - \frac{0.0592}{2} \log \frac{(4.00)}{(0.100)^2}$$

$$= 0.38\,V$$

Note for electronic calculator users: To solve the last equation, enter 4.00, square 0.100, and divide that into the 4.00. Press the $\boxed{=}$ key. Press the $\boxed{\log}$ key, then multiply that answer by 0.0592 and divide by 2. Change the sign of this result with the $\boxed{+/-}$ key and add in the 0.46.

14.6 PRACTICAL CELLS

We are used to the convenience of cells (more usually in the form of batteries) to power flashlights, cars, portable radios, and other devices requiring electric power. Many different kinds of cells are available, each with some advantages and some disadvantages. The desirable features of a practical cell include low cost, light weight, relatively constant potential, rechargeability, and long life. Not all of the desirable features can be incorporated into each cell.

The lead storage cell (six of which make up the lead storage battery commonly used in automobiles) will be discussed as an example of a practical cell. The cell, pictured in Fig. 14-2, consists of a lead electrode and a lead dioxide electrode immersed in relatively concentrated H_2SO_4 in a single container. When the cell delivers power (when it is used), the electrodes react as follows:

$$Pb + H_2SO_4 \longrightarrow PbSO_4(s) + 2H^+ + 2e^-$$

$$2H^+ + 2e^- + PbO_2 + H_2SO_4 \longrightarrow PbSO_4(s) + 2H_2O$$

The solid $PbSO_4$ formed in each reaction adheres to the electrode. The H_2SO_4 becomes diluted in two ways: Some of it is used up and some water is formed. The electrons going from anode to cathode do the actual work for which the electric power source is employed, such as starting the car. Both electrodes are placed in the same solution; no salt bridge is needed. The reason that this is possible is that both the oxidizing agent and the reducing agent, as well as the products of the oxidation and reduction, are solids, and they cannot migrate to the other electrode and react directly.

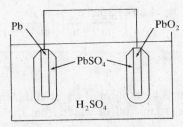

Fig. 14-2 Lead storage cell

The lead storage cell may be recharged by forcing electrons back the other way. Each electrode reaction is reversed, and Pb, PbO_2, and H_2SO_4 are produced again. In a car, the recharge reaction takes place when the battery is recharged at a gasoline station and even more often each time the car is driven and the alternator changes some of the mechanical energy of the engine into electrical energy and distributes it to the battery.

Solved Problems

ELECTRICAL UNITS

14.1. What factor label can be made from the current 6.00 A?

Ans.   6.00 C/s.

ELECTROLYSIS

14.2. Even if sodium metal were produced by the electrolysis of aqueous NaCl, what would happen to the sodium produced in the water?

    Ans. The sodium is so active that it would react immediately with the water. No elementary sodium should ever be expected from a water solution.

14.3. With 10 000 C of charge passing, how many grams of (*a*) Ag can be deposited from $AgNO_3$? (*b*) Pd can be deposited from $Pd(NO_3)_2$? (*c*) Since their atomic weights are so similar, why are the answers so different?

    Ans.

        (*a*)
$$Ag^+ + e^- \longrightarrow Ag$$
$$10\,000\,C\left(\frac{1\,\text{mol e}^-}{96\,500\,C}\right)\left(\frac{1\,\text{mol Ag}}{\text{mol e}^-}\right)\left(\frac{108\,\text{g Ag}}{\text{mol Ag}}\right) = 11.2\,\text{g Ag}$$

        (*b*)
$$Pd^{2+} + 2e^- \longrightarrow Pd$$
$$10\,000\,C\left(\frac{1\,\text{mol e}^-}{96\,500\,C}\right)\left(\frac{1\,\text{mol Pd}}{2\,\text{mol e}^-}\right)\left(\frac{106\,\text{g Pd}}{\text{mol Pd}}\right) = 5.49\,\text{g Pd}$$

    Since the Pd^{2+} has a double charge, it takes twice as many electrons to reduce it to the metal. A given number of electrons can reduce only half the number of moles of Pd as Ag.

14.4. Explain why Al cannot be produced from its salts in aqueous solution.

    Ans. Al is too active; the water will be reduced first.

14.5. Calculate the time required for 105 g of gold to be deposited from a gold(III) compound by a 5.00-A current.

    Ans.

$$105\,\text{g Au}\left(\frac{1\,\text{mol Au}}{197\,\text{g Au}}\right)\left(\frac{3\,\text{mol e}^-}{\text{mol Au}}\right)\left(\frac{96\,500\,C}{\text{mol e}^-}\right)\left(\frac{1\,s}{5.00\,C}\right) = 3.09 \times 10^4\,s$$
$$= 8.57\,\text{hr}$$

GALVANIC CELLS

14.6. Take a half cup of water from a bathtubful of water and measure its temperature. Assume that the value is 30 °C. Now take a cup full of water from the same bathtub. Is the temperature 30 °C or 60 °C?

    Ans. The temperature is 30 °C, because temperature, like potential, does not depend on the quantity of material present.

14.7. Calculate the standard potentials of the cells formed by combination of each of the following pairs of half-cells:

    (*a*) $Fe^{3+} + e^- \longrightarrow Fe^{2+}$ and $Fe^{2+} + 2e^- \longrightarrow Fe$

    (*b*) $Zn^{2+} + 2e^- \longrightarrow Zn$ and $Sn^{4+} + 2e^- \longrightarrow Sn^{2+}$

    (*c*) $I_2 + 2e^- \longrightarrow 2I^-$ and $Cl_2 + 2e^- \longrightarrow 2Cl^-$

    (*d*) $MnO_4^- + 8H^+ + 5e^- \longrightarrow Mn^{2+} + 4H_2O$ and $Fe^{3+} + e^- \longrightarrow Fe^{2+}$

    Ans.

        (*a*)

$2Fe^{3+} + 2e^- \longrightarrow 2Fe^{2+}$	0.77 V
$Fe \longrightarrow Fe^{2+} + 2e^-$	$+0.44$ V
$2Fe^{3+} + Fe \longrightarrow 3Fe^{2+}$	1.21 V

We used the data in Table 14-2. We doubled the species in the first chemical equation, but did not change the potential. We reversed the second equation, and changed the sign of the potential. We then added both the chemical equations and the potentials to get the answer.

(b)
$$Zn \longrightarrow Zn^{2+} + 2e^- \qquad\qquad +0.76\,V$$
$$Sn^{4+} + 2e^- \longrightarrow Sn^{2+} \qquad\qquad 0.13\,V$$
$$Sn^{4+} + Zn \longrightarrow Sn^{2+} + Zn^{2+} \qquad 0.89\,V$$

(c)
$$Cl_2 + 2e^- \longrightarrow 2Cl^- \qquad\qquad 1.36\,V$$
$$2I^- \longrightarrow I_2 + 2e^- \qquad\qquad -0.54\,V$$
$$Cl_2 + 2I^- \longrightarrow 2Cl^- + I_2 \qquad\qquad 0.82\,V$$

(d)
$$MnO_4^- + 8H^+ + 5e^- \longrightarrow Mn^{2+} + 4H_2O \qquad 1.51\,V$$
$$5Fe^{2+} \longrightarrow 5Fe^{3+} + 5e^- \qquad\qquad -0.77\,V$$
$$MnO_4^- + 8H^+ + 5Fe^{2+} \longrightarrow Mn^{2+} + 4H_2O + 5Fe^{3+} \qquad 0.74\,V$$

14.8. Calculate the potential of each of the following cells:

(a)
$$Sn^{4+} + Zn \longrightarrow Sn^{2+} + Zn^{2+}$$

(b)
$$Cl_2 + 2I^- \longrightarrow 2Cl^- + I_2$$

Ans. These are the same cells as are presented in Problem 14.7(b) and (c). All that you have to do is separate each one into its half-cells, and proceed as you did in that problem.

THE NERNST EQUATION

14.9. Calculate the potential of the following cell:

$$Sn^{4+}(1.50\,M) + Zn \longrightarrow Sn^{2+}(0.500\,M) + Zn^{2+}(2.00\,M)$$

Ans. The standard potential is 0.89 V, as calculated in Problem 14.7(b). The Nernst equation is used to calculate the actual potential:

$$\varepsilon = \varepsilon^\circ - \frac{0.0592}{2}\log\frac{[Sn^{2+}][Zn^{2+}]}{[Sn^{4+}]}$$

$$= 0.89 - 0.0296\log\frac{(0.500)(2.00)}{(1.50)} = 0.90\,V$$

14.10. Calculate the potential of the Daniell cell in which the copper(II) ion concentration and the zinc ion concentrations are both 0.100 M.

Ans. The Daniell cell has a standard potential of 1.10 V, as calculated in Example 14.5. The Nernst equation is used to calculate the actual potential.

$$\varepsilon = \varepsilon^\circ - \frac{0.0592}{2}\log\frac{[Zn^{2+}]}{[Cu^{2+}]}$$

$$= 1.10 - 0.0296\log\frac{(0.100)}{(0.100)} = 1.10\,V$$

Since both concentrations and both charges are the same, the potential is equal to the standard potential.

PRACTICAL CELLS

14.11. Explain why a Daniell cell cannot be placed in a single container, like the lead storage cell.

Ans. The copper(II) ions would migrate to the zinc electrode and be reduced to copper metal directly. The zinc electrode would become copper plated, and the cell would not function.

14.12. Can a Daniell cell be recharged?

Ans. No. If the Daniell cell were to be recharged, the Cu^{2+} ions would get into the zinc half-cell through the salt bridge. There, they would react directly with the zinc electrode, and the cell would be destroyed.

Supplementary Problems

14.13. What difference does it make to the conclusions about the chemical reaction that may occur in a cell if you reverse the wrong equation for a half-reaction?

Ans. If you reverse the upper half-reaction equation, the overall equation will be reversed and the potential will have the opposite sign. Since a positive cell potential means "tends to go to the right" and a negative potential means "tends to go to the left," the same information may be deduced either way. For example,

$$2\,Ag^+ + Cu \longrightarrow Cu^{2+} + 2\,Ag \qquad \varepsilon^\circ = 0.46\,V$$
$$Cu^{2+} + 2\,Ag \longrightarrow 2\,Ag^+ + Cu \qquad \varepsilon^\circ = -0.46\,V$$

Since its potential is positive, the first equation states that Ag^+ tends to react with Cu. Since its potential is negative, the second equation states that Ag^+ tends to react with Cu. (The reverse of the stated equation tends to proceed.) The same conclusion is drawn with either equation. You can always get a positive cell potential if you reverse the lower of the two half-cells from Table 14-2.

14.14. How many hours does it take to reduce 3.00 mol Fe^{3+} to Fe^{2+} with a 2.00-A current?

Ans. The net ionic equation for the reaction is

$$Fe^{3+} + e^- \longrightarrow Fe^{2+}$$

$$3.00\,\text{mol Fe}^{3+}\left(\frac{1\,\text{mol e}^-}{1\,\text{mol Fe}^{3+}}\right)\left(\frac{96\,500\,\text{C}}{\text{mol e}^-}\right)\left(\frac{1\,\text{s}}{2.00\,\text{C}}\right)\left(\frac{1\,\text{hr}}{3\,600\,\text{s}}\right) = 40.2\,\text{hr}$$

Despite the fact that we start with iron(III), it only takes 1 mol e^- for each mole of Fe^{3+}, as shown by the balanced chemical equation. We are not reducing the iron to elementary iron.

14.15. Explain why direct current (dc) rather than alternating current (ac) is used for electrolysis. Why is direct current used in cars?

Ans. In direct current, the electrons flow in the same direction all the time. In alternating current, the electrons flow one way for a short period of time (typically $\frac{1}{60}$ s) and then they flow the other way. To get any electrolysis that is not immediately undone, direct current is required. Direct current is also used in cars because cells generate direct current.

14.16. Calculate ε for the Ag/Cu cell containing 0.100 M Ag^+ and 0.100 M Cu^{2+}.

Ans. Using the equation from Example 14.7,

$$\varepsilon = 0.46 - \frac{0.0592}{2}\log\frac{(0.100)}{(0.100)^2}$$

$$= 0.43\,V$$

Despite the fact that the concentrations are the same (compare Problem 14.10), the potential is not equal to the standard potential, because the exponents in the Nernst equation are different.

14.17. What is the meaning of a positive sign for (*a*) a cell potential? (*b*) a half-cell potential?

Ans. (*a*) The reaction can proceed as written. (*b*) Nothing. No half-reaction can proceed alone no matter what the value of its potential.

Chapter 15

Equivalents and Normality

15.1 INTRODUCTION

Equivalents are measures of the quantity of a substance present, analogous to moles. Normality is a concentration unit, analogous to molarity. The similarities are so great that these two units should be very easy to learn to use, but many students have difficulty with them. Be sure that you understand the underlying definitions before you proceed.

15.2 EQUIVALENTS

The *equivalent* is defined in terms of a chemical reaction. It is defined in one of two different ways, depending on whether an oxidation-reduction reaction or an acid-base reaction is under discussion. For an oxidation-reduction reaction, an equivalent is the quantity of a substance that will react with or yield 1 mol of *electrons*. For an acid-base reaction, an equivalent is the quantity of a substance that will react with or yield 1 mol of *hydrogen ions or hydroxide ions*. Note that the equivalent is defined in terms of a reaction, not merely in terms of a formula. Thus, the same mass of the same compound undergoing different reactions can correspond to different numbers of equivalents. The ability to determine the number of equivalents per mole is the key to calculations in this chapter.

EXAMPLE 15.1. How many equivalents are there in 98.0 g of H_2SO_4 in each of the following reactions?

(*a*) $$H_2SO_4 + NaOH \longrightarrow NaHSO_4 + H_2O$$

(*b*) $$H_2SO_4 + 2NaOH \longrightarrow Na_2SO_4 + 2H_2O$$

Since 98.0 g of H_2SO_4 is 1.00 mol of H_2SO_4, we will concentrate on that quantity of H_2SO_4. Both of these reactions are acid-base reactions, and so we will define the number of equivalents of H_2SO_4 in terms of the number of moles of hydroxide ion with which it reacts. In the first equation, 1 mol of H_2SO_4 reacts with 1 mol of OH^-. By definition, that quantity of H_2SO_4 is 1 equivalent. For that equation, 1 equivalent of H_2SO_4 is equal to 1 mol of H_2SO_4. In the second equation, 1 mol of H_2SO_4 reacts with 2 mol OH^-. By definition, that quantity of H_2SO_4 is equal to 2 equivalents. Thus, in that equation, 2 equivalents of H_2SO_4 is 1 mol of H_2SO_4.

(*a*) $$1 \text{ equivalent} = 1 \text{ mol}$$

(*b*) $$2 \text{ equivalents} = 1 \text{ mol}$$

We can use these equalities as factors to change moles to equivalents or equivalents to moles, especially to calculate normalities (Sec. 15.3) or equivalent weights (Sec. 15.4).

EXAMPLE 15.2. How many equivalents are there in 98.0 g of H_2SO_4 in the following reaction?

$$8H^+ + H_2SO_4 + 4Zn \longrightarrow H_2S + 4Zn^{2+} + 4H_2O$$

This reaction is a redox reaction, and so we will define the number of equivalents of H_2SO_4 in terms of the number of moles of electrons with which it reacts. Since no electrons appear explicitly in an overall equation, we will write the half-reaction in which the H_2SO_4 appears:

$$8H^+ + H_2SO_4 + 8e^- \longrightarrow H_2S + 4H_2O$$

It is now apparent that 1 mol of H_2SO_4 reacts with 8 mol e^-, and by definition 8 equivalents of H_2SO_4 reacts with 8 mol e^-. Thus, 8 equivalents equals 1 mol in this reaction. Since 98.0 g H_2SO_4 is 1 mol, there are 8 equivalents in 98.0 g H_2SO_4.

Some instructors and texts ask the number of equivalents per mole of an acid or base without specifying a particular reaction. In that case merely assume that the substance undergoes an acid-base reaction as completely as possible. State that assumption in your answers on examinations.

EXAMPLE 15.3.   What is the number of equivalents in 2.00 mol H_2SO_4?

Assuming that the H_2SO_4 will react as an acid to replace both hydrogen atoms, we have reaction (b) from Example 15.1, and there are 2 equivalents per mole. In 2.00 mol H_2SO_4,

$$2.00 \text{ mol } H_2SO_4 \left(\frac{2 \text{ equivalents}}{\text{mol}} \right) = 4.00 \text{ equivalents}$$

The major use of equivalents stems from its definition. Once you define the number of equivalents in a certain mass of a substance, you do not need to write the equation for its reaction. That equation has already been used in defining the number of equivalents. Thus, a chemist can calculate the number of equivalents in a certain mass of substance, and his technicians can subsequently use that definition without knowing the details of the reaction.

One equivalent of one substance in a reaction always reacts with one equivalent of each of the other substances in that reaction.

EXAMPLE 15.4.   How many equivalents of NaOH does the 98.0 g of H_2SO_4 react with in reaction (b) of Example 15.1?

Since there are 2 equivalents of H_2SO_4, they must react with 2 equivalents of NaOH. Checking, we see that 2 equivalents of NaOH liberate 2 mol of OH^- and also react with 2 mol H^+, and thus there are 2 equivalents by definition also.

15.3 NORMALITY

Normality is defined as the number of equivalents of solute per liter of solution. Its unit is "normal," with the symbol N. Thus the normality of a solution may be "3.0 normal."

EXAMPLE 15.5.   What is the normality of a solution containing 7.00 equivalents in 5.00 L of solution?

$$\frac{7.00 \text{ equivalents}}{5.00 \text{ L}} = 1.40 \, N$$

"The normality of the solution is 1.40 normal," or "The solution is 1.40 normal."

Normality is some integral multiple of molarity, since there are always some integral number of equivalents per mole.

EXAMPLE 15.6.   What is the normality of 2.50 M H_2SO_4 in each of the following reactions?

(a) $H_2SO_4 + NaOH \longrightarrow NaHSO_4 + H_2O$

(b) $H_2SO_4 + 2 NaOH \longrightarrow Na_2SO_4 + 2 H_2O$

We found in Example 15.1 that there were 1 equivalent per mol in the first reaction and 2 equivalents per mol in the second. We use these factors to solve for normality:

(a) $2.50 \, M = \dfrac{2.50 \text{ mol}}{L} \left(\dfrac{1 \text{ equivalent}}{\text{mol}} \right) = \dfrac{2.50 \text{ equivalents}}{L} = 2.50 \, N$

(b) $2.50 \, M = \dfrac{2.50 \text{ mol}}{L} \left(\dfrac{2 \text{ equivalents}}{\text{mol}} \right) = \dfrac{5.00 \text{ equivalents}}{L} = 5.00 \, N$

EXAMPLE 15.7. What is the molarity of $4.00\,N$ H_2SO_4 in each of the following reactions?

(*a*) $$H_2SO_4 + NaOH \longrightarrow NaHSO_4 + H_2O$$

(*b*) $$H_2SO_4 + 2\,NaOH \longrightarrow Na_2SO_4 + 2\,H_2O$$

(*c*) $$8\,H^+ + H_2SO_4 + 8\,e^- \longrightarrow H_2S + 4\,H_2O$$

Again we use the factors found in Sec. 15.2:

(*a*) $$4.00\,N = \frac{4.00\,\text{equivalents}}{L}\left(\frac{1\,\text{mol}}{\text{equivalent}}\right) = \frac{4.00\,\text{mol}}{L} = 4.00\,M$$

(*b*) $$4.00\,N = \frac{4.00\,\text{equivalents}}{L}\left(\frac{1\,\text{mol}}{2\,\text{equivalents}}\right) = \frac{2.00\,\text{mol}}{L} = 2.00\,M$$

(*c*) $$4.00\,N = \frac{4.00\,\text{equivalents}}{L}\left(\frac{1\,\text{mol}}{8\,\text{equivalents}}\right) = \frac{0.500\,M}{L} = 0.500\,M$$

We can do concentration problems with normality just as we did them with molarity (Chap. 10). The quantities are expressed in equivalents, of course.

EXAMPLE 15.8. How many equivalents are present in $2.0\,L$ of $3.0\,N$ solution?

$$2.0\,L\left(\frac{3.0\,\text{equivalents}}{L}\right) = 6.0\,\text{equivalents}$$

EXAMPLE 15.9. How many liters of $4.00\,N$ H_3PO_4 does it take to hold 7.00 equivalents?

$$7.00\,\text{equivalents}\left(\frac{1\,L}{4.00\,\text{equivalents}}\right) = 1.75\,L$$

EXAMPLE 15.10. What is the final concentration of HCl if $2.0\,L$ of $3.0\,N$ HCl and $1.5\,L$ of $4.0\,N$ HCl are mixed and diluted to $5.0\,L$?

$$(2.0\,L)(3.0\,N) = 6.0\,\text{equivalents}$$

$$(1.5\,L)(4.0\,N) = 6.0\,\text{equivalents}$$

$$\text{total} = 12.0\,\text{equivalents}$$

$$\frac{12.0\,\text{equivalents}}{5.0\,L} = 2.4\,N$$

Equivalents are especially useful in dealing with stoichiometry problems in solution. Since 1 equivalent of one thing reacts with 1 equivalent of any other thing in the reaction, it is also true that the volume times the normality of the first thing is equal to the volume times the normality of the second.

EXAMPLE 15.11. What volume of $3.00\,N$ NaOH is required to neutralize $25.00\,mL$ of $2.00\,N$ H_2SO_4?

$$N_1V_1 = N_2V_2$$

$$V_2 = \frac{N_1V_1}{N_2} = \frac{(2.00\,N\ H_2SO_4)(0.02500\,L)}{3.00\,N\ NaOH} = 0.0167\,L = 16.7\,mL$$

Note that less volume of NaOH is required because its normality is greater.

15.4 EQUIVALENT WEIGHT

The *equivalent weight* of a substance is the mass in grams of 1 equivalent of the substance. The equivalent weight is often a useful value to characterize a substance.

EXAMPLE 15.12. A new solid acid was prepared in a laboratory; its formula weight was not known. It was titrated with standard base, and the number of moles of base was calculated. Without knowing the formula of the acid, can you tell how many moles of the acid was present in a certain mass of acid? Can you tell how many equivalents of acid was present?

Without knowing the formula, you cannot tell the number of moles, but you can tell the number of equivalents. For example, HX and H_2X_2 would both neutralize the same number of moles of NaOH per gram of acid. Suppose X had a formula weight of 24 g/mol. HX would have a molecular weight of 25; H_2X_2 would have a molecular weight of 50 g/mol. To neutralize 1.00 mol NaOH would take 1.00 mol HX or 0.500 mol H_2X_2:

$$HX + NaOH \longrightarrow NaX + H_2O$$

or

$$H_2X_2 + 2\,NaOH \longrightarrow Na_2X_2 + 2\,H_2O$$

One mole of HX is 25 g; 0.500 mol of H_2X_2 is also 25 g. You cannot tell from the mass which is the formula of the acid. In either case, the equivalent weight of the acid is 25 g/equivalent. The equivalent weight of H_2X_2 is given by

$$\frac{50\,g}{mol}\left(\frac{1\,mol}{2\,equivalents}\right) = \frac{25\,g}{equivalent}$$

To convert from formula weight to equivalent weight, use the same factors as were introduced in Sec. 15.2.

EXAMPLE 15.13. Calculate the equivalent weight of $H_2C_2O_4$ in its complete neutralization by NaOH.

$$H_2C_2O_4 + 2\,OH^- \longrightarrow C_2O_4^{2-} + 2\,H_2O$$

$$\frac{90.0\,g}{mol}\left(\frac{1\,mol}{2\,equivalents}\right) = \frac{45.0\,g}{equivalent}$$

EXAMPLE 15.14. Calculate the formula weight of an acid with three replaceable hydrogen ions and an equivalent weight of 20.1 g/equivalent, assuming complete neutralization.

$$\frac{20.1\,g}{equivalent}\left(\frac{3\,equivalents}{mol}\right) = \frac{60.3\,g}{mol}$$

Solved Problems

EQUIVALENTS

15.1. An equivalent is defined as that amount of a substance that reacts with or produces a mole of what?

Ans. In a redox reaction, electrons. In an acid-base reaction, hydrogen ions or hydroxide ions.

15.2. What is the number of equivalents per mole of HCl?

Ans. Assuming complete neutralization (or even oxidation of the Cl^- to Cl_2), we can see that 1 mol HCl will react with 1 mol NaOH (or 1 mol e^-); thus, 1 equivalent = 1 mol.

15.3. How many equivalents of NaOH react with the 98.0 g of H_2SO_4 in part (*a*) of Example 15.1?

Ans. 1 equivalent. (One OH^- can react with 1 mol H^+; 1 equivalent of H_2SO_4 is involved.)

15.4. What numbers of equivalents per mole are possible for H_3PO_4 in its acid-base reactions?

Ans. 1, 2, or 3, depending on how complete its reaction with base is.

15.5. How many equivalents are there per mole of the first reactant in each of the following equations?

(*a*) $2\,HCl + Ba(OH)_2 \longrightarrow BaCl_2 + 2\,H_2O$

(*b*) $H_2SO_4 + NaOH \longrightarrow NaHSO_4 + H_2O$

(*c*) $Zn \longrightarrow Zn^{2+} + 2\,e^-$

(*d*) $H_2SO_4 + 2\,H^+ + Cu \longrightarrow Cu^{2+} + SO_2 + 2\,H_2O$

(*e*) $MnO_4^- + 5\,e^- + 8\,H^+ \longrightarrow Mn^{2+} + 4\,H_2O$

(*f*) $2\,H_2SO_4 + Cu \longrightarrow CuSO_4 + SO_2 + 2\,H_2O$

(*g*) $NaH_2PO_4 + 2\,NaOH \longrightarrow Na_3PO_4 + 2\,H_2O$

(*h*) $NaHCO_3 + HCl \longrightarrow NaCl + CO_2 + H_2O$

(*i*) $NaHCO_3 + NaOH \longrightarrow Na_2CO_3 + H_2O$

Ans.   (*a*) 1, (*b*) 1, (*c*) 2, (*d*) 2, (*e*) 5, (*f*) 2 [same as (*d*) for the one H_2SO_4 molecule that is reduced], (*g*) 2, (*h*) 1, and (*i*) 1.

NORMALITY

15.6. If a bottle is labeled 4.6 *N* H_3PO_4, which reaction should be assumed for the definition of its normality?

Ans.

$$H_3PO_4 + 3\,NaOH \longrightarrow Na_3PO_4 + 3\,H_2O \qquad \text{(complete neutralization)}$$

15.7. Explain why volume times normality of one reagent is equal to volume times normality of the other reagents in the reaction.

Ans.   Volume times normality is the number of equivalents, and by definition, the number of equivalents of one reagent is the same as the number of equivalents of any other reagent in a given reaction.

15.8. Explain why normality is the number of milliequivalents per milliliter.

Ans.   By definition, normality is the number of equivalents per liter. Then

$$\frac{x\,\text{equivalents}}{L}\left(\frac{1\,000\,\text{mequiv}}{\text{equivalent}}\right)\left(\frac{1\,L}{1\,000\,mL}\right) = \frac{x\,\text{mequiv}}{mL}$$

15.9. What volume of 2.40 *N* H_2SO_4 is completely neutralized by 44.0 mL of (*a*) 1.22 *N* NaOH? (*b*) 1.22 *M* NaOH? (*c*) 1.22 *M* $Ba(OH)_2$? (*d*) 1.22 *N* $Ba(OH)_2$?

Ans.   (*a*)   The number of equivalents of NaOH is $0.0440\,L \times 1.22\,N = 0.0537$ equivalents. That also is the number of equivalents of H_2SO_4.

$$0.0537\,\text{equivalents}\left(\frac{1\,L}{2.40\,\text{equivalents}}\right) = 0.0224\,L = 22.4\,mL$$

(*b*)   Since 1.22 *M* NaOH is 1.22 *N* NaOH, the answer is the same as in part (*a*).

(*c*) $1.22\,M\,Ba(OH)_2 = \dfrac{1.22\,\text{mol}\,Ba(OH)_2}{L}\left(\dfrac{2\,\text{equivalents}}{\text{mol}}\right) = 2.44\,N\,Ba(OH)_2$

$$2.44\,N\,(0.0440\,L) = 0.107\,\text{equivalent}$$

The quantity of $Ba(OH)_2$ is 0.107 equivalent. The volume of H_2SO_4 is therefore

$$0.107\,\text{equivalent}\,H_2SO_4\left(\frac{1\,L}{2.40\,\text{equivalents}}\right) = 0.0446\,L = 44.6\,mL$$

(d)   1.22 N $Ba(OH)_2$ has the same neutralizing ability as 1.22 N NaOH, and the answer is the same as in part (a).

15.10. Calculate the number of milliequivalents per milliliter in 2.70 N H_2SO_4.

Ans.

$$2.70\,N = \frac{2.70\,\text{equivalents}}{L}\left(\frac{1\,000\,\text{mequiv}}{\text{equivalent}}\right)\left(\frac{1\,L}{1\,000\,\text{mL}}\right) = \frac{2.70\,\text{mequiv}}{\text{mL}}$$

Normality may de defined as the number of milliequivalents per milliliter as well as the number of equivalents per liter.

15.11. A bottle is marked 4.00 N H_2SO_4. What is its probable molarity?

Ans.   Assuming that the normality is for complete neutralization, the molarity is given by

$$\frac{4.00\,\text{equivalents}}{L}\left(\frac{1\,\text{mol}}{2\,\text{equivalents}}\right) = 2.00\,M$$

15.12. A bottle is marked 4.00 N H_2SO_4 for production of $NaHSO_4$. What is its molarity?

Ans.

$$H_2SO_4 + NaOH \longrightarrow NaHSO_4 + H_2O$$

$$\frac{4.00\,\text{equivalents}}{L}\left(\frac{1\,\text{mol}}{1\,\text{equivalent}}\right) = 4.00\,M$$

15.13. What volume of 1.23 N H_3PO_4 will 2.40 equivalents of a base neutralize completely? Explain why you did not need to know the formula of the base to answer this question.

Ans.   2.40 equivalents of base reacts with 2.40 equivalents of acid, no matter what the base.

$$2.40\,\text{equivalents}\ H_3PO_4\left(\frac{1\,L}{1.23\,\text{equivalents}}\right) = 1.95\,L$$

(The identity of the base was used to determine the number of equivalents of base; it is not necessary to know its identity here.)

EQUIVALENT WEIGHT

15.14. If 2.000 g of a solid acid is neutralized by 24.07 mL of 1.070 N NaOH, what is its equivalent weight?

Ans.

$$24.07\,\text{mL}\left(\frac{1.070\,\text{mequiv}}{\text{mL}}\right) = 25.75\,\text{mequiv}$$

$$\frac{2\,000\,\text{mg}}{25.75\,\text{mequiv}} = 77.67\,\text{mg/mequiv}$$

15.15. It takes 47.03 mL of a base to titrate 2.000 g of potassium hydrogen phthalate ($KHC_8H_4O_4$). What is the normality of the base?

Ans.

$$2.000\,g\left(\frac{1\,\text{equivalent}}{204\,g}\right) = 0.00980\,\text{equivalent}\ KHC_8H_4O_4$$

$$\frac{0.00980\,\text{equivalent base}}{0.04703\,L} = 0.208\,N$$

Supplementary Problems

15.16. A sample of 4.00 g of a solid acid was treated with 50.00 mL of 2.000 N NaOH, which dissolved it completely (by reacting with it). There was enough excess NaOH to require 10.07 mL of 1.000 N HCl to neutralize the excess base. What is the equivalent weight of the acid?

Ans.   The number of equivalents of base and HCl are as follows:

$$50.00\,\text{mL}\left(\frac{2.000\,\text{mequiv}}{\text{mL}}\right) = 100.0\,\text{mequiv NaOH}$$

$$10.07\,\text{mL}\left(\frac{1.000\,\text{mequiv}}{\text{mL}}\right) = 10.07\,\text{mequiv HCl}$$

The difference is the number of milliequivalents of the solid acid:

$$100.0\,\text{mequiv} - 10.07\,\text{mequiv} = 89.9\,\text{mequiv}$$

The equivalent weight is

$$\frac{4\,000\,\text{mg}}{89.9\,\text{mequiv}} = \frac{44.5\,\text{mg}}{\text{mequiv}} = \frac{44.5\,\text{g}}{\text{equivalent}}$$

15.17. Calculate the equivalent weight of each of the following acids toward complete neutralization. (*a*) H_3PO_4, (*b*) HCl, (*c*) H_2SO_4, (*d*) $H_2C_2O_4$, and (*e*) $HC_2H_3O_2$.

Ans.   (*a*) 32.7 g/equivalent, (*b*) 36.5 g/equivalent, (*c*) 49.0 g/equivalent, (*d*) 45.0 g/equivalent, and (*e*) 60.0 g/equivalent.

15.18. Write an equation for the half-reaction in which $H_2C_2O_4$ is oxidized to CO_2. What is the equivalent weight of the $H_2C_2O_4$?

Ans.

$$H_2C_2O_4 \longrightarrow 2\,CO_2 + 2\,H^+ + 2\,e^-$$

$$\frac{90.0\,\text{g}}{\text{mol}}\left(\frac{1\,\text{mol}}{2\,\text{equivalents}}\right) = \frac{45.0\,\text{g}}{\text{equivalent}}$$

15.19. What is the equivalent weight of $H_2C_2O_4$ in the following reaction?

$$6\,H^+ + 2\,MnO_4^- + 5\,H_2C_2O_4 \longrightarrow 10\,CO_2 + 8\,H_2O + 2\,Mn^{2+}$$

Ans.   The half-reaction is given in Problem 15.18. The 1 mol of acid liberates 2 mol of electrons, and so there are 2 equivalents of acid per mole of acid. If we wrote the half-reaction for the equation given, we would find that there are 10 mol of electrons involved in reducing 5 mol of acid, and so again we get

$$10\,\text{equivalents}/5\,\text{mol} = 2\,\text{equivalents}/\text{mol}$$

Still another method is to balance the equation by the oxidation state change method:

$$H_2\underset{\underset{\textstyle 2(3\rightarrow 4)\,=\,+2}{\rule{0pt}{0pt}}}{\underbrace{C_2}}O_4 \longrightarrow 2\,CO_2$$

The change in oxidation state is 2 per molecule of $H_2C_2O_4$, and so there are 2 equivalents per mole.

15.20. Calculate the oxidation number of carbon in (*a*) CH_2O and (*b*) CH_2Cl_2.

Ans.   (*a*) 0 and (*b*) 0.

15.21. (a) What is the concentration of NaOH if 43.17 mL of 2.073 N H_2SO_4 neutralizes 25.00 mL of the base?
(b) What is the concentration of NaOH if 43.17 mL of 2.073 M H_2SO_4 neutralizes 25.00 mL of the base?

Ans.

(a)
$$N_1 V_1 = N_2 V_2$$

$$N_2 = \frac{N_1 V_1}{V_2} = \frac{(2.073\,N)(43.17\,\text{mL})}{(25.00\,\text{mL})} = 3.580\,N$$

(b) The number of millimoles of acid is given by

$$43.17\,\text{mL}\left(\frac{2.073\,\text{mmol}}{\text{mL}}\right) = 89.49\,\text{mmol}\,H_2SO_4$$

The number of millimoles of base is determined from the 2 : 1 ratio in the balanced chemical equation:

$$89.49\,\text{mmol}\,H_2SO_4\left(\frac{2\,\text{mmol}\,NaOH}{\text{mmol}\,H_2SO_4}\right) = 179.0\,\text{mmol}\,NaOH$$

$$\frac{179.0\,\text{mmol}\,NaOH}{25.00\,\text{mL}\,NaOH} = 7.160\,M\ NaOH = 7.160\,N\,NaOH$$

Chapter 16

Solutions

16.1 QUALITATIVE CONCENTRATION TERMS

Solutions are mixtures, and therefore do not have definite compositions. For example, in a glass of water it is possible to dissolve 1 teaspoonful of sugar or 2 or 3 or more. However, for most solutions there is a limit to how much *solute* will dissolve in a given quantity of *solvent* at a given temperature. The maximum concentration of solute that will dissolve in contact with excess solute is called the *solubility* of the solute. Solubility depends on temperature. Most solids dissolve in liquids more at higher temperatures than at lower temperatures, while gases dissolve in cold liquids better than in hot liquids.

A solution in which the concentration of the solute is equal to the solubility is called a *saturated* solution. If the concentration is lower, the solution is said to be *unsaturated*. It is also possible to prepare a *supersaturated* solution, an unstable solution containing a greater concentration of solute than is present in a saturated solution. Such a solution deposits the excess solute if a crystal of the solute is added to it. It is prepared by dissolving solute at one temperature and carefully changing the temperature to a point where the solution is unstable.

EXAMPLE 16.1. A solution at 0 °C contains 119 g of sodium acetate per 100 g of water. If more sodium acetate is added, it does not dissolve, and no sodium acetate crystallizes from solution either. Describe the following solutions as saturated, unsaturated, or supersaturated. (*a*) 100 g sodium acetate in 100 g water at 0 °C. (*b*) 150 g sodium acetate in 100 g water at 0 °C. (*c*) 11.9 g sodium acetate in 10.0 g water at 0 °C.

(*a*) The solution is unsaturated; more solute could dissolve. (*b*) The solution is supersaturated; only 119 g is stable in water at this temperature. (*c*) The solution is saturated; the concentration is the same as that given in the statement of the problem.

EXAMPLE 16.2. How can the solution described in Example 16.1(*b*) be prepared?

The 150 g of sodium acetate is mixed with 100 g of water and heated to nearly 100 °C, where 170 g of solute would dissolve. The mixture is stirred until solution is complete, and then the solution is cooled until it gets to 0 °C. The solute would crystallize if it could, but this particular solute has difficulty doing so, and thus a supersaturated solution is formed. Adding a crystal of solid sodium acetate allows the excess solute to crystallize around the solid added, and the excess solute precipitates out of the solution, leaving the solution saturated.

16.2 MOLALITY

Both molarity (Chap. 10) and normality (Chap. 15) are defined in terms of a volume. Since the volume is temperature-dependent, so are the molarity and normality of the solution. Two units of concentration that are independent of temperature are introduced in this chapter. *Molality* is defined as the number of moles of solute per kilogram of solvent in a solution. The symbol for molality is *m*. Note the differences between molality and molarity:
1. Molality is defined in terms of kilograms, not liters.
2. Molality is defined in terms of solvent, not solution.
3. The symbol for molality is a lower case *m*, not capital *M*.
   Great care is necessary to make sure that you do not confuse molality and molarity.

EXAMPLE 16.3. Calculate the molality of a solution prepared by adding 0.200 mol of solute to 100 g of water.

$$\frac{0.200 \, \text{mol}}{0.100 \, \text{kg}} = 2.00 \, m$$

EXAMPLE 16.4. What mass of H_2O is required to make a 4.00-m solution with 2.06 mol NaCl?

$$2.06 \, \text{mol NaCl} \left(\frac{1000 \, \text{g} \, H_2O}{4.00 \, \text{mol NaCl}} \right) = 515 \, \text{g} \, H_2O$$

EXAMPLE 16.5. What mass of 3.00 m CH_3OH solution in water can be prepared with 50.0 g of CH_3OH?

$$50.0 \, \text{g} \, CH_3OH \left(\frac{1 \, \text{mol} \, CH_3OH}{32.0 \, \text{g} \, CH_3OH} \right) = 1.56 \, \text{mol} \, CH_3OH$$

$$1.56 \, \text{mol} \, CH_3OH \left(\frac{1.00 \, \text{kg solvent}}{3.00 \, \text{mol} \, CH_3OH} \right) = 0.520 \, \text{kg solvent}$$

$$50.0 \, \text{g} \, CH_3OH + 520 \, \text{g} \, H_2O = 570 \, \text{g solution}$$

Note the importance of keeping track of what materials you are considering, in addition to their units.

EXAMPLE 16.6. Calculate the molarity of a 1.40-m solution of NaCl in water if the density of the solution is 1.03 g/mL.

Since neither concentration depends on the quantity of solution under consideration, let us work with a solution containing 1.0000 kg H_2O. Then we have

$$1.40 \, \text{mol NaCl} \left(\frac{58.5 \, \text{g NaCl}}{\text{mol NaCl}} \right) = 81.9 \, \text{g NaCl}$$

The total mass of the solution is therefore 1 081.9 g, and its volume is

$$1\,081.9 \, \text{g} \left(\frac{1 \, \text{mL}}{1.03 \, \text{g}} \right) = 1\,050 \, \text{mL} = 1.05 \, \text{L}$$

The molarity is

$$\frac{1.40 \, \text{mol NaCl}}{1.05 \, \text{L}} = 1.33 \, M$$

16.3 MOLE FRACTION

The *mole fraction* of a substance in a solution is the ratio of the number of moles of that substance to the total number of moles in the solution. The symbol for mole fraction of A is usually X_A, although some texts use the symbol N_A. Thus, for a solution containing x mol of A, y mol of B, and z mol of C, the mole fraction of A is

$$X_A = \frac{x \, \text{mol A}}{x \, \text{mol A} + y \, \text{mol B} + z \, \text{mol C}}$$

EXAMPLE 16.7. What is the mole fraction of CH_3OH in a solution of 20.0 g CH_3OH and 20.0 g H_2O?

$$20.0 \, \text{g} \, CH_3OH \left(\frac{1 \, \text{mol} \, CH_3OH}{32.0 \, \text{g} \, CH_3OH} \right) = 0.625 \, \text{mol} \, CH_3OH$$

$$20.0 \, \text{g} \, H_2O \left(\frac{1 \, \text{mol} \, H_2O}{18.0 \, \text{g} \, H_2O} \right) = 1.11 \, \text{mol} \, H_2O$$

$$X_{CH_3OH} = \frac{0.625 \, \text{mol} \, CH_3OH}{0.625 \, \text{mol} \, CH_3OH + 1.11 \, \text{mol} \, H_2O} = 0.360$$

Since the mole fraction is a ratio of moles (of one substance) to moles (total), the units cancel out and the mole fraction has no units.

EXAMPLE 16.8. Show that the total of both mole fractions in a solution of two compounds is equal to 1.

$$X_A = \frac{x \text{ mol A}}{x \text{ mol A} + y \text{ mol B}}$$

$$X_B = \frac{y \text{ mol B}}{x \text{ mol A} + y \text{ mol B}}$$

$$X_A + X_B = \frac{x \text{ mol A}}{x \text{ mol A} + y \text{ mol B}} + \frac{y \text{ mol B}}{x \text{ mol A} + y \text{ mol B}}$$

$$= \frac{x \text{ mol A} + y \text{ mol B}}{x \text{ mol A} + y \text{ mol B}} = 1$$

Solved Problems

QUALITATIVE CONCENTRATION TERMS

16.1. Describe each of the solutions indicated as saturated, unsaturated, supersaturated, or impossible to tell. (*a*) More solute is added to a solution of that solute, and the additional solute all dissolves. Describe the original solution. (*b*) More solute is added to a solution of that solute, and the additional solute does not all dissolve. Describe the final solution. (*c*) A solution is left standing, and some of the solvent evaporates. After a time, some solute crystallizes out. Describe the final solution. (*d*) A hot saturated solution is cooled slowly, and no solid crystallizes out. Describe the cold solution. (*e*) A hot solution is cooled slowly, and after a time some solid crystallizes out. Describe the cold solution. (*f*) A hot saturated solution is cooled slowly, and no solid crystallizes out. The solute is a solid that is more soluble hot than cold. Describe the cold solution.

Ans. (*a*) The original solution was unsaturated, since it was possible to dissolve more solute at that temperature. (*b*) The final solution is saturated; it is holding all the solute that it can hold stably at that temperature, so not all the excess solute dissolves. (*c*) The final solution is saturated; it has solid solute in contact with the solution, and that solute does not dissolve. The solution must be holding as much as it can at this temperature. (*d*) It is impossible to tell with just this information [see part (*f*)]. The solute may get more soluble as the solution cools. (*e*) The solution is saturated. (*f*) The solution is supersaturated; it is holding more solute than is stable at the colder temperature.

16.2. The solubility of ethyl alcohol in water is said to be infinite. What does that mean?

Ans. It means that the alcohol is completely soluble in water no matter how much alcohol or how little water is present.

16.3. $Ce_2(SO_4)_3$ is unusual because it is a solid that dissolves in water better at lower temperatures than at higher temperatures. State how you might attempt to make a supersaturated solution of this compound in water at 50 °C.

Ans. Dissolve as much as possible in water at 0 °C. Then warm the solution carefully to 50 °C. Unless it has been done before, there is no guarantee that any particular compound will produce a supersaturated solution, but this is the way to try.

MOLALITY

16.4. Calculate the molality of each of the following solutions: (*a*) 10.0 g C_2H_5OH in 200 g H_2O and (*b*) 20.0 g NaCl in 100 g H_2O.

Ans.

(*a*) $10.0 \text{ g } C_2H_5OH \left(\dfrac{1 \text{ mol } C_2H_5OH}{46.0 \text{ g } C_2H_5OH} \right) = 0.217 \text{ mol } C_2H_5OH$

$$\dfrac{0.217 \text{ mol } C_2H_5OH}{0.200 \text{ kg } H_2O} = 1.09 \, m$$

(*b*) $20.0 \text{ g NaCl} \left(\dfrac{1 \text{ mol NaCl}}{58.5 \text{ g NaCl}} \right) = 0.342 \text{ mol NaCl}$

$$\dfrac{0.342 \text{ mol NaCl}}{0.100 \text{ kg } H_2O} = 3.42 \, m \text{ NaCl}$$

16.5. How many moles of solute are there in a 2.00-*m* solution containing 7.00 kg of solvent?

Ans.

$$7.00 \text{ kg solvent} \left(\dfrac{2.00 \text{ mol solute}}{\text{kg solvent}} \right) = 14.0 \text{ mol solute}$$

16.6. What mass of solvent is required to make a 3.00-*m* solution with 6.00 mol of solute?

Ans.

$$6.00 \text{ mol solute} \left(\dfrac{1 \text{ kg solvent}}{3.00 \text{ mol solute}} \right) = 2.00 \text{ kg solvent}$$

16.7. Calculate the total molality of all the ions in solution when 0.100 mol $Al(NO_3)_3$ is dissolved in 500 g H_2O.

Ans.   There are 0.100 mol Al^{3+} and 0.300 mol NO_3^- or a total of 0.400 mol ions in 0.500 kg of water. The total molality is

$$\dfrac{0.400 \text{ mol}}{0.500 \text{ kg}} = 0.800 \, m$$

16.8. Calculate the molality of the chloride ions in a solution containing 10.0 g $CaCl_2$ and 125 g H_2O.

Ans.

$$10.0 \text{ g } CaCl_2 \left(\dfrac{1 \text{ mol } CaCl_2}{111 \text{ g } CaCl_2} \right) = 0.0901 \text{ mol } CaCl_2$$

$$0.0901 \text{ mol } CaCl_2 \left(\dfrac{2 \text{ mol } Cl^-}{\text{mol } CaCl_2} \right) = 0.180 \text{ mol } Cl^-$$

$$\dfrac{0.180 \text{ mol } Cl^-}{0.125 \text{ kg } H_2O} = 1.44 \, m \text{ Cl}^-$$

MOLE FRACTION

16.9. What are the units of mole fraction?

Ans.   Mole fraction has no units. It is defined as one number of moles divided by another, and the units cancel.

16.10. Calculate the molality of a solution with mole fraction 0.100 of C_2H_5OH in water.

Ans. Assume 1.00 mol total, which contains:

$$0.100 \, mol \, C_2H_5OH$$

$$0.900 \, mol \, H_2O \left(\frac{18.0 \, g \, H_2O}{mol \, H_2O} \right) = 16.2 \, g \, H_2O$$

$$\frac{0.100 \, mol \, C_2H_5OH}{0.0162 \, kg \, H_2O} = 6.17 \, m$$

16.11. Calculate the molality and the mole fraction of each of the following solutions: (*a*) 1.00 mol CH_2O and 10.0 mol H_2O, (*b*) 2.00 mol CH_3OH and 500 g H_2O, and (*c*) 100 g C_2H_5OH and 500 g H_2O.

Ans.

(*a*)
$$10.0 \, mol \, H_2O \left(\frac{18.0 \, g \, H_2O}{mol \, H_2O} \right) = 180 \, g \, H_2O = 0.180 \, kg \, H_2O$$

$$molality = \frac{1.00 \, mol \, CH_2O}{0.180 \, kg \, H_2O} = 5.56 \, m$$

$$X_{CH_2O} = \frac{1.00 \, mol \, CH_2O}{1.00 \, mol \, CH_2O + 10.0 \, mol \, H_2O} = 0.0909$$

(*b*)
$$molality = \frac{2.00 \, mol \, CH_3OH}{0.500 \, kg \, H_2O} = 4.00 \, m$$

$$500 \, g \, H_2O \left(\frac{1 \, mol \, H_2O}{18.0 \, g \, H_2O} \right) = 27.8 \, mol \, H_2O$$

$$X_{CH_3OH} = \frac{2.00 \, mol \, CH_3OH}{29.8 \, mol \, total} = 0.0671$$

(*c*)
$$100 \, g \, C_2H_5OH \left(\frac{1 \, mol \, C_2H_5OH}{46.0 \, g \, C_2H_5OH} \right) = 2.17 \, mol \, C_2H_5OH$$

$$\frac{2.17 \, mol}{0.500 \, kg} = 4.34 \, m$$

From part (*b*) we have 27.8 mol H_2O.

$$X_{C_2H_5OH} = \frac{2.17 \, mol}{2.17 \, mol + 27.8 \, mol} = 0.0724$$

16.12. Calculate the mole fraction of the first component in each of the following solutions: (*a*) 50.0 g $C_6H_{12}O_6$ in 400 g H_2O and (*b*) 0.200-*m* solution of $C_6H_{12}O_6$ in water.

Ans.

(*a*)
$$50.0 \, g \, C_6H_{12}O_6 \left(\frac{1 \, mol \, C_6H_{12}O_6}{180 \, g \, C_6H_{12}O_6} \right) = 0.278 \, mol \, C_6H_{12}O_6$$

$$400 \, g \, H_2O \left(\frac{1 \, mol \, H_2O}{18.0 \, g \, H_2O} \right) = 22.2 \, mol \, H_2O$$

$$X_{C_6H_{12}O_6} = \frac{0.278 \, mol}{22.2 \, mol + 0.278 \, mol} = 0.0124$$

(*b*) In 1.00 kg H_2O there are 0.200 mol $C_6H_{12}O_6$ and

$$1000 \, g \, H_2O \left(\frac{1 \, mol \, H_2O}{18.0 \, g \, H_2O} \right) = 55.6 \, mol \, H_2O$$

The mole fraction is

$$X_{C_6H_{12}O_6} = \frac{0.200\,\text{mol}}{55.8\,\text{mol total}} = 0.00358$$

16.13. Explain why $X_A = 1.13$ cannot be correct.

Ans. The total of all mole fractions in a solution is 1, and no single mole fraction can exceed 1.

Supplementary Problems

16.14. Calculate the number of moles of CH_3OH in 5.00 L of 2.000 m solution if the density of the solution is 0.950 g/mL.

Ans.

$$5.00\,\text{L}\left(\frac{0.950\,\text{kg}}{\text{L}}\right) = 4.75\,\text{kg solution}$$

Per kilogram of water:

$$2.000\,\text{mol CH}_3\text{OH}\left(\frac{32.04\,\text{g}}{\text{mol}}\right) = 64.08\,\text{g CH}_3\text{OH} = 0.06408\,\text{kg CH}_3\text{OH}$$

$$1.000\,\text{kg H}_2\text{O} + 0.06408\,\text{kg CH}_3\text{OH} = 1.064\,\text{kg solution}$$

For the entire solution:

$$4.75\,\text{kg solution}\left(\frac{1\,\text{kg H}_2\text{O}}{1.064\,\text{kg solution}}\right) = 4.46\,\text{kg H}_2\text{O}$$

$$4.46\,\text{kg H}_2\text{O}\left(\frac{2.00\,\text{mol CH}_3\text{OH}}{\text{kg H}_2\text{O}}\right) = 8.92\,\text{mol CH}_3\text{OH}$$

16.15. The solubility of $Ba(OH)_2 \cdot 8H_2O$ in water at 15 °C is 5.6 g per 100 g water. (*a*) What is the molality of a saturated solution at 15 °C? (*b*) What is the molality of the hydroxide ions?

Ans.

$$(a) \qquad 5.6\,\text{g Ba(OH)}_2 \cdot 8\text{H}_2\text{O}\left(\frac{1\,\text{mol Ba(OH)}_2 \cdot 8\text{H}_2\text{O}}{315\,\text{g Ba(OH)}_2 \cdot 8\text{H}_2\text{O}}\right) = 0.018\,\text{mol Ba(OH)}_2 \cdot 8\text{H}_2\text{O}$$

$$\frac{0.018\,\text{mol}}{0.10\,\text{kg}} = 0.18\,m\,\text{compound}$$

$$(b) \qquad 0.18\,m\,\text{Ba(OH)}_2 \cdot 8\text{H}_2\text{O}\left(\frac{2\,\text{mol OH}^-}{\text{mol Ba(OH)}_2 \cdot 8\text{H}_2\text{O}}\right) = 0.36\,m\,\text{OH}^-$$

16.16. What is the total of the mole fractions of a three-component system?

Ans. 1.00.

Chapter 17

Electron Configuration of the Atom

17.1 INTRODUCTION

In Chap. 3 the elementary structure of the atom was introduced. The facts that protons, neutrons, and electrons are present in the atom and that electrons are arranged in shells allowed us to explain isotopes (Chap. 3), the octet rule for main group elements (Chap. 5), ionic and covalent bonding (Chap. 5), and much more. However, we still have not been able to deduce why the transition metal groups and inner transition metal groups arise, why many of the transition metals have ions of different charges, how the shapes of molecules are determined, and much more. In this chapter we introduce a more detailed description of the electronic structure of the atom which begins to answer some of these more difficult questions.

The modern theory of the electronic structure of the atom is based on experimental observations of the interaction of electricity with matter, studies of electron beams (cathode rays), studies of radioactivity, studies of the distribution of the energy emitted by hot solids, and studies of the wavelengths of light emitted by incandescent gases. A complete discussion of the experimental evidence for the modern theory of atomic structure is beyond the scope of this book. In this chapter only the results of the theoretical treatment will be described. These results will have to be memorized as "rules of the game," but they will be used so extensively throughout the general chemistry course that the notation used will soon become familiar.

17.2 BOHR THEORY

The first plausible theory of the electronic structure of the atom was proposed in 1914 by Niels Bohr (1885–1962), a Danish physicist. In order to explain the hydrogen spectrum (Fig. 17-1), he suggested that in each hydrogen atom, the electron revolves about the nucleus in one of several possible circular orbits, each having a definite radius corresponding to a definite energy for the electron. An electron in the orbit closest to the nucleus should have the lowest energy. With the electron in that orbit, the atom is said to be in its lowest energy state or *ground state*. If a discrete quantity of additional energy were absorbed by the atom in some manner, the electron would be able to move into another orbit having a higher energy. The hydrogen atom would then be in an *excited state*. An atom in the excited state will return to the ground state and give off its excess energy as light in the process. In returning to the ground state, the energy may be emitted all at once or it may be emitted in a stepwise manner as the electron drops from a higher allowed orbit to allowed orbits of lower and lower energy. Since each orbit corresponds to a definite energy level, the energy of the light emitted will correspond to the definite differences in energy between levels. Therefore, the light

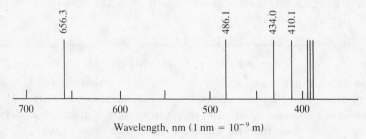

Fig. 17-1 Visible spectrum of hydrogen

251

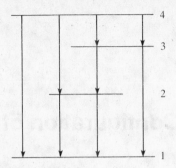

Fig. 17-2 Possible return paths for electron in orbit 4

$$4 \to 1 \qquad 4 \to 2 \to 1 \qquad 4 \to 3 \to 2 \to 1 \qquad 4 \to 3 \to 1$$

Only electron transitions down to the second orbit cause emission of visible light. Other transitions may involve infrared or ultraviolet light.

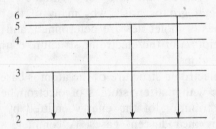

Fig. 17-3 The origin of the visible spectrum of hydrogen (not drawn to scale)

emitted as the atom returns to its ground state will have a definite energy or a definite set of energies (Fig. 17-2). The discrete amounts of energy emitted or absorbed by an atom or molecule are called *quanta* (singular, *quantum*). A quantum of light energy is called a *photon*.

The wavelength of a photon—a quantum of light—is inversely proportional to the energy of the light, and when the light is observed through a spectroscope, lines of different colors, corresponding to different wavelengths, are seen. The origin of the visible portion of the hydrogen spectrum is shown schematically in Fig. 17-3.

Bohr's original idea of orbits of discrete radii has been greatly modified, but the concept that the electron in the hydrogen atom occupies definite energy levels still applies. It can be calculated that an electron in a higher energy level is located on the average farther away from the nucleus than one in a lower energy level. It is customary to refer to the successive energy levels as electron *shells*. The terms *energy level* and *shell* are used interchangeably. The shells are often designated by capital letters, with K denoting the lowest energy level, as follows:

energy level: 1   2   3   4   5 ...

shell notation: K   L   M   N   O ...

The electrons in atoms other than hydrogen also occupy various energy levels. With more than one electron in each atom, the question of how many electrons can occupy a given level becomes important. The maximum number of electrons that can occupy a given shell depends on the shell number. For example, in any atom, the first shell can hold a maximum of only 2 electrons, the second shell can hold a maximum of 8 electrons, the third shell can hold a maximum of 18 electrons, and so forth. The maximum number of electrons that can occupy any particular shell is $2n^2$, where n is the shell number.

EXAMPLE 17.1.   What is the maximum number of electrons that can occupy the M shell?

The M shell in an atom corresponds to the third energy level ($n = 3$); hence the maximum number of electrons it can hold is

$$2n^2 = 2(3)^2 = 2 \times 9 = 18$$

The shell structure of the energy levels of various atoms is sometimes represented by diagrams such as are shown in Fig. 17-4 for the first 10 elements. It must be emphasized that these are *diagrams*, and are not pictures of atoms. (Electron dot diagrams represent the outermost of these electron shells.) Such diagrams are quite inadequate for depicting atoms of elements having atomic numbers beyond 20.

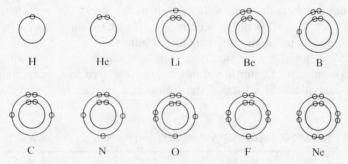

Fig. 17-4   Diagrams of shell structure for the first 10 elements

17.3   QUANTUM NUMBERS

The modern theory of the electronic structure of the atom stems from a complex mathematical equation (called the Schrödinger equation), which is beyond the mathematical requirements for the general chemistry course. We therefore take the results of the solution of this equation as postulates. Solution of the equation yields four *quantum numbers*, named and with limitations as shown in Table 17-1. (In some texts, the magnetic quantum number and the spin quantum number are symbolized m_l and m_s, respectively.) A given electron is specified in terms of four quantum numbers that govern its energy, its orientation in space, and its possible interactions with other electrons. Thus, listing the values of the four quantum numbers describes the probable location of the electron, somewhat analogously to listing the section, row, seat, and date on a ticket to a football game. To learn to express the electronic structure of an atom, it is necessary to learn (1) the names, symbols, and permitted values of the quantum numbers, and (2) the order of increasing energy of electrons as a function of their sets of quantum numbers.

Table 17-1   The Quantum Numbers

Name	Symbol	Limitations
Principal quantum number	n	Any positive integer
Angular momentum quantum number	l	$0, \ldots, (n-1)$ in integer steps
Magnetic quantum number	m	$-l, \ldots, 0, \ldots, +l$ in integer steps
Spin quantum number	s	$-\frac{1}{2}, +\frac{1}{2}$

The principal quantum number of an electron is denoted n. It is the most important quantum number in determining the energy of the electron. In general, the higher the principal quantum number, the higher the energy of the electron. Electrons with higher principal quantum numbers are also apt to be farther away from the nucleus than electrons with lower principal quantum numbers. The values of n can be any positive integer: 1, 2, 3, 4, 5, 6, 7, ... The first seven principal quantum numbers are the only important ones for electrons in ground states of atoms.

The angular momentum quantum number is denoted l. It also affects the energy of the electron, but in general not as much as the principal quantum number does. In the absence of an electric or magnetic field around the atom, only these two quantum numbers have any effect on the energy of the electron. The value of l can be 0 or any positive integer up to, but not including, the value of n for that electron.

The magnetic quantum number, denoted m, determines the orientation in space of the electron, but does not ordinarily affect the energy of the electron. Its values depend on the value of l for that electron, ranging from $-l$ through 0 to $+l$ in integral steps. Thus, for an electron with an l value of 2, the possible m values are -2, -1, 0, 1, and 2.

The spin quantum number, denoted s, is related to the "spin" of the electron on its "axis." It ordinarily does not affect the energy of the electron. Its possible values are $-\frac{1}{2}$ and $+\frac{1}{2}$. The value of s does not depend on the value of any other quantum number.

The permitted values for the other quantum numbers when $n = 2$ are shown in Table 17-2. The rules governing the assignment of quantum numbers are compared to a more familiar situation in Fig. 17-5. The following examples will illustrate the limitations on the values of the quantum numbers (Table 17-1).

Table 17-2 Permitted Values of Quantum Numbers when $n = 2$

n	2	2	2	2	2	2	2	2
l	0	0	1	1	1	1	1	1
m	0	0	-1	0	$+1$	-1	0	$+1$
s	$-\frac{1}{2}$	$+\frac{1}{2}$	$-\frac{1}{2}$	$-\frac{1}{2}$	$-\frac{1}{2}$	$+\frac{1}{2}$	$+\frac{1}{2}$	$+\frac{1}{2}$

Each vertical set of four quantum numbers represents one electron.

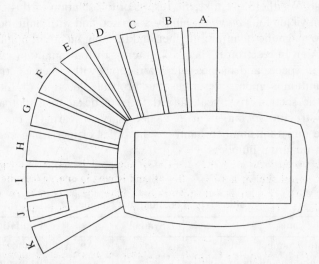

Fig. 17-5 Football stadium: Number of rows depends on section; number of seats depends on row; date is independent of all other factors.

EXAMPLE 17.2. What are the first seven permitted values for n?

　1, 2, 3, 4, 5, 6, and 7.

EXAMPLE 17.3. What values of l are permitted for an electron with principal quantum number $n = 3$?

　0, 1, and 2. l can have integral values from 0 up to $(n - 1)$.

EXAMPLE 17.4. What values are permitted for m in an electron in which the l value is 4?

　$-4, -3, -2, -1, 0, 1, 2, 3$, and 4. m can have integer values from $-l$ through 0 up to $+l$.

EXAMPLE 17.5. What values are permitted for s in an electron in which $n = 3$, $l = 2$, and $m = -1$?

　$-\frac{1}{2}$ and $+\frac{1}{2}$. It does not matter what values the other quantum numbers have; the s value must be either $-\frac{1}{2}$ or $+\frac{1}{2}$.

The angular momentum quantum numbers are often given letter designations, so that when they are stated along with principal quantum numbers, less confusion results. The letter designations of importance in the ground states of atoms are the following:

l Value	Letter Designation
0	s
1	p
2	d
3	f

17.4　QUANTUM NUMBERS AND ENERGIES OF ELECTRONS

The energy of the electrons in an atom is of paramount importance. The n and l quantum numbers determine the energy of each electron (apart from the effects of external electric and magnetic fields, which are most often not of interest in general chemistry courses). The energies of the electrons increase as the sum $(n + l)$ increases; the lower the value of $(n + l)$ for an electron in an atom, the lower is its energy. For two electrons with equal values of $(n + l)$, the one with the lower n value has lower energy. Thus, we can fill an atom with electrons starting with its lowest-energy electrons by starting with the electrons with the lowest sum $(n + l)$.

EXAMPLE 17.6. Arrange the electrons in the following list in order of increasing energy, lowest first:

	n	l	m	s
(a)	3	2	-1	$\frac{1}{2}$
(b)	4	0	0	$-\frac{1}{2}$
(c)	4	1	1	$-\frac{1}{2}$
(d)	2	1	-1	$\frac{1}{2}$

Electron (d) has the lowest value of $n + l$ ($2 + 1 = 3$), and so it is lowest in energy of the four electrons. Electron (b) has the next lowest sum of $n + l$ ($4 + 0 = 4$), and is next in energy (despite the fact that it does not have the next lowest n value). Electrons (a) and (c) both have the same sum of $n + l$ ($3 + 2 = 4 + 1 = 5$). Therefore, in this case, electron (a), the one with the lower n value, is lower in energy. Electron (c) is highest in energy.

The *Pauli exclusion principle* states that no two electrons in the same atom can have the same set of four quantum numbers. Along with the order of increasing energy, we can use this principle to deduce the order of filling of electron shells in atoms.

EXAMPLE 17.7. Use the Pauli principle and the $(n + l)$ rule to predict the sets of quantum numbers for the 12 electrons in the ground state of a magnesium atom.

We want the first electron to have the lowest energy possible. The lowest value of $(n + l)$ will have the lowest n possible and the lowest l possible. The lowest n permitted is 1 (Table 17-1). With that value of n, the only value of l permitted is 0. With $l = 0$, the value of m must be 0 $(-0 \ldots +0)$. The value of s can be either $-\frac{1}{2}$ or $+\frac{1}{2}$. Thus, the first electron can have either

$$n = 1, \qquad l = 0, \qquad m = 0, \qquad s = -\frac{1}{2}$$

or
$$n = 1, \qquad l = 0, \qquad m = 0, \qquad s = +\frac{1}{2}$$

The second electron also can have $n = 1$, $l = 0$, and $m = 0$. Its value of s can be either $+\frac{1}{2}$ or $-\frac{1}{2}$, but not the same as that for the first electron. If it were, this second electron would have the same set of four quantum numbers that the first electron has, which is not permitted by the Pauli principle. If we were to try to give the third electron the same values for the first three quantum numbers, we would be stuck when we came to assign the s value. Both $+\frac{1}{2}$ and $-\frac{1}{2}$ have already been used, and we would have a duplicate set of quantum numbers for two electrons, which is not permitted. We cannot use any other values for l or m with the value of $n = 1$, and so the third electron must have the next higher value, $n = 2$. The l values could be 0 or 1, and since 0 will give a lower $(n + l)$ sum, we choose that value for the third electron. Again the value of m must be 0 since $l = 0$, and s can have a value $-\frac{1}{2}$ (or $+\frac{1}{2}$). For the fourth electron, $n = 2$, $l = 0$, $m = 0$, and $s = +\frac{1}{2}$ (or $-\frac{1}{2}$ if the third were $+\frac{1}{2}$). The fifth electron can have $n = 2$ but not $l = 0$, since all combinations of $n = 2$ and $l = 0$ have been used. Therefore, $n = 2$, $l = 1$, $m = -1$, and $s = -\frac{1}{2}$ are assigned. The rest of the electrons in the magnesium atom are assigned quantum numbers as shown in Table 17-3.

Table 17-3 The Quantum Numbers of the Electrons of Magnesium

Electron	1	2	3	4	5	6	7	8	9	10	11	12
n	1	1	2	2	2	2	2	2	2	2	3	3
l	0	0	0	0	1	1	1	1	1	1	0	0
m	0	0	0	0	-1	0	1	-1	0	1	0	0
s	$-\frac{1}{2}$	$+\frac{1}{2}$	$-\frac{1}{2}$	$+\frac{1}{2}$	$-\frac{1}{2}$	$-\frac{1}{2}$	$-\frac{1}{2}$	$+\frac{1}{2}$	$+\frac{1}{2}$	$+\frac{1}{2}$	$-\frac{1}{2}$	$+\frac{1}{2}$

17.5 SHELLS, SUBSHELLS, AND ORBITALS

Electrons having the same value of n in an atom are said to be in the same *shell*. Electrons having the same value of n and the same value of l in an atom are said to be in the same *subshell*. (Electrons having the same values of n, l, and m in an atom are said to be in the same *orbital*.) Thus, the first two electrons of magnesium (Table 17-3) are in the first shell and also in the same subshell. The third and fourth electrons are in the same shell and subshell with each other. They are also in the same shell with the next six electrons (all have $n = 2$) but a different subshell ($l = 0$ rather than 1). With the letter designations of Sec. 17.3, the first two electrons of magnesium are in the $1s$ subshell, the next two electrons are in the $2s$ subshell, and the next six electrons are in the $2p$ subshell. The last two electrons occupy the $3s$ subshell.

Since the possible numerical values of l depend on the value of n, the number of subshells within a given shell is determined by the value of n. The number of subshells within a given shell is merely the value of n, the shell number. Thus, the first shell has one subshell; the second shell has two subshells, and so forth. These facts are summarized in Table 17-4. Even the atoms with the most electrons do not have enough electrons to completely fill the highest shells shown. The subshells that hold electrons in the ground states of the biggest atoms are **boldfaced**.

EXAMPLE 17.8. What are the values of n and l in each of the following subshells? (a) $3p$, (b) $4s$, (c) $6d$, and (d) $5f$.

(a) $n = 3$, $l = 1$, (b) $n = 4$, $l = 0$, (c) $n = 6$, $l = 2$, (d) $n = 5$, $l = 3$.

Table 17-4 Arrangement of Subshells in Electron Shells

Energy Level, n	Type of Subshell	Number of Subshells
1	s	1
2	s, p	2
3	s, p, d	3
4	s, p, d, f	4
5	s, p, d, f, g	5
6	s, p, d, f, g, h	6
7	s, p, d, f, g, h, i	7

EXAMPLE 17.9. Show that there can be only two electrons in any s subshell.

For any given value of n, there can be a value of $l = 0$, corresponding to an s subshell. For $l = 0$ there can be only one possible m value: $m = 0$. Hence, n, l, and m are all specified for a given s subshell. Electrons can then have spin values of $s = +\frac{1}{2}$ or $s = -\frac{1}{2}$. Thus, every possible set of four quantum numbers is used, and there are no other possibilities in that subshell. Each of the two electrons has the first three quantum numbers in common and has a different value of s. The two electrons are said to be *paired*.

Depending on the permitted values of the magnetic quantum number m, each subshell is further broken down into units called *orbitals*. The number of orbitals per subshell depends on the type of subshell but not on the value of n. Each orbital can hold a maximum of two electrons; hence, the maximum number of electrons that can occupy a given subshell is determined by the number of orbitals available. These relationships are presented in Table 17-5. The maximum number of electrons in any given energy level is thus determined by the subshells it contains. The first shell can contain 2 electrons; the second, 8 electrons; the third, 18 electrons; the fourth, 32 electrons; and so on.

Table 17-5 Occupancy of Subshells

Type of Subshell	Allowed Values of m	Number of Orbitals	Maximum Number of Electrons
s	0	1	2
p	$-1, 0, 1$	3	6
d	$-2, -1, 0, 1, 2$	5	10
f	$-3, -2, -1, 0, 1, 2, 3$	7	14

Suppose we want to write the electron configuration of scandium (atomic number 21). We can rewrite the first 12 electrons that we wrote above for magnesium, and then just keep going. As we added electrons, we filled the first shell of electrons first, then the second shell. When we are filling the third shell, we have to ask if the electrons with $n = 3$ and $l = 2$ will enter before the $n = 4$ and $l = 0$ electrons. Since $(n + l)$ for the former is 5 and that for the latter is 4, we must add the two electrons with $n = 4$ and $l = 0$ before the last 10 electrons with $n = 3$ and $l = 2$. In this discussion, the values of m and s tell us how many electrons can have the same set of n and l values, but do not matter as to which come first.

	13	14	15	16	17	18	19	20	21
n	3	3	3	3	3	3	4	4	3
l	1	1	1	1	1	1	0	0	2
m	-1	0	$+1$	-1	0	$+1$	0	0	-2
s	$-\frac{1}{2}$	$-\frac{1}{2}$	$-\frac{1}{2}$	$+\frac{1}{2}$	$+\frac{1}{2}$	$+\frac{1}{2}$	$-\frac{1}{2}$	$+\frac{1}{2}$	$-\frac{1}{2}$
$(n + l)$	4	4	4	4	4	4	4	4	5

Thus, an important development has occurred because of the $(n + l)$ rule. The fourth shell has started filling before the third shell has been completed. This is the origin of the transition series elements.

Thus, scandium, atomic number 21, has two electrons in its $1s$ subshell, two electrons in its $2s$ subshell, six electrons in its $2p$ subshell, two electrons in its $3s$ subshell, six electrons in its $3p$ subshell, two electrons in its $4s$ subshell, and its last electron in the $3d$ subshell.

We note in the electron configuration for electrons 13 through 20 for scandium that when the $(n + l)$ sum was 4 we added the $3p$ electrons before the $4s$ electrons. Each of these groups has an $(n + l)$ sum of 4. Since the $(n + l)$ values were the same, we added electrons having the lower n value first.

We conventionally use a more condensed notation for electron configurations, with the subshell notation and a superscript to denote the number of electrons in that subshell. To write the detailed electron configuration of any atom, showing how many electrons occupy each of the various subshells, one needs to know only the order of increasing energy of the subshells, given above, and the maximum number of electrons that will fit into each, given in Table 17-5. A convenient way to designate such a configuration is to write the shell and subshell designation, and add a superscript to denote the number of electrons occupying that subshell. For example, the electron configuration of the sodium atom is written as follows:

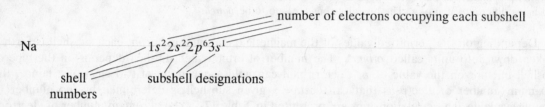

The shell number is represented by 1, 2, 3, and so forth, and the letters designate the subshells. The superscript numbers tell how many electrons occupy each subshell. Thus, in this example, there are two electrons in the $1s$ subshell, two electrons in the $2s$ subshell, six electrons in the $2p$ subshell, and only one electron in the $3s$ subshell. (The $3s$ subshell can hold a *maximum* of two electrons, but in this atom this subshell is not filled.) The total number of electrons in the atom can easily be determined by adding the numbers in all the subshells, that is, by adding all the superscripts. For sodium, this sum is 11, equal to the atomic number of sodium.

EXAMPLE 17.10. Write the electron configuration of magnesium.

$1s^2 2s^2 2p^6 3s^2$.

17.6 SHAPES OF ORBITALS

The *Heisenberg uncertainty principle* requires that, since the energy of the electron is known, its exact position cannot be known. It is possible to learn only the probable location of the electron in the vicinity of the atomic nucleus. The mathematical details of expressing the probability are quite complex, but it is possible to give an approximate description in terms of values of the quantum numbers n, l, and m. The shapes of the first few orbitals are shown in Fig. 17-6 for the case of the hydrogen atom. This figure shows that, in general, an electron in the $1s$ orbital is equally likely to be found in any direction about the nucleus. (The maximum probability is at a distance corresponding to the experimentally determined radius of the hydrogen atom.) In contrast, in the case of an electron in a $2p$ orbital, there are three possible values of the quantum number m. There are three possible regions in which the electron is most likely to be found. It is customary to depict these orbitals as being located along the cartesian (x, y, and z) axes of a three-dimensional graph. Hence, the three probability distributions are labeled p_x, p_y, and p_z, respectively.

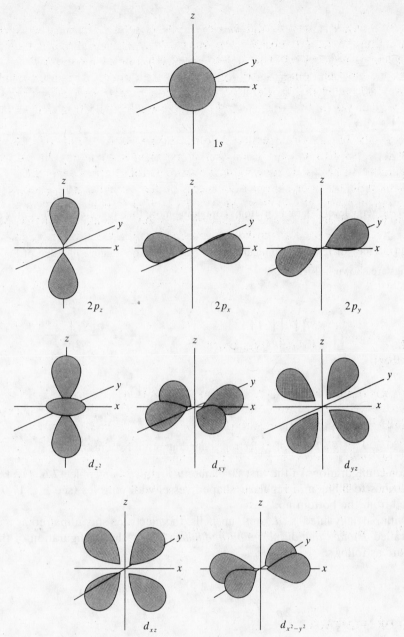

Fig. 17-6   Shapes of various orbitals

17.7   BUILDUP PRINCIPLE

With each successive increase in atomic number, a given atom has one more electron than the previous atom. Thus, it is possible to start with hydrogen and, adding one electron at a time, build up the electron configuration of each of the other elements. In the buildup of electronic structures of the atoms of the elements, the last electron added is added to the lowest-energy subshell possible. The relative order of the energies of all the important subshells in an atom is shown in Fig. 17-7. The energies of the various subshells are plotted along the vertical axis. The subshells are displaced left to right merely to avoid overcrowding. The order of increasing energy is as follows:

$$1s, 2s, 2p, 3s, 3p, 4s, 3d, 4p, 5s, 4d, 5p, 6s, 4f, 5d, 6p, 7s, 5f, 6d.$$

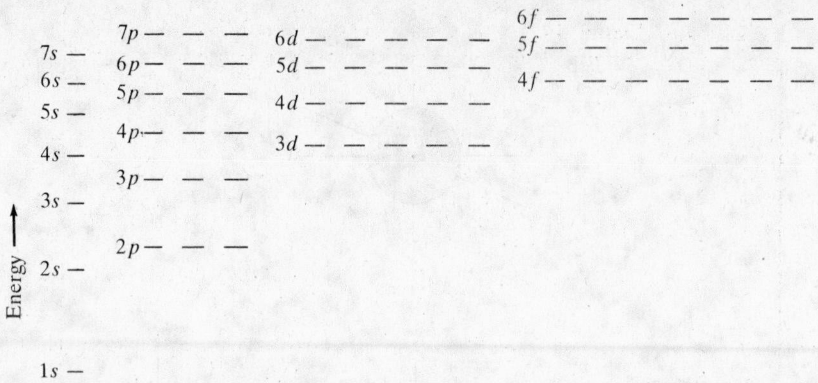

Fig. 17-7 Energy level diagram for atoms containing more than one electron (not drawn to scale)

EXAMPLE 17.11. Draw the electron configuration of Ne and Na on a figure like that of Fig. 17-7.

The drawings are shown in Fig. 17-8.

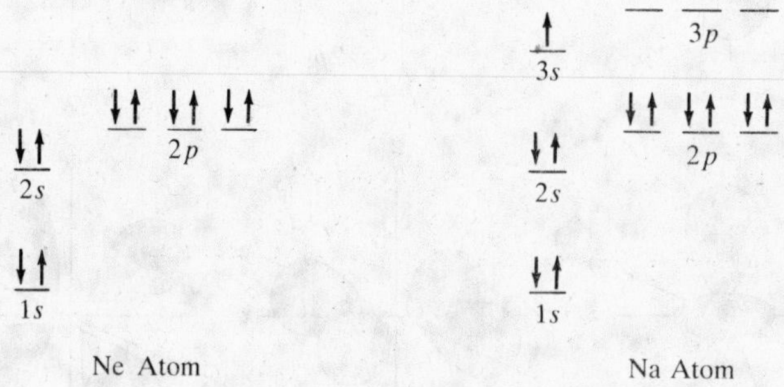

Ne Atom   Na Atom

Fig. 17-8 Electron configurations of Ne and Na

The electron configurations of the first 20 elements are given in Table 17-6. The buildup principle is crudely analogous to filling an irregularly shaped vessel with marbles (see Fig. 17-9). The available spaces are filled from the bottom up.

In atoms with partially filled p, d, or f subshells, the electrons stay unpaired as much as possible. This effect is called *Hund's rule of maximum multiplicity*. Thus the configuration of the nitrogen and oxygen atoms are as follows:

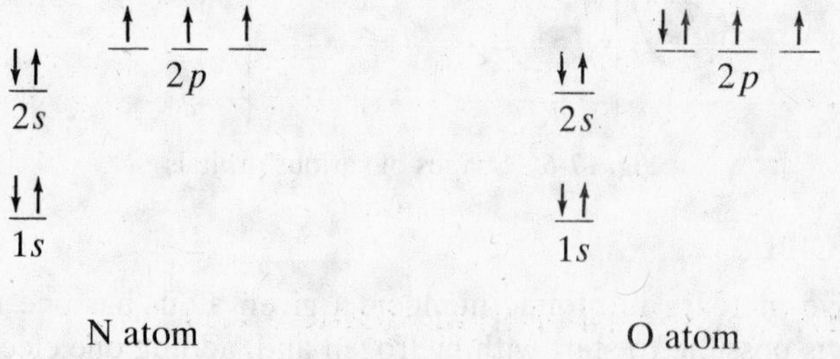

N atom   O atom

It turns out that fully filled or half-filled subshells have added stability compared with subshells having some other numbers of electrons. One effect of this added stability is the fact that some elements do not follow the $n + l$ rule exactly. For example, copper would be expected to have a configuration

$n + l$ configuration for Cu $1s^2 2s^2 2p^6 3s^2 3p^6 4s^2 3d^9$

the actual configuration for Cu $1s^2 2s^2 2p^6 3s^2 3p^6 4s^1 3d^{10}$

The actual configuration has two subshells of enhanced stability ($3d$ and $4s$) in contrast to the one ($4s$)

Table 17-6   Detailed Electron Configuration of the First 20 Elements

H	$1s^1$	1
He	$1s^2$	2
Li	$1s^2 2s^1$	3
Be	$1s^2 2s^2$	4
B	$1s^2 2s^2 2p^1$	5
C	$1s^2 2s^2 2p^2$	6
N	$1s^2 2s^2 2p^3$	7
O	$1s^2 2s^2 2p^4$	8
F	$1s^2 2s^2 2p^5$	9
Ne	$1s^2 2s^2 2p^6$	10
Na	$1s^2 2s^2 2p^6 3s^1$	11
Mg	$1s^2 2s^2 2p^6 3s^2$	12
Al	$1s^2 2s^2 2p^6 3s^2 3p^1$	13
Si	$1s^2 2s^2 2p^6 3s^2 3p^2$	14
P	$1s^2 2s^2 2p^6 3s^2 3p^3$	15
S	$1s^2 2s^2 2p^6 3s^2 3p^4$	16
Cl	$1s^2 2s^2 2p^6 3s^2 3p^5$	17
Ar	$1s^2 2s^2 2p^6 3s^2 3p^6$	18
K	$1s^2 2s^2 2p^6 3s^2 3p^6 4s^1$	19
Ca	$1s^2 2s^2 2p^6 3s^2 3p^6 4s^2$	20

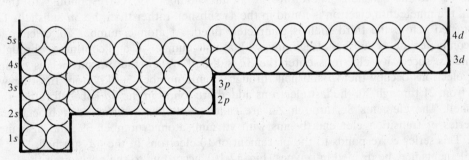

Fig. 17-9   Irregularly shaped container

of the expected configuration. (There are also some elements whose configurations do not follow the $n + l$ rule and which are not enhanced by the added stability of extra fully filled and half-filled subshells.)

17.8   ELECTRONIC STRUCTURE AND THE PERIODIC TABLE

The arrangement of electrons in successive energy levels in the atom provides an explanation of the periodicity of the elements, as found in the periodic table. The charges on the nuclei of the atoms increase in a regular manner as the atomic number increases. Therefore, the number of electrons surrounding the nucleus increases also. The number and arrangement of the electrons in the outermost shell of an atom varies in a periodic manner (compare Table 17-6). For example, all of the elements in Group IA—H, Li, Na, K, Rb, Cs, Fr—corresponding to the elements that begin a new row or period, have electron configurations with a single electron in the outermost shell, specifically an

s subshell.

$$\text{H} \qquad 1s^1$$

$$\text{Li} \qquad 1s^2 2s^1$$

$$\text{Na} \qquad 1s^2 2s^2 2p^6 3s^1$$

$$\text{K} \qquad 1s^2 2s^2 2p^6 3s^2 3p^6 4s^1$$

$$\text{Rb} \qquad 1s^2 2s^2 2p^6 3s^2 3p^6 4s^2 3d^{10} 4p^6 5s^1$$

$$\text{Cs} \qquad 1s^2 2s^2 2p^6 3s^2 3p^6 4s^2 3d^{10} 4p^6 5s^2 4d^{10} 5p^6 6s^1$$

$$\text{Fr} \qquad 1s^2 2s^2 2p^6 3s^2 3p^6 4s^2 3d^{10} 4p^6 5s^2 4d^{10} 5p^6 6s^2 4f^{14} 5d^{10} 6p^6 7s^1$$

The noble gases, located at the end of each period, have electron configurations of the type $ns^2 np^6$, where n represents the number of the outermost shell. Also, n is the number of the period in the periodic table in which the element is found.

Since atoms of all elements in a given group of the periodic table have analogous arrangements of electrons in their outermost shells and different arrangements from elements of other groups, it is reasonable to conclude that the outermost electron configuration of the atom is responsible for the chemical characteristics of the element. Elements with similar arrangements of electrons in their outer shells will have similar properties. For example, the formulas of their oxides will be of the same type. The electrons in the outermost shells of the atoms are referred to as *valence electrons*. The outermost shell is often called the *valence shell*.

As the atomic numbers of the elements increase, the arrangements of electrons in successive energy levels vary in a periodic manner. As shown in Fig. 17-7, the energy of the $4s$ subshell is lower than that of the $3d$ subshell. Therefore, at atomic number 19, corresponding to the element potassium, the nineteenth electron is found in the $4s$ subshell rather than the $3d$ subshell. The fourth shell is started before the third shell is completely filled. At atomic number 20, calcium, a second electron completes the $4s$ subshell. Beginning with atomic number 21 and continuing through the next nine elements, successive electrons enter the $3d$ subshell. When the $3d$ subshell is complete, the following electrons occupy the $4p$ subshell through atomic number 36, krypton. In other words, for elements from 21 through 30, the last electrons added are found in the $3d$ subshell rather than in the valence shell. The elements Sc through Zn are called *transition elements* or d block elements. A second series of transition elements begins with yttrium, atomic number 39, and also includes 10 elements. This series corresponds to the placement of 10 electrons in the $4d$ subshell.

The elements may be divided into types (Fig. 17-10), according to the position of the last electron added to those present in the preceding element. In the first type, the last electron added enters the valence shell. These elements are called the *main group elements*. In the second type, the last electron enters a d subshell in the next to last shell. These elements are the *transition elements*. The third type

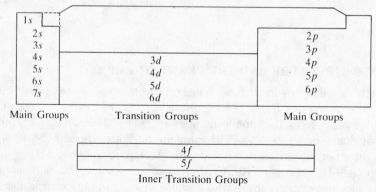

Fig. 17-10   Periodic table as an aid to assigning electron configurations

of elements has the last electron enter the f subshell in the $(n-2)$ shell—the second shell below the valence shell. These elements are the *inner transition elements*.

An effective way to determine the detailed electron configuration of any element is to use the periodic table to determine which subshell to fill next. Each s subshell holds a maximum of 2 electrons; each p subshell holds a maximum of 6 electrons; each d subshell holds a maximum of 10 electrons; and each f subshell holds a maximum of 14 electrons (Table 17-5). These numbers match the numbers of elements in a given period in the various blocks. To get the electron configuration, start at hydrogen (atomic number = 1) and continue in order of atomic number, using the periodic table of Fig. 17-10.

EXAMPLE 17.12.   Determine the detailed electron configuration of sodium.

Starting at hydrogen, we put two electrons into the $1s$ subshell, then two more electrons into the $2s$ subshell. We continue (at atomic number = 5) with the $2p$ subshell, and enter six electrons there, corresponding to the six elements (elements 5 to 10, inclusive) in that p block of the periodic table. We have one more electron to put into the $3s$ subshell, which is next. Thus, we always start at hydrogen, and we end at the element required. The number of electrons that we add to each subshell is equal to the number of elements in the block of the periodic table. In this case, we added electrons from hydrogen to sodium, following the atomic numbers in order, and we got a configuration $1s^2 2s^2 2p^6 3s^1$.

EXAMPLE 17.13.   Write the detailed electron configurations for K, S, and Rb.

$$K \qquad 1s^2 2s^2 2p^6 3s^2 3p^6 4s^1$$
$$S \qquad 1s^2 2s^2 2p^6 3s^2 3p^4$$
$$Rb \qquad 1s^2 2s^2 2p^6 3s^2 3p^6 4s^2 3d^{10} 4p^6 5s^1$$

In each case, the superscripts total to the atomic number of the element.

EXAMPLE 17.14.   Determine the detailed electron configuration of Gd, atomic number 64.

$$Gd \quad 1s^2 2s^2 2p^6 3s^2 3p^6 4s^2 3d^{10} 4p^6 5s^2 4d^{10} 5p^6 6s^2 5d^1 4f^7$$

We note that the $5d$ subshell started before the $4f$ subshell, but only one electron entered that shell before the $4f$ subshell started. Indeed, the periodic table predicts this correct configuration for Gd better than the $n+l$ rule or other common memory aids.

Instead of writing out the entire electron configuration of an atom, especially an atom with many electrons, we sometimes abbreviate the configuration by using the configuration of the previous noble gas and represent the rest of the electrons explicitly. For example, the full configuration of iron can be given as

$$Fe \quad 1s^2 2s^2 2p^6 3s^2 3p^6 4s^2 3d^6$$

Alternately, we can use the configuration of the previous noble gas, Ar, and add the extra electrons:

$$Fe \quad [Ar]4s^2 3d^6$$

To determine the electron configuration in this manner, start with the noble gas of the previous period and use the subshell notation from only the period of the required element. Thus, for Fe, the notation for Ar (the previous noble gas) is included in the square brackets, and the $4s^2 3d^6$ is obtained across the fourth period. It is suggested that you do not use this notation until you have mastered the full notation. Also, on examinations, use the full notation unless the question or the instructor indicates that the shortened notation is acceptable.

17.9 ELECTRON CONFIGURATIONS OF IONS

To get the electron configuration of ions, a new rule is followed. We first write the electron configuration of the neutral atom. Then, for positive ions, we remove the electrons in the subshell with highest principal quantum number first. Note that these electrons might not have been added last, because of the $n + l$ rule. Nevertheless, the electrons from the shell with highest principal quantum number are removed first. For negative ions, we add electrons to the shell of highest principal quantum number. (That shell has the electrons added last by the $n + l$ rule.)

EXAMPLE 17.15. What is the electron configuration of each of the following? (a) Na^+, (b) S^{2-}, (c) Fe^{2+}, and (d) Fe^{3+}.

(a) The configuration of Na is $1s^2 2s^2 2p^6 3s^1$. For the ion Na^+, the outermost ($3s$) electron is removed, yielding

$$Na^+ \quad 1s^2 2s^2 2p^6 3s^0 \quad \text{or} \quad 1s^2 2s^2 2p^6$$

(b) The configuration of S is $1s^2 2s^2 2p^6 3s^2 3p^4$. For the anion, we add two electrons for the two extra charges:

$$S^{2-} \quad 1s^2 2s^2 2p^6 3s^2 3p^6$$

(c) The configuration of Fe is $1s^2 2s^2 2p^6 3s^2 3p^6 4s^2 3d^6$. For Fe^{2+}, we remove the two $4s$ electrons! They are the electrons in the outermost shell, despite the fact that they were not the last electrons added. The configuration is

$$Fe^{2+} \quad 1s^2 2s^2 2p^6 3s^2 3p^6 4s^0 3d^6$$

or

$$1s^2 2s^2 2p^6 3s^2 3p^6 3d^6$$

(d) The configuration of Fe^{3+} is $1s^2 2s^2 2p^6 3s^2 3p^6 3d^5$. An inner electron is removed after both $4s$ electrons are removed.

Being able to write correct electron configurations for transition metal ions becomes very important in discussions of coordination compounds (Chemistry 2).

Solved Problems

BOHR THEORY

17.1. Draw a picture of the electron jump corresponding to the first line in the visible emission spectrum of hydrogen according to the Bohr theory.

> *Ans.* The electron falls from the third orbit to the second. In hydrogen, only jumps to or from the second orbit are in the visible region of the spectrum. The picture is shown as the leftmost arrow of Fig. 17-3.

17.2. When electrons fall to lower energy levels, light is given off. What energy effect is expected when an electron jumps to a higher energy orbit?

> *Ans.* Absorption of energy is expected. The energy may be light energy (of the same energies as are given off in emission), or it may be heat or other types of energy.

17.3. What energy changes can an electron make in going from its fifth orbit to its ground state?

> *Ans.* It can go from (a) $5 \rightarrow 4 \rightarrow 3 \rightarrow 2 \rightarrow 1$, (b) $5 \rightarrow 4 \rightarrow 3 \rightarrow 1$, (c) $5 \rightarrow 4 \rightarrow 2 \rightarrow 1$, (d) $5 \rightarrow 4 \rightarrow 1$, (e) $5 \rightarrow 3 \rightarrow 2 \rightarrow 1$, (f) $5 \rightarrow 3 \rightarrow 1$, (g) $5 \rightarrow 2 \rightarrow 1$, or (h) $5 \rightarrow 1$.

QUANTUM NUMBERS

17.4. What values are permitted for m in an electron in which the l value is 0?

> *Ans.* 0. (The value of m may range from -0 to $+0$; that is, it must be 0.)

17.5. What are the possible values of m for an electron with $l = 1$?

   Ans. $-1, 0$, and $+1$.

17.6. What are the permitted values of s for an electron with $n = 2$, $l = 1$, and $m = -1$?

   Ans. $s = -\frac{1}{2}$ or $+\frac{1}{2}$, no matter what values the other quantum numbers have.

17.7. In a football stadium, can two tickets have the same set of section, row, seat, and date? How many of these must be different to have a legal situation? (*b*) In an atom, can two electrons have the same set of n, l, m, and s? How many of these must be different to have a "legal" situation?

   Ans.   (*a*) At least one of the four must be different. (*b*) At least one of the four must be different.

17.8. Two baseball fans happen to discover that they have tickets to the July 4 doubleheader baseball game in New York, and further that' each is sitting in Section A, Row 5, Seat 11. Explain how this could be legal.

   Ans.   One has a ticket to Yankee Stadium and the other to Shea Stadium. (Unlikely, but legal.) Similarly, two electrons can have the same set of four quantum numbers if they are in different atoms.

17.9. In this chapter, to what two uses is the letter s put?

   Ans.   The letter s represents the fourth quantum number—the spin quantum number. It also represents the subshell with an l value of 0.

17.10. What is the maximum number of electrons that can occupy the K shell? the L shell?

   Ans.   The K shell, the first shell, can hold a maximum of two electrons. The L shell can hold a maximum of eight electrons.

17.11. What is the maximum number of electrons that can occupy the N shell?

   Ans.   The N shell in an atom corresponds to the fourth energy level ($n = 4$); hence, the maximum number of electrons it can hold is

$$2n^2 = 2(4)^2 = 2 \times 16 = 32$$

17.12. What is the maximum number of electrons that can occupy the N shell before the start of the O shell?

   Ans.   The maximum number of electrons in any outermost shell (except the first shell) is 8. The fifth shell starts before the d subshell of the fourth shell starts.

QUANTUM NUMBERS AND ENERGIES OF ELECTRONS

17.13. Arrange the electrons in the following list in order of increasing energy, lowest first:

	n	l	m	s
(*a*)	4	2	-1	$\frac{1}{2}$
(*b*)	5	0	0	$-\frac{1}{2}$
(*c*)	4	1	1	$-\frac{1}{2}$
(*d*)	4	1	-1	$\frac{1}{2}$

   Ans.   Electrons (*c*) and (*d*) have the lowest value of $n + l$ ($4 + 1 = 5$) and the lowest n, and so they are lowest in energy of the four electrons. Electron (*b*) also has an equal sum of $n + l$ ($5 + 0 = 5$), but its n value is higher than those of electrons (c) and (d). Electron (b) is therefore next in energy (despite the fact that it has the highest n value). Electron (a) has the highest sum of $n + l$ ($4 + 2 = 6$) and is highest in energy.

17.14. How many electrons are permitted in the O shell? Explain why no atom in its ground state has that many electrons in that shell.

   Ans.   The O shell corresponds to $n = 5$, and so there could be as many as $2(5)^2 = 50$ electrons in that shell. However, there are only a few more than 100 electrons in even the biggest atom. By the time you put 2 electrons in the first shell, 8 in the second, 18 in the third, and 32 in the fourth, you have already accounted for 60 electrons. Moreover, the fifth shell cannot completely fill until overlying shells start to fill. There are just not that many electrons in any actual atom.

17.15. If the n and l quantum numbers are the ones affecting the energy of the electron, why are the m and s quantum numbers important?

   Ans.   Their permitted values tell us how many electrons there can be with that given energy. That is, they tell us how many electrons are allowed in a given subshell.

SHELLS, SUBSHELLS, AND ORBITALS

17.16. What is the difference between (a) the 3s subshell and a 3s orbital? (b) the 3p subshell and a 3p orbital?

   Ans.   (a) Since the s subshell contains only one orbital, the 3s orbital *is* the 3s subshell. (b) The 3p subshell contains three 3p orbitals—known as $3p_x$, $3p_y$, and $3p_z$.

17.17. (a) How many orbitals are there in the third shell of an atom? (b) How many electrons can be held in the third shell of an atom? (c) How many electrons can be held in the third shell of an atom before the fourth shell starts to fill?

   Ans.   (a) 9 (one s orbital, three p orbitals, and five d orbitals). (b) 18 [$2(3)^2$]. (c) 8. (The 4s subshell starts to fill before the 3d subshell. Thus, only the 3s and the 3p subshells are filled before the fourth shell starts.)

17.18. How many electrons are permitted in each of the following subshells?

	n	l
(a)	4	2
(b)	5	0
(c)	4	1
(d)	4	0

   Ans.   (a) 10. This is a d subshell, with five orbitals corresponding to m values of -2, -1, 0, 1, 2. Each orbital can hold a maximum of 2 electrons, and so the subshell can hold $5 \times 2 = 10$ electrons. (b) 2. This is an s subshell. (c) 6. This is a p subshell (with three orbitals). (d) 2. This is an s subshell. The principal quantum number does not matter. [Compare part (b).]

17.19. How many electrons are permitted in each of the following subshells? (a) 2s, (b) 4d, and (c) 6p.

   Ans.   (a) 2, (b) 10, and (c) 6. Note that the principal quantum number does not affect the number of orbitals and thus the maximum number of electrons. The angular momentum quantum number is the only criterion of that.

17.20. In Chap. 3 why were electron dot diagrams drawn with four areas of electrons? Why are at least two of the electrons paired if at least two are shown?

   Ans.   The four areas represent the four s and p orbitals of the outermost shell. If there are at least two electrons in the outermost shell, the first two are paired because they are in the s subshell. The other electrons do not pair up until all have at least one electron in each area (orbital).

SHAPES OF ORBITALS

17.21. Draw an outline of the shape of the $1s$ orbital.

   Ans.   See Fig. 17-6.

17.22. How do the three $2p$ orbitals differ from each other?

   Ans.   They are oriented differently in space. The $2p_x$ orbital lies along the x axis; the $2p_y$ orbital lies along the y axis; the $2p_z$ orbital lies along the z axis.

17.23. How many different drawings, like those in Fig. 17-6, are there for the $4f$ orbitals?

   Ans.   There are seven, corresponding to the seven f orbitals in a subshell (Table 17-5). General chemistry students are never asked to draw them, however.

17.24. Which $2p$ orbital of a given atom would be expected to have the most interaction with another atom lying along the y axis of the first atom?

   Ans.   The $2p_y$ orbital. That one is oriented along the direction toward the second atom.

17.25. Toward which direction is the $1s$ orbital aligned?

   Ans.   None; it is spherically symmetric.

BUILDUP PRINCIPLE

17.26. Write detailed electron configurations for the following atoms: (*a*) Ca, (*b*) V, (*c*) Cl, and (*d*) As.

   Ans.   (*a*)   Ca $1s^2 2s^2 2p^6 3s^2 3p^6 4s^2$.
   (*b*)   V $1s^2 2s^2 2p^6 3s^2 3p^6 4s^2 3d^3$.
   (*c*)   Cl $1s^2 2s^2 2p^6 3s^2 3p^5$.
   (*d*)   As $1s^2 2s^2 2p^6 3s^2 3p^6 4s^2 3d^{10} 4p^3$.

ELECTRONIC STRUCTURE AND THE PERIODIC TABLE

17.27. Write detailed electron configurations for (*a*) F, (*b*) Cl, (*c*) Br, and (*d*) I.

   Ans.   (*a*) $1s^2 2s^2 2p^5$.
   (*b*) $1s^2 2s^2 2p^6 3s^2 3p^5$.
   (*c*) $1s^2 2s^2 2p^6 3s^2 3p^6 4s^2 3d^{10} 4p^5$.
   (*d*) $1s^2 2s^2 2p^6 3s^2 3p^6 4s^2 3d^{10} 4p^6 5s^2 4d^{10} 5p^5$.

17.28. What neutral atom is represented by each of the following configurations? (*a*) $1s^2 2s^2 2p^6$, (*b*) $1s^2 2s^2$, and (*c*) $1s^2 2s^2 2p^6 3s^2 3p^6$.

   Ans.   (*a*) Ne, (*b*) Be, and (*c*) Ar.

17.29. Write the detailed electron configuration for Kr in shortened form.

   Ans.   Kr $[\text{Ar}]4s^2 3d^{10} 4p^6$.

ELECTRON CONFIGURATIONS OF IONS

17.30. Write detailed electron configurations for the following ions: (*a*) Ca^{2+}, (*b*) V^{3+}, (*c*) Cl^-, and (*d*) As^{3-}.

   Ans.   (*a*) Ca^{2+} $1s^2 2s^2 2p^6 3s^2 3p^6$.
   (*b*) V^{3+} $1s^2 2s^2 2p^6 3s^2 3p^6 3d^2$.
   (*c*) Cl^- $1s^2 2s^2 2p^6 3s^2 3p^6$.
   (*d*) As^{3-} $1s^2 2s^2 2p^6 3s^2 3p^6 4s^2 3d^{10} 4p^6$.

17.31. What is the electron configuration of H^+?

   Ans.   There are no electrons left in H^+, and so the configuration is $1s^0$. This is not really a stable chemical species. H^+ is a convenient abbreviation for a more complicated ion, H_3O^+.

17.32. What positive ion with a single charge is represented by each of the following configurations? (*a*) $1s^2 2s^2 2p^6$, (*b*) $1s^2$, and (*c*) $1s^2 2s^2 2p^6 3s^2 3p^6$.

   Ans.   (*a*) Na^+, (*b*) Li^+, and (*c*) K^+.

17.33. What is the electron configuration of Pb^{2+}?

   Ans.   The configuration for Pb is

$$Pb \quad 1s^2 2s^2 2p^6 3s^2 3p^6 4s^2 3d^{10} 4p^6 5s^2 4d^{10} 5p^6 6s^2 4f^{14} 5d^{10} 6p^2$$

The configuration for Pb^{2+} is the same except for the loss of the outermost two electrons. There are four electrons in the sixth shell; the two in the last subshell of that shell are lost first:

$$Pb^{2+} \quad 1s^2 2s^2 2p^6 3s^2 3p^6 4s^2 3d^{10} 4p^6 5s^2 4d^{10} 5p^6 6s^2 4f^{14} 5d^{10} 6p^0$$

In shortened terminology:

$$Pb \qquad [Xe]6s^2 4f^{14} 5d^{10} 6p^2$$

$$Pb^{2+} \qquad [Xe]6s^2 4f^{14} 5d^{10} 6p^0$$

17.34. What positive ion with a double charge is represented by each of the following configurations? (*a*) $1s^2 2s^2 2p^6$, (*b*) $1s^2$, and (*c*) $1s^2 2s^2 2p^6 3s^2 3p^6$.

   Ans.   (*a*) Mg^{2+}, (*b*) Be^{2+}, and (*c*) Ca^{2+}.

17.35. What ion with a single negative charge is represented by each of the following configurations? (*a*) $1s^2 2s^2 2p^6$, (*b*) $1s^2$, and (*c*) $1s^2 2s^2 2p^6 3s^2 3p^6$.

   Ans.   (*a*) F^-, (*b*) H^-, and (*c*) Cl^-.

17.36. What is the difference in the electron configurations of Fe and Ni^{2+}? (They both have 26 electrons.)

   Ans.

$$Fe \qquad 1s^2 2s^2 2p^6 3s^2 3p^6 4s^2 3d^6$$

$$Ni \qquad 1s^2 2s^2 2p^6 3s^2 3p^6 4s^2 3d^8$$

$$Ni^{2+} \qquad 1s^2 2s^2 2p^6 3s^2 3p^6 4s^0 3d^8$$

The Ni atom has two more $3d$ electrons; the Ni^{2+} ion has lost its $4s$ electrons. Thus, the Ni^{2+} ion has two more $3d$ electrons and two fewer $4s$ electrons than the Fe atom has.

Supplementary Problems

17.37. State the octet rule in terms described in this chapter.

   Ans.   A state of great stability is a state in which the outermost s and p subshells are filled and no other subshell of the outermost shell has any electrons.

17.38. List all the ways given in this chapter to determine the order of increasing energy of subshells.

   Ans.   The $n + l$ rule, the energy level diagram (Fig. 17-7), the periodic table (and some mnemonics given by other texts).

17.39. What are the advantages of the periodic table over the other ways of determining the order of increasing energy in subshells given in Problem 17.38?

Ans.   The periodic table is generally available on examinations, it is easy to use after a little practice, it allows starting at any noble gas, it reminds us of the $5d$ electron added before the $4f$ electrons (at La), and it tells us by the number of elements in a block how many electrons are in the subshell.

17.40. Write the electron configuration of K and Cu. What difference is there that could explain why K is so active and Cu so relatively inactive?

Ans.

$$K \qquad 1s^2 2s^2 2p^6 3s^2 3p^6 4s^1$$
$$Cu \qquad 1s^2 2s^2 2p^6 3s^2 3p^6 4s^1 3d^{10}$$

The difference is the extra ten $3d$ electrons, plus the extra ten protons in the nucleus that go along with them. Adding the protons and adding the electrons in an inner subshell makes the outermost electron more tightly bound to the nucleus.

17.41. Starting at the first electron added to an atom, (*a*) what is the number of the first electron in the second shell of the atom? (*b*) What is the atomic number of the first element of the second period? (*c*) What is the number of the first electron in the third shell of the atom? (*d*) What is the atomic number of the first element of the third period? (*e*) What is the number of the first electron in the fourth shell of the atom? (*f*) What is the atomic number of the first element of the fourth period?

Ans.   (*a*) and (*b*), 3; (*c*) and (*d*), 11; and (*e*) and (*f*), 19. There obviously is some relationship between the electron configuration and the periodic table.

17.42. Check Table 17-3 to ensure that no two electrons have the same set of four quantum numbers.

Chapter 18

Thermochemistry

18.1 INTRODUCTION

The energy involved in chemical reactions is often as important as the chemical products. For example, the fuel used in home furnaces and in automobiles is used solely for their energy content, and not for the chemical products of their combustion. The measurement of energy is discussed in Sec. 18.2, heat capacity is discussed in Sec. 18.3, the energy involved in changes of phase is treated in Sec. 18.4, and the energy involved in physical and chemical processes is taken up in Sec. 18.5.

18.2 ENERGY CHANGE, HEAT, AND WORK

Energy is the ability to do work. A basic principle of science is that energy cannot be created or destroyed. This statement is known as the *law of conservation of energy* or as the *first law of thermodynamics*. Energy exists in a variety of forms, including heat, light, and electrical and mechanical energy. A variety of energy types is presented in Table 18-1, which is to be used for reference and not necessarily memorized. Although energy cannot be created or destroyed, the various forms of energy can be converted into one another.

Table 18-1 Energy Classification (Reference Table)

A. Heat
B. Work
Chemical energy
Light energy
Solar energy
Radio waves
Microwaves
Mechanical energy
Potential energy (energy of position)
Kinetic energy (energy of motion)
Electrical energy
Nuclear (atomic) energy
Sound

EXAMPLE 18.1. Name a common device that changes (*a*) chemical energy to electrical energy and (*b*) electrical energy to heat.

One or two examples of the many possible are given for each: (*a*) a battery and (*b*) a toaster or electric space heater.

Heat is different than all the other types of energy. All other types of energy can be completely changed into heat, but the reverse is not true; heat cannot be changed *completely* into any other kind of energy. In this chapter we will classify energy as heat and work, where work includes all other types of energy besides heat (Table 18-1).

EXAMPLE 18.2. A solution of HCl reacts with a solution of Na_2CO_3 in a general chemistry lab, yielding H_2O, $CO_2(g)$, and NaCl. Some heat and some work are also produced. State how you can recognize that heat has been produced, and identify the form of the work produced.

You can recognize that heat is produced because the final solution is warmer than the original solutions. The heat liberated by the chemical reaction has warmed up the solution. (Note that temperature is not heat, however. See Sec. 18.3.) The work done is energy of motion. The gaseous CO_2 liberated has to push back the atmosphere.

The total energy of a sample of matter is not particularly interesting and it would be almost impossible to measure. The *change in energy* in the sample is the quantity of interest in this section. Change in energy is denoted ΔE, where Δ is the Greek letter delta, meaning "change in." ΔE is defined as $E_2 - E_1$. In most texts, it is stated to be equal to

$$\Delta E = q + w$$

where q stands for the heat involved in the process and w stands for the work involved. By definition, heat *added* to a system is called positive and work *added* to a system is called positive. The word *system* refers to the portion of matter under investigation. (In some texts, work *removed* from a system is defined as positive, and in those texts the equation above is changed to $\Delta E = q - w$.)

EXAMPLE 18.3. Calculate the energy change in a system if 500 J of heat and 200 J of work are added to the system.

$$\Delta E = q + w = 500\,J + 200\,J = 700\,J$$

Since the value of ΔE is positive, there is more energy in the system after the changes than was there before the changes. E_2, the energy at the end of the process, is greater than E_1, the energy at the beginning.

EXAMPLE 18.4. Calculate the energy change in a system if 500 J of heat is added to the system and 200 J of work is done by the system.

Here the work is negative, $-200\,J$, because the work done *by* the system is removed from the system.

$$\Delta E = q + w = 500\,J + (-200\,J) = 300\,J$$

Warning: In this chapter especially, the student must be conscious of the signs of measured quantities and also of the units involved. The signs are an especially error-prone topic because energy put into a system is defined as positive and energy removed from a system is defined as negative. Thus, the sign of a quantity of heat in a word problem is not indicated by the words "plus" or "minus" but by the words "added" or "removed" or "done on a system" or "done by the system," for example. Moreover, some texts use the equation for change in energy

$$\Delta E = q - w$$

with w defined as work done *by* the system. The units are also error-prone because tables are usually given with joules for certain quantities and kilojoules for other quantities. Also, heat capacities are expressed in both gram and mole quantities. Be especially aware of the signs and units in this chapter!

EXAMPLE 18.5. What is the energy change in the system if 2.50 kJ of heat is added to the system and 1 500 J of work is done by the system?

$$\Delta E = q + w = 2.50\,kJ + (-1\,500\,J)\left(\frac{1\,kJ}{1\,000\,J}\right) = 1.00\,kJ$$

In this case, the energies in different units had to be converted to the same unit before addition. The work done by the system, a loss of energy to the system, is negative.

In the reaction of Na_2CO_3 with HCl, Example 18.2, some heat was produced and some work was also produced. The quantity of heat produced can be measured rather easily by the quantity of materials and the temperature rise, as will be illustrated. The quantity of work does not affect the temperature, and so is not of immediate interest to us. A quantity called *enthalpy change*, abbreviated ΔH, is equal to the heat produced in a process as long as the process is carried out under constant pressure and as long as no other work except expansion (or contraction) against the atmosphere is considered. Since these conditions are precisely the conditions normally used on an open laboratory bench in the general chemistry laboratory, very often $\Delta H = q$. Thus, we may use ΔH in our discussions. (The advantage to using ΔH rather than q is that in more advanced situations where they are not equal, ΔH is more useful than is q.)

18.3   HEAT CAPACITY

When heat is added to a system, in the absence of a chemical reaction the system may warm up or a change of phase may occur. In this section the warming process only will be considered. Phase changes will be taken up in Sec. 18.4.

Temperature and heat are not the same. Temperature is a measure of the intensity of the energy in a system. Consider the following experiment: Hold a lit candle under a pail of water with one-half inch of water in the bottom. Hold an identical candle, also lit, under a pail full of water for the same length of time. To which sample of water is more heat added? Which sample of water gets hotter?

The same quantity of heat is added to each pail, since identical candles were used for the same lengths of time. However, the water in the pail with less water in it is heated to a higher temperature. The greater quantity of water would require more heat to get it to the same higher temperature.

The *heat capacity* of a substance is defined as the quantity of energy required to change a certain quantity of the substance 1 °C. The *specific heat capacity* of a substance is the quantity of heat required to heat 1 g of the substance 1 °C. Specific heat capacity is often called *specific heat*. The *molar heat capacity* of a substance is the heat required to raise the temperature of 1 mol of the substance 1 °C. For example, the specific heat of water is 4.184 J/g · deg. (The abbreviation deg is used to represent degrees Celsius.) This means that 4.184 J will warm 1 g of water 1 °C. To warm 2 g of water 1 °C requires twice as much energy, 8.368 J. To warm 1 g of water 2 °C requires 8.368 J of energy also. In general, the heat required to effect a certain change in temperature in a certain sample of a given material is calculated with one of the following related equations:

$$\text{heat required} = (\text{mass})(\text{specific heat})(\text{change in temperature}) = (m)(\text{SH})(\Delta t)$$

$$\text{heat required} = (\text{moles})(\text{molar heat capacity})(\text{change in temperature}) = (n)(C)(\Delta t)$$

Heat capacities may be used as factors in factor-label method solutions to problems. Be aware that there are two units in the denominator, mass (or moles) and temperature change. Thus, to get energy, one must multiply the heat capacity by both mass (or moles) and temperature change.

EXAMPLE 18.6.   How much heat does it take to raise the temperature of 100.0 g of water 17.0 °C?

$$\text{heat} = (m)(\text{SH})(\Delta t) = (100.0\,\text{g})\left(\frac{4.184\,\text{J}}{\text{g} \cdot \text{deg}}\right)(17.0\,\text{deg}) = 7110\,\text{J} = 7.11\,\text{kJ}$$

EXAMPLE 18.7.   How much heat does it take to raise the temperature of 100.0 g of water from 10.0 °C to 27.0 °C?

This is the same problem as Example 18.6. In that problem the temperature change was specified. In this example, the initial and final temperatures are given, but the temperature change is the same 17.0 °C. The answer is again 7.11 kJ.

EXAMPLE 18.8.   What is the final temperature after 5 100 J of heat is added to 50.0 g of water at 44.0 °C?

$$\Delta t = \frac{\Delta H}{(m)(\text{SH})} = \frac{5\,100\,\text{J}}{(50.0\,\text{g})(4.184\,\text{J/g} \cdot \text{deg})} = 24.4\,\text{deg}$$

Note that the problem is to find the final temperature; the 24.4 °C is the *temperature change*.

$$t_{final} = t_{initial} + \Delta t = 44.0\,°C + 24.4\,°C = 68.4\,°C$$

EXAMPLE 18.9. What is the specific heat of a metal if 202 J is required to heat 44.0 g of the metal from 22.0 °C to 33.6 °C?

$$SH = \frac{\Delta H}{(m)(\Delta t)} = \frac{202\,J}{(44.0\,g)(33.6\,°C - 22.0\,°C)} = \frac{0.396\,J}{g \cdot deg}$$

EXAMPLE 18.10. What is the final temperature of 2.25 mol of water initially at 17.1 °C from which 1 910 J of heat is removed? The molar heat capacity of water is 75.38 J/mol · deg.

$$\Delta t = \frac{\Delta H}{nC} = \frac{-1910\,J}{(2.25\,mol)(75.38\,J/mol \cdot deg)} = -11.3\,°C$$

$$t_{final} = t_{initial} + \Delta t = 17.1\,°C + (-11.3\,°C) = 5.8\,°C$$

We note several things about this example. First, the number of moles of water and the molar heat capacity were used. Second, since the heat was removed, the value used in the equation was negative. The final temperature is obviously lower than the initial temperature, since heat was removed.

EXAMPLE 18.11. How much heat must be added to 170 g of a metal with atomic weight 120 g/mol to raise its temperature 11.1 °C? Its molar heat capacity is 25.5 J/mol · deg.

$$\Delta H = (m)(SH)(\Delta t) \qquad or \qquad \Delta H = (n)(C)(\Delta t)$$

$$= (170\,g)\left(\frac{1\,mol}{120\,g}\right)\left(\frac{25.5\,J}{mol \cdot deg}\right)(11.1\,deg) = 401\,J$$

Which of the equations given for ΔH is used here? Both. You can see from the factor-label method solution that the atomic weight divided into the molar heat capacity is the specific heat capacity while the mass divided by atomic weight is the number of moles. Thus, we have either moles times molar heat capacity times change in temperature or mass times specific heat times change in temperature.

$$\Delta H = \underbrace{(170\,g)\overbrace{\left(\frac{1\,mol}{120\,g}\right)}^{\substack{\text{number of}\\\text{moles}}}\underbrace{\left(\frac{25.5\,J}{mol \cdot deg}\right)}_{\text{specific heat}}}_{\text{mass}}\overset{\substack{\text{molar heat}\\\text{capacity}}}{}(11.1\,deg) = 401\,J$$

Assuming no chemical reaction, what will happen if we place a piece of metal at one temperature into water at another temperature? The metal will gain energy from or lose energy to the water until their temperatures are the same. Whichever was at the higher temperature initially will lose energy to the other. In the absence of any energy loss to any other body, the quantity of heat removed from one will be added to the other. We can use this fact to measure specific heats of substances, or to predict the final temperatures to which such combinations will arrive. The principle involved may be summarized in the equation:

$$heat\ gained = -(heat\ lost)$$

The minus sign stems from the fact that heat loss is defined as negative and heat gain as positive. Without the minus sign, we would have a positive quantity equal to a negative quantity.

EXAMPLE 18.12. A 100-g piece of metal at 55.0 °C is placed into 200 g of water at 20.0 °C. The final temperature is 22.5 °C. What is the specific heat of the metal?

$$SH_{H_2O} = 4.184\,J/g \cdot deg$$

We first ask what the temperature change of the water is, and what the temperature change of the metal is.

$$\Delta t_{water} = 22.5\,°C - 20.0\,°C = 2.5\,°C$$

$$\Delta t_{metal} = 22.5\,°C - 55.0\,°C = -32.5\,°C$$

We use these values of temperature change to determine the specific heat:

$$heat\ gained\ by\ water = -(heat\ lost\ by\ metal)$$

$$(m)(SH)(\Delta t) = -(m)(SH_{metal})(\Delta t)$$

$$(200\,g)\left(\frac{4.184\,J}{g\cdot°C}\right)(2.5\,°C) = -(100\,g)(SH_{metal})(-32.5\,°C)$$

$$SH_{metal} = 0.64\,J/g\cdot°C$$

EXAMPLE 18.13.   Calculate the final temperature if a 50.0-g piece of metal ($SH = 0.444\,J/g\cdot deg$) at 44.9 °C is placed into 150.0 g of water at 19.7 °C.

$$heat\ gained\ by\ water = -(heat\ lost\ by\ metal)$$

$$(m)(SH)(\Delta t) = -(m)(SH_{metal})(\Delta t)$$

In this example, the final temperature is unknown, and will be represented as t_f. The changes in temperature are

$$\Delta t_{water} = t_f - 19.7\,°C$$

$$\Delta t_{metal} = t_f - 44.9\,°C$$

$$(150.0\,g)\left(\frac{4.184\,J}{g\cdot°C}\right)(t_f - 19.7\,°C) = -(50.0\,g)\left(\frac{0.444\,J}{g\cdot°C}\right)(t_f - 44.9\,°C)$$

$$(627.6)(t_f - 19.7\,°C) = -(22.2)(t_f - 44.9\,°C)$$

$$627.6 t_f - (1.24 \times 10^4)\,°C = -22.2 t_f + 997\,°C$$

$$649.8 t_f = 13\,400\,°C$$

$$t_f = 20.6\,°C$$

The *law of Dulong and Petit* states that the molar heat capacity of crystalline elements is approximately 25 J/mol · deg. With this law, we can calculate approximate atomic weights from heat capacity data.

EXAMPLE 18.14.   An element has a specific heat of 0.123 J/g · deg. Calculate its approximate atomic weight.

The molar heat capacity is about 25 J/mol · deg, and the specific heat is 0.123 J/g · deg. Therefore, the atomic weight is approximately

$$\frac{25\,J/mol\cdot deg}{0.123\,J/g\cdot deg} = \frac{200\,g}{mol} = \frac{2.0 \times 10^2\,g}{mol}$$

This law was very important in establishing atomic weights for metallic elements in the 1800s. Once you know an atomic weight, you can calculate an empirical formula from composition data (Sec. 4.7). Conversely, if you know a formula, you can calculate atomic weights. However, at least one of these needs to be known. The availability of approximate atomic weights and exact combining weights allowed calculation of exact atomic weights and formulas.

EXAMPLE 18.15.   If 6.90 g of an element, X, combines with 35.5 g of chlorine, does the element have atomic weight 6.90 and a chloride of formula XCl or does it have atomic weight 13.8 and a chloride with formula XCl_2?

Both possibilities (and others) exist with the data available. If the formula is XCl, then 1 mol of metal bonds to 1 mol of chlorine, and the mole of metal has a mass of 6.90 g. If the formula is XCl_2, then 1 mol of metal bonds to 2 mol of Cl atoms, or 1 mol of Cl bonds to 0.5 mol of metal. The 0.5 mol of metal has a mass of 6.90 g, and the atomic weight (the mass of 1 mol) is 13.8 g.

18.4 PHASE CHANGES

A *phase change* is a change from solid to liquid, solid to gas, liquid to gas, or the reverse of these. Examples of phase changes include all the following: The boiling of water causes liquid water to be converted to gas. The *sublimation* of carbon dioxide (dry ice) causes the carbon dioxide to change from solid directly into the gas phase. The freezing of water causes liquid water to change to solid.

When a pure substance changes phase because it is heated or cooled, its temperature remains constant until the entire phase change is complete. For example, if you add heat to a cube of pure ice at 0 °C, it will melt. The temperature of the water-ice mixture will remain at 0 °C until all the ice is melted. Only then, if heating is continued, will the temperature rise. To summarize, heating a pure solid or liquid at a temperature different from its melting point or boiling point will cause its temperature to rise; heating it at its melting point or boiling point will cause it to change phase at constant temperature.

A heating curve or a cooling curve may be used to emphasize the constant temperature of phase changes. In such a curve, temperature is plotted on the vertical axis against heat added on the horizontal axis (Fig. 18-1).

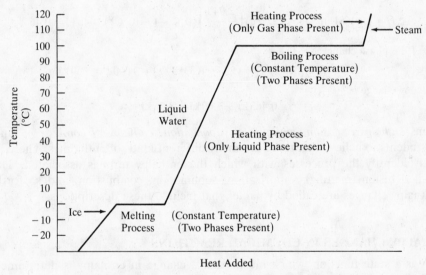

Fig. 18-1 Heating curve for pure water (not to scale)

The equations involving the enthalpy of phase changes include no term for change in temperature, because there is no temperature change for phase changes. The typical equation thus involves multiplying the mass times the enthalpy change per gram or the number of moles times the enthalpy change per mole.

EXAMPLE 18.16. Calculate the heat necessary to melt 40.0 g of ice at 0 °C to water at 0 °C. The enthalpy of fusion of ice is 335 J/g.

$$\text{heat} = (40.0\,\text{g})(335\,\text{J/g}) = 13\,400\,\text{J} = 13.4\,\text{kJ}$$

Note particularly that the equation has no temperature change in it and that the enthalpy of fusion term has no unit of temperature in it.

A table of enthalpies of phase changes is given as Table 18-2. The enthalpies of the reverse processes are merely the negatives of the values given. For example, the freezing of water has an enthalpy change equal to the negative of the enthalpy for the melting process, that is, $-335\,\text{J/g}$.

EXAMPLE 18.17. How much heat is liberated when 44.0 g of benzene is frozen at its freezing point?

$$(44.0\,\text{g})(-127\,\text{J/g}) = -5\,590\,\text{J} = -5.59\,\text{kJ}$$

The negative value indicates that heat is lost by the benzene during the freezing process.

Table 18-2   Enthalpies of Phase Change (at Normal Phase Change Temperatures)

	Melting (Fusion)	Boiling (Vaporization)	Sublimation
Water	335 J/g	2 260 J/g	
Carbon dioxide			368 J/g
Benzene	127 J/g	550 J/g	

A process that involves both a phase change and a temperature change may have its total enthalpy change calculated as the sum of the separate processes. For example, if an ice cube at $-10\,°C$ is warmed to liquid water at $0\,°C$, the enthalpy change may be calculated in two parts: the calculation involving specific heat for the warming of solid water at $-10\,°C$ to solid water at $0\,°C$ and the calculation involving the phase change at $0\,°C$.

EXAMPLE 18.18.   Calculate the enthalpy change accompanying the heating of 22.7 g of ice at $-10\,°C$ to liquid water at $0\,°C$. The specific heat of ice is 2.04 J/g·deg.

The enthalpy of the warming process is given by

$$\Delta H = (m)(SH)(\Delta t) = (22.7\,g)\left(\frac{2.04\,J}{g\cdot°C}\right)(0\,°C - (-10\,°C)) = 463\,J = 0.463\,kJ$$

The enthalpy of the phase change is

$$\Delta H = (m)(335\,J/g) = (22.7\,g)(335\,J/g) = 7\,600\,J = 7.60\,kJ$$

The total enthalpy change is

$$0.463\,kJ + 7.60\,kJ = 8.06\,kJ$$

The terms *enthalpy of fusion*, *enthalpy of vaporization*, *enthalpy of combustion*, and many more cause some students to believe that there are many different kinds of enthalpies. There are not. These names merely identify the processes with which the enthalpy term is associated. Thus, there are processes called fusion (melting), vaporization, sublimation, combustion, and so forth. The corresponding enthalpy changes are called by names that include these descriptions.

18.5   ENTHALPY CHANGES IN CHEMICAL REACTIONS

Enthalpy is a state function, which means that a change in enthalpy is determined only by the initial and final states of the system undergoing change, and not by the path of the change. This principle enables us to calculate the enthalpy change during a chemical reaction by subtracting the enthalpy of formation of the reactants from the enthalpy of formation of the products. The *enthalpy of formation* of a compound is defined as the enthalpy change that occurs when a compound is formed from its elements in their standard states. The enthalpy of formation of an element in its standard state is 0 by definition, because there is no change necessary to get the element from its standard state to the same element in its standard state. A table of standard enthalpies of formation is given in Table 18-3.

Table 18-3   Standard Enthalpies of Formation (ΔH_f) (kJ/mol)

CH_4	-74.85	CuO	-157
C_2H_4	52.28	HCl	-92.13
C_2H_6	-84.68	NaCl	-411.0
$H_2O(l)$	-285.9	Na_2CO_3	$-1\,430.1$
CO	-110.5		
CO_2	-393.5		

To get the enthalpy of a reaction, merely subtract the enthalpies of formation of the reactants from the enthalpies of formation of the products.

EXAMPLE 18.19.   Calculate the enthalpy of the following reaction:

$$C_2H_4 + 3O_2 \longrightarrow 2CO_2 + 2H_2O$$

The total of the enthalpies of formation of the products, from Table 18-3, is

$$2\,\text{mol CO}_2\left(\frac{-393.5\,\text{kJ}}{\text{mol}}\right) = -787.0\,\text{kJ}$$

$$2\,\text{mol H}_2\text{O}\left(\frac{-285.9\,\text{kJ}}{\text{mol}}\right) = -571.8\,\text{kJ}$$

$$\text{total} = -1\,358.8\,\text{kJ}$$

The total of the enthalpies of formation of the reactants is

$$1\,\text{mol C}_2\text{H}_4\left(\frac{52.28\,\text{kJ}}{\text{mol}}\right) = 52.28\,\text{kJ}$$

$$3\,\text{mol O}_2\left(\frac{0\,\text{kJ}}{\text{mol}}\right) = 0\,\text{kJ}$$

$$\text{total} = 52.28\,\text{kJ}$$

The difference, equal to the enthalpy change for burning 1 mol of C_2H_4, is

$$-1\,358.8\,\text{kJ} - (52.28\,\text{kJ}) = -1\,411.1\,\text{kJ}$$

EXAMPLE 18.20.   How much heat will be emitted if 2.50 mol of C_2H_4 is burned to CO_2 and water?

In Example 18.19, we calculated the enthalpy change for 1 mol of C_2H_4. In this example, all we need do is multiply that value by 2.50 mol:

$$2.50\,\text{mol C}_2\text{H}_4\left(\frac{-1\,411.1\,\text{kJ}}{\text{mol C}_2\text{H}_4}\right) = -3\,530\,\text{kJ}$$

Solved Problems

ENERGY CHANGE, HEAT, AND WORK

18.1.   What is the difference between dissolving and melting?

   Ans.   Both involve changing from solid to liquid, but melting is caused by addition of energy and dissolving is caused by the addition of a solvent. For example, an ice cube melts when left in a room above 0 °C. It dissolves in alcohol even if the temperature is below 0 °C. In both cases, a liquid results.

18.2.   Name a common device that will make the following transformations: (*a*) electrical energy into kinetic energy, (*b*) electrical energy into chemical energy, and (*c*) kinetic energy into electrical energy.

   Ans.   (*a*) An electric motor, (*b*) a battery charger, and (*c*) a generator or alternator.

18.3.   Why does a bicycle tire get hot at the valve when pumped up by a tire pump?

   Ans.   When work is done on the system, the energy of the system rises. The added energy causes a rise in temperature.

18.4.   If 100 J of heat is added to a system and 0.100 kJ of work is done on the system, what is the value of ΔE?

   Ans.

$$\Delta E = 100\,\text{J} + 0.100\,\text{kJ}\left(\frac{1\,000\,\text{J}}{\text{kJ}}\right) = 200\,\text{J}$$

18.5. If 100 J of heat is added to a system and 0.100 kJ of work is done by the system, what is the value of ΔE?

    Ans.

$$\Delta E = 100 \,\text{J} - 0.100 \,\text{kJ} \left(\frac{1000 \,\text{J}}{\text{kJ}} \right) = 0 \,\text{J}$$

18.6. Two solutions, each at 25 °C, are mixed. A chemical reaction takes place, producing a final solution at 35 °C. (*a*) Does the reaction give off heat or take up heat? (*b*) Does the solution give off heat or take up heat?

    Ans.   (*a*) We simplify this process by imagining that it takes place in two consecutive steps. The first step is the chemical reaction, which liberates heat. (ΔH is negative.) This heat is the cause of the temperature rise. (*b*) The second step is the addition of that heat to the solution, causing the temperature rise. The solution takes in the heat. ΔH is positive for this step. If no heat escapes to the surroundings, the overall ΔH is 0.

HEAT CAPACITY

18.7. A system is initially at 25 °C. What will be the final temperature if the system (*a*) is warmed 35 °C? (*b*) is warmed to 35 °C? (*c*) What is the difference between final temperature and temperature change?

    Ans.   (*a*) 60 °C and (*b*) 35 °C. (*c*) The temperature difference is the final temperature minus the initial temperature. The final temperature is merely a single temperature. Be sure to read the problems carefully so that you do not mistake temperature change for initial or final temperature.

18.8. (*a*) Calculate the Kelvin equivalent of 0 °C and 10 °C. (*b*) Calculate the difference in these Kelvin temperatures. (*c*) Explain why the difference between these temperatures is the same in Celsius and Kelvin.

    Ans.   (*a*) 273 K and 283 K. (*b*) 283 K − 273 K = 10 K. (*c*) (10 °C + 273) − (0 °C + 273) = 10 °C. The 273° cancels out when two temperatures are subtracted.

18.9. The student political science society decides to take a trip to Washington, D.C. The hotel rate is $15 per student each night. How much will it cost for 23 students to stay for three nights?

    Ans.

$$23 \,\text{students} \left(\frac{15 \,\text{dollars}}{\text{student} \cdot \text{night}} \right) 3 \,\text{nights} = 1035 \,\text{dollars}$$

18.10. How much heat is required to raise the temperature of 2.29 mol of iron 11.3 °C? (SH = 0.447 J/g · °C)

    Ans.

$$\text{heat} = nC(\Delta t) = (2.29 \,\text{mol}) \left(\frac{55.85 \,\text{g}}{\text{mol}} \right) \left(\frac{0.447 \,\text{J}}{\text{g} \cdot \text{deg}} \right) (11.3 \,\text{deg})$$

$$= 646 \,\text{J}$$

18.11. How much heat is required to raise 133 g of iron from 20.0 °C to 31.3 °C? (SH = 0.447 J/g · °C)

    Ans.

$$\Delta t = 31.3 \,°\text{C} - 20.0 \,°\text{C} = 11.3 \,°\text{C}$$

$$\text{heat} = (m)(\text{SH})(\Delta t) = (133 \,\text{g}) \left(\frac{0.447 \,\text{J}}{\text{g} \cdot °\text{C}} \right) (11.3 \,°\text{C}) = 672 \,\text{J}$$

18.12. How much heat is required to raise 2.29 mol of iron from 20.0 °C to 31.3 °C? (SH = 0.447 J/g · °C)

Ans. This is the same as Problem 18.10 except that the initial and final temperatures are given instead of the temperature change.

18.13. How much heat is required to raise 133 g of iron from 20.0 °C to 31.3 °C? (C = 25.0 J/mol · °C)

Ans.

$$\text{heat} = 133\,\text{g}\left(\frac{1\,\text{mol}}{55.85\,\text{g}}\right)\left(\frac{25.0\,\text{J}}{\text{mol}\cdot\text{deg}}\right)(11.3\,\text{deg}) = 673\,\text{J}$$

18.14. A piece of metal at 50 °C is added to water at 20 °C. The final temperature is 23 °C. (*a*) What is the change in temperature of the water? (*b*) of the metal?

Ans. (*a*) 3 °C and (*b*) −27 °C.

18.15. A donor gives a certain sum of money to help the political science society visit Washington, D.C., and the UN in New York. The donation covers hotel expenses only. The donor specifies that equal amounts are to be used for the two trips. The New York stay will last 4 nights; the Washington stay will last from noon on the nineteenth to noon on the twenty-second of the month. The rate in New York is \$23.01/person · night, and that in Washington is \$15.34/person · night. If 14 students visit Washington, how many can go to New York?

Ans.

$$\text{Washington cost} = \text{New York cost}$$

$$14\,\text{students}\left(\frac{15.34\,\text{dollars}}{\text{student}\cdot\text{night}}\right)3\,\text{nights} = x\,\text{students}\left(\frac{23.01\,\text{dollars}}{\text{student}\cdot\text{night}}\right)4\,\text{nights}$$

$$x = 7$$

18.16. Calculate the final temperature when 55.0 g of iron (SH = 0.447 J/g · °C) at 53.3 °C is added to 145.0 g of water at 20.0 °C.

Ans.

$$\text{heat gained by water} = -(\text{heat lost by metal})$$

$$(m)(\text{SH})(\Delta t) = -(m)(\text{SH}_{\text{metal}})(\Delta t)$$

In this example, the final temperature is unknown, and will be represented as t_f. The changes in temperature are

$$\Delta t_{\text{water}} = t_f - 20.0\,°\text{C}$$

$$\Delta t_{\text{metal}} = t_f - 53.3\,°\text{C}$$

$$(145.0\,\text{g})\left(\frac{4.184\,\text{J}}{\text{g}\cdot°\text{C}}\right)(t_f - 20.0\,°\text{C}) = -(55.0\,\text{g})\left(\frac{0.447\,\text{J}}{\text{g}\cdot°\text{C}}\right)(t_f - 53.3\,°\text{C})$$

$$606.7(t_f - 20.0) = 24.6(53.3 - t_f)$$

$$631.3 t_f = 1.34 \times 10^4$$

$$t_f = 21.2\,°\text{C}$$

18.17. The specific heat of a certain element is 0.119 J/g · deg. Using the law of Dulong and Petit, calculate its approximate atomic weight.

Ans.

$$\frac{25\,\text{J/mol}\cdot\text{deg}}{0.119\,\text{J/g}\cdot\text{deg}} = \frac{210\,\text{g}}{\text{mol}}$$

18.18. Calculate the approximate specific heat of aluminum.

Ans.   According to the law of Dulong and Petit, the molar heat capacity of aluminum is approximately 25 J/mol · deg. Hence,

$$\frac{25 \text{ J/mol} \cdot \text{deg}}{27 \text{ g/mol}} = 0.93 \text{ J/g} \cdot \text{deg}$$

18.19. What factor can be used to transform molar heat capacity into specific heat?

Ans.   The reciprocal of atomic weight (or formula weight). For example,

$$\frac{20 \text{ J}}{\text{mol} \cdot \text{deg}} \left(\frac{1 \text{ mol}}{100 \text{ g}} \right) = \frac{0.20 \text{ J}}{\text{g} \cdot \text{deg}}$$

Molar	Reciprocal	Specific
Heat	of Atomic	Heat
Capacity	Weight	

18.20. Calculate the final temperature of 50.0 g of water at 45.5 °C which is mixed with 25.0 g of water at 0.0 °C.

Ans.

$$\text{heat gained} = -(\text{heat lost})$$
$$(m)(\text{SH})(\Delta t) = -(m)(\text{SH})(\Delta t)$$

Since both samples are water, the specific heats are the same and they cancel out.

$$(25.0 \text{ g})(t_f - 0.0 \text{ °C}) = -(50.0 \text{ g})(t_f - 45.5 \text{ °C})$$
$$t_f = 30.3 \text{ °C}$$

18.21. Calculate the approximate atomic weight of lead using its specific heat, 0.12 J/g · deg, and the law of Dulong and Petit.

Ans.

$$\frac{25 \text{ J/mol} \cdot \text{deg}}{0.12 \text{ J/g} \cdot \text{deg}} = 210 \text{ g/mol}$$

PHASE CHANGES

18.22. Under what circumstances does addition of heat to a system cause no rise in temperature?

Ans.   If the system is a pure substance changing phase, no rise in temperature occurs.

18.23. Describe the process called (*a*) vaporization and (*b*) fusion. To what do (*c*) ΔH_{vap} and (*d*) ΔH_{fus} refer?

Ans.   (*a*) The change of a liquid into a gas. (*b*) The change of a solid into a liquid. (*c*) The enthalpy change accompanying a vaporization process. (*d*) The enthalpy change accompanying a melting (fusion) process.

18.24. (*a*) How much heat does it take to just melt 100 g of ice at 0 °C? (*b*) What will be the final temperature of 100 g of ice at 0 °C after 30 kJ of heat is added? The heat of fusion of ice is 335 J/g.

Ans.

(*a*) $(100 \text{ g})(335 \text{ J/g}) = 3.35 \times 10^4 \text{ J} = 33.5 \text{ kJ}$

(*b*)   If only 30 kJ of heat is added, not all the ice will melt. The final state will contain both solid and liquid, and the temperature will still be 0 °C.

18.25. If 50.0 kJ of energy is added to 100 g of ice at 0 °C, (a) how much energy is required to melt the ice? The heat of fusion of ice is 335 J/g. (b) How much of the original 50.0 kJ is left over to warm the water? (c) What is the final temperature of the water?

Ans.

(a) $(100\,g)(335\,J/g) = 3.35 \times 10^4\,J = 33.5\,kJ$

(b) $50.0\,kJ - 33.5\,kJ = 16.5\,kJ$

(c) $\Delta t = \dfrac{16\,500\,J}{(100\,g)(4.184\,J/g \cdot {}^\circ C)} = 39.4\,{}^\circ C$

The final temperature is 39.4 °C, since the temperature started at 0 °C.

18.26. What is the final temperature after 50.0 kJ is added to 100 g of ice at 0 °C?

Ans. This problem is the same as Problem 18.25.

ENTHALPY CHANGES IN CHEMICAL REACTIONS

18.27. Calculate the value of ΔH for the reaction of 77.0 g of CH_4 according to the equation

$$3\,CuO + CH_4 \longrightarrow CO + 2\,H_2O + 3\,Cu$$

Ans. ΔH_f values are taken from Table 18-3. For each mole of CH_4,

$$\Delta H = \Delta H_{f(products)} - \Delta H_{f(reactants)}$$
$$= (-110.5\,kJ) + 2(-285.9\,kJ) + 3(0) - 3(-157\,kJ) - (-74.85\,kJ)$$
$$= -136\,kJ$$

For 77.0 g CH_4,

$$77.0\,g\,CH_4\left(\frac{1\,mol\,CH_4}{16.0\,g\,CH_4}\right)\left(\frac{-136\,kJ}{mol\,CH_4}\right) = -654\,kJ$$

18.28. The heat capacity of CH_4 is 36.0 J/mol · deg. Calculate ΔH_{total} for forming 1.00 mol of CH_4 at 25 °C from its elements at 25 °C, then warming the CH_4 10 °C.

Ans.

$$\Delta H_{total} = \Delta H_f + \Delta H_{warming}$$
$$= -74.85\,kJ + 1.00\,mol\left(\frac{36.0\,J}{mol \cdot deg}\right)10\,deg$$
$$= -74.85\,kJ + 0.36\,kJ = -74.49\,kJ$$

This example illustrates the necessity of watching the signs and units carefully.

Supplementary Problems

18.29. If 24.32 g of a crystalline element reacts with exactly 15.999 g of oxygen, what is the value of the atomic weight of the element if (a) the metal atoms and oxygen atoms react in a 1:1 ratio? (b) the metal atoms and oxygen atoms react in a 1:2 ratio? (c) the metal atoms and oxygen atoms react in a 2:1 ratio? (d) If the specific heat of the element is determined to be 1.025 J/g · deg, which of the atomic weights is correct?

Ans. (a) 24.32 g/mol, (b) 48.64 g/mol, and (c) 12.16 g/mol. (d) Using the law of Dulong and Petit, we can calculate

$$\frac{25\,J/mol \cdot deg}{1.025\,J/g \cdot deg} = \frac{24\,g}{mol}$$

The atomic weight in part (a) is close to that determined from the law of Dulong and Petit, so that the assumption that the atoms react in a 1 : 1 ratio is correct.

18.30. If $1\,000\,J$ of electrical energy is added with an immersion heater to 55.7 g of water at 20.0 °C, (a) how much heat is added? (b) how much work is done on the system? (c) how much does the temperature rise?

    Ans.   (a) No heat. (b) $1\,000\,J$ of work. The work added will raise the temperature of the water as much as $1\,000\,J$ of heat would raise it.

       (c) $$\Delta H = (m)(SH)(\Delta t)$$

$$\Delta t = \frac{1\,000\,J}{(55.7\,g)(4.184\,J/g \cdot °C)} = 4.29\,°C$$

18.31. Under what circumstances can the temperature of a system be raised (with no chemical reaction) without any addition of heat?

    Ans.   Addition of work (any form of energy other than heat) may cause a temperature rise.

18.32. Calculate the approximate specific heat of lead.

    Ans.

$$SH = \frac{25\,J}{mol \cdot deg}\left(\frac{1\,mol}{208\,g}\right) = 0.12\,J/g \cdot deg$$

18.33. Calculate ΔH for the combustion of 100 g of CO.

    Ans.

$$2\,CO + O_2 \longrightarrow 2\,CO_2$$

Per 2 mol of CO:

$$\Delta H = 2\Delta H_{f(CO_2)} - 2\Delta H_{f(CO)}$$

$$= 2(-393.5\,kJ) - 2(-110.5\,kJ) = -566.0\,kJ$$

Per 100 g of CO:

$$100\,g\,CO\left(\frac{1\,mol\,CO}{28.0\,g\,CO}\right)\left(\frac{-566.0\,kJ}{2\,mol\,CO}\right) = -1\,010\,kJ = -1.01 \times 10^3\,kJ$$

Chapter 19

Rates and Equilibrium

19.1 INTRODUCTION

We have learned (Chap. 7) that some reactions occur under one set of conditions while an opposite reaction occurs under another set of conditions. For example, we learned that sodium and chlorine combine when treated with each other, but that molten NaCl decomposes when treated with electricity:

$$2\,Na + Cl_2 \longrightarrow 2\,NaCl$$

$$2\,NaCl(molten) \xrightarrow{\text{electricity}} 2\,Na + Cl_2$$

However, some sets of reactants can undergo both a forward and a reverse reaction under the same set of conditions. This circumstance leads to a state called *chemical equilibrium*. Before we take up equilibrium, however, we have to learn about the factors that affect the rate of a chemical reaction.

19.2 RATES OF CHEMICAL REACTION

Some chemical reactions proceed very slowly, others with explosive speed, and still others somewhere in between. The "dissolving" of underground limestone deposits by water containing carbon dioxide to form caverns is an example of a slow reaction; it can take centuries. The explosion of TNT is an example of a very rapid reaction.

The *rate of a reaction* is defined as the change in concentration of any of its reactants or products per unit time. There are six factors that affect the rate of a reaction:

1. The nature of the reactants. Carbon tetrachloride (CCl_4) does not burn in oxygen, but methane (CH_4) burns very well indeed. In fact, CCl_4 used to be used in fire extinguishers, while CH_4 is the major component of natural gas. This factor is least controllable by the chemist, and so is of least interest here.
2. Temperature. In general, the higher the temperature of a system, the faster the chemical reaction will proceed. A rough rule of thumb is that a 10 °C rise in temperature will about double the rate of a reaction.
3. The presence of a catalyst. A *catalyst* is a substance that can accelerate (or slow down) a chemical reaction without undergoing a permanent change in its own composition. For example, the decomposition of $KClO_3$ by heat is accelerated by the presence of a small quantity of MnO_2. After the reaction, the $KClO_3$ has been changed to KCl and O_2, but the MnO_2 is still MnO_2.
4. The concentration of the reactants. In general, the higher the concentration of the reactants, the faster the reaction.
5. The pressure of gaseous reactants. In general, the higher the pressure of gaseous reactants, the faster the reaction. This factor is merely a corollary of factor 4, since the higher pressure is in effect a higher concentration.
6. State of subdivision. The smaller the pieces of a solid reactant—the smaller the state of subdivision—the faster the reaction. Wood shavings burn faster than solid wood, for example, because they have more surface area in contact with the oxygen with which they are combining (for a given mass of wood). In a sense, this is also a corollary of factor 4.

Most of the factors that affect the rate of a reaction are qualitative or semiquantitative, but the dependency of the rate on concentration (or pressure, which is a measure of concentration) may be stated as a *rate law expression*. For the reaction

$$a\,A + b\,B \longrightarrow products$$

the rate law generally has the following form:

$$rate = k[A]^x[B]^y$$

The proportionality constant k, called the *rate constant*, has a constant value for a given reaction at a given temperature. The terms in square brackets are concentration terms (compare Chap. 14), and x and y are exponents which are often integral. The exponent x is called the *order with respect to* A, and y is called the *order with respect to* B. The sum $x + y$ is called the *overall order* of the reaction. The values for x and y can be 0, 1, 2, 3 or 0.5, 1.5, or 2.5, but never more than 3. These values must be determined by experiment, and do not necessarily equal the values of a and b in the chemical equation.

EXAMPLE 19.1. In a certain reaction, doubling the initial concentration of the only reactant doubles the initial rate of the reaction. What is the order of the reaction?

Since the rate doubles as the concentration doubles, the reaction must be first order.

$$rate = k[A]^x$$

$$2 = \frac{rate_2}{rate_1} = \frac{k[A]_2^x}{k[A]_1^x} = \left(\frac{[A]_2}{[A]_1}\right)^x = 2^x$$

x must equal 1.

EXAMPLE 19.2. In a certain reaction, doubling the initial concentration of the only reactant quadruples the initial rate of the reaction. What is the order of the reaction?

Since the rate quadruples as the concentration doubles, the reaction must be second order.

$$rate = k[A]^x$$

$$4 = \frac{rate_2}{rate_1} = \frac{k[A]_2^x}{k[A]_1^x} = 2^x$$

x must equal 2.

EXAMPLE 19.3. The rate of a given reaction changes as the reaction proceeds. Explain why.

The concentrations of the reactants change as the reaction progresses, and so the rate changes because it depends on the concentrations. An illustration of the effect of time on the rate of a first-order process is the decay of a radioactive substance, considered in Sec. 22.3.

EXAMPLE 19.4. Consider the following data about the reaction

$$A + B \longrightarrow products$$

Run	Initial Concentration of A	Initial Concentration of B	Initial Rate
1	0.10 M	1.0 M	$2.1 \times 10^{-3}\,M/s$
2	0.20 M	1.0 M	$8.4 \times 10^{-3}\,M/s$
3	0.20 M	2.0 M	$8.4 \times 10^{-3}\,M/s$

Determine the order with respect to A and with respect to B, as well as the overall order.

$$rate = k[A]^x[B]^y$$

When the initial concentration of A doubles and that of B stays the same (compare runs 1 and 2), the rate increases by a factor of 4. The value of x must be 2. When the concentration of B doubles with constant A concentration (as in runs 2 and 3), the rate does not change. The value of y is 0.

$$4 = \frac{\text{rate}_2}{\text{rate}_1} = \frac{k[A]_2^x}{k[A]_1^x} = 2^x \qquad x \text{ must equal } 2$$

$$1 = \frac{\text{rate}_3}{\text{rate}_2} = \frac{k[B]_3^y}{k[B]_2^y} = 2^y \qquad y \text{ must equal } 0$$

The overall order is $2 + 0 = 2$.

EXAMPLE 19.5. What can you determine about the reaction from just runs 1 and 3 of the table in Example 19.4?

You cannot tell if the change in rate stems from the change in A concentration, the change in B concentration, or both. Keeping all but one of the factors constant during a series of reactions, and using the effect of that one change, is the characteristic of a *controlled experiment*.

EXAMPLE 19.6. From the following data, calculate the value of k for the reaction

$$A + B \longrightarrow \text{products}$$

Run	Initial Concentration of A (M)	Initial Concentration of B (M)	Initial Rate (M/s)
1	0.40	1.5	1.1×10^{-4}
2	0.80	1.5	2.2×10^{-4}
3	0.80	3.0	4.4×10^{-4}

Doubling each concentration, with the other constant, doubles the rate, and so the reaction is first order with respect to each reactant. The rate law then has the form

$$\text{rate} = k[A][B]$$

Substituting the values from run 1 of this set of data yields

$$1.1 \times 10^{-4} \, M/s = k(0.40 \, M)(1.5 \, M)$$

$$k = \frac{1.1 \times 10^{-4} \, M/s}{0.60 \, M^2} = \frac{1.8 \times 10^{-4}}{M \cdot s}$$

The rate law equation gives some insight into the actual steps by which the reaction takes place—the reaction mechanism.

19.3 CHEMICAL EQUILIBRIUM

Many chemical reactions convert practically all the reactant(s) (at least the limiting quantity) into products under a given set of conditions. These reactions are said to *go to completion*. In other reactions, as the products are formed they in turn react to form the original reactants again. This situation—two opposing reactions occurring at the same time—leads to formation of some products, but none of the reactants is completely converted to products. A state in which two exactly opposite reactions are occurring at the same rate is called *chemical equilibrium*. (In fact, all chemical reactions are equilibrium reactions, at least theoretically.) For example, nitrogen and hydrogen gases react with each other at 500 °C and high pressure to form ammonia; under the same conditions, ammonia decomposes to produce hydrogen and nitrogen:

$$3 H_2 + N_2 \longrightarrow 2 NH_3$$
$$2 NH_3 \longrightarrow 3 H_2 + N_2$$

To save effort, we often write these two exactly opposite equations as one, with double arrows:

$$3H_2 + N_2 \rightleftarrows 2NH_3 \quad \text{or} \quad 2NH_3 \rightleftarrows 3H_2 + N_2$$

We call the reagents on the right of the chemical equation as it is written the *products* and those on the left the *reactants*, despite the fact that we can write the equation with either set of reagents on the left side.

With the reaction just above, if you start with a mixture of nitrogen and hydrogen and allow it to come to 500 °C at 200 atm pressure, some nitrogen and hydrogen combine to form ammonia. If you heat ammonia to 500 °C at 200 atm pressure, some of it decomposes to nitrogen and hydrogen. Both reactions can occur in the same vessel at the same time.

What happens when we first place hydrogen and nitrogen in a container at 500 °C and allow them to react? At first, there is no ammonia present, and so the only reaction that occurs is the combination of the two elements. As time passes, there is less and less nitrogen and hydrogen, and the combination reaction therefore slows down (factor 4 or 5, Sec. 19.2). Meanwhile, the concentration of ammonia is building up, and the decomposition of the ammonia therefore increases in rate. There comes a time when both the combination reaction and the decomposition reaction occur at the same rate. When that happens, the concentration of ammonia will not change any more. Apparently, the reaction stops. However, in reality both the combination reaction and the decomposition reaction continue to occur; their effects merely cancel each other. A state of *equilibrium* has been achieved.

Le Châtelier's Principle

If we change the conditions on the N_2, H_2, NH_3 system at equilibrium, such as changing the temperature, we can get some further reaction—either combination or decomposition. Soon, however, the system will achieve a new equilibrium at the new set of conditions.

Le Châtelier's principle states that if a stress is applied to a system *at equilibrium*, the equilibrium will shift in a tendency to reduce that stress. A *stress* is something done *to* the system (not *by* the equilibrium reaction). The stresses that we consider are change of temperature, change of pressure, change of concentration(s), and addition of a catalyst. Let us consider the effect on a typical equilibrium by each of these stresses.

Temperature Change

To consider the effect of temperature change, let us rewrite one of the equations including the heat involved:

$$3H_2 + N_2 \rightleftarrows 2NH_3 + \text{heat}$$

How do we get the temperature of the system to rise? By adding heat. When we add heat, this equilibrium system reacts to reduce that stress, that is, to use up some of the added heat. It can use up heat in the reverse reaction, the decomposition of ammonia to hydrogen and nitrogen. When the substances written as products of the reaction (on the right side of the equation) react to produce more reactants (on the left side of the equation), we say that the reaction has *shifted to the left*. When the opposite process occurs, we say that the equilibrium has *shifted to the right*. Thus, raising the temperature on this system already at equilibrium causes a shift to the left; some of the ammonia decomposes without being replaced.

EXAMPLE 19.7. What would happen to the above system at equilibrium if the temperature were lowered?

We lower the temperature of a system by removing heat. The equilibrium tries to minimize that change by restoring some heat; it can do that by shifting to the right. Thus, more nitrogen and hydrogen are converted to ammonia.

The Effect of Pressure

Pressure affects the gases in a system much more than it affects the liquid or solids. We will investigate the same ammonia, hydrogen, nitrogen system discussed above. If the system is at equilibrium, what will an increase in pressure by the chemist do to the equilibrium? The system will shift to try to reduce the stress, as required by Le Châtelier's principle. How can this system reduce its own pressure? By reducing the total number of moles present. It can shift to the right to produce 2 mol of gas for every 4 mol used up:

$$3H_2 + N_2 \rightleftharpoons 2NH_3$$

Of course, it need not shift much. For example, if 0.0030 mol H_2 reacts with 0.0010 mol N_2 to produce 0.0020 mol NH_3, the total number of moles will have been reduced (by 0.0020 mol), and the pressure will therefore have been reduced.

EXAMPLE 19.8.   What will be the effect of increased pressure on the following system at equilibrium?

$$2NH_3 \rightleftharpoons 3H_2 + N_2$$

Again some nitrogen and hydrogen will be converted to ammonia. This time, since the equation is written in the reverse of the way it was written above, the equilibrium will shift to the left. Of course, the same physical effect is produced: More ammonia is formed. But the answer in terms of the direction of the shift is different since the equation is written "backward."

EXAMPLE 19.9.   What effect would a decrease in volume have on the following system at equilibrium at 500 °C, where all the reagents are gases?

$$H_2 + I_2 \rightleftharpoons 2HI$$

The decrease in volume would increase the pressure on each of the gases (Chap. 11). The equilibrium would not shift, however, because there are equal numbers of moles of gases on the two sides. Neither possible shift would cause a reduction of pressure.

The Effect of Concentration

An increase in concentration of one of the reactants or products of the equilibrium will cause the equilibrium to shift to try to reduce that concentration increase.

EXAMPLE 19.10.   How will addition of hydrogen gas affect the following equilibrium system?

$$3H_2 + N_2 \rightleftharpoons 2NH_3$$

Addition of hydrogen will at first increase the concentration of hydrogen. The equilibrium will therefore shift to reduce some of that increased concentration; it will shift right. Some of the added hydrogen will react with some of the nitrogen originally present to produce more ammonia. Note especially that the hydrogen concentration will be above the original hydrogen concentration but below the concentration it would have if no shift had taken place.

EXAMPLE 19.11.   Suppose that, under a certain set of conditions, a mixture of nitrogen, hydrogen, and ammonia is at equilibrium. The concentration of hydrogen is 0.250 mol/L; that of nitrogen is 0.100 mol/L, and that of ammonia is 0.200 mol/L. Now 0.003 mol of hydrogen is added to 1.00 L of the mixture. What is the widest possible range of hydrogen concentration in the new equilibrium?

Before addition of the extra hydrogen, its concentration was 0.250 mol/L. After addition of the extra hydrogen, but before any equilibrium shift could take place, there would be 0.253 mol/L. The shift of the equilibrium would use up some but not all of the added hydrogen, and so the final hydrogen concentration must be above 0.250 mol/L and below 0.253 mol/L. Some nitrogen has been used up, and its final concentration must be less than its original concentration, 0.100 mol/L. Some additional ammonia has been formed, and so its final concentration has been increased over 0.200 mol/L. Notice that Le Châtelier's principle does not tell us how much of a shift there will be, but only qualitatively in which direction a shift will occur.

Presence of a Catalyst

Addition of a catalyst to a system at equilibrium will not cause any change in the position of the equilibrium; it will shift neither left nor right. Addition of the catalyst speeds up both the forward reaction and the reverse reaction equally.

19.4   EQUILIBRIUM CONSTANTS

Although Le Châtelier's principle does not tell us how much an equilibrium will be shifted, there is a way to determine the position of an equilibrium once data have been determined for the equilibrium experimentally. The ratio of concentrations of products to reactants, each raised to a suitable power, is constant for a given equilibrium reaction. The letters A, B, C, and D are used here to stand for general chemical species. Thus, for a chemical reaction in general,

$$a\,\text{A} + b\,\text{B} \rightleftharpoons c\,\text{C} + d\,\text{D}$$

it is true that the following ratio always has the same value at a given temperature:

$$K = \frac{[\text{C}]^c[\text{D}]^d}{[\text{A}]^a[\text{B}]^b}$$

Here the square brackets indicate the concentration of the chemical species within the bracket. That is, [A] means the concentration of A, and so forth. $[\text{A}]^a$ means the concentration of A raised to the a power, where a is the value of the coefficient of A in the balanced equation for the chemical equilibrium. The value of the ratio of concentration terms is symbolized by the letter K, called the *equilibrium constant*. For example, for the reaction of nitrogen and hydrogen referred to in Sec. 19.3,

$$3\,\text{H}_2 + \text{N}_2 \rightleftharpoons 2\,\text{NH}_3$$

the ratio is

$$K = \frac{[\text{NH}_3]^2}{[\text{H}_2]^3[\text{N}_2]}$$

The exponents 2 and 3 are the coefficients of ammonia and hydrogen, respectively, in the balanced equation.

Please note the following points about such an equation:

1. This is a mathematical equation. Its values are numbers, involving the concentrations of the chemicals.
2. Each concentration is raised to the correct power, given by the coefficient in the chemical equation. (Contrast this exponent with that in Sec. 19.2.)
3. The concentrations of the products of the reaction are written in the numerator of the right-hand side of the equilibrium constant expression; the concentrations of the reactants are in the denominator.
4. The terms are multiplied together, not added.
5. Each equilibrium constant expression is associated with a particular chemical reaction written in a given direction.

EXAMPLE 19.12.   Write the equilibrium constant expression for the reaction

$$2\,\text{NH}_3 \rightleftharpoons 3\,\text{H}_2 + \text{N}_2$$

For this equation, the terms involving the concentrations of the elements are in the numerator, because they are the products:

$$K = \frac{[\text{H}_2]^3[\text{N}_2]}{[\text{NH}_3]^2}$$

There are a great many types of equilibrium problems. We will try to start with the easiest and work to the harder ones as we go.

EXAMPLE 19.13. Calculate the value of the equilibrium constant for the following reaction if at equilibrium the concentration of A is 2.00 M, that of B is 3.50 M, that of C is 0.500 M, and that of D is 1.50 M.

$$A + B \rightleftharpoons C + D$$

Since all the coefficients in the balanced equation are equal to 1, the value of the equilibrium constant is

$$K = \frac{[C][D]}{[A][B]}$$

We merely substitute the equilibrium concentrations into this equation to determine the value of the equilibrium constant:

$$K = \frac{(0.500\ M)(1.50\ M)}{(2.00\ M)(3.50\ M)} = 0.107$$

One way to make Example 19.13 harder is by giving numbers of moles and a volume instead of concentrations at equilibrium. Since the equilibrium constant is defined in terms of concentrations, we must first convert the numbers of moles and volume to concentrations. Note especially that the volume of all the reactants is the same, since they are all in the same system.

EXAMPLE 19.14.

$$A + B \rightleftharpoons C + D$$

Calculate the value of the equilibrium constant if at equilibrium there are 1.00 mol A, 1.75 mol B, 0.250 mol C, and 0.750 mol D in 500 mL of solution.

Since numbers of moles and a volume are given, it is easy to calculate the equilibrium concentrations of the species. In this case, [A] = 2.00 M, [B] = 3.50 M, [C] = 0.500 M, and [D] = 1.50 M, which are exactly the concentrations given in the last example. Thus, this problem has exactly the same answer.

It is somewhat more difficult to determine the value of the equilibrium constant if some initial concentrations are given instead of equilibrium concentrations.

EXAMPLE 19.15. Calculate the value of the equilibrium constant in the following reaction if 1.00 mol of A and 2.00 mol of B are placed in 1.00 L of solution and allowed to come to equilibrium. The equilibrium concentration of C is found to be 0.20 M.

$$A + B \rightleftharpoons C + D$$

To determine the equilibrium concentrations of all the reactants and products, we must deduce the changes that have occurred. We can assume by the wording of the problem that no C or D was added by the chemist, that some A and B have been used up, and that some D has also been produced. It is perhaps easiest to tabulate the various concentrations. We will use the chemical equation as our table headings, and enter the values that we know now:

	A	B	$\rightleftharpoons$	C	D
Initial concentrations	1.00	2.00		0.00	0.00
Changes produced by the reaction					
Equilibrium concentrations				0.20	

We deduce that to produce 0.20 M C it takes 0.20 M A and 0.20 M B. Moreover, we know that 0.20 M D was also produced. **The magnitudes of the values in the second row of the table—the changes produced by the reaction—are always in the same ratio as the coefficients in the balanced chemical equation.**

	A	B	$\rightleftharpoons$	C	D
Initial concentrations	1.00	2.00		0.00	0.00
Changes produced by the reaction	−0.20	−0.20		+0.20	+0.20
Equilibrium concentrations				0.20	

We merely add the columns to find the rest of the equilibrium concentrations:

	A	B	⇌	C	D
Initial concentrations	1.00	2.00		0.00	0.00
Changes produced by the reaction	−0.20	−0.20		+0.20	+0.20
Equilibrium concentrations	0.80	1.80		0.20	0.20

Now that we have calculated the equilibrium concentrations we can substitute these values into the equilibrium constant expression:

$$K = \frac{[C][D]}{[A][B]} = \frac{(0.20)(0.20)}{(0.80)(1.80)} = 0.028$$

When the chemical equation is more complex, the equilibrium constant expression is also more complex and the deductions about the equilibrium concentrations of reactants and products are more involved too.

EXAMPLE 19.16.   Calculate the value of the equilibrium constant for the following reaction if 1.00 mol of A and 2.00 mol of B are placed in 1.00 L of solution and allowed to come to equilibrium. The equilibrium concentration of C is found to be 0.20 M.

$$2A + B \rightleftharpoons C + D$$

The equilibrium constant expression for this equation is

$$K = \frac{[C][D]}{[A]^2[B]}$$

	2A	B	⇌	C	D
Initial concentrations	1.00	2.00		0.00	0.00
Changes produced by the reaction					
Equilibrium concentrations				0.20	

The changes brought about by the chemical reaction are a little different in this case. Twice as many moles per liter of A are used up as moles per liter of C are produced. Note that the magnitudes in the middle row of this table and the coefficients in the balanced chemical equation are in the same ratio.

	2A	B	⇌	C	D
Initial concentrations	1.00	2.00		0.00	0.00
Changes produced by the reaction	−0.40	−0.20		+0.20	+0.20
Equilibrium concentrations				0.20	

Adding the columns:

	2A	B	⇌	C	D
Initial concentrations	1.00	2.00		0.00	0.00
Changes produced by the reaction	−0.40	−0.20		+0.20	+0.20
Equilibrium concentrations	0.60	1.80		0.20	0.20

The equilibrium values are substituted into the equilibrium constant expression:

$$K = \frac{[C][D]}{[A]^2[B]} = \frac{(0.20)(0.20)}{(0.60)^2(1.80)} = 0.062$$

The next type of problem gives the initial concentrations of the reactants (and products) plus the value of the equilibrium constant and requires calculation of one or more equilibrium concentrations. We use algebraic quantities (such as x) to represent at least one equilibrium concentration, and solve for the others if necessary in terms of that quantity.

EXAMPLE 19.17. For the reaction

$$A + B \rightleftharpoons C + D$$

1.50 mol of A and 2.25 mol B are placed in a 1.00-L vessel and allowed to come to equilibrium. The value of the equilibrium constant is 0.0010. Calculate the concentration of C at equilibrium.

	A	+	B	$\rightleftharpoons$	C	+	D
Initial concentrations	1.50		2.25		0.00		0.00
Changes produced by the reaction							
Equilibrium concentrations					x		

The changes due to the chemical reaction are as easy to determine as before, and therefore so are the equilibrium concentrations:

	A	+	B	$\rightleftharpoons$	C	+	D
Initial concentrations	1.50		2.25		0.00		0.00
Changes produced by the reaction	$-x$		$-x$		x		x
Equilibrium concentrations	$1.50 - x$		$2.25 - x$		x		x

We could substitute these equilibrium concentrations in the equilibrium constant expression, and solve using the quadratic equation. However, it is more convenient to attempt to approximate the equilibrium concentrations by neglecting a small quantity (x) *when added to or subtracted from* a larger quantity (such as 2.25 or 1.50). We do not neglect small quantities unless they are added to or subtracted from larger quantities! Thus, we approximate the equilibrium concentrations as $[A] \cong 1.50 \, M$ and $[B] \cong 2.25 \, M$. The equilibrium constant expression is thus

$$K = \frac{[C][D]}{[A][B]} = \frac{(x)(x)}{(1.50)(2.25)} = \frac{x^2}{3.38} = 0.0010$$

$$x = 0.058 \, M$$

Next we must check whether our approximation was valid. Is it true that $1.50 - 0.058 \cong 1.50$ and $2.25 - 0.058 \cong 2.25$? Considering the limits of accuracy of the equilibrium constant expression, results within 5% accuracy are considered valid for the general chemistry course. These results are thus valid. The equilibrium concentrations are

$$[A] = 1.50 - 0.058 = 1.44 \, M$$
$$[B] = 2.25 - 0.058 = 2.19 \, M$$
$$[C] = [D] = 0.058 \, M$$

If the approximation had caused an error of 10% or more, you would not be able to use it. You would have to solve by a more rigorous method, such as the quadratic equation.

EXAMPLE 19.18. Repeat the last example, but with an equilibrium constant value of 0.100.

The problem is done in exactly the same manner until you find that the value of x is 0.581 M.

$$\frac{x^2}{3.38} = 0.100$$

$$x = 0.581$$

When that value is subtracted from 1.50 or 2.25, the answer is not nearly the original 1.50 or 2.25. The approximation is not valid. Thus, we must use the quadratic equation.

$$K = \frac{(x)^2}{(1.50 - x)(2.25 - x)} = 0.100$$

$$x^2 = (0.100)(1.50 - x)(2.25 - x) = 0.100(x^2 - 3.75x + 3.38)$$

$$0.900x^2 + 0.375x - 0.338 = 0$$

$$x = \frac{-b \pm \sqrt{(b^2 - 4ac)}}{2a} = \frac{-0.375 \pm \sqrt{(0.375)^2 - 4(0.900)(-0.338)}}{2(0.900)}$$

Two solutions for x, one positive and one negative, can be obtained from this equation, but only one will have physical meaning. There cannot be any negative concentrations.

$$x = 0.439\, M = [C] = [D]$$

The concentrations of A and B are then

$$[A] = 1.50 - 0.439 = 1.06\, M$$
$$[B] = 2.25 - 0.439 = 1.81\, M$$

The value from the equilibrium constant expression is therefore equal to the value of K given in the problem.

$$K = \frac{(0.439)^2}{(1.06)(1.81)} = 0.100$$

Solved Problems

RATES OF CHEMICAL REACTION

19.1. Under what conditions does a glowing lump of charcoal react faster: in air or in pure oxygen? Explain.

Ans. It will react faster in pure oxygen, since the concentration of oxygen is greater there.

19.2. Does sugar dissolve faster in hot coffee or in lukewarm coffee? Explain.

Ans. In hot coffee. The higher the temperature, the faster the process.

19.3. Does lump sugar or granular sugar dissolve faster in water, all other factors being equal? Explain.

Ans. Granular sugar dissolves faster; it has more surface area in contact with the liquid.

19.4. At the same temperature, which sample of a gas has more molecules per unit volume—one at high pressure or one at lower pressure? Which sample would react faster with a solid substance, all other factors being equal?

Ans. The high pressure gas has more molecules per unit volume and therefore reacts faster.

19.5. Write a rate law equation for each of the following: (*a*) A reaction first order with respect to A and second order with respect to B. (*b*) A reaction zero order with respect to A and second order with respect to B. (*c*) A reaction first order with respect to A and first order with respect to B. (*d*) A reaction second order with respect to B, its only reactant.

Ans.

   (*a*) $\text{rate} = k[A][B]^2$

   (*b*) $\text{rate} = k[B]^2$

   (*c*) $\text{rate} = k[A][B]$

   (*d*) $\text{rate} = k[B]^2$

19.6. Calculate the value of k for part (*a*) of the last problem if the rate is $1.6 \times 10^{-4}\ \text{mol}/\text{L} \cdot \text{s}$ when $[A] = 1.00\, M$ and $[B] = 2.00\, M$.

Ans.

$$1.6 \times 10^{-4}\ \text{mol}/\text{L} \cdot \text{s} = k(1.00\,\text{mol}/\text{L})(2.00\,\text{mol}/\text{L})^2$$

$$k = 4.0 \times 10^{-5}\ \text{L}^2/\text{mol}^2 \cdot \text{s}$$

19.7. What are the units of k for part (b) of Problem 19.5?

Ans. L/mol · s. The rate is always in M/s (or mol/L · s). The units of k change depending on the overall order of the reaction. Check:

$$\text{rate} = k[\text{B}]^2$$
$$\text{mol/L} \cdot \text{s} = (\text{L/mol} \cdot \text{s})(\text{mol}^2/\text{L}^2) = \text{mol/L} \cdot \text{s}$$

CHEMICAL EQUILIBRIUM

19.8. Write an "equation" for the addition of heat to a water-ice mixture at $0\,°\text{C}$ to produce more liquid water at $0\,°\text{C}$. Which way does the equilibrium shift when you try to raise the temperature?

Ans.

$$\text{H}_2\text{O(s)} + \text{heat} \rightleftharpoons \text{H}_2\text{O(l)}$$

The equilibrium shifts toward liquid water as you add heat in an attempt to raise the temperature. (However, the temperature does not change until all the ice is melted and this equilibrium system is destroyed.)

19.9. What does Le Châtelier's principle state about the effect of an addition of NH_3 on a mixture of N_2 and H_2 before it attains equilibrium?

Ans. Nothing. Le Châtelier's principle applies only to a system already at equilibrium.

19.10. What effect would a decrease in volume have on the following system at equilibrium at $500\,°\text{C}$?

$$2\,\text{C(s)} + \text{O}_2 \rightleftharpoons 2\,\text{CO}$$

Ans. The equilibrium would shift to the left. The decrease in volume would increase the pressure of each of the gases, but not of the carbon, which is a solid. The number of moles of *gas* would be decreased by the shift to the left.

19.11. For the reaction

$$\text{A} + \text{B} \rightleftharpoons \text{C} + \text{heat}$$

does a rise in temperature increase or decrease the rate of (a) the forward reaction? (b) the reverse reaction? (c) Which effect is greater?

Ans. (a) Increase and (b) increase. (An increase in temperature increases all rates.) (c) The added heat shifts the equilibrium to the left. That means that the reverse reaction was speeded up more than the forward reaction was speeded up.

19.12. Hydrogen gas is added to an equilibrium system of hydrogen, nitrogen, and ammonia. The equilibrium shifts to reduce the stress of the added hydrogen. How much hydrogen will be present at the new equilibrium compared with the old equilibrium—more, less, or the same concentration?

Ans.

$$\text{N}_2 + 3\,\text{H}_2 \rightleftharpoons 2\,\text{NH}_3 + \text{heat}$$

There will be more hydrogen and more ammonia present at the new equilibrium, and less nitrogen. The concentration of hydrogen is not as great as the original concentration plus that which would have resulted from the addition of more hydrogen, however. Some of the total has been used up in the equilibrium shift.

EQUILIBRIUM CONSTANTS

19.13. The balanced chemical equation

$$3\,\text{H}_2 + \text{N}_2 \rightleftharpoons 2\,\text{NH}_3$$

means which one(s) of the following: (a) One can place H_2 and N_2 in a reaction vessel only in the ratio 3 mol to 1 mol. (b) If one places NH_3 in a reaction vessel, one cannot put any N_2

and/or H_2 in with it. (c) If one puts 3 mol of H_2 and 1 mol of N_2 into a reaction vessel, it will produce 2 mol of NH_3. (d) For every 1 mol of N_2 that reacts, 3 mol of H_2 will also react and 2 mol of NH_3 will be produced. (e) For every 2 mol of NH_3 that decomposes, 3 mol of H_2 and 1 mol of N_2 are produced.

Ans.   Only parts (d) and (e) are correct. The balanced equation governs the reacting ratios only. It cannot determine how much of any chemical may be placed in a vessel—(a) and (b)—or if a reaction will go to completion—(c).

19.14. Write equilibrium constant expressions for the following equations. Tell how they are related.

(a) $$N_2O_4 \rightleftharpoons NO_2 + NO_2$$

(b) $$N_2O_4 \rightleftharpoons 2NO_2$$

Ans.

(a) $$K = \frac{[NO_2][NO_2]}{[N_2O_4]}$$

(b) $$K = \frac{[NO_2]^2}{[N_2O_4]}$$

The equilibrium constant expressions are the same because the chemical equations are the same. It is easy to see why the coefficients of the chemical equation are used as exponents in the equilibrium constant expression by writing out the equation and expression as in part (a).

19.15. Write an equilibrium constant expression for each of the following:

(a) $$PCl_3(g) + Cl_2(g) \rightleftharpoons PCl_5(g)$$

(b) $$2NO + O_2 \rightleftharpoons 2NO_2$$

(c) $$SO_2 + Cl_2 \rightleftharpoons SO_2Cl_2$$

(d) $$2NH_3 \rightleftharpoons N_2 + 3H_2$$

(e) $$2HBr(g) \rightleftharpoons H_2 + Br_2(g)$$

Ans.

(a) $$K = \frac{[PCl_5]}{[PCl_3][Cl_2]}$$

(b) $$K = \frac{[NO_2]^2}{[NO]^2[O_2]}$$

(c) $$K = \frac{[SO_2Cl_2]}{[SO_2][Cl_2]}$$

(d) $$K = \frac{[N_2][H_2]^3}{[NH_3]^2}$$

(e) $$K = \frac{[H_2][Br_2]}{[HBr]^2}$$

19.16. Write an equilibrium constant expression for each of the following:

(a) $$SO_2 + \tfrac{1}{2}O_2 \rightleftharpoons SO_3$$

(b) $$2NO_2 + \tfrac{1}{2}O_2 \rightleftharpoons N_2O_5$$

(c) $$CH_3COCH_2COCH_3 \rightleftharpoons CH_3C(OH)=CHCOCH_3$$

Ans.

(*a*)
$$K = \frac{[SO_3]}{[SO_2][O_2]^{1/2}}$$

(*b*)
$$K = \frac{[N_2O_5]}{[NO_2]^2[O_2]^{1/2}}$$

(*c*)
$$K = \frac{[CH_3C(OH)\!=\!CHCOCH_3]}{[CH_3COCH_2COCH_3]}$$

19.17. Write equilibrium constant expressions for the following equations. Tell how these expressions are related.

(*a*) $2\,NO_2 \rightleftharpoons N_2O_4$

(*b*) $N_2O_4 \rightleftharpoons 2\,NO_2$

(*c*) $NO_2 \rightleftharpoons \frac{1}{2}\,N_2O_4$

Ans.

(*a*)
$$K = \frac{[N_2O_4]}{[NO_2]^2}$$

(*b*)
$$K = \frac{[NO_2]^2}{[N_2O_4]}$$

(*c*)
$$K = \frac{[N_2O_4]^{1/2}}{[NO_2]}$$

The K in part (*b*) is the reciprocal of that in part (*a*). The K in part (*c*) is the square root of that in part (*a*).

19.18. In the reaction

$$X + Y \rightleftharpoons Z$$

1.0 mol of X, 1.3 mol of Y, and 2.2 mol of Z are found at equilibrium in a 1.0-L reaction mixture. (*a*) Calculate the value for K. (*b*) If the same mixture had been found in a 2.0-L reaction mixture, would the value of K have been the same? Explain.

Ans.

(*a*)
$$K = \frac{[Z]}{[X][Y]} = \frac{(2.2)}{(1.0)(1.3)} = 1.7$$

(*b*)
$$K = \frac{(1.1)}{(0.50)(0.65)} = 3.4$$

The value is not the same. The value of K is related to the *concentrations* of the reagents, not to their numbers of moles.

19.19.

$$A + B \rightleftharpoons C + D$$

If 1.00 mol of A and 2.00 mol of B are placed in a 1.00-L vessel and allowed to achieve equilibrium, 0.20 mol of C is found. In order to determine the value of the equilibrium constant, determine (*a*) the quantity of C produced, (*b*) the quantity of D produced, (*c*) the quantities of A and B used up, (*d*) the quantity of A remaining at equilibrium, by considering the initial quantity and the quantity used up, (e) the quantity of B remaining at equilibrium, and (f) the value of the equilibrium constant.

Ans. (a), (b), and (c) 0.20 mol each.

(d)
$$1.00 - 0.20 = 0.80 \text{ mol}$$

(e)
$$2.00 - 0.20 = 1.80 \text{ mol}$$

(f)
$$K = \frac{[C][D]}{[A][B]} = \frac{(0.20)^2}{(0.80)(1.80)} = 2.8 \times 10^{-2}$$

19.20. In which line of the table used to calculate the equilibrium concentrations are the values in the ratio of the coefficients of the balanced chemical equation? Are other lines?

Ans. The terms in the second line are in that ratio, since it describes the changes made by the reaction. The terms in the other lines are not generally in that ratio.

19.21.

$$2A + B \rightleftharpoons C + 2D$$

If 0.75 mol of A and 1.30 mol of B are placed in a 1.00-L vessel and allowed to achieve equilibrium, 0.15 mol of C is found. Use a table such as shown in Example 19.15 to determine the value of the equilibrium constant.

Ans.

	2A	+ B	$\rightleftharpoons$	C	+ 2D
Initial	0.75	1.30		0	0
Change	-0.30	-0.15		$+0.15$	$+0.30$
Equilibrium	0.45	1.15		0.15	0.30

$$K = \frac{[C][D]^2}{[A]^2[B]} = \frac{(0.15)(0.30)^2}{(0.45)^2(1.15)} = 5.8 \times 10^{-2}$$

19.22. For the following reaction, $K = 1.0 \times 10^{-5}$.

$$W + Q \rightleftharpoons R + Z$$

If 1.0 mol of W and 3.0 mol of Q are placed in a 1.0-L vessel and allowed to come to equilibrium, calculate the equilibrium concentration of Z using the following steps: (a) If the equilibrium concentration of Z is equal to x, how much Z was produced by the chemical reaction? (b) How much R was produced by the chemical reaction? (c) How much W and Q were used up by the reaction? (d) How much W is left at equilibrium? (e) How much Q is left at equilibrium? (f) With the value of the equilibrium constant given, will x (equal to the Z concentration at equilibrium) be significant when subtracted from 1.0? (g) Approximately what concentrations of W and Q will be present at equilibrium? (h) What is the value of x? (i) What is the concentration of R at equilibrium? (j) Is the answer to part (f) justified?

Ans. (a) x. (There was none present originally, so all that is there must have come from the reaction.) (b) x. (The Z and R are produced in equal mole quantities, and they are in the same volume.) (c) x each. (The reacting ratio of W to Q to R to Z is 1:1:1:1.) (d) $1.0 - x$. (The original concentration minus the concentration used up.) (e) $3.0 - x$. (The original concentration minus the concentration used up.) (f) No. (We will try it first and see that this is true.) (g) 1.0 mol/L and 3.0 mol/L, respectively.

(h)
$$K = \frac{[R][Z]}{[W][Q]} = \frac{x^2}{(1.0)(3.0)} = 1.0 \times 10^{-5}$$
$$x^2 = 3.0 \times 10^{-5}$$
$$x = 5.5 \times 10^{-3}$$

(i)
$$[R] = x = 5.5 \times 10^{-3} \text{ mol/L}$$

(j) Yes. $1.0 - (5.5 \times 10^{-3}) \cong 1.0$.

19.23. For the following reaction of gases, $K = 1.0 \times 10^{-9}$:

$$W + 2Q \rightleftharpoons 3R + Z$$

Determine the concentration of Z at equilibrium by repeating each of the steps in Problem 19.22. Calculate the equilibrium concentration of R if 1.0 mol of W and 3.0 mol of Q are placed in a 1.0-L vessel and allowed to attain equilibrium.

Ans. (a) x; (b) $3x$; (c) W: x, Q: $2x$; (d) $1.0 - x$; (e) $3.0 - 2x$; (f) no; (g) 1.0 mol/L and 3.0 mol/L, respectively.

(h)
$$K = \frac{[R]^3[Z]}{[W][Q]^2} = \frac{(3x)^3(x)}{(1.0)(3.0)^2} = 3.0x^4 = 1.0 \times 10^{-9}$$

$$x = 4.3 \times 10^{-3} = [Z]$$

(i)
$$[R] = 3x = 1.3 \times 10^{-2} \text{ mol/L}$$

(j) Yes.

Supplementary Problems

19.24. Identify or explain each of the following terms: (a) equilibrium, (b) rate of reaction, (c) catalyst, (d) completion, (e) Le Châtelier's principle, (f) stress, (g) shift, (h) shift to the right or left, (i) equilibrium constant, and (j) equilibrium constant expression.

Ans. (a) Equilibrium is a state in which two exactly opposite processes occur at equal rates. No apparent change takes place at equilibrium. (b) Rate of reaction is the number of moles per liter of reactant that reacts per unit time. (c) A catalyst is a substance that alters the rate of a chemical reaction without permanent change in its own composition. (d) A reaction goes to *completion* when one or more of the reactants is entirely used up. In contrast, a reaction at equilibrium has some of each reactant formed again from products, and no reactant or product concentration goes to 0. (e) Le Châtelier's principle states that when a stress is applied to a system at equilibrium, the equilibrium tends to shift to reduce the stress. (f) Stress is a change of conditions imposed on a system at equilibrium. (g) Shift is a change in the concentrations of reactants and products as a result of a stress. (h) Shift to the right is production of more products with the using up of reactants; shift to the left is exactly the opposite. (i) An equilibrium constant is a set value of the ratio of concentrations of products to concentrations of reactants, each raised to the appropriate power and multiplied together. (j) An equilibrium constant expression is the equation that relates the equilibrium constant to the concentration ratio.

19.25. Compare and contrast the function of a catalyst in a chemical reaction with that of a marriage broker—a matchmaker.

19.26. Nitrogen and sulfur oxides undergo the following rapid reactions in oxygen. Determine the overall reaction. What is the role of NO?

$$2NO + O_2 \longrightarrow 2NO_2$$
$$NO_2 + SO_2 \longrightarrow NO + SO_3$$

Ans. Combining the first reaction with twice the second yields the following reaction. That reaction will proceed only slowly in the absence of the NO and NO_2.

$$2SO_2 + O_2 \longrightarrow 2SO_3$$

The NO is a catalyst. It speeds up the conversion of SO_2 to SO_3 and is not changed permanently in the process.

19.27. What is the best set of temperature and pressure conditions for the Haber process—the industrial process to convert hydrogen and nitrogen to ammonia?

Ans. To get the equilibrium to shift as much as possible toward ammonia, you should use a high pressure and a low temperature. However, the lower the temperature, the slower the reaction. Therefore, a high pressure, 200 atm, and an intermediate temperature, 500 °C, are actually used in the industrial process. Naturally, a catalyst to speed up the reaction is also used.

19.28. What happens to the position of the equilibrium in each of the following cases?

$$N_2 + 3H_2 \rightleftharpoons 2NH_3 + \text{heat}$$

(a) Nitrogen is added and the volume is reduced. (b) Heat is added and ammonia is added. (c) Ammonia is taken out and a catalyst is added. (d) Nitrogen and ammonia are both added.

Ans.   (a) Each stress—addition of nitrogen and increase in pressure (by reduction of the volume)—causes a shift to the right, and so the equilibrium will shift to the right when both stresses are applied. (b) Addition of ammonia and addition of heat will each cause the equilibrium to shift to the left; addition of both will also cause shift to the left. (c) Addition of a catalyst will cause no shift, but removal of ammonia will cause a shift to the right, and so these stresses together will cause a shift to the right. (d) You cannot tell, because addition of the nitrogen would cause a shift to the right and addition of ammonia would cause a shift to the left. Since no data are given and Le Châtelier's principle is only qualitative, no conclusion is possible.

19.29. Consider a change in total volume of the equilibria described. In each case, calculate K for a 1.0-L total volume and again for a 2.0-L total volume, and explain in each case why there is or is not a difference.
(a) In the reaction

$$B \rightleftharpoons C + D$$

1.3 mol of B, 2.4 mol of C, and 2.4 mol of D are found in a 1.0-L reaction mixture at equilibrium. Calculate the value of K. If the same mixture were found in a 2.0-L reaction mixture at equilibrium, would K be different?
(b) In the reaction

$$A + B \rightleftharpoons C + D$$

1.0 mol of A, 1.3 mol of B, 2.4 mol of C, and 2.4 mol of D are found in a 1.0-L reaction mixture at equilibrium. Calculate the value of K. If the same mixture were found in a 2.0-L reaction mixture at equilibrium, would K be different? Compare this answer and that in part (a), and explain.

Ans.   (a) In the two reactions we have

$$K_1 = \frac{(2.4)(2.4)}{(1.3)} = 4.4 \qquad K_2 = \frac{(1.2)(1.2)}{(0.65)} = 2.2$$

and the values are different. (b) In the two reactions we have

$$K_1 = \frac{(2.4)(2.4)}{(1.00)(1.3)} = 4.4 \qquad K_2 = \frac{(1.2)(1.2)}{(0.50)(0.65)} = 4.4$$

The values are the same. In part (b), 2.0 L appears twice in the numerator and twice in the denominator, so that it cancels out. In part (a), it appears only once in the denominator and twice in the numerator, and so the constant in the 2.0-L volume would be half as large as that in the 1.0-L volume. Note that in part (a), the same mixture would *not* appear in the 2.0-L mixture, because the value of K does not change with volume.

19.30. Calculate the units of K in Problem 19.17(a) and (c). Can you tell from the units which of the equations is referred to?

Ans.   The units for part (a) are L/mol; those for part (c) are $(L/mol)^{1/2}$. If a value is given in L/mol for the reaction, the equation in part (a) is the one to be used. If the units are the square root of those, the equation of part (c) is to be used.

19.31. What is the sequence of putting values in the table in the solution of Problem 19.21?

Ans.   The superscript numbers before the table entries give the sequence.

	2A	+	B	$\rightleftharpoons$	C	+	2D
Initial	[1] 0.75		[1] 1.30		[1] 0		[1] 0
Change	[3] −0.30		[3] −0.15		[2] +0.15		[3] +0.30
Equilibrium	[4] 0.45		[4] 1.15		[1] 0.15		[4] 0.30

19.32. For the reaction

$$A + B \rightleftharpoons C + \text{heat} \qquad K = 1.0 \times 10^{-4}$$

would an increase in temperature raise or lower the value of K?

Ans.   According to Le Châtelier's principle, raising the temperature would shift this equilibrium to the left. That means that there would be less C and more A and B present at the new equilibrium temperature. The value of the equilibrium constant at that temperature would therefore be lower than the one at the original temperature.

$$K = \frac{[C]}{[A][B]}$$

19.33. Calculate the value of the equilibrium constant for the reaction below if there are present at equilibrium 7.0 mol of N_2, 9.0 mol of O_2, and 0.130 mol of NO_2 in 2.0 L.

$$N_2 + 2O_2 \rightleftharpoons 2NO_2$$

Ans.

$$[N_2] = 7.0\,\text{mol}/2.0\,\text{L} = 3.5\,\text{mol}/\text{L}$$

$$[O_2] = 9.0\,\text{mol}/2.0\,\text{L} = 4.5\,\text{mol}/\text{L}$$

$$[NO_2] = 0.130\,\text{mol}/2.0\,\text{L} = 0.065\,\text{mol}/\text{L}$$

$$K = \frac{[NO_2]^2}{[N_2][O_2]^2} = \frac{(0.065)^2}{(3.5)(4.5)^2} = 6.0 \times 10^{-5}$$

19.34. For the reaction

$$2NO_2 \rightleftharpoons N_2O_4$$

calculate the value of the equilibrium constant if initially 1.00 mol of NO_2 and 1.00 mol of N_2O_4 are placed in a 1.00-L vessel and at equilibrium 0.75 mol of N_2O_4 is left in the vessel.

Ans.

	$2NO_2 \rightleftharpoons$	N_2O_4
Initial	1.00	1.00
Change	0.50	−0.25
Equilibrium	1.50	0.75

$$K = \frac{[N_2O_4]}{[NO_2]^2} = \frac{(0.75)}{(1.50)^2} = 0.33$$

19.35. Calculate the number of moles of Cl_2 produced at equilibrium in a 10.0-L vessel when 1.00 mol of PCl_5 is heated to 250 °C. $K = 0.041\,\text{mol}/\text{L}$.

Ans.

$$PCl_5 \rightleftharpoons PCl_3 + Cl_2$$

$$K = 0.041 = \frac{[PCl_3][Cl_2]}{[PCl_5]} = \frac{x^2}{0.100 - x}$$

$$x^2 = 0.0041 - 0.041x$$

$$x^2 + 0.041x - 0.0041 = 0$$

$$x = \frac{-0.041 + \sqrt{(0.041)^2 + 4(0.0041)}}{2} = 0.047$$

$$(0.047\,\text{mol}/\text{L})(10.0\,\text{L}) = 0.47\,\text{mol}$$

19.36. In a 10.0-L vessel at 448 °C, H_2 and $I_2(g)$ react to form HI. $K = 50$. If 0.500 mol of each reactant is used, how many moles of I_2 remain at equilibrium?

Ans. The initial concentration of each reagent is 0.500 mol/10.0 L = 0.0500 mol/L. Let the equilibrium concentration of HI be $2x$ for convenience.

$$H_2 \quad + \quad I_2 \quad \rightleftharpoons 2\,HI$$

Initial	0.0500	0.0500	0
Change	$-x$	$-x$	$2x$
Equilibrium	$0.0500 - x$	$0.0500 - x$	$2x$

$$K = \frac{[HI]^2}{[H_2][I_2]} = \frac{(2x)^2}{(0.0500 - x)^2} = 50$$

Taking the square root of each side of the last equation yields

$$\frac{2x}{0.0500 - x} = 7.1$$

$$2x = 7.1(0.0500) - 7.1x$$

$$9.1x = 7.1(0.0500)$$

$$x = 0.039$$

$$[I_2] = 0.0500 - 0.039 = 0.011 \, \text{mol/L}$$

In 10.0 L, there is then 0.11 mol I_2.

19.37. Calculate the equilibrium CO_2 concentration if 1.00 mol of CO and 1.00 mol of H_2O are placed in a 1.00-L vessel and allowed to come to equilibrium at 975 °C.

$$CO + H_2O(g) \rightleftharpoons CO_2 + H_2 \qquad K = 0.63$$

Ans.

$$K = \frac{[CO_2][H_2]}{[CO][H_2O]} = 0.63$$

$$K = \frac{x^2}{(1.00 - x)^2} = 0.63$$

Taking the square root of this expression yields:

$$\frac{x}{1.00 - x} = 0.79$$

$$x = 0.79 - 0.79x$$

$$1.79x = 0.79$$

$$x = 0.44 \, \text{mol/L}$$

19.38. What effect would a decrease in volume have on the following system at equilibrium at 500 °C?

$$H_2 + I_2 \rightleftharpoons 2\,HI$$

Ans. The equilibrium would not shift. The decrease in volume would increase the pressure of each of the gases, but at 500 °C, all of these reagents are gases. The total number of moles of gas would not be affected by either shift, so no shift takes place.

19.39. What is the difference between the concentration ratio in the equilibrium constant expression and that in the Nernst equation (Chap. 14)?

Ans. In a reaction at equilibrium, the ratio can have only one value at any given temperature. In the Nernst equation, the value can change, since the reaction can be stopped short of equilibrium simply by disconnecting a wire or the salt bridge.

19.40. From the following data, calculate the order of the reaction

$$A + B \longrightarrow \text{products}$$

Run	Initial Concentration of A (M)	Initial Concentration of B (M)	Initial Rate (M/s)
1	1.00	2.00	4.00×10^{-5}
2	1.00	3.00	6.00×10^{-5}
3	3.00	2.00	1.20×10^{-4}
4	9.00	2.00	3.60×10^{-4}

Ans. The reaction is first order with respect to each reactant. Since the 1.50 factor increase in [B] causes a 1.50 factor increase in rate, the reaction is first order with respect to B. Since the 3.00-fold increase in A concentration causes a 3.00-fold increase in rate, it is also first order with respect to A.

19.41. Substitute the values from runs 2 and 3 in Example 19.6 and compare the results for k with the result from run 1.

Ans.

$$k = \frac{2.2 \times 10^{-4}\,M/s}{(0.80\,M)(1.5\,M)} = \frac{4.4 \times 10^{-4}\,M/s}{(0.80\,M)(3.0\,M)} = \frac{1.8 \times 10^{-4}}{M \cdot s}$$

The answers are the same each time.

Chapter 20

Acid-Base Theory

20.1 INTRODUCTION

Thus far, we have used the *Arrhenius theory* of acids and bases (Secs. 6.4 and 7.3) in which acids are defined as hydrogen-containing compounds that react with bases. Bases are compounds containing OH^- ions or that form OH^- ions when they react with water. Bases react with acids to form salts and water. Metallic hydroxides and ammonia are the most familiar bases to us.

The *Brønsted theory* expands the definition of acids and bases to allow us to explain much more of solution chemistry. For example, the Brønsted theory allows us to explain why a solution of ammonium chloride tests acidic and a solution of sodium acetate tests basic. Most of the substances that we consider acids in the Arrhenius theory are also acids in the Brønsted theory, and the same is true of bases. In both theories, *strong acids* are those that react completely with water to form ions. *Weak acids* ionize only slightly. We can now explain this partial ionization as an equilibrium reaction of the ions, the weak acid, and the water. A similar statement can be made about weak bases:

$$HC_2H_3O_2 + H_2O \rightleftharpoons C_2H_3O_2^- + H_3O^+$$

$$NH_3 + H_2O \rightleftharpoons NH_4^+ + OH^-$$

20.2 THE BRØNSTED THEORY

In the Brønsted theory, an *acid* is defined as a substance that donates a proton to another substance. In this sense, a proton is a hydrogen atom that has lost its electron; it has nothing to do with the protons in the nuclei of other atoms. (The nuclei of ^2H are also considered protons; they are also hydrogen ions.) A base is a substance that accepts a proton from another substance. The reaction of an acid and a base produces another acid and base. The following reaction is thus an acid-base reaction according to Brønsted:

$$\underset{\text{acid}}{HC_2H_3O_2} + \underset{\text{base}}{H_2O} \rightleftharpoons \underset{\text{base}}{C_2H_3O_2^-} + \underset{\text{acid}}{H_3O^+}$$

The $HC_2H_3O_2$ is an acid because it donates its proton to the H_2O to form $C_2H_3O_2^-$ and H_3O^+. The H_2O is a base because it accepts that proton. But this is an equilibrium reaction, and the $C_2H_3O_2^-$ reacts with H_3O^+ to form $HC_2H_3O_2$ and H_2O. The $C_2H_3O_2^-$ is a base because it accepts the proton from H_3O^+; the H_3O^+ is an acid because it donates a proton. H_3O^+ is called the *hydronium ion*. It is the combination of a proton and a water molecule, and is the species that we have been abbreviating H^+ thus far in this book. (H^+ does not really exist; it does not have the configuration of a noble gas; see Sec. 5.3.)

The acid on the left of this equation is related to the base on the right; they are said to be *conjugates* of each other. The $HC_2H_3O_2$ is the conjugate acid of the base $C_2H_3O_2^-$. Similarly, the H_2O is the conjugate base of the H_3O^+. *Conjugates differ in each case by H^+.*

EXAMPLE 20.1. Write an equilibrium equation for the reaction of NH_3 and H_2O, and label each of the conjugate acids and bases.

$$\overset{\displaystyle \text{conjugates}}{\overbrace{\underset{\text{base}}{NH_3} + \underset{\text{acid}}{H_2O} \rightleftharpoons \underset{\text{acid}}{NH_4^+} + \underset{\text{base}}{OH^-}}}$$

conjugates

The NH_3 is a base because it accepts a proton from water, which is therefore an acid. The NH_4^+ is an acid because it can donate a proton to OH^-, a base.

We have labeled water as both an acid and a base. It is useful to think of water as either, because ~ally has no more properties of one than the other. Water is sometimes referred to as *amphiprotic*. ~eacts as an acid in the presence of bases and it reacts as a base in the presence of acids. Various acids have different strengths. Some acids are strong; that is, they react with water ~pletely to form their conjugate bases. Other acids are weaker, and they form conjugate bases that ~ronger than the conjugate bases of strong acids. In fact, the stronger the acid, the weaker its ~te base. Some acids are *feeble*; they do not donate protons at all. We can classify the acids and ~njugate bases as follows:

Conjugate Acid	Conjugate Base
Strong	Feeble
Weak	Weak
Feeble	Strong

~ng applies to bases which are molecules and their conjugate acids (which are ions). ~ that weak acids do not have strong conjugate bases, as stated in some texts. For ~ is weak, and its conjugate base, the acetate ion, is certainly not strong. It is even ~n acetic acid is as an acid.

~sify the following acids and bases according to their strength: HCl, $HC_2H_3O_2$, NaOH,

~; $HC_2H_3O_2$ is a weak acid; NaOH is a strong base; NH_3 is a weak base.

~fy the conjugates of the species in the last example according to their strength.

~ $C_2H_3O_2^-$ is a weak base; Na^+ is a feeble acid; NH_4^+ is a weak acid.

~ndicates that Cl^- and Na^+ have no tendency to react with water to form their ~ does react with water to some extent to form $HC_2H_3O_2$ and OH^-. NH_4^+ ~ small extent to form H_3O^+ and NH_3. Consider the following equation:

$$HCl + H_2O \rightleftharpoons Cl^- + H_3O^+$$
$$\text{acid} \qquad \text{base} \qquad \text{base} \qquad \text{acid}$$

~with water 100%, the Cl^- does not react with H_3O^+ at all. If the chloride ion ~rom the hydronium ion, it certainly cannot take the proton from water, which is ~han the hydronium ion is.

~ is the difference between the reaction of $HC_2H_3O_2$ with H_2O and with OH^-?

~es place to a slight extent; $HC_2H_3O_2$ is a weak acid. The second reaction goes almost ~eact almost completely with OH^-.

~olution is determined by the hydronium ion concentration of the solution. The ~ore acidic the solution; the lower $[H_3O^+]$, the more basic the solution. Other ~, OH^-, affect the acidity of a solution by affecting the concentration of H_3O^+. ~ in water in greater concentration than H_3O^+ makes the solution basic. If the ~s are reversed, the solution is acidic.

~ain why $NaC_2H_3O_2$ tests basic in water solution.

~t react with water at all; it is feeble. The $C_2H_3O_2^-$ reacts with water to a slight extent:

$$C_2H_3O_2^- + H_2O \rightleftharpoons HC_2H_3O_2 + OH^-$$

reacts with the H_3O^+ present in water. The excess OH^- makes the solution basic.

20.3 ACID-BASE EQUILIBRIUM

Equilibrium constants can be written for the ionization of weak acids and weak bases, just as for any other equilibria. For the equation

$$HC_2H_3O_2 + H_2O \rightleftharpoons C_2H_3O_2^- + H_3O^+$$

we would ordinarily (Chap. 19) write:

$$K = \frac{[C_2H_3O_2^-][H_3O^+]}{[HC_2H_3O_2][H_2O]}$$

However, in dilute aqueous solution, the concentration of H_2O is practically constant, and its concentration is conventionally built into the value of the equilibrium constant. The new constant, variously called K_a or K_i for acids (K_b or K_i for bases), does not have the water concentration term in the denominator:

$$K_a = \frac{[C_2H_3O_2^-][H_3O^+]}{[HC_2H_3O_2]}$$

EXAMPLE 20.6. Calculate the value of K_a for $HC_2H_3O_2$ in 0.100 M solution if the H_3O^+ concentration of the solution is found to be 1.34×10^{-3} M.

$$HC_2H_3O_2 \quad + H_2O \rightleftharpoons C_2H_3O_2^- + \quad H_3O^+$$

Initial	0.100	0	0
Change			
Equilibrium			1.34×10^{-3}

$$HC_2H_3O_2 \quad + H_2O \rightleftharpoons C_2H_3O_2^- + \quad H_3O^+$$

Initial	0.100	0	0
Change	-1.34×10^{-3}	1.34×10^{-3}	1.34×10^{-3}
Equilibrium			1.34×10^{-3}

$$HC_2H_3O_2 \quad + H_2O \rightleftharpoons C_2H_3O_2^- + \quad H_3O^+$$

Initial	0.100	0	0
Change	-1.34×10^{-3}	1.34×10^{-3}	1.34×10^{-3}
Equilibrium	0.099	1.34×10^{-3}	1.34×10^{-3}

The value of the equilibrium constant is given by

$$K_a = \frac{[C_2H_3O_2^-][H_3O^+]}{[HC_2H_3O_2]} = \frac{(1.34 \times 10^{-3})(1.34 \times 10^{-3})}{0.099} = 1.81 \times 10^{-5}$$

EXAMPLE 20.7. Calculate the hydronium ion concentration of a 0.200 M solution of acetic acid, using the equilibrium constant of Example 20.6.

In this example, we will use x for the unknown concentration of hydronium ions. We will solve in terms of x.

$$HC_2H_3O_2 + H_2O \rightleftharpoons C_2H_3O_2^- + H_3O^+$$

Initial	0.200	0	0
Change	$-x$	x	x
Equilibrium	$0.200 - x$	x	x

$$K_a = \frac{[C_2H_3O_2^-][H_3O^+]}{[HC_2H_3O_2]} = \frac{(x)(x)}{0.200 - x} = 1.81 \times 10^{-5}$$

We can solve this equation most easily by assuming that x is small enough to neglect when *added to* or *subtracted from* a greater concentration. Therefore,

$$K_a = \frac{(x)(x)}{0.200} = 1.81 \times 10^{-5}$$

$$x^2 = 3.62 \times 10^{-6}$$

$$x = 1.90 \times 10^{-3} = [H_3O^+]$$

In cases where x is too large to neglect from the concentration from which it is subtracted, a more exact method is required. The quadratic formula is an exact method to determine the value of x in a equation of the form

$$ax^2 + bx + c = 0$$

The solution has the form

$$x = \frac{-b \pm \sqrt{b^2 - 4ac}}{2a}$$

Two solutions for x (one with the plus sign and one with the minus sign) can be obtained from this formula, but only one will have physical meaning. For example, there can be no negative concentrations. Solving this problem exactly yields

$$K_a = \frac{(x)(x)}{0.200 - x} = 1.81 \times 10^{-5}$$

$$x^2 = 1.81 \times 10^{-5}(0.200 - x) = 3.62 \times 10^{-6} - 1.81 \times 10^{-5}x$$

$$x^2 + 1.81 \times 10^{-5}x - 3.62 \times 10^{-6} = 0$$

$$x = \frac{-1.81 \times 10^{-5} + \sqrt{(1.81 \times 10^{-5})^2 - 4(-3.62 \times 10^{-6})}}{2}$$

$$x = 1.89 \times 10^{-3}$$

In this case, the approximate solution gives almost the same answer as the exact solution. In general, you should use the approximate method and check your answer to see that it is reasonable; only if it is not should you use the quadratic equation.

20.4   AUTOIONIZATION OF WATER

Since water is defined as both an acid and a base (Sec. 20.2), it is not surprising to find that water can react with itself, even though only to a very limited extent, in a reaction called *autoionization*:

$$\underset{\text{acid}}{H_2O} + \underset{\text{base}}{H_2O} \rightleftharpoons \underset{\text{acid}}{H_3O^+} + \underset{\text{base}}{OH^-}$$

An equilibrium constant for this reaction, called K_w, does not have terms for the concentration of water; otherwise it is like the other equilibrium constants considered so far.

$$K_w = [H_3O^+][OH^-]$$

The value for this constant in dilute aqueous solution at 25°C is 1.0×10^{-14}. Thus, water ionizes very little when it is pure, and even less in acidic or basic solution.

The equation for K_w means that there is always some H_3O^+ and always some OH^- in any aqueous solution. Their concentrations are inversely proportional. A solution is acidic if the H_3O^+ concentration exceeds the OH^- concentration; it is neutral if the two concentrations are equal; and it is basic if the OH^- concentration exceeds the H_3O^+ concentration:

$$[H_3O^+] > [OH^-] \quad \text{acidic}$$

$$[H_3O^+] = [OH^-] \quad \text{neutral}$$

$$[H_3O^+] < [OH^-] \quad \text{basic}$$

EXAMPLE 20.8.   Calculate the hydronium ion concentration in pure water at 25°C.

$$2H_2O \rightleftharpoons H_3O^+ + OH^-$$

$$K_w = [H_3O^+][OH^-] = 1.0 \times 10^{-14}$$

Both ions must be equal in concentration since they were produced in equal molar quantities by the autoionization reaction. Let the concentration of each be equal to x.

$$x^2 = 1.0 \times 10^{-14}$$
$$x = 1.0 \times 10^{-7} = [H_3O^+] = [OH^-]$$

EXAMPLE 20.9. Calculate the hydronium ion concentration in 0.10 M NaOH.

$$K_w = [H_3O^+][OH^-] = 1.0 \times 10^{-14}$$
$$[OH^-] = 0.10 \ M$$
$$[H_3O^+](0.10) = 1.0 \times 10^{-14}$$
$$[H_3O^+] = 1.0 \times 10^{-13}$$

The OH^- from the NaOH has caused the autoionization reaction of water to shift to the left, yielding only a tiny fraction of the H_3O^+ that is present in pure water.

20.5   pH

The pH scale was invented to reduce the necessity for using exponential numbers to report acidity. The pH is defined as

$$pH = -\log [H_3O^+]$$

EXAMPLE 20.10. Calculate the pH of (a) 0.10 M H_3O^+ solution and (b) 0.10 M HCl solution.

(a)
$$[H_3O^+] = 0.10 = 1.0 \times 10^{-1}$$
$$pH = -\log [H_3O^+] = -\log (1.0 \times 10^{-1}) = 1.00$$

(b) Since HCl is a strong acid, it reacts completely with water to give 0.10 M H_3O^+ solution. Thus, the pH is the same as in part (a).

EXAMPLE 20.11. Calculate the pH of 0.10 M NaOH and of 0.10 M Ba(OH)$_2$.

In 0.10 M NaOH,

$$[OH^-] = 0.10 = 1.0 \times 10^{-1}$$
$$[H_3O^+] = 1.0 \times 10^{-13} \qquad \text{(see Example 20.9)}$$
$$pH = -\log (1.0 \times 10^{-13}) = +13.00$$

In 0.10 M Ba(OH)$_2$, $[OH^-] = 0.20 \ M$. (There are two OH^- ions per formula unit.)

$$[OH^-] = 0.20 = 2.0 \times 10^{-1}$$
$$[H_3O^+] = (1.0 \times 10^{-14})/(2.0 \times 10^{-1}) = 5.0 \times 10^{-14}$$
$$pH = 13.30$$

Special note for electronic calculator users: To determine the negative of the logarithm of a quantity, enter the quantity, press the $\boxed{\text{LOG}}$ key, then press the change sign key, $\boxed{+/-}$. Do not use the minus key.

EXAMPLE 20.12. What is the difference in calculating the pH of a solution of a weak acid and that of a strong acid?

For a strong acid, the H_3O^+ concentration can be determined directly from the concentration of the acid. For a weak acid, the H_3O^+ concentration must be determined first from an equilibrium constant calculation (Sec. 20.3); then the pH is calculated.

EXAMPLE 20.13. Calculate the pH of (a) 0.010 M HCl and (b) 0.010 M $HC_2H_3O_2$ ($K_a = 1.8 \times 10^{-5}$).

(a)
$$[H_3O^+] = 0.010 = 1.0 \times 10^{-2}$$
$$pH = 2.00$$

(b)
$$HC_2H_3O_2 + H_2O \rightleftharpoons H_3O^+ + C_2H_3O_2^-$$

Initial	0.010	0	0
Change	$-x$	x	x
Equilibrium	$0.010 - x$	x	x

$$\frac{x^2}{0.010} = 1.8 \times 10^{-5}$$
$$x = 4.2 \times 10^{-4}$$
$$pH = 3.37$$

From the way pH is defined and the value of K_w, we can deduce that solutions with pH = 7 are neutral, those with pH less than 7 are acidic, and those with pH greater than 7 are basic.

$$[H_3O^+] > [OH^-] \qquad \text{acidic} \qquad pH < 7$$
$$[H_3O^+] = [OH^-] \qquad \text{neutral} \qquad pH = 7$$
$$[H_3O^+] < [OH^-] \qquad \text{basic} \qquad pH > 7$$

20.6 BUFFER SOLUTIONS

A *buffer solution* is a solution of a weak acid and its conjugate base or a weak base and its conjugate acid. The main property of a buffer solution is its resistance to changes in its pH despite addition of small quantities of strong acid or strong base. The student must know the following three things about buffer solutions:
1. How they are prepared.
2. What they do.
3. How they do it.

A buffer solution may be prepared by addition of a weak acid to a salt of that acid or addition of a weak base to a salt of that base. For example, a solution of acetic acid and sodium acetate is a buffer solution. The weak acid ($HC_2H_3O_2$) plus its conjugate base ($C_2H_3O_2^-$, from the sodium acetate) constitute a buffer solution. There are other ways to make such a combination of weak acid plus conjugate base (see Problem 20.26).

A buffer solution resists change in its acidity. For example, a certain solution of acetic acid and sodium acetate has a pH of 4. When a small quantity of NaOH is added, the pH goes up to 4.2. If that quantity of NaOH had been added to the same volume of an unbuffered solution at pH 4, the pH would have gone up to 12.

The buffer solution works on the basis of Le Châtelier's principle. Consider the equation for the reaction of acetic acid with water:

$$\underset{\text{excess}}{HC_2H_3O_2} + \underset{\substack{\text{huge} \\ \text{excess}}}{H_2O} \rightleftharpoons \underset{\text{excess}}{C_2H_3O_2^-} + \underset{\substack{\text{limited} \\ \text{quantity}}}{H_3O^+}$$

The solution of $HC_2H_3O_2$ and $C_2H_3O_2^-$ in H_2O results in the relative quantities of each of the species in the equation as shown under the equation. If H_3O^+ is added to the equilibrium system, the equilibrium shifts to use up some of the added H_3O^+. If the acetate ion were not present to take up the added H_3O^+, the pH would drop. Since the acetate ion reacts with much of the added H_3O^+, there is little increase in H_3O^+ and little drop in pH. If OH^- is added to the solution, it reacts with the H_3O^+ present. But the removal of that H_3O^+ is a stress, which causes this equilibrium to shift to the right, replacing much of the H_3O^+ removed by the OH^-. The pH does not rise nearly as much in the buffered solution as it would have in an unbuffered solution.

Calculations may be made as to how much the pH is changed by addition of strong acid or base. You should first determine how much of each conjugate would be present if that strong acid or base reacted completely with the weak acid or conjugate base originally present. Use the results as the initial values of concentrations for the equilibrium calculations.

EXAMPLE 20.14.   A solution containing 0.100 mol $HC_2H_3O_2$ and 0.200 mol $NaC_2H_3O_2$ in 1.00 L is treated with 0.010 mol NaOH. Assume that there is no change in the volume of the solution. (*a*) What was the original pH of the solution? (*b*) What is the pH of the solution after the addition of the NaOH? $K_a = 1.8 \times 10^{-5}$.

(*a*)
$$HC_2H_3O_2 + H_2O \rightleftharpoons C_2H_3O_2^- + H_3O^+$$

$$K_a = \frac{[C_2H_3O_2^-][H_3O^+]}{[HC_2H_3O_2]} = 1.8 \times 10^{-5}$$

$$HC_2H_3O_2 + H_2O \rightleftharpoons H_3O^+ + C_2H_3O_2^-$$

Initial	0.100	0	0.200
Change	$-x$	x	x
Equilibrium	$0.100 - x$	x	$0.200 + x$

Assuming x to be negligible when added to or subtracted from larger quantities yields

$$K_a = \frac{(0.200)x}{(0.100)} = 1.8 \times 10^{-5}$$

$$x = 9.0 \times 10^{-6}$$

The assumption was valid.

$$pH = 5.05$$

(*b*) We assume that the 0.010 mol of NaOH reacted with 0.010 mol of $HC_2H_3O_2$ to produce 0.010 mol more of $C_2H_3O_2^-$ (Sec. 8.3). That gives us

$$HC_2H_3O_2 + H_2O \rightleftharpoons H_3O^+ + C_2H_3O_2^-$$

Initial	0.090	0	0.210
Change	$-x$	x	x
Equilibrium	$0.090 - x$	x	$0.210 + x$

$$K_a = \frac{(0.210)x}{(0.090)} = 1.8 \times 10^{-5}$$

$$x = 7.7 \times 10^{-6}$$
$$pH = 5.11$$

The pH has risen from 5.05 to 5.11 by the addition of 0.010 mol NaOH. (That much NaOH would have raised the pH of an unbuffered solution originally at the same pH to a final pH value of 12.00.)

Solved Problems

THE BRØNSTED THEORY

20.1.   Explain the difference between the strength of an acid and its concentration.

> *Ans.*   Strength refers to the extent the acid or base will ionize in water. Concentration is a measure of the quantity of the acid or base in a certain volume of solution.

20.2.   Is NH_4Cl solution in water acidic or basic? Explain.

> *Ans.*   It is acidic. The feeble Cl^- does not react with water. The NH_4^+ reacts somewhat to form H_3O^+, making the solution somewhat acidic.

$$\underset{\text{acid}}{NH_4^+} + \underset{\text{base}}{H_2O} \rightleftharpoons \underset{\text{base}}{NH_3} + \underset{\text{acid}}{H_3O^+}$$

Another way to explain this effect is to say that NH_4^+ is a weak conjugate acid and Cl^- is a feeble conjugate base. The stronger of these makes the solution slightly acidic.

20.3. Draw an electron dot diagram for the hydronium ion, and explain why it is expected to be more stable than the hydrogen ion, H^+.

Ans.

$$H:\overset{\displaystyle ..}{\underset{\displaystyle ..}{\ddot{O}}}:H^+$$
$$H$$

Since each hydrogen atom has two electrons and the oxygen atom has eight electrons, the octet rule is satisfied. In H^+, the atom would have no electrons, which is not the electronic configuration of a noble gas.

20.4. H_3BO_3 is a much weaker acid than is $HC_2H_3O_2$. The anion of which one is the stronger base?

Ans. $H_2BO_3^-$ is a stronger base than $C_2H_3O_2^-$ since it is the conjugate of the weaker acid.

20.5. (*a*) In a 0.200 *M* solution of HA (a weak acid), 1.0% of the HA is ionized. What are the actual concentrations of HA, H_3O^+, and A^- in a solution prepared by adding 0.200 mol HA to sufficient water to make 1.00 L of solution? (*b*) What are the actual concentrations of HA, H_3O^+, and A^- in a solution prepared by adding 0.200 mol A^- (from NaA) plus 0.200 mol H_3O^+ (from HCl) to sufficient water to make 1.00 L of solution?

Ans. (*a*) HA ionizes 1.0% to yield:

$$[HA] = 0.198\ M \qquad (99.0\%\ of\ 0.200\ M)$$
$$[H_3O^+] = 0.0020\ M \qquad (1.0\%\ of\ 0.200\ M)$$
$$[A^-] = 0.0020\ M$$

(*b*) The H_3O^+ reacts with the A^- to give HA and H_2O:

$$H_3O^+ + A^- \rightleftharpoons HA + H_2O$$

This reaction is exactly the reverse of the ionization reaction of HA, so that if the ionization reaction goes 1.0%, this reaction goes 99.0%. Thus, after this reaction there are:

$$[HA] = 0.198\ M \qquad (99.0\%\ of\ 0.200\ M)$$
$$[H_3O^+] = 0.0020\ M \qquad (1.0\%\ of\ 0.200\ M)$$
$$[A^-] = 0.0020\ M$$

20.6. Write the net ionic equation for the reaction of HCl(aq) and $NaC_2H_3O_2$. What relationship is there between this equation and the equation for the ionization of acetic acid?

Ans. HCl(aq) is really a solution of H_3O^+ and Cl^-:

$$HCl + H_2O \longrightarrow H_3O^+ + Cl^-$$

The net ionic equation of interest is then

$$H_3O^+ + C_2H_3O_2^- \longrightarrow HC_2H_3O_2 + H_2O$$

This equation is the reverse of the ionization equation for acetic acid.

20.7. How is an Arrhenius salt defined in the Brønsted system?

Ans. In the Brønsted system, an Arrhenius salt is defined as a combination of two conjugates—the conjugate acid of the original base and the conjugate base of the original acid. For example, the salt NH_4ClO is a combination of the conjugate acid (NH_4^+) of NH_3 and the conjugate base (ClO^-) of HClO.

ACID-BASE EQUILIBRIUM

20.8. The $[H_3O^+]$ in 0.150 M benzoic acid ($HC_7H_5O_2$) is 3.0×10^{-3} M. Calculate the value for K_a.

   Ans.

$$HC_7H_5O_2 + H_2O \rightleftharpoons H_3O^+ + C_7H_5O_2^-$$

$$[H_3O^+] = [C_7H_5O_2^-] = 3.0 \times 10^{-3}$$

$$[HC_7H_5O_2] = 0.150 - (3.0 \times 10^{-3}) = 0.147$$

$$K_a = \frac{[H_3O^+][C_7H_5O_2^-]}{[HC_7H_5O_2]} = \frac{(3.0 \times 10^{-3})^2}{(0.147)} = 6.1 \times 10^{-5}$$

20.9. K_a for formic acid, $HCHO_2$, is 1.80×10^{-4}. Calculate $[H_3O^+]$ in 0.00100 M $HCHO_2$.

   Ans.

$$HCHO_2 + H_2O \rightleftharpoons H_3O^+ + CHO_2^-$$

$$K_a = \frac{[H_3O^+][CHO_2^-]}{[HCHO_2]} = 1.80 \times 10^{-4}$$

	$HCHO_2$	$+ H_2O \rightleftharpoons H_3O^+$	$+ CHO_2^-$
Initial	0.00100	0	0
Change	$-x$	x	x
Equilibrium	$0.00100 - x$	x	x

Neglecting x when subtracted from 0.00100:

$$K_a = \frac{(x)(x)}{(0.00100)} = 1.80 \times 10^{-4}$$

$$x^2 = 1.80 \times 10^{-7}$$

$$x = 4.24 \times 10^{-4}$$

This value of x is 42.4% of the concentration from which it was subtracted. The approximation is wrong. The quadratic formula must be used:

$$K_a = \frac{(x)(x)}{(0.00100 - x)} = 1.80 \times 10^{-4}$$

$$x^2 = -1.80 \times 10^{-4}x + 1.80 \times 10^{-7}$$

$$x^2 + 1.80 \times 10^{-4}x - 1.80 \times 10^{-7} = 0$$

$$x = \frac{-b \pm \sqrt{b^2 - 4ac}}{2a}$$

$$= \frac{-1.80 \times 10^{-4} + \sqrt{(1.80 \times 10^{-4})^2 - 4(-1.80 \times 10^{-7})}}{2}$$

$$x = 3.44 \times 10^{-4} = [H_3O^+]$$

   Check:

$$\frac{(3.44 \times 10^{-4})^2}{[0.00100 - (3.44 \times 10^{-4})]} = 1.8 \times 10^{-4}$$

20.10. What concentration of acetic acid, $HC_2H_3O_2$, is needed to give a hydronium ion concentration of 4.8×10^{-3} M? $K_a = 1.8 \times 10^{-5}$.

Ans.

$$HC_2H_3O_2 + H_2O \rightleftharpoons H_3O^+ + C_2H_3O_2^-$$

$$K_a = \frac{[H_3O^+][C_2H_3O_2^-]}{[HC_2H_3O_2]} = 1.8 \times 10^{-5}$$

$$HC_2H_3O_2 \; + H_2O \rightleftharpoons H_3O^+ + C_2H_3O_2^-$$

	$HC_2H_3O_2$	H_3O^+	$C_2H_3O_2^-$
Initial	x	0	0
Change			
Equilibrium		0.0048	

$$HC_2H_3O_2 \; + H_2O \rightleftharpoons H_3O^+ + C_2H_3O_2^-$$

	$HC_2H_3O_2$	H_3O^+	$C_2H_3O_2^-$
Initial	x	0	0
Change	-0.0048	0.0048	0.0048
Equilibrium	$x - 0.0048 \cong x$	0.0048	0.0048

$$K_a = \frac{(0.0048)^2}{x} = 1.8 \times 10^{-5}$$

$$x = 1.3 \; M$$

20.11. What is the CN^- concentration in 0.020 M HCN? $K_a = 6.2 \times 10^{-10}$.

Ans.

$$HCN + H_2O \rightleftharpoons CN^- + H_3O^+$$

$$K_a = \frac{[CN^-][H_3O^+]}{[HCN]} = 6.2 \times 10^{-10}$$

$$HCN \quad + H_2O \rightleftharpoons CN^- + H_3O^+$$

	HCN	CN^-	H_3O^+
Initial	0.020	0	0
Change	$-x$	x	x
Equilibrium	$0.020 - x$	x	x

$$K_a = \frac{x^2}{0.020} = 6.2 \times 10^{-10}$$

$$x = 3.5 \times 10^{-6} = [CN^-]$$

AUTOIONIZATION OF WATER

20.12. Calculate the hydronium ion concentration in 0.10 M NaCl.

Ans.   Since both Na^+ and Cl^- are feeble, neither reacts with water at all. Thus, the hydronium ion concentration in this solution is the same as that in pure water, 1.0×10^{-7} M.

20.13. According to Le Châtelier's principle, what does the H_3O^+ concentration from the ionization of an acid (strong or weak) do to the ionization of water?

Ans.

$$2H_2O \rightleftharpoons H_3O^+ + OH^-$$

The presence of H_3O^+ from an acid will repress this water ionization. The water will ionize less than it does in neutral solution, and the H_3O^+ generated by the water will be negligible in all but the most dilute acid. The OH^- generated by the water will be equally small, but since it is the only OH^- present, it still has to be considered.

20.14. What does a 0.10 M HCl solution contain?

Ans.   It contains 0.10 M H_3O^+, 0.10 M Cl^-, H_2O, and a slight concentration of OH^- from the ionization of the water.

20.15. What does a 0.10 M $HC_2H_3O_2$ solution contain?

Ans.   It contains nearly 0.10 M $HC_2H_3O_2$, a little H_3O^+ and $C_2H_3O_2^-$, H_2O, and a slight concentration of OH^- from the ionization of the water.

pH

20.16. Write an equilibrium constant expression for each of the following:

(a) $$HC_2H_3O_2 + H_2O \rightleftharpoons H_3O^+ + C_2H_3O_2^-$$

(b) $$2H_2O \rightleftharpoons H_3O^+ + OH^-$$

Ans.

(a) $$K_a = \frac{[H_3O^+][C_2H_3O_2^-]}{[HC_2H_3O_2]}$$

(b) $$K_w = [H_3O^+][OH^-]$$

20.17. Calculate the pH of 0.100 M NH_3. $K_b = 1.8 \times 10^{-5}$.

Ans.

$$NH_3 + H_2O \rightleftharpoons NH_4^+ + OH^-$$

$$K_b = \frac{[NH_4^+][OH^-]}{[NH_3]} = 1.8 \times 10^{-5}$$

	NH_3	$+ H_2O \rightleftharpoons$	NH_4^+	$+ OH^-$
Initial	0.100		0	0
Change	$-x$		x	x
Equilibrium	$0.100 - x$		x	x

$$K_b = \frac{(x)(x)}{(0.100)} = 1.8 \times 10^{-5}$$

$$x^2 = 1.8 \times 10^{-6}$$

$$x = 1.3 \times 10^{-3} = [OH^-]$$

$$[H_3O^+] = \frac{K_w}{[OH^-]} = \frac{1.0 \times 10^{-14}}{1.3 \times 10^{-3}} = 7.7 \times 10^{-12}$$

$$pH = 11.11$$

20.18. Calculate the pH of each of the following solutions: (a) 0.100 M HCl, (b) 0.100 M NaOH, (c) 0.100 M Ba(OH)$_2$, and (d) 0.100 M NaCl.

Ans.   (a) 1.000, (b) 13.000, (c) 13.301, and (d) 7.000.

20.19. Calculate the pH, where applicable, of every solution mentioned in Problems 20.8 through 20.14.

Ans.   Merely take the logarithm of the *hydronium ion concentration* and change the sign: Problem 20.8, 2.52; Problem 20.9, 3.46; Problem 20.10, 2.32; Problem 20.11, 5.46; Problem 20.12, 7.00; and Problem 20.14, 1.00.

BUFFER SOLUTIONS

20.20. Is it true that when one places a weak acid and its conjugate base in the same solution they react with each other? That is, do they both appear on the same side of the equation?

> *Ans.* They do not react with each other in that way. The conjugate acts as a stress as required by Le Châtelier's principle, and does not allow the weak acid to ionize as much as it would if the conjugate were not present.

20.21. One way to get acetic acid and sodium acetate into the same solution is to add them both to water. State another way to get both these reagents in a solution.

> *Ans.* One can add acetic acid and neutralize some of it with sodium hydroxide (Sec. 8.3). The excess acetic acid remains in the solution, and the quantity that reacts with the sodium hydroxide will form sodium acetate. Thus, both acetic acid and sodium acetate will be present in the solution without any sodium acetate having been added directly. See the next problem.

20.22. State another way that a solution can be made containing acetic acid and sodium acetate, besides the way indicated in Problem 20.21.

> *Ans.* One can add sodium acetate and hydrochloric acid. The reverse of the ionization reaction occurs, yielding acetic acid:
>
> $$C_2H_3O_2{}^- + H_3O^+ \rightleftharpoons HC_2H_3O_2 + H_2O$$
>
> If a smaller number of moles of hydrochloric acid is added than moles of sodium acetate present, some of the sodium acetate will be in excess (Sec. 8.3), and will be present in the solution with the acetic acid formed in this reaction.

20.23. Calculate the hydronium ion concentration of a 0.200 M acetic acid solution also containing 0.150 M sodium acetate. $K_a = 1.81 \times 10^{-5}$.

> *Ans.*
>
> $$HC_2H_3O_2 + H_2O \rightleftharpoons H_3O^+ + C_2H_3O_2{}^-$$
>
> | | | | |
> |---|---|---|---|
> | Initial | 0.200 | 0 | 0.150 |
> | Change | $-x$ | x | x |
> | Equilibrium | $0.200 - x$ | x | $0.150 + x$ |
>
> $$K_a = \frac{[C_2H_3O_2{}^-][H_3O^+]}{[HC_2H_3O_2]} = \frac{(x)(0.150 + x)}{0.200 - x} = 1.81 \times 10^{-5}$$
>
> $$K_a \cong \frac{(x)(0.150)}{0.200} \cong 1.81 \times 10^{-5}$$
>
> $$x = 2.41 \times 10^{-5} = [H_3O^+]$$

20.24. According to Le Châtelier's principle, what is the effect of $NaC_2H_3O_2$ on a solution of $HC_2H_3O_2$?

> *Ans.* The added $C_2H_3O_2{}^-$ will cause the ionization of $HC_2H_3O_2$ to be repressed. The acid will not ionize as much as if the acetate were not present.

20.25. A solution contains 0.10 mol of HA (a weak acid) and 0.20 mol of NaA (a salt of the acid). Another solution contains 0.10 mol of HA to which 0.20 mol of NaA is added. The final volume of each solution is 1.0 L. What difference, if any, is there between the two solutions?

> *Ans.* There is no difference between the two solutions. The wording of the problems is different, but the final result is the same.

20.26. What are the concentrations in 1.00 L of solution after 0.100 mol of NaOH is added to 0.200 mol of $HC_2H_3O_2$ but before any equilibrium reaction is considered? Of what use would these be to calculate the pH of this solution?

Ans.   The balanced equation is

$$NaOH + HC_2H_3O_2 \longrightarrow NaC_2H_3O_2 + H_2O$$

The limiting quantity is NaOH, and so 0.100 mol NaOH reacts with 0.100 mol $HC_2H_3O_2$ to produce 0.100 mol $NaC_2H_3O_2$ + 0.100 mol H_2O. There is 0.100 mol excess $HC_2H_3O_2$ left in the solution. So far, this problem is exactly the same as Problem 8.71. Now we can start the equilibrium calculations with these initial quantities. We have a buffer solution problem.

20.27. Calculate the pH of a solution containing 0.100 M NH_3 plus 0.100 M NH_4Cl. $K_b = 1.8 \times 10^{-5}$.

Ans.

$$NH_3 + H_2O \rightleftharpoons NH_4^+ + OH^-$$

$$K_b = \frac{[NH_4^+][OH^-]}{[NH_3]} = 1.8 \times 10^{-5}$$

	NH_3	$+ H_2O \rightleftharpoons$	NH_4^+	$+ OH^-$
Initial	0.100		0.100	0
Change	$-x$		x	x
Equilibrium	$0.100 - x$		$0.100 + x$	x

$$K_b = \frac{(x)(0.100)}{(0.100)} = 1.8 \times 10^{-5}$$

$$x = 1.8 \times 10^{-5} = [OH^-]$$

$$[H_3O^+] = \frac{K_w}{[OH^-]} = \frac{1.0 \times 10^{-14}}{1.8 \times 10^{-5}} = 5.6 \times 10^{-10}$$

$$pH = 9.25$$

Supplementary Problems

20.28. Identify each of the following terms: (*a*) hydronium ion, (*b*) Brønsted theory, (*c*) proton (Brønsted sense), (*d*) acid (Brønsted sense), (*e*) base (Brønsted sense), (*f*) conjugate, (*g*) strong, (*h*) acid dissociation constant, (*i*) ionization constant, (*j*) base dissociation constant, (*k*) autoionization, (*l*) pH, and (*m*) K_w.

Ans.   (*a*) The hydronium ion is H_3O^+, which is a proton added to a water molecule. (*b*) The Brønsted theory defines acids as proton donors and bases as proton acceptors. (*c*) A proton in the Brønsted sense is a hydrogen nucleus—a hydrogen ion. (*d*) An acid is a proton donor. (*e*) A base in the Brønsted sense is a proton acceptor. (*f*) A conjugate is the product of a proton loss or gain from a Brønsted acid or base. (*g*) Of an acid or a base, reacting completely with water to form H_3O^+ or OH^- ions. The common strong acids are HCl, $HClO_3$, $HClO_4$, HBr, HI, HNO_3, and H_2SO_4 (first proton). All soluble hydroxides are strong bases.

20.29. Fluoroacetic acid, $HC_2H_2FO_2$, has a K_a of 2.6×10^{-3}. What concentration of the acid is required to get $[H_3O^+] = 1.5 \times 10^{-3}$?

Ans.

$$HC_2H_2FO_2 + H_2O \rightleftharpoons C_2H_2FO_2^- + H_3O^+$$

$$K_a = \frac{[C_2H_2FO_2^-][H_3O^+]}{[HC_2H_2FO_2]} = 2.6 \times 10^{-3}$$

$$HC_2H_2FO_2 + H_2O \rightleftharpoons C_2H_2FO_2^- + H_3O^+$$

	$HC_2H_2FO_2$	$C_2H_2FO_2^-$	H_3O^+
Initial	x	0	0
Change	-0.0015	0.0015	0.0015
Equilibrium	$x - 0.0015$	0.0015	0.0015

If you try to neglect 0.0015 with respect to x, you get a foolish answer. This problem must be solved exactly.

$$K_a = \frac{(0.0015)^2}{x - 0.0015} = 2.6 \times 10^{-3}$$

$$2.2 \times 10^{-6} = (2.6 \times 10^{-3})x - 3.9 \times 10^{-6}$$

$$x = 2.3 \times 10^{-3} = [HC_2H_2FO_2]$$

20.30. What is the effect of $NaC_2H_3O_2$ on a solution of $HC_2H_3O_2$?

Ans.   The added $C_2H_3O_2^-$ will cause the ionization of $HC_2H_3O_2$ to be repressed. The acid will not ionize as much as if the acetate were not present. This is the same as Problem 20.24, whether or not Le Châtelier's principle is mentioned.

20.31. What chemicals are left in solution after 0.100 mol of NaOH is added to 0.200 mol of NH_4Cl?

Ans.   The balanced equation is

$$NaOH + NH_4Cl \longrightarrow NH_3 + NaCl + H_2O$$

The limiting quantity is NaOH, so that 0.100 mol NaOH reacts with 0.100 mol NH_4Cl to produce 0.100 mol NH_3 + 0.100 mol NaCl + 0.100 mol H_2O. There is also 0.100 mol excess NH_4Cl left in the solution. This is a buffer solution of NH_3 plus NH_4^+ in which the Na^+ and Cl^- are inert.

20.32. For the reaction

$$W + X \rightleftharpoons Y + Z$$

(*a*) Calculate the value of the equilibrium constant if 0.100 mol of W and 55.40 mol of X were placed in a 1.00-L vessel and allowed to come to equilibrium, at which point 0.010 mol of Z was present. (*b*) Using the value of the equilibrium constant calculated in (*a*), determine the concentration of Y at equilibrium if 0.200 mol of W and 55.40 mol of X were placed in a 1.00-L vessel and allowed to come to equilibrium. (*c*) Calculate the ratio of the concentration of W at equilibrium to that initially present in part (*b*). Calculate the same ratio for X. (*d*) Calculate the value of

$$K' = \frac{[Y][Z]}{[W]}$$

from the data of part (*a*). (*e*) Using the equation of part (*d*) and the data of part (*b*), calculate the concentration of Y at equilibrium. (*f*) Explain why the answer to part (*e*) is the same as that to part (*b*).

Ans.

(*a*) $$K = \frac{[Y][Z]}{[W][X]} = \frac{(0.010)^2}{(0.090)(55.39)} = 2.0 \times 10^{-5}$$

(*b*) $$\frac{x^2}{(0.200)(55.40)} = 2.0 \times 10^{-5}$$

$$x^2 = 2.2 \times 10^{-4}$$

$$x = 1.5 \times 10^{-2}$$

(*c*) $$\frac{0.185}{0.200} = 0.925 \qquad \frac{55.38}{55.40} = 0.9996 \qquad \text{(practically no change for X)}$$

(*d*) $$K' = \frac{(0.010)^2}{(0.090)} = 1.1 \times 10^{-3}$$

(*e*) $$\frac{x^2}{0.200} = 1.1 \times 10^{-3}$$

$$x^2 = 2.2 \times 10^{-4}$$

$$x = 1.5 \times 10^{-2}$$

(f) The concentration of X is essentially constant throughout the reaction [see part (c)]. The equilibrium constant of part (a) is amended to include that constant concentration; the new constant is K'. That constant is just as effective in calculating the equilibrium concentration of Y as is the original constant, K. This is the same effect as using K_a, not K and $[H_2O]$, for weak acid and weak base equilibrium calculations.

20.33. Consider the ionization of the general acid HA:

$$HA + H_2O \rightleftharpoons H_3O^+ + A^-$$

Calculate the hydronium ion concentration of a 0.100 M solution of acid for (a) $K_a = 1.0 \times 10^{-9}$ and (b) $K_a = 1.0 \times 10^{-1}$.

Ans.   (a) Let $[H_3O^+] = x$

$$K_a = \frac{[H_3O^+][A^-]}{[HA]} = 1.0 \times 10^{-9}$$

$$\frac{x^2}{(0.10 - x)} = 1.0 \times 10^{-9}$$

$$x^2 = 1.0 \times 10^{-10}$$

$$x = 1.0 \times 10^{-5}$$

(b) Let $[H_3O^+] = x$

$$K_a = \frac{[H_3O^+][A^-]}{[HA]} = 1.0 \times 10^{-1}$$

$$\frac{x^2}{(0.10 - x)} = 1.0 \times 10^{-1}$$

$$x^2 = 1.0 \times 10^{-1}(0.10 - x)$$

Here the x cannot be neglected when subtracted from the 10^{-1}, because the equilibrium constant is too big. The problem is solved using the quadratic formula.

$$x = 6.2 \times 10^{-2}$$

20.34. (a) What ions are present in a solution of sodium acetate? (b) According to Le Châtelier's principle, what effect would the addition of sodium acetate have on a solution of 0.200 M acetic acid? (c) Compare the hydronium ion concentrations of Example 20.7 and Problem 20.23 to check your answer to part (b).

Ans.   (a) Na^+ and $C_2H_3O_2^-$. (b) The acetate ion should repress the ionization of the acetic acid, so that there would be less hydronium ion concentration in the presence of the sodium acetate. (c) The hydronium ion concentration dropped from 1.9×10^{-3} to 2.4×10^{-5} as a result of the addition of the sodium acetate.

Chapter 21

Organic Chemistry

21.1 INTRODUCTION

Historically, the term *organic chemistry* has been associated with the study of compounds obtained from plants and animals. However, more than 150 years ago it was found that typical organic compounds could be prepared in the laboratory without the use of any materials derived directly from living organisms. Indeed, today great quantities of synthetic materials, having properties as desirable as, or more desirable than, those of natural products, are produced commercially. These materials include fibers, perfumes, medicines, paints, pigments, rubber, and building materials.

In modern terms, an organic compound is one that contains at least one carbon-to-carbon and/or carbon-to-hydrogen bond. (Urea and thiourea are compounds considered to be organic which do not fit this description.) In addition to carbon and hydrogen, the elements that may be present in organic compounds are oxygen, nitrogen, phosphorus, sulfur, and the halogens (fluorine, chlorine, bromine, and iodine). With just these few elements, literally millions of organic compounds are known and thousands of new compounds are synthesized every year.

That an entire branch of chemistry can be based on such a relatively small number of elements can be attributed to the fact that carbon atoms have the ability to link together to form long chains, rings, and a variety of combinations of branched chains and fused rings.

In this chapter, some basic concepts of organic chemistry will be described. The objectives of the discussions will be to emphasize the systematic relationships which exist in simple cases. Extension of the concepts presented will be left to more advanced texts.

21.2 BONDING IN ORGANIC COMPOUNDS

The elements that are commonly part of organic compounds are all located in the upper right corner of the periodic table. They are all nonmetals. The bonds between atoms of these elements are essentially covalent. (Some organic molecules may form ions; nevertheless, the bonds *within* each organic ion are covalent. For example, the salt sodium acetate consists of sodium ions, Na^+, and acetate ions, $C_2H_3O_2^-$. Despite the charge, the bonds *within* the acetate ion are all covalent.)

The covalent bonding in organic compounds can be described by means of the electron dot notation (Chap. 5). The carbon atoms has four electrons in its outermost shell:

$$\cdot \dot{\underset{\cdot}{C}} \cdot$$

In order to complete its octet, each carbon atom must share a total of four electron pairs. The order of a bond is the number of electron pairs shared in that bond. The total number of shared pairs is called the *total bond order* of an atom. Thus, carbon must have a total bond order of four (except in CO). A *single bond* is a sharing of one pair; a *double bond*, two; and a *triple bond*, three. Therefore, in organic compounds, each carbon atom forms either four single bonds, a double bond and two single bonds, a triple bond and a single bond, or two double bonds. As shown in the table below, each of these possibilities corresponds to a total bond order of 4.

Number and Types of Bonds	Total Bond Order	
Four single bonds	4×1	$= 4$
One double bond and two single bonds	$(1 \times 2) + (2 \times 1)$	$= 4$
One triple bond and one single bonds	$(1 \times 3) + (1 \times 1)$	$= 4$
Two double bonds	2×2	$= 4$

317

A hydrogen atom has only one electron in its outermost shell, and can accommodate a maximum of two electrons in its outermost shell. Hence, in any molecule, each hydrogen atom can form only one bond—a single bond. The oxygen atom, with six electrons in its outermost shell, can complete its octet by forming either two single bonds or one double bond, for a total bond order of 2 (except in CO). The total bond orders of the other elements usually found in organic compounds can be deduced in a similar manner. The results are given in Table 21-1.

Table 21-1   Total Bond Orders in Organic Compounds

Element	Symbol	Total Bond Order
Carbon	C	4
Hydrogen	H	1
Oxygen	O	2
Nitrogen	N	3
Sulfur	S	2
Halogen	X	1
Phosphorus	P	3

With the information given in Table 21-1, it is possible to write an electron dot structure for an organic molecule.

EXAMPLE 21.1.   Write an electron dot structure for each of the following molecular formulas: (a) CH_4, (b) CH_2O, (c) CH_4O, (d) C_2Cl_2, and (e) CH_5N.

$$(a) \quad H:\overset{\displaystyle H}{\underset{\displaystyle H}{\overset{..}{C}}}:H \qquad (b) \quad H:\overset{\displaystyle H}{\overset{..}{C}}::\overset{..}{\underset{..}{O}}: \qquad (c) \quad H:\overset{\displaystyle H}{\underset{\displaystyle H}{\overset{..}{C}}}:\overset{..}{\underset{..}{O}}:H$$

$$(d) \quad :\overset{..}{\underset{..}{C}l}:C:::C:\overset{..}{\underset{..}{C}l}: \qquad (e) \quad H:\overset{\displaystyle H}{\underset{\displaystyle H}{\overset{..}{C}}}:\overset{\displaystyle H}{\underset{\displaystyle H}{N}}:H$$

Each element in this example is characterized by a total bond order corresponding to that listed in Table 21-1.

21.3   STRUCTURAL AND LINE FORMULAS

Electron dot formulas are useful for deducing the structures of organic molecules, but it is more convenient to use simpler representations—structural or graphic formulas—in which a line is used to denote a shared pair of electrons. Because each pair of electrons shared between two atoms is equivalent to a total bond order of 1, each shared pair can be represented by a line between the symbols of the elements. Unshared electrons on the atoms are usually not shown in this kind of representation. The resulting representations of molecules are called *graphic formulas* or *structural formulas*. The structural formulas for the compounds (a) to (e) described in Example 21.1 may be written as follows:

$$(a) \quad H-\overset{\displaystyle H}{\underset{\displaystyle H}{\overset{|}{\underset{|}{C}}}}-H \qquad (b) \quad H-\overset{\displaystyle H}{\overset{|}{C}}=O \qquad (c) \quad H-\overset{\displaystyle H}{\underset{\displaystyle H}{\overset{|}{\underset{|}{C}}}}-O-H$$

$$(d) \quad Cl-C\equiv C-Cl \qquad (e) \quad H-\overset{\displaystyle H}{\underset{\displaystyle H}{\overset{|}{\underset{|}{C}}}}-\overset{\displaystyle }{\underset{\displaystyle H}{\overset{}{\underset{|}{N}}}}-H$$

For even more convenience in representing the structures of organic compounds, particularly in printed material, *line formulas* are used, so-called because they are printed on one line. In line formulas, each carbon atom is written on a line adjacent to the symbols for the other elements to which it is bonded. Line formulas show the general sequence in which the carbon atoms are attached, but in order to interpret them properly, the permitted total bond orders of all the respective atoms must be kept in mind. Again referring to the compounds (*a*) to (*e*) described above, the line formulas are as follows:

(*a*) CH_4 (*b*) CH_2O (*c*) CH_3OH (*d*) $CClCCl$ or $ClCCCl$ (*e*) CH_3NH_2

If the permitted total bond orders of the respective atoms are remembered, it is apparent that the line formula CH_4 cannot represent such structures as

$$H—C—H—H \qquad \text{(incorrect)}$$
$$\overset{|}{H}$$

which has a total bond order of 2 for one of the hydrogen atoms and a total bond order of only 3 for the carbon atom. Similarly, the formula CH_2O cannot represent the structure $H—C—O—H$ or $H—C{=}O—H$, because in either of these cases, the total bond order of the carbon atom is below 4 and in $H—C{=}O—H$ the total bond order of the oxygen is above 2. Accordingly, line formulas must be interpreted in terms of the permitted total bond orders.

EXAMPLE 21.2.   Write structural formulas for the molecules represented by the following formulas: (*a*) C_2H_4 and (*b*) CH_3COCH_3.

(*a*) Since the hydrogen atoms can have only a total bond order of 1, the two carbon atoms must be linked together. In order for each carbon atom to have a total bond order of 4, the two carbon atoms must be linked to each other by a double bond and also be bonded to two hydrogen atoms each.

$$\overset{\displaystyle H \quad H}{\underset{}{\overset{|\qquad|}{H—C{=}C—H}}}$$

(*b*) The line formula CH_3COCH_3 implies that two of the three carbon atoms each have three hydrogen atoms attached. This permits them to form one additional single bond, to the middle carbon atom. The middle carbon atom, with two single bonds to carbon atoms, must complete its total bond order of 4 with a double bond to the oxygen atom.

$$\overset{\displaystyle H \quad O \quad H}{\overset{|\quad\;\|\quad\;|}{H—C—C—C—H}}$$
$$\underset{\displaystyle H \qquad\quad H}{\underset{|\qquad\quad|}{}}$$

21.4 HYDROCARBONS

Compounds consisting of only carbon and hydrogen have the simplest compositions of all organic compounds. These compounds are called *hydrocarbons*. It is possible to classify the hydrocarbons into four series, based on the characteristic structures of the molecules in each series. These series are known as (1) the alkane series, (2) the alkene series, (3) the alkyne series, and (4) the aromatic series. There are many subdivisions of each series, and it is also possible to have molecules that could be classified as belonging to more than one series.

The *alkane series* is also called the *saturated hydrocarbon series* because the molecules of this class have carbon atoms connected by single bonds only, and therefore have the maximum number of hydrogen atoms possible for the number of carbon atoms. These substances may be represented by the general formula C_nH_{2n+2} and molecules of successive members of the series differ from each other by only a CH_2 unit. The line formulas and names of the first 10 members of the series, given in Table 21-2, should be memorized because these names form the basis for naming many other organic compounds. It should be noted that the first parts of the names of the later members listed are the

Table 21-2   The Simplest Alkanes

Number of C Atoms	Molecular Formula	Line Formula	Name
1	CH_4	CH_4	Methane
2	C_2H_6	CH_3CH_3	Ethane
3	C_3H_8	$CH_3CH_2CH_3$	Propane
4	C_4H_{10}	$CH_3CH_2CH_2CH_3$	Butane
5	C_5H_{12}	$CH_3CH_2CH_2CH_2CH_3$	Pentane
6	C_6H_{14}	$CH_3CH_2CH_2CH_2CH_2CH_3$	Hexane
7	C_7H_{16}	$CH_3CH_2CH_2CH_2CH_2CH_2CH_3$	Heptane
8	C_8H_{18}	$CH_3CH_2CH_2CH_2CH_2CH_2CH_2CH_3$	Octane
9	C_9H_{20}	$CH_3CH_2CH_2CH_2CH_2CH_2CH_2CH_2CH_3$	Nonane
10	$C_{10}H_{22}$	$CH_3CH_2CH_2CH_2CH_2CH_2CH_2CH_2CH_2CH_3$	Decane

Greek or Latin prefixes that denote the number of carbon atoms. Also note that the characteristic ending of each name is -ane. Names of other organic compounds are derived from these names by dropping the -ane ending and adding other endings. At room temperature, the first four members of this series are gases; the remainder of those listed in Table 21-2 are liquids. Higher members of the series, having more than 13 carbon atoms, are solids at room temperature.

Compounds in the alkane series are chemically rather inert. Aside from burning in air or oxygen to produce carbon dioxide and water (or carbon monoxide and water), the most characteristic reaction they undergo is reaction with halogen molecules. The latter reaction is initiated by light. Using pentane as a typical alkane hydrocarbon and chlorine as a typical halogen, the above reactions are represented by the following equations:

$$C_5H_{12} + 8O_2 \longrightarrow 5CO_2 + 6H_2O$$

$$C_5H_{12} + Cl_2 \xrightarrow{\text{light}} C_5H_{11}Cl + HCl$$

Because of their limited reactivity, the saturated hydrocarbons are also called the *paraffins*. This term is derived from the Latin words meaning "little affinity."

The _alkene series_ of hydrocarbons is characterized by having one double bond in the carbon chain. The characteristic formula for members of the series is C_nH_{2n}. Since there must be at least two carbon atoms present to have a carbon-to-carbon double bond, the first member of this series is ethene, C_2H_4, also known as ethylene. Propene (propylene), C_3H_6, and butene (butylene), C_4H_8, are the next two members of the series. Note that the systematic names of these compounds denote the number of carbon atoms in the chain with the name derived from that of the alkane having the same number of carbon atoms (Table 21-2). Note also that the characteristic ending -ene is part of the name of each of these compounds.

Owing to the presence of the double bond, the alkenes are said to be *unsaturated* and are more reactive than the alkanes. The structure of propene, for example, is $CH_2{=}CHCH_3$. The alkenes may react with hydrogen gas in the presence of a catalyst to produce the corresponding alkane; they may react with halogens or with hydrogen halides at relatively low temperatures to form compounds containing only single bonds. These possibilities are illustrated in the following equations in which ethylene is used as a typical alkene:

$$CH_2{=}CH_2 + H_2 \xrightarrow{\text{catalyst}} CH_3CH_3$$

$$CH_2{=}CH_2 + Br_2 \longrightarrow CH_2BrCH_2Br$$

$$CH_2{=}CH_2 + HCl \longrightarrow CH_3CH_2Cl$$

The _alkyne series_ of hydrocarbons is characterized by having molecules with one triple bond each. They have the general formula C_nH_{2n-2}. Like other unsaturated hydrocarbons, the alkynes are quite

ORGANIC CHEMISTRY

reactive. Ethyne is commonly known as *acetylene*. It is the most important member of the series commercially, being widely used as a fuel in acetylene torches and also as a raw material in the manufacture of synthetic rubber and other industrial chemicals.

The *aromatic hydrocarbons* are characterized by molecules containing six-membered rings of carbon atoms with each carbon atom attached to a maximum of one hydrogen atom. The simplest member of the series is benzene, C_6H_6. Using the total bond order rules discussed above, the structural formula of benzene can be written as follows:

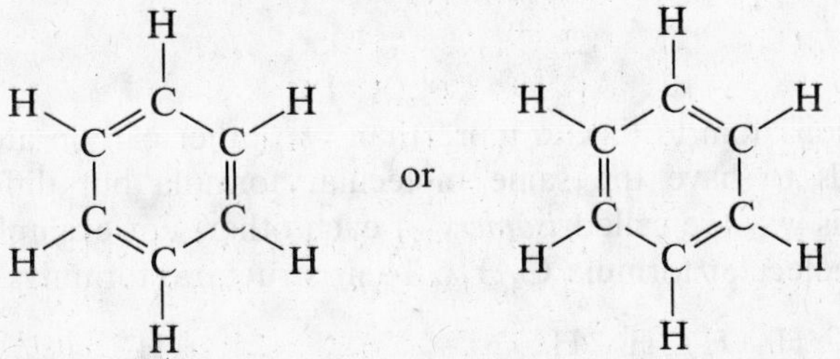

Such a molecule, containing alternating single and double bonds, would be expected to be quite reactive. Actually, benzene is quite unreactive, and its chemical properties resemble those of the alkanes much more than those of the unsaturated hydrocarbons. For example, the characteristic reaction of benzene with halogens resembles that of the reaction of the alkanes:

$$C_6H_6 + Br_2 \xrightarrow{Fe} C_6H_5Br + HBr$$

Unlike the alkanes, however, the reaction of benzene with the halogens is catalyzed by iron. The relative lack of reactivity in aromatic hydrocarbons is attributed to *delocalized double bonds*. That is, the second pair of electrons in each of the three possible carbon-to-carbon double bonds is shared by all six carbon atoms rather than by any two specific carbon atoms. Two ways of writing structural formulas which indicate this type of bonding in the benzene molecules are as follows:

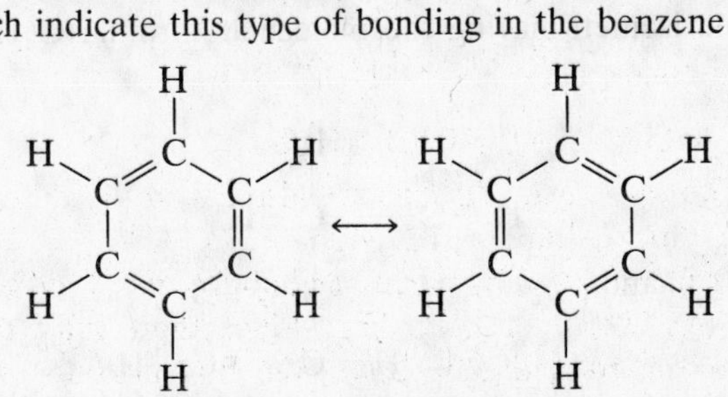

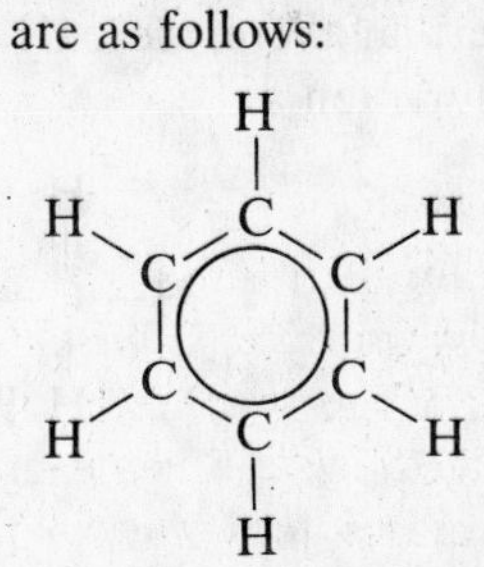

Because of its great stability, the six-membered ring of carbon atoms persists in most reactions. For simplicity, the ring is sometimes represented as a hexagon, each corner of which is assumed to be occupied by a carbon atom with a hydrogen atom attached (unless some other atom is explicitly indicated at that point). The delocalized electrons are indicated by a circle within the hexagon. The following representations illustrate these rules:

benzene, C_6H_6 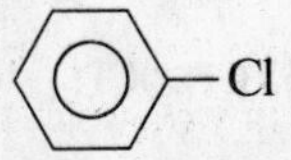   chlorobenzene, C_6H_5Cl

Other aromatic hydrocarbons include naphthalene, $C_{10}H_8$, and anthracene, $C_{14}H_{10}$, whose structures are shown as follows:

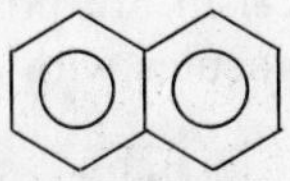

naphthalene 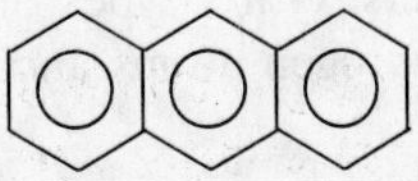   anthracene

Derivatives of aromatic hydrocarbons may have chains of carbon atoms substituted on the aromatic ring, as for example toluene, $C_6H_5CH_3$, and styrene, $C_6H_5CH{=}CH_2$:

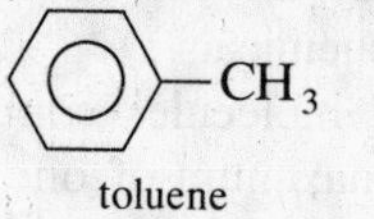

toluene
(methyl benzene)

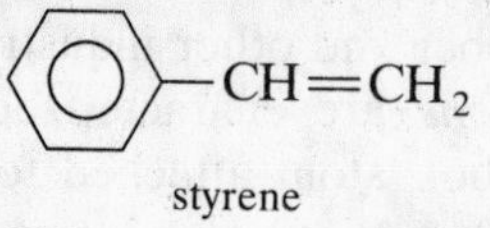

styrene

21.5 ISOMERISM

The ability of a carbon atom to link to more than two other carbon atoms makes it possible for two or more compounds to have the same molecular formula but different structures. Sets of compounds related in this way are called *isomers* of each other. For example, there are two different compounds having the molecular formula C_4H_{10}. Their structural formulas are as follows:

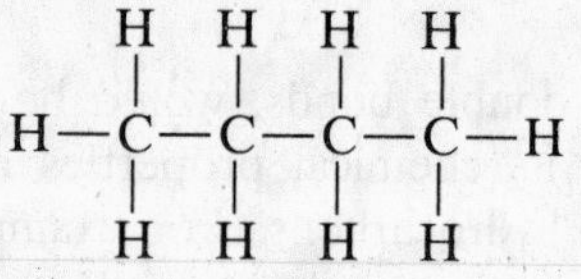

butane
(normal boiling point 1°C)

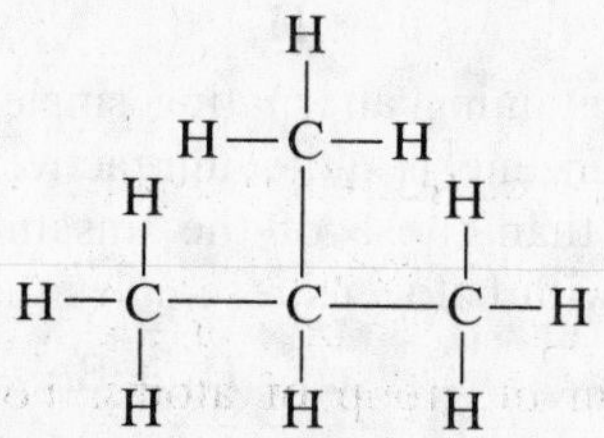

methylpropane
(normal boiling point -10°C)

These are two distinctly different compounds, with different chemical and physical properties; for example, they have different boiling points.

Note that there are only two possible isomers having the molecular formula C_4H_{10}. Structural formulas written as

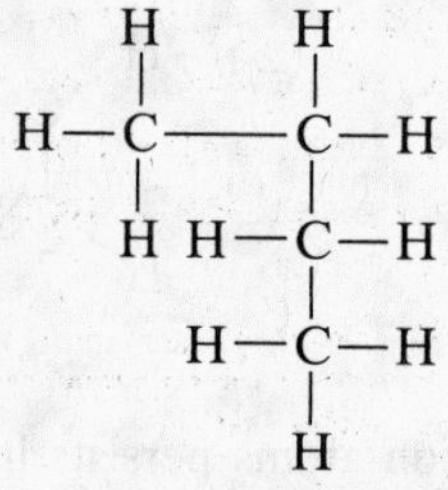

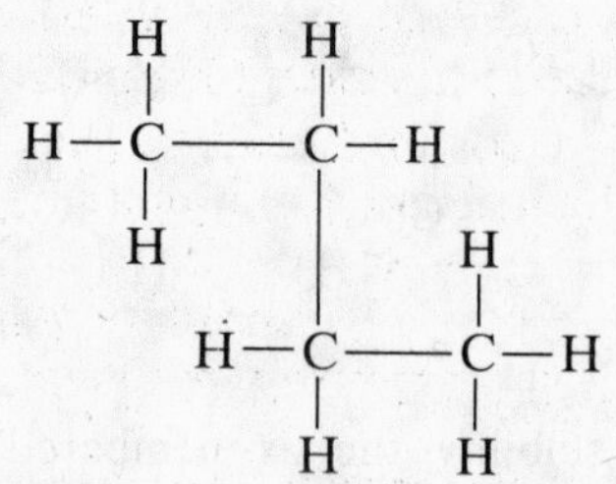

are the same as butane. They have a continuous chain of four carbon atoms and are completely saturated with hydrogen atoms; they have no carbon branches.

EXAMPLE 21.3.   Write the line formulas for butane and for methylpropane, both C_4H_{10}.

$CH_3CH_2CH_2CH_3$ $CH_3CH(CH_3)CH_3$ or $CH_3CH(CH_3)_2$ or $CH(CH_3)_3$ or $(CH_3)_3CH$

butane   methylpropane

Similarly, three isomers of pentane, C_5H_{12}, exist. As the number of carbon atoms in a molecule increases, the number of possible isomers increases markedly. Theoretically, for the formula $C_{20}H_{42}$ there are 366 319 possible isomers. Other types of isomerism are possible in molecules that contain atoms other than carbon and hydrogen atoms and in molecules with double bonds or triple bonds in the chain of carbon atoms.

Compounds called *cycloalkanes*, having molecules with no double bonds but having a cyclic or ring structure, are isomeric with alkenes whose molecules contain the same number of carbon atoms. For example, cyclopentane and 2-pentene have the same molecular formula, C_5H_{10}, but have completely different structures:

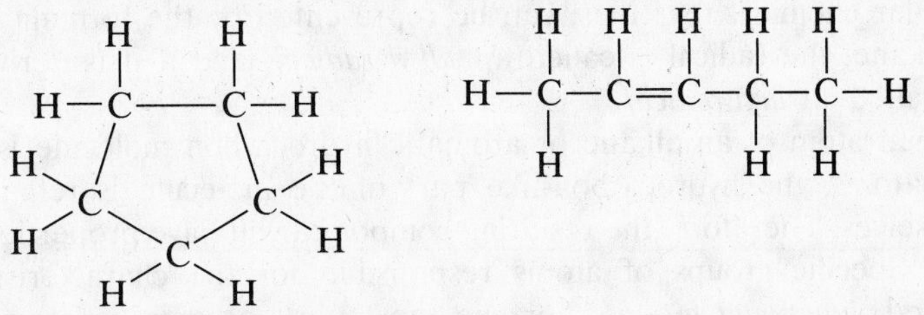

Cyclopentane has the low chemical reactivity which is typical of saturated hydrocarbons, while 2-pentene is much more reactive. Similarly, ring structures containing double bonds, called cyclo-alkenes, can be shown to be isomeric with alkynes.

Literally thousands of isomers can exist. Even relatively simple molecules can have many isomers. Thus, the phenomenon of isomerism accounts in part for the enormous number and variety of compounds of carbon.

21.6 RADICALS AND FUNCTIONAL GROUPS

The millions of organic compounds other than hydrocarbons can be regarded as *derivatives* of hydrocarbons, where one (or more) of the hydrogen atoms on the parent molecule is replaced by another kind of atom or group of atoms. For example, if one hydrogen atom in a molecule of methane, CH_4, is replaced by an —OH group, the resulting compound is methyl alcohol, also called methanol, CH_3OH. (This replacement is often not easy to perform in the laboratory, and is meant in the context used here as a mental exercise.) In many cases, the compound is named in a manner that designates the hydrocarbon parent from which it was derived Thus, the word methyl is derived from the word methane. The hydrocarbon part of the molecule is often called the *radical*, and is denoted R in formulas. The ending of the parent hydrocarbon name is changed from *-ane* to *-yl*. The names of some common radicals are listed in Table 21-3. It should be noted that the radical derived from benzene, C_6H_5—, is called the phenyl radical. Some other radicals are also given names that are not derived from the names of the parent hydrocarbons, but these other cases will not be discussed here. Since, in many reactions, the hydrocarbon part of the organic compound is not changed and does not affect the nature of the reaction, it is useful to generalize many reactions by using the symbol R— to

Table 21-3   Some Common Radicals

Parent Hydrocarbon		Radical		
Name	Line Formula	Name	Line Formula	Structural Formula
Methane	CH_4	Methyl	CH_3—	
Ethane	CH_3CH_3	Ethyl	CH_3CH_2—	
Propane	$CH_3CH_2CH_3$	Propyl	$CH_3CH_2CH_2$—	
Benzene	C_6H_6	Phenyl	C_6H_5—	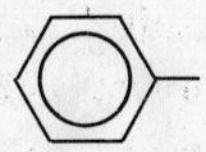

denote any radical. Thus, the compounds CH_3Cl, CH_3CH_2Cl, $CH_3CH_2CH_2Cl$, and so forth, all of which undergo similar chemical reactions, can be represented by the formula RCl. If the radical is derived from an alkane, the radical is called an *alkyl radical*, and if it is derived from an aromatic hydrocarbon, it is called an *aryl radical*.

When a hydrogen atom of an alkane or aromatic hydrocarbon molecule is replaced by another atom or group of atoms, the hydrocarbon-like part of the molecule is relatively inert, like these hydrocarbons themselves. Therefore, the resulting compound will have properties characteristic of the substituting group. Specific groups of atoms responsible for the characteristic properties of the compound are called *functional groups*. For the most part, organic compounds can be classified according to the functional group they contain. The most important classes of such compounds include (1) alcohols, (2) ethers, (3) aldehydes, (4) ketones, (5) acids, (6) amines, and (7) esters. The general formulas for these classes are given in Table 21-4, where the symbol for a radical, R, is written for the hydrocarbon part of the molecule.

Table 21-4   Formulas for Functional Groups

Type	Characteristic Functional Group	Example
Alcohol	—OH	R—OH
Ether	—O—	R—O—R′
Aldehyde	—C=O | H	R—C=O | H
Ketone	—C— || O	R—C—R′ || O
Acid	—C—OH || O	R—C—OH || O
Ester	—C—O— || O	R—C—OR′ || O
Amine	—NH$_2$, ⟩NH, —⟩N	R—NH$_2$, RNHR′, RR′R″N

The radicals labeled R′ or R″ may be the same as or different from the radicals labeled R in the same compounds.

There are several other types of function groups, which will not be described here. Also, molecules of some compounds contain more than one functional group, and there may be even more than one kind of functional group in each molecule. The purpose of this discussion is merely to describe some of the possible compounds. Therefore, the methods of preparation of the various classes of compounds will not be given here for every class, nor will more than a few of their properties be described.

21.7 ALCOHOLS

Compounds containing the functional group —OH are called *alcohols*. The —OH group is covalently bonded to a carbon atom in an alcohol molecule, and the molecules do *not* ionize in water solution to give OH⁻ ions. On the contrary, they react with metallic sodium to liberate hydrogen in a reaction analogous to that of sodium with water.

$$2\,ROH + 2\,Na \longrightarrow H_2 + 2\,NaOR$$

$$2\,HOH + 2\,Na \longrightarrow H_2 + 2\,NaOH$$

The simplest alcohol is methanol, CH_3OH, also called methyl alcohol in a less systematic system of naming. Methanol is also known as wood alcohol. Ethanol, CH_3CH_2OH, also known as ethyl alcohol or grain alcohol, is the principal constituent of intoxicating beverages. Other alcohols of importance are included in Table 21-5. Note that the systematic names of alcohols characteristically end in -ol.

Table 21-5   Common Alcohols

Systematic Name	Formula	Common Name	Familiar Source or Use
Methanol	CH_3OH	Methyl alcohol	Wood alcohol
Ethanol	CH_3CH_2OH	Ethyl alcohol	Grain alcohol
1-Propanol	$CH_3CH_2CH_2OH$	Propyl alcohol	Rubbing alcohol
2-Propanol	$CH_3CHOHCH_3$	Isopropyl alcohol	Rubbing alcohol
1,2-Ethanediol	CH_2OHCH_2OH	Ethylene glycol	Antifreeze
1,2,3-Propanetriol	$CH_2OHCHOHCH_2OH$	Glycerine	Animal fat

When the number of carbon atoms in an alcohol molecule is greater than 2, several isomers are possible, depending on the location of the —OH group as well as on the nature of the carbon chain. For example, the structural formulas of four isomeric alcohols with the formula C_4H_9OH are given in Table 21-6.

Table 21-6   Isomeric Alcohols

Systematic Name	Formula	Common Name
1-Butanol		Butyl alcohol (primary)
2-Butanol		Secondary butyl alcohol
2-Methyl-1-Propanol		Isobutyl alcohol
2-Methyl-2-Propanol		Tertiary butyl alcohol

21.8   ETHERS

A chemical reaction that removes a molecule of water from two alcohol molecules results in the formation of an *ether*.

$$CH_3CH_2OH + HOCH_2CH_3 \longrightarrow CH_3CH_2OCH_2CH_3 + H_2O$$

This reaction is run by heating the alcohol in the presence of concentrated sulfuric acid, a good dehydrating agent. Diethyl ether, $CH_3CH_2OCH_2CH_3$, was formerly used as an anesthetic in surgery, but it caused undesirable side effects in many patients, and was replaced by other types of agents.

The two radicals of the ether molecule need not be identical. A mixed either is named after both radicals.

EXAMPLE 21.4. Name the following ethers: (a) $CH_3OCH_2CH_2CH_2CH_3$, (b) $C_6H_5OCH_2CH_3$, and (c) CH_3OCH_3.

(a) Methyl butyl ether, (b) phenyl ethyl ether, and (c) dimethyl ether.

21.9 ALDEHYDES AND KETONES

Aldehydes are produced by the mild oxidation of primary alcohols.

$$
\begin{array}{cc}
\underset{\text{ethyl alcohol}}{\overset{\displaystyle H-\underset{\underset{H}{|}}{\overset{\overset{H}{|}}{C}}-\underset{\underset{H}{|}}{\overset{\overset{H}{|}}{C}}-O-H}{}} & \xrightarrow{\text{mild oxidation}} & \underset{\text{acetaldehyde}}{\overset{\displaystyle H-\underset{\underset{H}{|}}{\overset{\overset{H}{|}}{C}}-\overset{\overset{H}{|}}{C}=O}{}}
\end{array}
$$

The mild oxidation of a secondary alcohol produces a *ketone*.

$$
\underset{\text{isopropyl alcohol}}{H-\overset{H}{\underset{H}{C}}-\overset{H}{\underset{O}{\underset{|}{\underset{H}{|}}}{C}}-\overset{H}{\underset{H}{C}}-H} \xrightarrow{\text{oxidation}} \underset{\text{acetone}}{H-\overset{H}{\underset{H}{C}}-\overset{}{\underset{O}{C}}-\overset{H}{\underset{H}{C}}-H}
$$

Each of these groups is characterized by having a *carbonyl group*, $\diagdown C{=}O$, but the aldehyde has the $\diagdown C{=}O$ group on one end of the carbon chain, and the ketone has the $\diagdown C{=}O$ group on a carbon other than one on the end. Acetone is a familiar solvent for varnishes and lacquers. As such, it is used as a nail polish remover.

21.10 ACIDS AND ESTERS

Acids can be produced by the oxidation of aldehydes or primary alcohols. For example, the souring of wine results from the oxidation of the ethyl alcohol in wine to acetic acid:

$$
CH_3CH_2OH \xrightarrow{\text{oxidation}} \underset{\text{acetic acid}}{CH_3-\overset{}{\underset{\underset{O}{\|}}{C}}-O-H}
$$

A solution of acetic acid formed in this manner is familiar as vinegar. Acids are also obtained from natural sources. A few examples are listed in Table 21-7.

Acids react with alcohols to produce *esters*:

$$
\underset{\text{acetic acid}}{CH_3COOH} + \underset{\text{ethyl alcohol}}{HOCH_2CH_3} \longrightarrow \underset{\substack{\text{ethyl acetate} \\ \text{(an ester)}}}{CH_3COOCH_2CH_3} + H_2O
$$

Esters have a pleasant, fruity odor. Indeed, the odors of many fruits and oils are due to esters. Esters

Table 21-7 Common Organic Acids

Name	Formula	Source
Formic	$\underset{\displaystyle \overset{\|}{O}}{HCOH}$ or HCOOH	Ants and some other insects
Acetic	$\underset{\displaystyle \overset{\|}{O}}{CH_3COH}$	Vinegar
Butyric	$\underset{\displaystyle \overset{\|}{O}}{CH_3CH_2CH_2COH}$	Rancid butter
Stearic	$\underset{\displaystyle \overset{\|}{O}}{CH_3(CH_2)_{16}COH}$	Animal fat
Lactic	$\underset{\displaystyle \overset{\|}{O}}{CH_3CHOHCOH}$	Sour milk

are named by combining the radical name of the alcohol with that of the negative ion of the acid. The ester in the equation above is an example.

EXAMPLE 21.5.   Name the following esters: (a) $C_4H_9\overset{\displaystyle O}{\overset{\displaystyle \|}{O}}CH$ and (b) $C_6H_5\overset{\displaystyle O}{\overset{\displaystyle \|}{O}}CCH_3$.

(a) Butyl formate and (b) phenyl acetate.

The formation of an ester from an alcohol and an acid is an equilibrium reaction. The reverse reaction can be promoted by removing the acid from the reaction mixture, for example by treating it with NaOH. Animal fats are converted to soaps (fatty acid salts) and glycerine (a trialcohol) in this manner.

$$3\,C_{17}H_{35}COOH \;+\; \begin{array}{c} CH_2OH \\ | \\ CHOH \\ | \\ CH_2OH \end{array} \;\rightleftharpoons\; \begin{array}{c} CH_2OCOC_{17}H_{35} \\ | \\ CHOCOC_{17}H_{35} \\ | \\ CH_2OCOC_{17}H_{35} \end{array} +\,3\,H_2O$$

$$\text{fatty acid} \qquad \text{glycerine} \qquad\qquad \text{animal fat}$$

$$+$$

$$3\,NaOH$$

$$\downarrow$$

$$3\,Na^+C_{17}H_{35}COO^-$$

$$\text{a soap}$$

$$+$$

$$3\,H_2O$$

Nitroglycerine, a powerful explosive, is an ester of the inorganic acid HNO_3 and glycerol (glycerine).

$$\begin{array}{c} H_2C-OH \\ | \\ HC-OH \\ | \\ H_2C-OH \end{array} +3\,HNO_3 \;\xrightarrow[H_2SO_4]{}\; 3\,H_2O + \begin{array}{c} H_2C-O-NO_2 \\ | \\ HC-O-NO_2 \\ | \\ H_2C-O-NO_2 \end{array}$$

$$\text{glycerine} \qquad\qquad\qquad\qquad \text{nitroglycerine}$$

21.11   AMINES

Amines can be considered derivatives of ammonia, NH_3, in which one or more hydrogen atoms have been replaced by organic radicals. For example, replacing one hydrogen atom on the nitrogen results in a *primary amine*, RNH_2. A *secondary amine* has a formula of the type R_2NH, and a *tertiary amine* has no hydrogen atoms on the nitrogen atom in its molecules, and has the formula R_3N. Like ammonia, amines react as Brønsted bases:

$$RNH_2 + H_2O \rightleftharpoons RNH_3^+ + OH^-$$

Aromatic amines are of considerable importance commercially. The simplest aromatic amine, aniline, $C_6H_5NH_2$, is used in the production of various dyes and chemicals for color photography.

Solved Problems

INTRODUCTION

21.1.   Name one type of reaction that practically all organic compounds undergo.

  Ans.   Combustion.

BONDING IN ORGANIC COMPOUNDS

21.2.   Write an electron dot diagram for CH_5N.

  Ans.

$$\begin{array}{c} H \quad \overset{\cdot\cdot}{} \\ H\!:\!\overset{\cdot\cdot}{C}\!:\!\overset{\cdot\cdot}{N}\!:\!H \\ H \quad H \end{array}$$

21.3.   What is the total bond order of carbon in (*a*) CO_2? (*b*) CO? (*c*) $H_2C{=}O$?

  Ans.   (*a*) 4, (*b*) 3, and (*c*) 4.

21.4.   What is the total bond order of sulfur in CH_3SCH_3?

  Ans.   2.

21.5.   Draw electron dot diagrams for C_3H_8 and C_2H_4.

  Ans.

$$\begin{array}{c} H \; H \; H \\ H\!:\!\overset{\cdot\cdot}{C}\!:\!\overset{\cdot\cdot}{C}\!:\!\overset{\cdot\cdot}{C}\!:\!H \\ H \; H \; H \end{array} \qquad \begin{array}{c} H \quad H \\ H\!:\!\overset{\cdot\cdot}{C}\!:\!:\!\overset{\cdot\cdot}{C}\!:\!H \end{array}$$

STRUCTURAL AND LINE FORMULAS

21.6.   Draw electron dot diagrams and structural formulas for molecules with the following molecular formulas: (*a*) CH_3N and (*b*) CHN.

  Ans.

  (*a*)

$$\begin{array}{c} H\!:\!\overset{}{C}\!:\!:\!\overset{\cdot\cdot}{N}\!:\!H \\ \overset{\cdot\cdot}{H} \end{array} \qquad \begin{array}{c} H{-}C{=}N{-}H \\ | \\ H \end{array}$$

  (*b*)

$$H\!:\!C\!:\!:\!:\!N\!: \qquad H{-}C{\equiv}N$$

21.7.   Describe the difference, if any, between the compound(s) represented by the formulas CH_3CH_2OH and $HOCH_2CH_3$.

  Ans.   There is no difference.

21.8. Write the structural formula for $P(CH_3)_3$. What is the total bond order of phosphorus in this compound?

Ans.

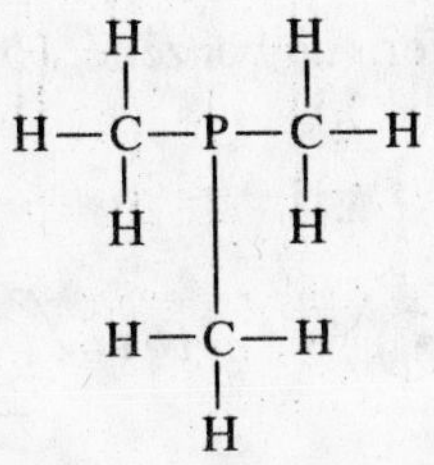

$$
\begin{array}{ccc}
 & H & \quad\ H \\
 & | & \quad\ | \\
H- & C-P-C & -H \\
 & | & \quad\ | \\
 & H & \quad\ H \\
 & & | \\
 & & H-C-H \\
 & & | \\
 & & H
\end{array}
$$

The total bond order of P is 3.

21.9. Explain the meaning in organic formulas of a pair of parentheses with no subscript behind it, such as in $CH_3CH_2CH(CH_3)C_3H_7$.

Ans.   The parentheses designate a side chain.

21.10. Write structural formulas for (*a*) CH_3OCH_3 and (*b*) $(CH_3)_2NH$.

Ans.

(*a*)

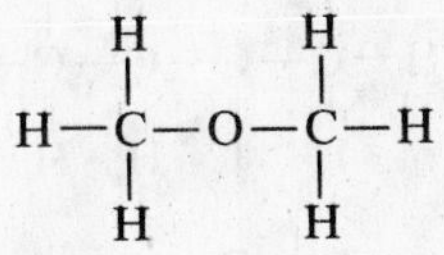

$$
\begin{array}{ccc}
H & & H \\
| & & | \\
H-C-O-C-H \\
| & & | \\
H & & H
\end{array}
$$

(*b*)

$$
\begin{array}{ccc}
H & & H \\
| & & | \\
H-C-N-C-H \\
| & | & | \\
H & H & H
\end{array}
$$

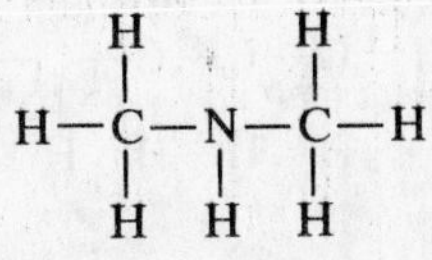

21.11. Draw as many representations as are introduced in this chapter for (*a*) butane and (*b*) benzene.

Ans.

(*a*) C_4H_{10}

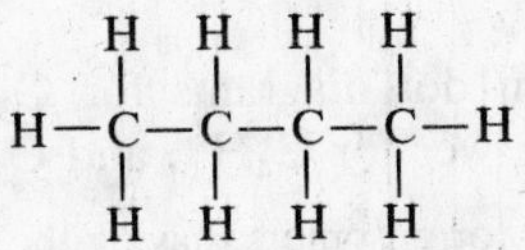

$$
\begin{array}{cccc}
H & H & H & H \\
| & | & | & | \\
H-C-C-C-C-H \\
| & | & | & | \\
H & H & H & H
\end{array}
$$

$CH_3CH_2CH_2CH_3$

(*b*)

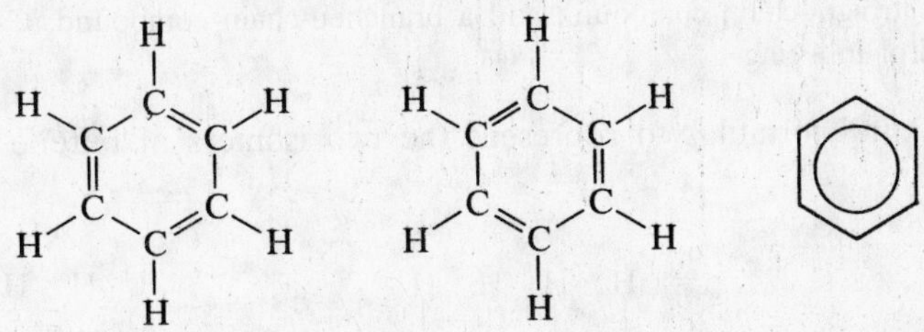

HYDROCARBONS

21.12. (*a*) Explain why alkenes have two fewer hydrogen atoms per molecule than alkanes with the same number of carbon atoms. (*b*) Explain why cycloalkanes have two fewer hydrogen atoms per molecule than alkanes with the same number of carbon atoms. (*c*) Explain why the cycloalkanes react more like alkanes than alkenes react like alkanes, despite the similarity in molecular formula of alkenes and cycloalkanes.

Ans.   (*a*) The extra pair of electrons between two carbon atoms leaves two fewer electrons to bond to hydrogen atoms. (*b*) The extra bond between the two "end" carbon atoms leaves two fewer electrons to bond to hydrogen atoms. (*c*) There are no multiple bonds in alkanes and cycloalkanes.

21.13. (a) List the common names of the first three members of the alkene series and (b) the first member of the alkyne series.

   Ans.   (a) Ethylene, propylene, and butylene. (b) Acetylene.

21.14. Write structural formulas for (a) benzene, (b) toluene (methyl benzene), and (c) chlorobenzene.

   Ans.

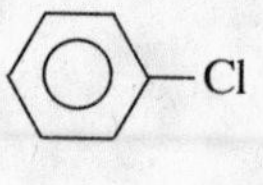

ISOMERISM

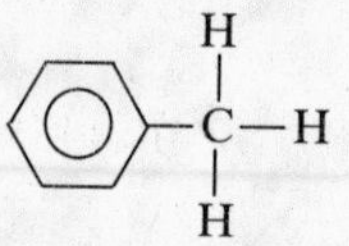

21.15. Write formulas which contain only one multiple bond and no rings for all possible isomers of butadiene. Butadiene is $CH_2=CHCH=CH_2$.

   Ans. $HC\equiv C-CH_2-CH_3$ and $CH_3-C\equiv C-CH_3$.

21.16. Write structural formulas for the three isomers of pentane.

   Ans.

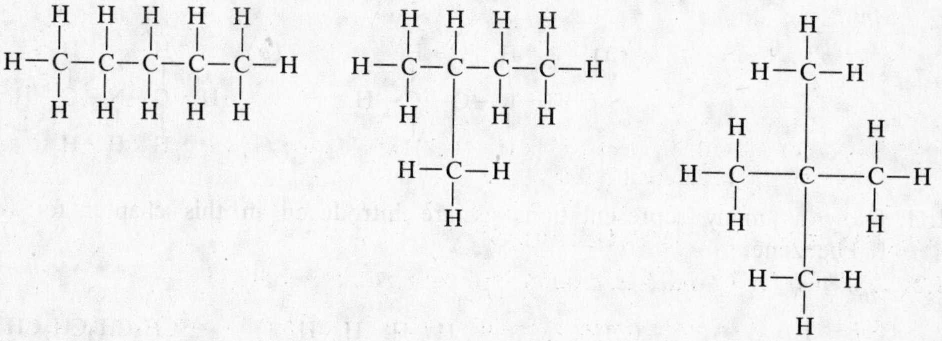

21.17. Problem 21.5 required electron dot diagrams for C_3H_8 and C_2H_4. Explain why the same question would be harder to answer for C_4H_{10} and C_3H_6.

   Ans.   These compounds can each form isomers (having the same molecular formulas). C_4H_{10} can occur as a straight-chain compound and a branched-chain compound. C_3H_6 can occur with a double bond or in a ring.

21.18. Draw structural formulas to represent the two isomers of butene which contain continuous carbon chains.

   Ans.

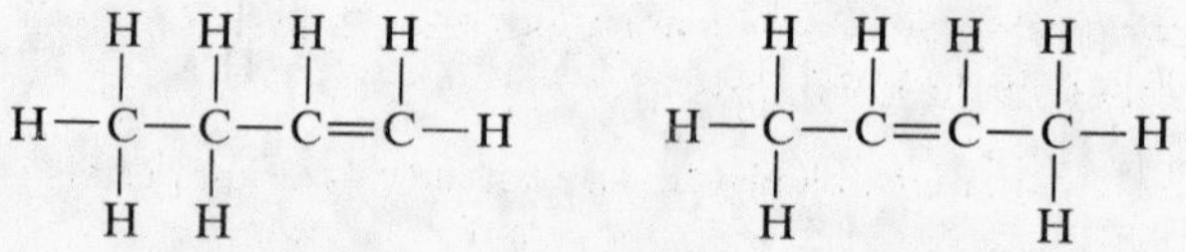

21.19. Explain why there can be no 2-carbon branch in a compound that contains a 4-carbon longest continuous chain.

   Ans.   A branch cannot start at the end of a carbon chain, and so it must be from one of the two center carbon atoms in the present compound. If a 2-carbon chain is added there, the 2-carbon chain becomes part of the longest chain, and a 1-carbon side chain is left.

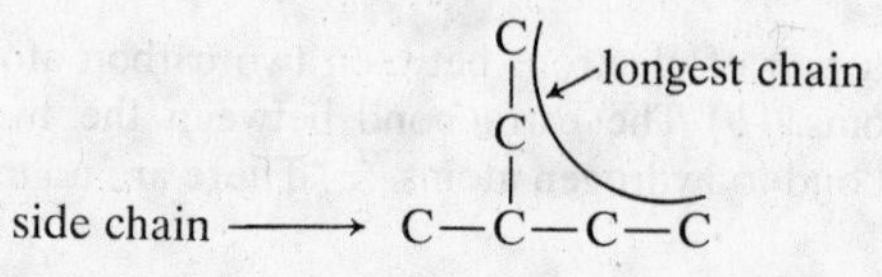

21.20. Write carbon skeletons for the four isomers of octane which each contain three 1-carbon branches.

Ans.

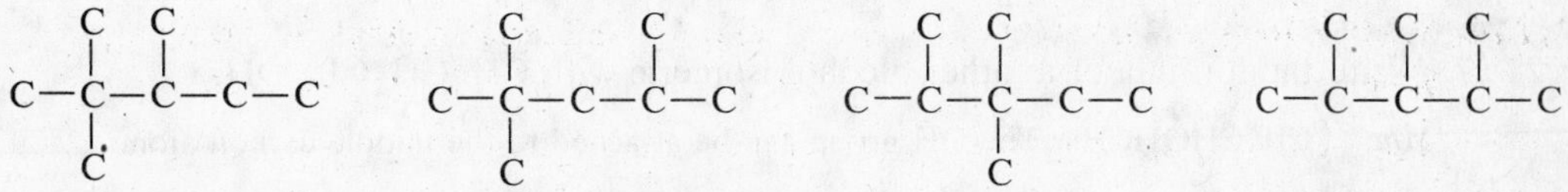

21.21. Which ones of the following are identical to each other?

Ans.

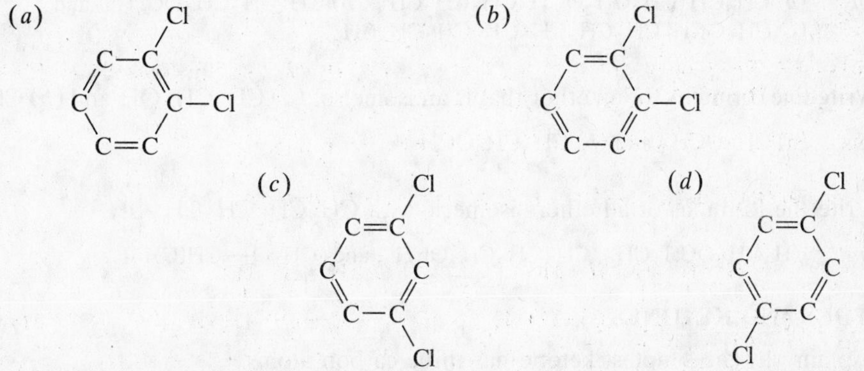

(a)

(b)

(c)

(d)

Ans.   (a) and (b) are identical because the double bonds are really delocalized. (c) and (d) have the Cl atoms on different carbon atoms.

RADICALS AND FUNCTIONAL GROUPS

21.22. Is it true that isomers can occur only with hydrocarbons?

Ans.   No. Compounds with functional groups can form isomers even more extensively than can hydrocarbons (given the same number of carbon atoms).

21.23. What usefulness has the concept of oxidation number in deciding whether two classes of compounds can be isomers of each other?

Ans.   In order to be isomeric, two classes of compounds with only one functional group each can be isomeric only if the carbon atoms of the functional group (or attached to the functional group) have the same oxidation number. For example, aldehydes and ketones can be isomers because the carbon atoms of their functional groups have the same oxidation number (0).

21.24. List the pairs of organic functional groups that can be isomeric with each other if the numbers of carbon atoms in their compounds are the same.

Ans.   Alcohols and ethers; aldehydes and ketones; acids and esters.

21.25. Identify the radical(s) and the functional group in each of the following molecules:
(a) CH_3CH_2OH, (b) CH_3CH_2COOH, (c) CH_3CH_2CHO, and (d) $CH_3CH_2OCH_3$.

Ans.

	Radical(s)	Functional Group
(a)	CH_3CH_2-	$-OH$
(b)	CH_3CH_2-	$-COOH$
(c)	CH_3CH_2-	$-CHO$
(d)	CH_3CH_2- and $-CH_3$	$-O-$

ALCOHOLS

21.26. Explain why an alcohol, ROH, is not a base.

Ans. The OH group is not ionic.

21.27. Write the formula of another alcohol isomeric with $CH_3CH_2CH_2OH$.

Ans. $CH_3CHOHCH_3$. The OH group can be attached to the middle carbon atom.

ETHERS

21.28. Write line formulas for (a) dibutyl ether, (b) propyl phenyl ether, and (c) methyl octyl ether.

Ans. (a) $CH_3CH_2CH_2CH_2OCH_2CH_2CH_2CH_3$, (b) $CH_3CH_2CH_2OC_6H_5$, and
(c) $CH_3OCH_2CH_2CH_2CH_2CH_2CH_2CH_3$.

21.29. Write line formulas for an ether that is an isomer of (a) CH_3CH_2OH and (b) $CH_3CH_2CH_2OH$.

Ans. (a) CH_3OCH_3 and (b) $CH_3CH_2OCH_3$.

21.30. Write the formulas of all ethers isomeric with $CH_3CH_2CH_2CH_2OH$.

Ans. $CH_3CH_2OCH_2CH_3$, $CH_3CH_2CH_2OCH_3$, and $CH_3O-CH(CH_3)_2$.

ALDEHYDES AND KETONES

21.31. Explain why the simplest ketone has three carbon atoms.

Ans. A ketone has a doubly bonded oxygen atom on a carbon that is *not* an end carbon atom. The smallest carbon chain that has a carbon not on an end has three carbon atoms.

21.32. Write balanced chemical equations for the reduction by hydrogen gas to the corresponding alcohol for (a) CH_3CHO and (b) CH_3COCH_3.

Ans.

(a) $$CH_3CHO + H_2 \xrightarrow{\text{catalyst}} CH_3CH_2OH$$

(b) $$CH_3COCH_3 + H_2 \xrightarrow{\text{catalyst}} CH_3CHOHCH_3$$

In each case, the hydrogen molecule adds across the $C=O$ double bond.

ACIDS AND ESTERS

21.33. Contrast the following two formulas for acetic acid: $HC_2H_3O_2$ and CH_3COOH. Explain the advantages of each. Which hydrogen atom is lost upon ionization of acetic acid in water?

Ans. The first formula is easily identified as an acid, with H written first. It is that hydrogen atom which ionizes in water. The second formula shows the bonding. The hydrogen atom attached to the oxygen atom ionizes in water.

21.34. Explain why an organic acid, RCOOH, has acidic properties.

Ans. The hydrogen atom attached to the oxygen atom is ionizable.

21.35. If pentanoic acid is C_4H_9COOH, what is the formula for hexanoic acid?

Ans. $C_5H_{11}COOH$.

21.36. Explain why oxygen is not usually used to oxidize an alcohol to the corresponding acid.

Ans.   It is too easy to oxidize the compound to CO_2 and H_2O.

21.37. The mild and moderate oxidation of CH_3CH_2OH are described in the text. What are the products of the vigorous oxidation of CH_3CH_2OH with excess oxygen gas?

Ans. CO_2 and H_2O.

21.38. Write balanced chemical equations for the reaction of CH_3COOH with (*a*) CH_3OH, (*b*) CH_3CH_2OH, (*c*) $CH_3CH_2CH_2OH$, and (*d*) $CH_3CH_2CH_2CH_2OH$. (*e*) Explain the utility of the symbol R.

Ans.

$$(a) \qquad CH_3OH + HOCOCH_3 \longrightarrow CH_3OCOCH_3 + H_2O$$

$$(b) \qquad CH_3CH_2OH + HOCOCH_3 \longrightarrow CH_3CH_2OCOCH_3 + H_2O$$

$$(c) \qquad CH_3CH_2CH_2OH + HOCOCH_3 \longrightarrow CH_3CH_2CH_2OCOCH_3 + H_2O$$

$$(d) \qquad CH_3CH_2CH_2CH_2OH + HOCOCH_3 \longrightarrow CH_3CH_2CH_2CH_2OCOCH_3 + H_2O$$

(*e*)   Each of these reactions corresponds to the equation below. The fact that all four can be represented by this single equation makes R a useful representation.

$$ROH + HOCOCH_3 \longrightarrow ROCOCH_3 + H_2O$$

21.39. Ethyl acetate is reduced by hydrogen in the presence of a catalyst to yield one organic compound, which contains oxygen but is not an ether. What is the compound?

Ans. CH_3CH_2OH.

AMINES

21.40. Write the formulas for (*a*) ammonium chloride, (*b*) methyl ammonium chloride, (*c*) dimethyl ammonium chloride, (*d*) trimethyl ammonium chloride, and (*e*) tetramethyl ammonium chloride.

Ans.   (*a*) NH_4Cl.
       (*b*) CH_3NH_3Cl. (One H atom has been replaced by a CH_3 group.)
       (*c*) $(CH_3)_2NH_2Cl$. (Two H atoms have been replaced by CH_3 groups.)
       (*d*) $(CH_3)_3NHCl$. (Three H atoms have been replaced by CH_3 groups.)
       (*e*) $(CH_3)_4NCl$. (All four H atoms have been replaced by CH_3 groups.)

21.41. Write balanced chemical equations for the reaction with H_3O^+ (excess) of aqueous (*a*) NH_3, (*b*) CH_3NH_2, and (*c*) $NH_2CH_2CH_2NH_2$.

Ans.

$$(a) \qquad H_3O^+ + NH_3 \longrightarrow NH_4^+ + H_2O$$

$$(b) \qquad H_3O^+ + CH_3NH_2 \longrightarrow CH_3NH_3^+ + H_2O$$

$$(c) \qquad 2H_3O^+ + NH_2CH_2CH_2NH_2 \longrightarrow {}^+NH_3CH_2CH_2NH_3^+ + 2H_2O$$

Supplementary Problems

21.42. Define or identify each of the following terms:

organic chemistry   delocalized double bond
total bond order   isomerism
graphic formula   cycloalkane
structural formula   radical
line formula   functional group
hydrocarbon   alcohol
alkane   ether
alkene   aldehyde
alkyne   ketone
aromatic hydrocarbon   carbonyl group
saturated   ester

21.42. What is the smallest number of carbon atoms possible in a molecule of (*a*) a ketone? (*b*) an aldehyde? (*c*) an aromatic hydrocarbon?

Ans. (*a*) 3. (A molecule cannot have a "middle" carbon atom unless it has at least three carbon atoms.) (*b*) 1. (*c*) 6.

21.44. What is the total bond order of oxygen in all aldehydes, alcohols, ketones, acids, and ethers?

Ans. 2.

21.45. In which classes of compounds can the R group(s) NOT be a hydrogen atom?

Ans. Ethers and esters (R on the oxygen atom cannot be H).

21.46. Which classes of compounds contain the element grouping —O—H?

Ans. Alcohols and acids.

21.47. Which of the following is apt to be the stronger acid—ClCOOH or HCOOH? Explain.

Ans. The electrons in the OH bond are attracted away from the hydrogen atom more by the Cl atom in ClCOOH than by the H atom in HCOOH, since Cl is more electronegative than H. The H atom on the O atom in ClCOOH is therefore easier to remove; that is, it is more acidic.

21.48. Determine the oxidation state of the carbon atom in the simplest member of each of the following groups: (*a*) alcohol, (*b*) aldehyde, and (*c*) acid.

Ans. (*a*) CH_3OH, -2; (*b*) CH_2O, 0; and (*c*) HCOOH, $+2$.

21.49. Using toothpicks and marshmallows or gumdrops, show that one part of an organic molecule can rotate about a carbon-to-carbon single bond with respect to another part, but not about a carbon-to-carbon double bond.

21.50. Using toothpicks and marshmallows or gumdrops, make a model of pentane. The angles between the toothpicks must be 109.5°. Show that the following represent the same compound:

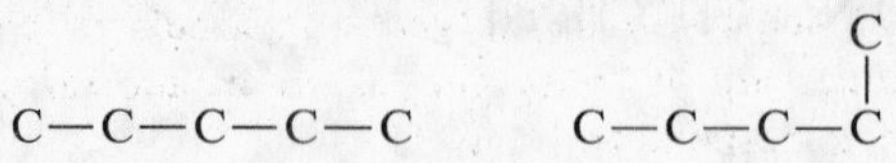

$$C—C—C—C—C \qquad \begin{matrix} & & & C \\ & & & | \\ C—C—C—C \end{matrix}$$

21.51. Build a ball-and-stick model of butane, and twist the model into a shape corresponding to each of the following representations.

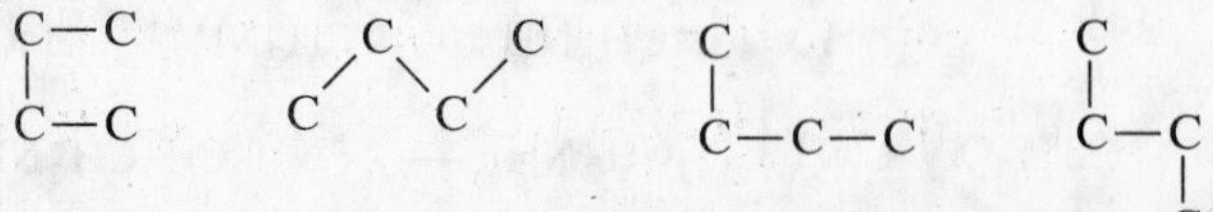

21.52. Give the name of the class of organic compound represented by each of the following: (a) CH_3COCH_3, (b) CH_3CH_2OH, (c) CH_3CHO, (d) $CH_2{=}CHCH_3$, (e) CH_3COOH, (f) CH_3COOCH_3, (g) $CH_3OCH_2CH_3$, (h) CH_3CH_3, and (i) CH_3NH_2.

Ans.   (a) ketone, (b) alcohol, (c) aldehyde, (d) alkene, (e) acid, (f) ester, (g) ether, (h) alkane, and (i) amine.

21.53. Burning of which compound, CH_3CH_2OH or CH_3CH_3, is apt to provide more heat? Explain why you made your choice.

Ans. CH_3CH_3. The CH_3CH_2OH is already partially oxidized.

21.54. Calculate the concentration of hydroxide ion in 1.00 L of a solution containing 0.100 mol of $CH_3CH_2NH_2$ ($K_b = 1.6 \times 10^{-11}$).

Ans.

$$RNH_2 + H_2O \rightleftharpoons RNH_3^+ + OH^-$$

Let $x = [OH^-]$

$$\frac{x^2}{0.100} = 1.6 \times 10^{-11}$$

$$x = 1.3 \times 10^{-6} = [OH^-]$$

21.55. What functional group is most likely to be found in a polyester shirt?

Ans.   Ester, $-\overset{\displaystyle \|}{\underset{\displaystyle O}{C}}-O-$.

21.56. Ethyl acetate is reduced by hydrogen in the presence of a catalyst to yield one oxygen-containing organic compound, but propyl acetate yields two. What are the two compounds?

Ans. CH_3CH_2OH and $CH_3CH_2CH_2OH$.

21.57. The formation of which ones of the functional groups described results in the simultaneous formation of water?

Ans.   Ether and ester.

21.58. In this question consider organic molecule with no rings or carbon-to-carbon double or triple bonds. (a) Show that ethers can be isomers of alcohols but not of aldehydes. (b) Show that aldehydes and ketones can be isomers of each other, but not of acids or alcohols. (c) Write the structural formula for an alcohol isomeric with diethyl ether.

Ans.   (a) These alcohols have molecular formulas $C_nH_{2n+2}O$. Ethers can be thought of as $C_nH_{2n+1}OC_{n'}H_{2n'+1}$ or $C_{n+n'}H_{2(n+n')+2}O$. If $n + n' = n''$, the general formula of the ether would be $C_{n''}H_{2n''+2}O$. Since the n in the general formula of the alcohol and the n'' in the general formula of the ether are arbitrary, they could be the same. The ether can be an isomer of an alcohol. (b) Both correspond to $C_nH_{2n}O$.

(c)

$$H-\overset{\overset{\displaystyle H}{|}}{\underset{\underset{\displaystyle H}{|}}{C}}-\overset{\overset{\displaystyle H}{|}}{\underset{\underset{\displaystyle H}{|}}{C}}-\overset{\overset{\displaystyle H}{|}}{\underset{\underset{\displaystyle H}{|}}{C}}-\overset{\overset{\displaystyle H}{|}}{\underset{\underset{\displaystyle H}{|}}{C}}-O-H \quad \text{or} \quad H-\overset{\overset{\displaystyle H}{|}}{\underset{\underset{\displaystyle H}{|}}{C}}-\overset{\overset{\displaystyle H}{|}}{\underset{\underset{\displaystyle H}{|}}{C}}-\overset{\overset{\displaystyle H}{|}}{\underset{\underset{\displaystyle O-H}{|}}{C}}-\overset{\overset{\displaystyle H}{|}}{\underset{\underset{\displaystyle H}{|}}{C}}-H$$

$$\text{or} \qquad \underset{\underset{\displaystyle\underset{\displaystyle H}{|}}{\overset{\displaystyle|}{\underset{\displaystyle H}{\overset{\displaystyle|}{C}}}}{\overset{\displaystyle H}{\underset{\displaystyle H}{\underset{|}{\overset{|}{C}}}}} \qquad \qquad \text{or} \qquad \cdots$$

H—C—C—C—O—H   or   H—C—C—C—H

21.59. In the formula $CH_3CH_2OCOCH_3$, which oxygen atom is double-bonded?

  Ans.   The double-bonded oxygen is conventionally written after the carbon atom to which it is attached, so the second oxygen is double-bonded to the carbon. The structural formula is

$$H-\underset{\underset{\displaystyle H}{|}}{\overset{\overset{\displaystyle H}{|}}{C}}-\underset{\underset{\displaystyle H}{|}}{\overset{\overset{\displaystyle H}{|}}{C}}-O-\underset{\underset{\displaystyle O}{\|}}{C}-\underset{\underset{\displaystyle H}{|}}{\overset{\overset{\displaystyle H}{|}}{C}}-H$$

21.60. Urea and thiourea have formulas NH_2CONH_2 and NH_2CSNH_2. (*a*) Explain why they are considered to be organic compounds. (*b*) Explain why they do not fit the general definition given in Sec. 21.1.

  Ans.   (*a*) Urea is a product of animal metabolism and thiourea is its sulfur analog. (*b*) Both hydrogen atoms of their "parent," formaldehyde, H_2CO, have been replaced with amino groups, and there are no C—C or C—H bonds left.

Chapter 22

Nuclear Reactions

22.1 INTRODUCTION

In all the processes discussed so far, the nucleus of every atom remained unchanged. In this chapter, the effect of changes in the nucleus will be considered. Reactions involving changes in the nuclei are called *nuclear reactions*. They involve great quantities of energy, and this energy is referred to as *atomic energy* or more precisely as *nuclear energy*. There are two types of nuclear reactions—spontaneous and induced. We will consider them in that order.

Nuclear reactions differ from ordinary chemical reactions in the following ways:
1. Atomic numbers change.
2. Although there is no change in the total of the mass numbers, the quantity of matter does change significantly. Some matter is changed to energy.
3. Reactions are those of specific isotopes rather than the naturally occurring mixtures of isotopes.
4. The quantities usually used in calculations are atoms rather than moles of atoms.

22.2 NATURAL RADIOACTIVITY

The nuclei of the atoms of some elements are inherently unstable, and they disintegrate over time to yield other nuclei. The process is called *radioactive decay*, and the decay of each nucleus is called an *event*. There is nothing that humans can do about this type of radioactivity; as long as these elements exist, the nuclei will decompose. Depending on the isotope involved, some number of existing atoms decompose over a period of time (see Sec. 22.3). Three types of small particles are emitted from the nuclei during natural radioactive decay; they are named after the first three letters of the Greek alphabet. Their names and properties are listed in Table 22-1. When an *alpha* or *beta particle* is emitted from the nucleus, the identity of the element is changed because the atomic number is changed.

A stream of alpha particles is sometimes called an alpha ray. A stream of beta particles is called a beta ray. A *gamma ray* is composed of a stream of gamma particles.

We can denote the charge and mass number of these small particles as we denote atomic numbers and mass numbers in Chap. 3.

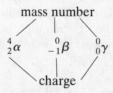

The superscripts refer to the mass numbers of the particles; the subscripts refer to their charges.

Table 22-1 Products of Natural Radioactivity

Symbol	Name	Mass Number	Charge	Identity
α	Alpha particle	4	2+	Helium nucleus
β	Beta particle	0	1−	High-energy electron
γ	Gamma ray	0	0	High-energy light

Nuclear equations are written with both the total charge and the total of the mass numbers unchanged from reactants to products. That is, the total of the subscripts of the reactants equals the total of the subscripts of the products and the total of the superscripts of the reactants equals the total of the superscripts of the products. The subscripts of isotopes may be omitted because the symbol of the element gives the atomic number.

EXAMPLE 22.1. Show that the mass number and the total charge are both conserved in the natural disintegration of $^{238}_{92}\text{U}$:

$$^{238}\text{U} \longrightarrow {}^{234}\text{Th} + \alpha$$

The equation may be rewritten including all atomic numbers and mass numbers:

$$^{238}_{92}\text{U} \longrightarrow {}^{234}_{90}\text{Th} + {}^{4}_{2}\text{He} \qquad \text{or} \qquad {}^{238}_{92}\text{U} \longrightarrow {}^{234}_{90}\text{Th} + {}^{4}_{2}\alpha$$

Adding the $234 + 4$ superscripts of the products gives the total superscript of the reactant. Adding the $90 + 2$ subscripts of the products gives the total subscript of the reactant. The nuclear equation is balanced.

EXAMPLE 22.2. Complete the following nuclear equation:

$$^{234}_{90}\text{Th} \longrightarrow {}^{234}_{91}\text{Pa} + ?$$

Inserting the proper subscript and superscript indicates that the product is a particle with -1 charge and 0 mass number:

$$^{234}_{90}\text{Th} \longrightarrow {}^{234}_{91}\text{Pa} + {}^{0}_{-1}?$$

The missing particle is a beta particle (Table 22-1).

$$^{234}_{90}\text{Th} \longrightarrow {}^{234}_{91}\text{Pa} + {}^{0}_{-1}\beta$$

Note that a beta particle has been emitted from the nucleus. This change has been accompanied by the increase in the number of protons by 1 and a decrease in the number of neutrons by 1. In effect, a neutron has been converted into a proton and an electron, and the electron has been ejected from the nucleus.

The emission of a gamma particle causes no change in the charge or mass number of the original particle. (It does cause a change in its internal energy, however.) For example,

$$^{119}_{50}\text{Sn} \longrightarrow {}^{119}_{50}\text{Sn} + {}^{0}_{0}\gamma$$

The same $^{119}_{50}\text{Sn}$ isotope is produced, but it has a lower energy after the emission of the gamma particle.

22.3   HALF-LIFE

Not all of the nuclei of a given sample of a radioactive isotope disintegrate at the same time; the nuclei disintegrate over a period of time. The number of radioactive disintegrations per unit time that occur in a given sample of a naturally radioactive isotope is directly proportional to the quantity of that isotope present. The more nuclei present, the more will disintegrate per second (or per year, etc.).

EXAMPLE 22.3. Sample A of $^{238}_{92}\text{U}$ has a mass of 1.00 kg; sample B of the same isotope has a mass of 2.00 kg. Compare the rate of decay (the number of disintegrations per second) in the two samples.

There will be twice the number of disintegrations per second in sample B (the 2.00-kg sample) as in sample A, because there were twice as many $^{238}_{92}\text{U}$ atoms there to start. The number of disintegrations per gram per second is a constant, because both samples are the same isotope—$^{238}_{92}\text{U}$.

After a certain period of time, sample B of Example 22.3 will have disintegrated so much that there will be only 1.00 kg of $^{238}_{92}\text{U}$ left. (Products of its decay will also be present.) How would the decay rate of this 1.00 kg of $^{238}_{92}$U compare with that of the 1.00 kg of $^{238}_{92}$U originally present in sample A of Example 22.3? The rates should be the same, since each contains 1.00 kg of $^{238}_{92}$U. That means that sample B is disintegrating at a slower rate as its number of $^{238}_{92}$U atoms decreases. Half of *this* sample will disintegrate in a time equal to the time it took half the original sample B to disintegrate. The period in which half a naturally radioactive sample disintegrates is called its *half-life* because that is the time required for half of *any given sample* of the isotope to disintegrate (Fig. 22-1). The half-lives of several isotopes are given in Table 22-2.

Table 22-2   Half-Lives of Some Nuclei

Isotope	Half-Life	Radiation[a]
^{238}U	4.5×10^9 years	Alpha
^{235}U	7.1×10^8 years	Alpha
^{237}Np	2.2×10^6 years	Alpha
^{14}C	5 730 years	Beta
^{90}Sr	19.9 years	Beta
^{3}H	12.3 years	Beta
^{140}Ba	12.5 days	Beta
^{131}I	8.0 days	Beta
^{15}O	118 s	Beta
^{94}Kr	1.4 s	Beta

[a]In most of these processes, gamma radiation is also emitted.

If 1.0 kg takes 1 half-life to undergo the process shown for sample A, it also should take the same 1 half-life to undergo the process shown for sample B, after sample gets down to 1.0 kg.

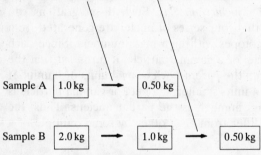

Fig. 22-1   A half-life example

EXAMPLE 22.4.   A certain isotope has a half-life of 2.00 days. How much of a 4.00-g sample of this isotope will remain after 6.00 days?

After the first 2.00 days, half the 4.00-g sample will still be the same isotope. After 2.00 more days, half of the 2.00 g remaining will still be the original isotope. That is, 1.00 g will remain. After the third 2.00-day period, half of this 1.00 g will remain, or 0.500 g, of the original isotope will have its original identity.

EXAMPLE 22.5.   A certain isotope has a half-life of 2.00 days. How much of a 4.00-g sample of this isotope will have decomposed after 6.00 days?

We calculated in Example 22.4 that 0.500 g would remain. That means that 4.00 g − 0.500 g = 3.50 g of the isotope will have disintegrated. (Most of the 3.50 g will produce some other isotope.) Notice the difference in the wording of Examples 22.4 and 22.5. How much remains is determined from the half-life; how much decomposes is determined from the original quantity and how much remains.

Half-life problems may be solved with the equation

$$\log\left(\frac{N}{N_0}\right) = \left(\frac{-0.301}{t_{1/2}}\right)t$$

where N is the number of radioactive atoms of the original isotope remaining at time t, N_0 is the number of those atoms at the beginning of the process, and $t_{1/2}$ is the half-life.

EXAMPLE 22.6.   Calculate the time required for a radioactive sample to lose one-third of the atoms of its parent isotope. The half-life is 33 min.

If one-third of the atoms are lost, two-thirds remain. Therefore,

$$\log\left(\frac{2}{3}\right) = \left(\frac{-0.301}{33 \text{ min}}\right)t$$

$$t = \frac{(33 \text{ min})(-0.176)}{-0.301} = 19 \text{ min}$$

In 19 min, one-third of the atoms will disintegrate. Note that this interval of time is less than the half-life, in which one-half the atoms would disintegrate, which is reasonable.

22.4   RADIOACTIVE SERIES

When an isotope disintegrates spontaneously, the products include one of the small particles from Table 22-1 and a large nucleus. The new nucleus is not necessarily stable; it might itself decompose spontaneously, yielding another small particle and still another nucleus. A disintegrating nucleus is called a *parent nucleus*, and the product is called the *daughter nucleus*. Such disintegrations can continue until a stable nucleus is produced. It turns out that four series of nuclei are generated from the disintegration of four different naturally occurring isotopes with high mass numbers. Since each nucleus can produce only an alpha particle, a beta particle, or a gamma particle, it turns out that the mass number of the daughter nucleus can be different from the parent nucleus by 4 units or 0 units. If an alpha particle is emitted, the daughter nucleus will be 4 units smaller; if a beta or gamma particle is produced, the daughter nucleus will be the same in mass number as the parent nucleus. Thus, the mass numbers of all the members of a given series can differ from each other by some multiple of 4 units.

EXAMPLE 22.7.   When a ^{238}U nucleus disintegrates, the following series of alpha and beta particles is emitted: alpha, beta, beta, alpha, alpha, alpha, alpha, alpha, beta, alpha, beta, beta, beta, alpha. (Since emission of gamma particles accompanies practically every disintegration and since gamma particles do not change the atomic number or mass number of an isotope, they are not listed.) Show that each isotope produced has a mass number that differs from 238 by some multiple of 4.

Each alpha particle loss lowers the mass number of the product nucleus by 4. Each beta particle loss lowers the mass number by 0. The mass numbers thus go down as follows:

$$238 \xrightarrow{\alpha} 234 \xrightarrow{\beta} 234 \xrightarrow{\beta} 234 \xrightarrow{\alpha} 230 \xrightarrow{\alpha} 226 \xrightarrow{\alpha} 222 \xrightarrow{\alpha} 218$$
$$\downarrow{\alpha}$$
$$206 \xleftarrow{\alpha} 210 \xleftarrow{\beta} 210 \xleftarrow{\beta} 210 \xleftarrow{\beta} 210 \xleftarrow{\alpha} 214 \xleftarrow{\beta} 214$$

There are four such series of naturally occurring isotopes. The series in which all mass numbers are evenly divisible by 4 is called the "$4n$ series." The series with mass numbers 1 more than the corresponding $4n$ series members is called the "$4n + 1$ series." Similarly, there are a "$4n + 2$ series" and a "$4n + 3$ series." Since the mass numbers change by 4 or 0, no member of any series can produce a product in a different series.

22.5   NUCLEAR FISSION AND FUSION

Nuclear fission refers to splitting a (large) nucleus into two smaller ones, not including the tiny particles listed in Table 22-3. *Nuclear fusion* refers to the combination of small nuclei to make a larger one. Both of these types of processes are included in the term artificial transmutation.

Transmutation means converting one element to another (by changing the nucleus). The first artificial transmutation was the bombardment of $^{14}_{7}$N by alpha particles in 1919 by Lord Rutherford.

$$^{14}_{7}\text{N} + ^{4}_{2}\text{He} \longrightarrow ^{17}_{8}\text{O} + ^{1}_{1}\text{H}$$

The alpha particles could be obtained from a natural decay process. At present, a variety of particles can be used to bombard nuclei (Table 22-3), some of which are raised to high energies in "atom smashing" machines. Again, nuclear equations can be written, in which the net charge and the total of the mass numbers on one side must be the same as their counterparts on the other side.

Table 22-3 Nuclear Projectiles and Products[a]

Name	Symbol	Identity	Nuclear Rest Mass (amu)
Proton	^{1}H or p	Hydrogen nucleus	1.00728
Deuteron	^{2}H or d	Heavy hydrogen nucleus	2.0135
Tritium	^{3}H	Tritium nucleus	3.01550
Helium-3	^{3}He	Light helium nucleus	3.01493
Neutron	^{1}n or n	Free neutron	1.008665
Alpha	α	Helium nucleus	4.001503
Beta	β	High-energy electron	0.00054858
Gamma	γ	High-energy light	0.0
Positron	$_{+}\beta$	Positive electron	0.00054858

[a]Larger projectiles are identified by their regular isotopic symbols, such as $^{12}_{6}$C.

EXAMPLE 22.8. What small particle(s) must be produced with the other products of the reaction of a neutron with a $^{235}_{92}$U nucleus by the following reaction?

$$^{235}_{92}\text{U} + ^1_0\text{n} \longrightarrow ^{140}_{56}\text{Ba} + ^{94}_{36}\text{Kr} + ?$$

In order to get the subscripts and the superscripts in the equation to balance, the reaction must produce two neutrons:

$$^{235}_{92}\text{U} + ^1_0\text{n} \longrightarrow ^{140}_{56}\text{Ba} + ^{94}_{36}\text{Kr} + 2^1_0\text{n}$$

This reaction is an example of a nuclear *chain reaction*, in which the products of the reaction cause more of the same reaction to proceed. The two neutrons can, if they do not escape from the sample first, cause two more such reactions. The four neutrons produced from these reactions can cause four more such reactions, and so forth. Soon, a large number of nuclei are converted, and simultaneously a small amount of matter is converted into a great deal of energy. Atomic bombs and nuclear energy plants both run on this principle.

EXAMPLE 22.9. If each neutron in a certain nuclear reaction can produce two new neutrons, and each reaction takes 1 s, how many neutrons can be produced theoretically in the twentieth second?

Assuming that no neutrons escaped, the number of neutrons produced during the twentieth second is

$$2^{20} = 1\,048\,576$$

The number produced in the sixtieth second is 1.15×10^{18}.

These nuclear reactions actually take place in much less than 1 s each, and the number of reactions can exceed 10^{18} within much less than a minute. Since the energy of each "event" is relatively great, a large amount of energy is available.

Such nuclear reactions are controllable by keeping the sample size small. Most of the neutrons escape from the sample instead of causing further reactions. The smallest mass of sample that can cause a sustained nuclear reaction, called a chain reaction, is called the *critical mass*. Another way to control the nuclear reaction is to insert control rods into the nuclear fuel. The rods absorb some of the neutrons and prevent a runaway reaction.

When a positron is emitted from a nucleus, it can combine with an electron to produce energy. Show that the following equations, when combined, yield exactly the number of electrons required for the product nucleus.

$$^{22}\text{Na} \longrightarrow ^{22}\text{Ne} + _{+1}\beta$$

$$\text{e}^- + _{+1}\beta \longrightarrow \text{energy}$$

One of the 11 electrons outside the Na nucleus is annihilated in the second reaction, leaving 10 electrons for the Ne nucleus.

22.6   NUCLEAR ENERGY

Nuclear energy in almost inconceivable quantities can be obtained from nuclear fission and fusion reactions according to Einstein's famous equation.

$$E = mc^2$$

The E in this equation is the energy of the process. The m is the mass of the matter that is converted to energy—the change in rest mass. Note well that it is *not* the total mass of the reactant nucleus, but only the mass of the matter that is converted to energy. Sometimes the equation is written as

$$E = (\Delta m)c^2$$

The c in the equation is the velocity of light, 3.00×10^8 m/s. The constant c^2 is so large that conversion of a very tiny quantity of mass produces a huge quantity of energy.

EXAMPLE 22.10. Calculate the amount of energy produced when 1.00 g of matter is converted to energy. (Note: More than 1.00 g of isotope is used in this reaction.)

$$1 \text{ J} = 1 \text{ kg} \cdot \text{m}^2/\text{s}^2 = 10^3 \text{ g} \cdot \text{m}^2/\text{s}^2$$

$$E = (\Delta m)c^2$$

$$= (1.00 \text{ g})(3.00 \times 10^8 \text{ m/s})^2$$

$$= 9.00 \times 10^{16} \text{ g} \cdot \text{m}^2/\text{s}^2 = 9.00 \times 10^{13} \text{ J}$$

$$= 9.00 \times 10^{10} \text{ kJ}$$

Ninety billion kilojoules of energy is produced by the conversion of 1 g of matter to energy! The tremendous quantities of energy available in the atomic bomb and the hydrogen bomb stem from the large value of the constant c^2 in Einstein's equation. Conversion of a tiny portion of matter yields a huge quantity of energy. To produce electricity commercially, nuclear plants also rely on this type of energy.

Nuclear fusion reactions involve combinations of nuclei. The fusion reaction of the "hydrogen bomb" involves the fusing of deuterium, $_1^2\text{H}$, in Li ^{2}H:

$$_1^2\text{H} + _1^2\text{H} \longrightarrow _2^3\text{He} + _0^1\text{n}$$

$$_1^2\text{H} + _1^2\text{H} \longrightarrow _1^3\text{H} + _1^1\text{H}$$

The ^{3}H produced (along with that produced from the fission of ^{6}Li) can react further, yielding even greater energy per event.

$$_1^3\text{H} + _1^2\text{H} \longrightarrow _2^4\text{He} + _0^1\text{n}$$

These fusion processes must be started at extremely high temperatures—on the order of tens of millions of degrees Celsius—which are achieved on earth by fission reactions. That is, the hydrogen bomb is triggered by an atomic bomb. Nuclei have to get very close for a fusion reaction to occur, and the strong repulsive force between two positively charged nuclei tends to keep them apart. Very high temperatures give the nuclei enough kinetic energy to overcome this repulsion. No such problem exists with fission, since the neutron projectile has no charge and can easily get close to the nuclear target. (A report was made in 1989 of a "cold fusion" reaction—a fusion reaction at ordinary temperatures with relatively little energy input—but that report has not yet been confirmed.) The stars get their energy from fusion reactions at extremely high temperatures.

The mass of matter at rest is referred to as *rest mass*. When matter is put into motion, its mass increases corresponding to its increased energy. The extra mass is given by

$$E = mc^2$$

When a nuclear event takes place, some rest mass is converted to extra mass of the product particles because of their high speed or to the mass of photons of light. While the total mass is conserved in the process, some rest mass (that is, some matter) is converted to energy.

Nuclear *binding energy* is the energy equivalent (in $E = mc^2$) of the difference between the mass of the nucleus of an atom and the sum of the masses of its uncombined protons and neutrons. For example, the mass of a ^4_2He nucleus is 4.0026 amu. The mass of a free proton is 1.00728 amu, and that of a free neutron is 1.00866 amu. The free particles exceed the nucleus in mass by

$$2(1.00728) + 2(1.00866) - 4.0015 = 0.0304 \text{ amu}$$

This mass has an energy equivalent of 4.54×10^{-12} J for each He nucleus. You would have to put in that much energy into the combined nucleus to get the free particles; that is why that energy is called the binding energy.

The difference in binding energies of the reactants and products of a nuclear reaction can be used to calculate the energy which the reaction will provide.

EXAMPLE 22.11. The mass of a ^7Li nucleus is 7.0154 amu. Using this value and those given above, calculate the energy given off in the reaction of 1 mol of ^7Li:

$$^7\text{Li} + {}^1\text{H} \longrightarrow 2\,{}^4\text{He}$$

The reactants have a mass of 7.0154 g + 1.00728 g = 8.0227 g. The products have a mass of 2(4.0015 g) = 8.0030 g. The difference in mass, 0.0197 g, has an energy equivalent

$$E = mc^2 = \left(1.97 \times 10^{-5} \text{ kg}\right)\left(3.00 \times 10^8 \text{ m/s}\right)^2$$

$$= 1.77 \times 10^{12} \text{ J}.$$

Over 1 billion kilojoules of energy is produced!

Solved Problems

INTRODUCTION

22.1. What is the difference between C and ^{12}C?

   Ans.   The first symbol stands for the naturally occurring mixture of isotopes of carbon; the second stands for only one isotope—the most common one.

22.2. The alchemists of the middle ages spent years and years trying to convert base metals like lead to gold. Is such a change possible? Why did they not succeed?

   Ans.   Base metals can be changed to gold in nuclear reactions, but not in ordinary chemical reactions, since both are elements The alchemists never succeeded because they could not perform nuclear reactions. (Indeed, they never dreamed of their existence.)

NATURAL RADIOACTIVITY

22.3. What is the difference between α, $^4_2\alpha$, and $^4_2\text{He}^{2+}$?

   Ans.   There is no difference. All three are representations of an alpha particle (or a helium nucleus).

22.4. In a nuclear equation, the subscripts are often omitted, but the superscripts are not. Where can you look to find the subscripts? Why can you *not* look there for the superscripts?

   Ans.   You can look at a periodic table for the atomic numbers, which are the subscripts. You cannot look at the periodic table for the superscripts, because mass numbers are not generally there. (Mass numbers for the few elements that do not occur naturally are provided in parentheses in most periodic tables.) The atomic weights that are given can give a clue to the mass number of the most abundant isotope in many cases, but not all. For example, the atomic weight of Br is nearest 80, but the two isotopes that represent the naturally occurring mixture have mass numbers 79 and 81.

22.5. Complete the following nuclear equations:

$$^{233}_{91}\text{Pa} \longrightarrow ^{233}_{92}\text{U} + ?$$

$$^{221}_{87}\text{Fr} \longrightarrow ^{217}_{85}\text{At} + ?$$

$$^{213}_{83}\text{Bi} \longrightarrow ? + ^{4}_{2}\alpha$$

$$^{213}_{83}\text{Bi} \longrightarrow ^{213}_{84}\text{Po} + ?$$

Ans.

$$^{233}_{91}\text{Pa} \longrightarrow ^{233}_{92}\text{U} + ^{0}_{-1}\beta$$

$$^{221}_{87}\text{Fr} \longrightarrow ^{217}_{85}\text{At} + ^{4}_{2}\alpha$$

$$^{213}_{83}\text{Bi} \longrightarrow ^{209}_{81}\text{Tl} + ^{4}_{2}\alpha$$

$$^{213}_{83}\text{Bi} \longrightarrow ^{213}_{84}\text{Po} + ^{0}_{-1}\beta$$

22.6. Complete the following nuclear equations:

$$^{214}_{83}\text{Bi} \longrightarrow ^{210}_{81}\text{Tl} + ?$$

$$^{214}_{83}\text{Bi} \longrightarrow ^{214}_{84}\text{Po} + ?$$

$$? \longrightarrow ^{211}_{83}\text{Bi} + ^{0}_{-1}\beta$$

$$^{234}_{90}\text{Th} \longrightarrow ^{234}_{91}\text{Pa} + ?$$

Ans.

$$^{214}_{83}\text{Bi} \longrightarrow ^{210}_{81}\text{Tl} + ^{4}_{2}\alpha$$

$$^{214}_{83}\text{Bi} \longrightarrow ^{214}_{84}\text{Po} + ^{0}_{-1}\beta$$

$$^{211}_{82}\text{Pb} \longrightarrow ^{211}_{83}\text{Bi} + ^{0}_{-1}\beta$$

$$^{234}_{90}\text{Th} \longrightarrow ^{234}_{91}\text{Pa} + ^{0}_{-1}\beta$$

22.7. Complete the following nuclear equations:

$$? \longrightarrow ^{119}\text{Sn} + ^{0}\gamma$$

$$^{239}\text{Np} \longrightarrow ^{239}\text{Pu} + ?$$

$$^{22}\text{Na} \longrightarrow ? + ^{0}_{+1}\beta$$

Ans.

$$^{119}_{50}\text{Sn} \longrightarrow ^{119}_{50}\text{Sn} + ^{0}_{0}\gamma$$

$$^{239}_{93}\text{Np} \longrightarrow ^{239}_{94}\text{Pu} + ^{0}_{-1}\beta$$

$$^{22}_{11}\text{Na} \longrightarrow ^{22}_{10}\text{Ne} + ^{0}_{+1}\beta \quad \text{(positron)}$$

22.8. Consider the equation

$$^{238}_{92}\text{U} \longrightarrow ^{234}_{90}\text{Th} + ^{4}_{2}\text{He}$$

Does this equation refer to the nuclei of these elements or to atoms of the elements as a whole?

Ans.   Nuclear equations are written to describe nuclear changes. However, since the correct number of electrons is available for the products, the equations can also be regarded as describing the reactions of complete atoms as well as just their nuclei.

HALF-LIFE

22.9. A certain isotope has a half-life of 3.00 s. How much of a 20.0 g sample of this isotope will remain after 6.00 s?

Ans.   After the first 3.00 s, half the 20.0-g sample will still be the same isotope. After 3.00 s more, half of the 10.0 g remaining will still be the original isotope. That is, 5.00 g will remain.

22.10. One-fourth of a certain sample of a radioactive isotope is present 16.0 min after its original weighing. How much will be present after 8.00 min more?

Ans.   The 16.0 min represents two half-lives, since the number of atoms is reduced to one-fourth the original quantity. (One-half disintegrated in the first half-life, and one-half of those left disintegrated in the second, leaving one-fourth of the original number at the end of the 16.0 min.) The half-life is therefore 8.00 min, and the sample is reduced to one-eighth of its original quantity after 8.00 min more. That is, half of the one-fourth number of atoms remain after one more half-life.

22.11. Draw a graph of the mass of the radioactive atoms left in the decomposition of a 1 200-g sample of a radioactive isotope with a half-life of 10.0 days. Extend the graph to allow readings up to 50 days. Use the vertical axis for mass and the horizontal axis for time.

Ans.

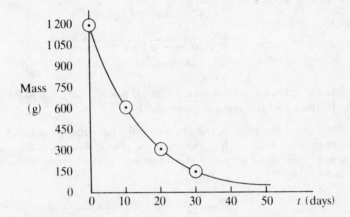

22.12. From the graph of Problem 22.11, estimate how many grams of the isotope will remain after 5.00 days.

Ans.   850 g.

RADIOACTIVE SERIES

22.13. Match the end product and the parent of each of the four radioactive series without consulting any reference tables or other data.

Parent	End Product
^{232}Th	^{206}Pb
^{237}Np	^{208}Pb
^{238}U	^{207}Pb
^{235}U	^{209}Bi

Ans.

^{232}Th	^{208}Pb
^{237}Np	^{209}Bi
^{238}U	^{206}Pb
^{235}U	^{207}Pb

In each case, the final product must differ from the original parent by some multiple of 4 mass numbers. For example, the ^{208}Pb differs in mass number from ^{232}Th by $24 = 4 \times 6$. There must have been six alpha particles emitted in this decay series, with a reduction of four mass numbers each. (The beta and gamma particles emitted do not affect the mass number.)

NUCLEAR FISSION AND FUSION

22.14. Complete the following nuclear equations:

$$^{14}_{7}N + ? \longrightarrow ^{17}_{8}O + ^{1}_{1}H$$

$$^{27}_{13}Al + ^{4}_{2}\alpha \longrightarrow ^{30}_{15}P + ?$$

$$^{235}_{92}U + ? \longrightarrow ^{140}_{56}Ba + ^{94}_{36}Kr + 2\,^{1}_{0}n$$

$$^{235}_{92}U + ^{1}_{0}n \longrightarrow ? + ^{143}_{54}Xe + 3\,^{1}_{0}n$$

Ans.

$$^{14}_{7}N + ^{4}_{2}\alpha \longrightarrow ^{17}_{8}O + ^{1}_{1}H$$

$$^{27}_{13}Al + ^{4}_{2}\alpha \longrightarrow ^{30}_{15}P + ^{1}_{0}n$$

$$^{235}_{92}U + ^{1}_{0}n \longrightarrow ^{140}_{56}Ba + ^{94}_{36}Kr + 2\,^{1}_{0}n$$

$$^{235}_{92}U + ^{1}_{0}n \longrightarrow ^{90}_{38}Sr + ^{143}_{54}Xe + 3\,^{1}_{0}n$$

22.15. What is the difference between (*a*) the mass of an ^{1}H nucleus and the mass of an ^{1}H atom? (*b*) the mass of an ^{1}H atom and the mass number of ^{1}H?

Ans.   (*a*) The difference is the mass of the electron. (*b*) The actual mass of an atom is nonintegral. (Calculations involving mass, such as those using $E = mc^2$, should use the actual mass.) The mass number is an integer, equal to the number of protons plus neutrons in the nucleus. In this case, the mass number is 1.

NUCLEAR ENERGY

22.16. Calculate the energy of the reaction that the free neutron undergoes if it does not encounter a nucleus to react with.

$$n \longrightarrow p + e^-$$

Ans.   The data from Table 22-3 are used. The sum of the masses of the products minus the mass of the neutron, converted to energy with Einstein's equation, yields the energy produced.

$$\Delta m = \left[1.008665 - (1.00728 + 0.00054858)\right] \text{ amu} = 0.00084 \text{ amu}$$

This mass is changed to kilograms, in order to calculate the energy in joules:

$$0.00084 \text{ amu} \left(\frac{1 \text{ g}}{6.022 \times 10^{23} \text{ amu}} \right) \left(\frac{1 \text{ kg}}{1\,000 \text{ g}} \right) = 1.4 \times 10^{-30} \text{ kg}$$

The energy is given by

$$E = mc^2 = \left(1.4 \times 10^{-30} \text{ kg}\right)\left(3.00 \times 10^8 \text{ m/s}\right)^2 = 1.3 \times 10^{-13} \text{ J}$$

22.17. Calculate the energy of the reaction of a positron with an electron.

$$_{+}\beta + e^- \longrightarrow \text{energy}$$

Ans.   Data from Table 22-3 are used. The sum of the masses of the reactants is converted to energy with Einstein's equation.

$$\Delta m = 2(0.00054858 \text{ amu}) = 0.0010972 \text{ amu}$$

$$0.0010972 \text{ amu} \left(\frac{1 \text{ g}}{6.022 \times 10^{23} \text{ amu}} \right) \left(\frac{1 \text{ kg}}{1\,000 \text{ g}} \right) = 1.822 \times 10^{-30} \text{ kg}$$

The energy is given by

$$E = mc^2 = \left(1.822 \times 10^{-30} \text{ kg} \right) \left(3.000 \times 10^8 \text{ m/s} \right)^2 = 1.640 \times 10^{-13} \text{ J}$$

22.18. Calculate the energy of the reaction of one atom of ^{14}C to yield ^{14}N and a beta particle. The *atomic* masses are ^{14}C = 14.003241 amu and ^{14}N = 14.003074 amu.

Ans.   Since the extranuclear electrons are not affected, the nuclear reaction may be represented by

$$^{14}\text{C} \longrightarrow {}^{14}\text{N}^+ + \beta$$

The six extranuclear electrons from the carbon plus the one corresponding to the beta particle are exactly the seven needed for the nitrogen atom. Thus, the energy may be calculated from the difference between the *atomic masses*:

$$\Delta m = 14.003241 \text{ amu} - 14.003074 \text{ amu} = 0.000167 \text{ amu}$$

$$0.000167 \text{ amu} \left(\frac{1 \text{ g}}{6.022 \times 10^{23} \text{ amu}} \right) \left(\frac{1 \text{ kg}}{1\,000 \text{ g}} \right) = 2.77 \times 10^{-31} \text{ kg}$$

The energy is given by

$$E = mc^2 = \left(2.77 \times 10^{-31} \text{ kg} \right) \left(3.00 \times 10^8 \text{ m/s} \right)^2 = 2.49 \times 10^{-14} \text{ J}$$

Supplementary Problems

22.19. What is the difference between β and $_+\beta$?

Ans. β is a high-energy electron. $_+\beta$ is a positron—a particle with the same properties as the electron except for the sign of its charge, which is positive.

22.20. Without consulting any references, text, or tables, add the correct symbol corresponding to each pair of subscript and superscript. (Example: $^{12}_{6}$? is $^{12}_{6}$C.) (a) $^{1}_{0}$?, (b) $^{1}_{1}$?, (c) $^{0}_{+}$?, (d) $^{0}_{-}$?, (e) $^{0}_{0}$?, (f) $^{4}_{2}$?, and (g) $^{2}_{1}$?.

Ans.   (a) $^{1}_{0}$n, (b) $^{1}_{1}$p or $^{1}_{1}$H, (c) $^{0}_{+}\beta$, (d) $^{0}_{-}\beta$ or $^{0}_{-}$e, (e) $^{0}_{0}\gamma$, (f) $^{4}_{2}\alpha$ or $^{4}_{2}$He, and (g) $^{2}_{1}$d or $^{2}_{1}$H.

22.21. What is the difference between radioactive decay processes and other types of nuclear reactions?

Ans.   Other types of reactions require a small particle to react with a nucleus to produce a nuclear reaction; radioactive decay processes are spontaneous with only the one nucleus as reactant.

2.22. (a) How many alpha particles are emitted in the radioactive decay series starting with $^{235}_{92}$U and ending with $^{207}_{82}$Pb? (b) How many beta particles are emitted? (c) Can you tell the order of these emissions without consulting reference data? (d) Can you tell how many gamma particles are emitted?

Ans.   (a) The mass number changes by 28 units in this series, corresponding to seven alpha particles ($7 \times 4 = 28$). (b) The seven alpha particles emitted would have reduced the atomic number by 14 units if no beta particles had been emitted; since the atomic number is reduced by only 10 units, four ($14 - 10$) beta particles are also emitted. (c) It is impossible to tell from these data alone what the order of emission is. (d) It is impossible to tell the number of gamma particles emitted, since emission of a gamma particle does not change either the mass number or the atomic number.

22.23. In a certain type of nuclear reaction, one neutron is a projectile (a reactant) and two neutrons are produced. Assume that each process takes 1 s. If every product neutron causes another event, how many neutrons will be produced (and not be used up again) (*a*) in 3 s? (*b*) in 10 s?

Ans.

（*a*）
$$n \xrightarrow[1\,s]{} 2n \xrightarrow[1\,s]{} 4n \xrightarrow[1\,s]{} 8n$$

In 3 s, $2^3 = 8$ neutrons are produced. (*b*) In 10 s, $2^{10} = 1\,024$ neutrons will be produced.

22.24. In a certain type of nuclear reaction, one neutron is a projectile (a reactant) and two neutrons are produced. Assume that each process takes 1 s. Suppose that half of all the product neutrons cause another event each, and the other half escape from the sample. How many neutrons will be produced in the third second?

Ans.   Two.

$$
\begin{array}{ccccc}
 & \text{one escapes} & & \text{one escapes} & \\
 & \nearrow & & \nearrow & \\
n \longrightarrow 2n & & 2n & & 2n \\
 & \text{one reacts} & & \text{one reacts} &
\end{array}
$$

The reaction continues with the same number of neutrons being produced as started the reaction in the first place.

22.25. If 1.00 mol of C is burned to CO_2 in an ordinary chemical reaction, 393×10^3 J of energy is liberated. (*a*) If 12.0 g of C could be totally converted to energy, how much energy could be liberated? (*b*) If 0.00100% of the mass could be converted to energy, how much energy could be liberated?

Ans.

（*a*）
$$E = mc^2 = (12.0\ \text{g})(3.00 \times 10^8\ \text{m/s})^2 = 1.08 \times 10^{18}\ \text{g} \cdot \text{m}^2/\text{s}^2$$
$$= 1.08 \times 10^{15}\ \text{kg} \cdot \text{m}^2/\text{s}^2 = 1.08 \times 10^{15}\ \text{J}$$

（*b*）
$$0.0000100 \times 1.08 \times 10^{15}\ \text{J} = 1.08 \times 10^{10}\ \text{J}$$

Over 25 000 times more energy would be liberated by converting 0.00100% of the carbon mass to energy than by burning all of it chemically.

22.26. How many beta particles are emitted in the decomposition series from $^{238}_{92}\text{U}$ to $^{206}_{82}\text{Pb}$?

Ans.   The number of alpha particles, calculated from the loss of mass number, is 8, because the mass number was lowered by 32. The number of beta particles is equal to twice the number of alpha particles minus the difference in atomic numbers of the two isotopes:

$$(2 \times 8) - 10 = 6$$

Six beta particles are emitted. (See Example 22.7 and Problem 22.22.)

22.27. Can the half-life of an isotope be affected by changing the compound it is in?

Ans.   No. The chemical environment has a negligible effect on the nuclear properties of the atom.

Glossary

A: (1) symbol for mass number. (2) symbol for ampere.

absolute temperature: temperature on the Kelvin scale.

absolute zero: 0 K = −273.15 °C; the coldest temperature theoretically possible.

acetylene: ethyne; $CH\equiv CH$.

acid: (1) a compound containing ionizable hydrogen atoms. (2) a proton donor (Brønsted theory).

acid, organic: a compound of the general type RCOOH.

acid salt: a salt produced by partial neutralization of an acid containing more than one ionizable hydrogen atom, for example, $NaHSO_4$.

activity: reactivity; tendency to react.

alcohol: an organic compound with molecules containing a covalently bonded —OH group on the radical; R—OH.

aldehyde: an organic compound of the general type RCHO.

alkali metal: a metal of periodic group IA: Li, Na, K, Rb, Cs, or Fr.

alkaline earth metal: an element of periodic group IIA.

alkane: a hydrocarbon containing only single bonds.

alkene: a hydrocarbon containing one double bond.

alkyl radical: a hydrocarbon radical from the alkane series.

alkyne: a hydrocarbon containing one triple bond.

alpha particle: a ^4He nucleus ejected from a larger nucleus in a spontaneous radioactive reaction.

amine: an organic compound of the general type RNH_2, R_2NH, or R_3N.

ammonia: NH_3.

ammonium ion: NH_4^+.

ampere: unit of electric current. 1 A = 1 C/s.

amphiprotic: the ability of a substance to react with itself to produce both a conjugate acid and base. Water is amphiprotic, producing H_3O^+ and OH^-.

-ane: name ending for the alkane series of hydrocarbons.

anhydrous: without water. The waterless salt capable of forming a hydrate, such as $CuSO_4$ which reacts with water to form $CuSO_4\cdot 5H_2O$.

anion: a negative ion.

anode: electrode at which oxidation takes place, in either a galvanic cell or an electrolysis cell.

aqueous solution: water solution.

aromatic hydrocarbon: a hydrocarbon containing at least one benzene ring, perhaps with hydrogen atoms replaced by other groups.

349

Arrhenius theory: theory of acids and bases in which acids are defined as hydrogen-containing compounds that react with bases.

aryl radical: a hydrocarbon radical derived from the aromatic series.

atmosphere: the air surrounding the earth.

atmosphere, standard: a unit of pressure, equal to 760 torr.

atmospheric pressure: the pressure of the atmosphere; barometric pressure.

atom: the smallest particle of an element that retains the composition of the element.

atomic energy: energy from the nuclei of atoms; nuclear energy.

atomic number: the number of protons in the nucleus of the atom.

atomic theory: Dalton's postulates, based on experimental evidence, which proposed that all matter is composed of atoms.

atomic weight: the relative mass, compared to ^{12}C, of an average atom of an element.

autoionization: reaction of a substance with itself to produce ions.

auto-oxidation: reaction of a substance with itself to produce a product with a lower oxidation number and another with a higher oxidation number; disproportionation.

Avogadro's number: a mole; 6.02×10^{23} units; the number of atomic mass units per gram.

balancing an equation: adding coefficients to make the numbers of atoms of each element the same on both sides of an equation.

balancing a nuclear equation: ensuring that the totals of the charges and the totals of the mass numbers are the same on both sides of a nuclear equation.

barometer: a device for measuring air pressure.

barometric pressure: the pressure of the atmosphere.

base: (1) a compound containing OH^- ions or which reacts with water to form OH^- ions. (2) a proton acceptor (Brønsted theory).

base of an exponential number: the number or unit which is multiplied by itself. For example, the base in 2.0×10^4 is 10, and in 5 cm^3 it is cm.

battery: a combination of two or more galvanic cells.

benzene: a cyclic compound with molecular formula C_6H_6; the base for the aromatic hydrocarbon series.

beta particle: a high-energy electron ejected from a nucleus in a nuclear reaction.

binary: composed of two elements.

binding energy: the energy equivalent of the difference between the mass of a nucleus and the sum of the masses of the (uncombined) protons and neutrons that make it up.

Bohr theory: the first theory of atomic structure which involved definite internal energy levels for electrons.

Boltzmann constant, k: a constant equal to the gas law constant divided by Avogadro's number: $k = 1.38 \times 10^{-23}$ J/molecule $\cdot$ K.

bonding: the chemical attraction of atoms for each other within chemical compounds.

Boyle's law: the volume of a given sample of gas at constant temperature is inversely proportional to the pressure of the gas: $V = k/P$.

Brønsted theory: a theory of acids and bases that defines acids as proton donors and bases as proton acceptors.

buffer solution: a solution of a weak acid and its conjugate base or a weak base and its conjugate acid. The solution resists change in pH even on addition of small quantities of strong acid or base.

buret: a calibrated tube used to deliver exact volumes of liquid.

c: the velocity of light in a vacuum: 3.00×10^8 m/s.

carbonyl group: the $\diagdown C = O$ group in an organic compound.

catalyst: a substance that alters the rate of a chemical reaction without undergoing permanent change in its own composition.

cathode: electrode at which reduction takes place, in either a galvanic cell or an electrolysis cell.

cation: a positive ion.

Celsius temperature scale: a temperature scale with 0° defined as the freezing point of pure water and 100° defined as the normal boiling point of pure water.

centi-:   prefix meaning 0.01.

chain reaction: a self-sustaining series of reactions in which the products of one reaction, such as neutrons, initiate more of the same reaction. One such particle can therefore start a whole series of reactions.

Charles' law: for a given sample of gas at constant pressure, the volume is directly proportional to the *absolute* temperature: $V = kT$.

circuit: a complete path necessary for an electric current.

classification of matter: the grouping of matter into elements, compounds, and mixtures.

coefficient: (1) the number placed before the formula of a reactant or product to balance an equation. (2) the decimal value in a number in standard exponential notation.

coinage metal: an element of periodic group IB: Cu, Ag, or Au.

combination reaction: a reaction in which two elements, an element and a compound, or two compounds combine to form a single compound.

combined gas law: for a given sample of gas, the volume is inversely proportional to the pressure and directly proportional to the absolute temperature:

$$\frac{P_1 V_1}{T_1} = \frac{P_2 V_2}{T_2}$$

combustion: reaction with oxygen gas.

completion: complete consumption of at least one of the reactants in a chemical reaction. A reaction goes to completion if its limiting quantity is used up.

compound: a chemical combination of elements.

concentration: the number of parts of solute in a given quantity of solution (molarity or normality) or of solvent (molality).

conduct: allow passage of electricity by movement of electrons in a wire or ions in a liquid.

conjugates: an acid or base plus the product of the reaction of that acid or base with water. The product is a base or acid that differs from the original substance by H^+.

"control" of electrons: assignment of the electrons in a covalent bond for oxidation number purposes to the more electronegative atom sharing them.

controlled experiment: a series of individual experiments in which all factors except one are held constant so that the effect of that factor on the outcome can be determined. Another series of experiments can be performed for each additional factor to be tested.

coulomb: unit of electric charge; 96 500 C is equal to the charge on 1 mol of electrons.

covalent bonding: bonding by shared electron pairs.

critical mass: the smallest mass of a sample that will sustain a chain reaction. Smaller masses will lose neutrons or other projectile particles from their bulk and there will not be sufficient projectile particles to keep the chain going.

current: (1) the flow of electrons or ions; (2) passage of a certain number of coulombs per second.

cycloalkane: a hydrocarbon containing a ring of carbon atoms and only single bonds.

cycloalkene: a hydrocarbon containing a ring of carbon atoms and one double bond.

dalton: a unit of atomic weight; 1 amu.

Dalton's law of partial pressures: the total pressure of a mixture of gases is equal to the sum of the partial pressures of the components.

Daniell cell: a galvanic cell composed of copper/copper(II) ion and zinc/zinc ion half-cells.

daughter nucleus: the large nucleus (as opposed to an alpha particle) that results from the spontaneous disintegration of a (parent) nucleus.

decay: radioactive disintegration.

decomposition reaction: a reaction in which a compound decomposes to yield two elements, an element and a compound, or two compounds.

definite proportions: having the same ratio of masses of individual elements in every sample of the compound.

delocalized double bonds: double bonds that are not permanently located between two specific atoms. The electron pairs can be written equally well between one pair of atoms or another, as in benzene.

Δ: Greek letter delta, meaning "change of."

density: mass divided by volume. A body of lower density will float in a liquid of higher density.

derivative: a compound of a hydrocarbon with at least one hydrogen atom replaced by a functional group.

deuterium: the isotope of hydrogen with a mass number of 2. Also called heavy hydrogen.

deuteron: the nucleus of a deuterium atom.

diatomic: composed of two atoms.

diffusion: the passage of one gas through another.

dimensional analysis: factor-label method.

direct proportion: as the value of one variable rises, the value of the other rises by the same factor. Directly proportional variables have a constant quotient, for example, $V/T = k$ for a given sample of a gas at constant pressure.

disintegration: spontaneous emission from a radioactive nucleus of an alpha, a beta, or a gamma particle.

disproportionation: reaction of a reactant with itself to produce a product with a lower oxidation number and another with a higher oxidation number.

double bond: a covalent bond with two shared pairs of electrons.

double decomposition: double substitution

double replacement: double substitution.

double-substitution reaction: a reaction of (ionic) compounds in which the reactant cations swap anions. Also called double replacement, double decomposition, or metathesis.

Dulong and Petit, law of: the molar heat capacities of crystalline elements are approximately 25 J/mol · deg.

effusion: the escape of gas molecules through tiny openings in the container holding the gas.

Einstein's equation: $E = mc^2$.

elastic collision: a collision in which the total kinetic energy of the colliding particles does not change.

electric current: the concerted (nonrandom) movement of charged particles such as electrons in a wire or ions in a solution.

electrode: a solid conductor of electricity used to connect a current-carrying wire to a solution in an electrolysis cell or a galvanic cell.

electrolysis: a process in which an electric current produces a chemical reaction.

electron: a negatively charged particle that occurs outside the nucleus of the atom and is chiefly responsible for the bonding between atoms.

electron dot diagram: a scheme for representing valence electrons in an atom with dots (or crosses or circles).

electronegativity: the relative attraction for electrons of atoms involved in covalent bonding. The higher the attraction, the higher the electronegativity.

electronic charge: the magnitude of the charge on an electron; 1.60×10^{-19} C.

element: a substance that cannot be broken down into simpler substances by ordinary chemical means.

empirical formula: formula for a compound that contains the simplest whole number ratio of atoms of the elements.

end point: the point in a titration at which the indicator changes color permanently and the titration is stopped.

-ene: name ending for the alkene series of hydrocarbons.

energy: the capacity to produce change or the ability to do work.

energy change: the energy of a system following a change minus the energy of the system before the change: $\Delta E = E_2 - E_1$.

enthalpy change: a thermodynamic property. The enthalpy change is equal to the heat added for a system under constant pressure in which no work other than expansion work is done.

enthalpy of combustion: the enthalpy change accompanying a combustion process.

enthalpy of formation: the enthalpy change in the process of making a compound from its elements in their standard states.

enthalpy of fusion: the enthalpy change accompanying a melting (fusion) process.

enthalpy of vaporization: the enthalpy change accompanying a vaporization process.

equation: notation for a chemical reaction containing formulas of each reactant and product, with coefficients to make the numbers of atoms of each element equal on both sides.

equilibrium: state in which two opposing processes occur at equal rates, causing no apparent change.

equilibrium constant: a constant equal to the ratio of concentrations of products to reactants, each raised to a suitable power, which is dependent for a given reaction on temperature only.

equivalent: (1) in a redox reaction, the quantity of a substance that reacts with or produces 1 mol of electrons. (2) in an acid-base reaction, the quantity of a substance that reacts with or produces 1 mol of H^+ or OH^- ions.

equivalent weight: the number of grams per equivalent of a substance.

ester: an organic compound of the general type RCOOR'.

ether: an organic compound of the general type ROR'.

ethylene: ethene; $CH_2{=}CH_2$.

event: the reaction of one nucleus (plus a projectile, if any) in a nuclear reaction.

excess quantity: more than sufficient of one reagent to ensure that another reagent reacts completely.

excited state: the state of an atom with the electron(s) in higher energy levels than the lowest possible.

exponent: a superscript telling how many times the coefficient is multiplied by the base. For example, the exponent in 2.0×10^3 is 3; the 2.0 is multiplied by 10 three times: $2.0 \times 10^3 = 2.0 \times 10 \times 10 \times 10$.

exponential number: a number expressed with a coefficient times a power of ten, for example, 1.0×10^3.

extrapolation: reading a graph beyond the experimental points.

factor-label method: a method of problem solving that uses the units to indicate which algebraic operation to do for quantities that are directly proportional; dimensional analysis.

faraday: the charge on 1 mol of electrons; 96 500 C.

feeble acid or base: an acid or base that has practically no tendency to react with water.

first law of thermodynamics: the law of conservation of energy; the increase in the energy of a system is equal to the (algebraic sum of the) heat added to the system and the work added to the system.

fission: the process in which a nucleus is split into two more or less equal parts by bombardment with a projectile particle such as a neutron or a proton.

fluid: a gas or liquid.

formula: a combination of symbols with proper subscripts that identifies a compound or molecule.

formula unit: the material represented (1) by the simplest formula of an ionic compound, (2) by a molecule, or (3) by an uncombined atom.

formula weight: the sum of the atomic weights of all the atoms in a formula.

functional group: the reactive part of an organic molecule.

fusion: (1) melting. (2) combining two nuclei in a nuclear reaction.

galvanic cell: a cell in which a chemical reaction produces an electric potential.

gamma ray: a stream of gamma particles, essentially photons of high-energy light with zero rest mass and zero charge.

Graham's law: the rate of effusion or diffusion of a gas is inversely proportional to the square root of its molecular weight.

gram: the basic unit of mass in the metric system.

graphic formula: a formula in which all atoms are shown with their bonding electron pairs represented by lines; structural formula.

ground state: the lowest energy state of an atom.

half-life: the time it takes for one-half of the nuclei of any given sample of a particular radioactive isotope to disintegrate spontaneously.

half-reaction: the oxidation or reduction half of a redox reaction.

halogen: an element of periodic group VIIA: F, Cl, Br, I, (or At).

heat: a form of energy; the only form of energy that cannot be completely converted into another form.

heat capacity: the energy required to change a certain quantity of a substance by a certain temperature.

heating curve: a graph of the temperature of a body versus the quantity of heat added to the body.

heavy hydrogen: deuterium.

Heisenberg uncertainty principle: the location and the energy of a small particle such an an electron cannot both be known precisely at any given time.

heterogeneous: having distinguishable parts.

homogeneous: alike throughout; having parts that are indistinguishable even with an optical microscope.

Hund's rule of maximum multiplicity: when electrons partially but not fully fill a subshell, they remain as unpaired as possible.

hydrate: a compound composed of a stable salt plus some number of molecules of water, for example, $CuSO_4 \cdot 5H_2O$.

hydrocarbon: a compound of carbon and hydrogen only.

hydronium ion: H_3O^+; the product of reaction of a proton and water.

hypothesis: a proposed explanation of observable results.

ideal gas law: $PV = nRT$.

indicator: a substance that has an intense color in acid or base solution and another color in the other type of solution. Indicators are used to determine when the end point of a titration has been reached.

initial concentration: the concentration in an equilibrium system before the equilibrium reaction has proceeded at all.

inner transition series: the two series of elements at the bottom of the periodic table. The series containing elements 58–71 and 90–103, arising from the filling of $4f$ and $5f$ subshells.

interpolation: reading a graph between the experimental points.

inverse proportion: as the value of one variable rises, the value of the other goes down by the same factor. Inversely proportional variables have a constant product, for example, $PV = k$ for a given sample of gas at constant temperature.

ion: a charged atom or group of atoms.

ion-electron method: a method of balancing redox equations using ions in half-reactions.

ionic bond: the attraction between oppositely charged ions.

ionization constant: the equilibrium constant for the reaction of a weak acid or base with water.

isomerism: existence of isomers.

isomers: different compounds having the same molecular formula.

isotopes: two or more atoms of the same element with different numbers of neutrons in their nuclei.

joule: a unit of energy. 4.184 J raises the temperature of 1.00 g of water 1.00 °C.

k: Boltzmann constant: $k = R/N = 1.38 \times 10^{-23}$ J/molecule · K.

K_a: acid ionization constant.

K_b: base ionization constant.

K_i: acid or base ionization constant.

K_w: ionization constant for the autoionization of water.

kelvin: the unit of the Kelvin temperature scale.

Kelvin temperature scale: a temperature scale with its 0 at the lowest theoretically possible temperature (0 K = -273 °C) and with temperature differences the same as those on the Celsius scale.

ketone: an organic compound of the type RCOR'.

kilo: prefix meaning 1 000.

kilogram: the legal standard of mass in the United States; 1 000 g.

kinetic energy: energy of motion, as in a 10-ton Mack truck going 50 miles per hour: $KE = \frac{1}{2}mv^2 = \frac{1}{2}$ (10 ton) (50 miles/hour)^2 (where m is mass and v is velocity).

kinetic molecular theory: a theory that explains the properties of gases in terms of the actions of their molecules.

law: an accepted generalization of observable facts and experiments.

law of conservation of energy: energy cannot be created or destroyed.

law of conservation of mass: mass is neither created nor destroyed during any process.

law of conservation of matter: matter is neither created nor destroyed during any ordinary chemical reaction. This law differs from the law of conservation of mass because rest mass can be converted into energy (which has its own mass associated with it). Some matter is converted to energy in nuclear reactions.

law of definite proportions: all samples of a given compound, no matter what their source, have the same percentage of each of the elements.

law of Dulong and Petit: the molar heat capacity of crystalline elements is approximately 25 J/mol · deg.

law of multiple proportions: for a given mass of one element, the two masses of each other element in two compounds of the same elements are in a ratio of small whole numbers.

lead storage cell: a cell composed of $Pb/PbSO_4$ and $PbO_2/PbSO_4$ electrodes in an H_2SO_4 electrolyte.

Le Châtelier's principle: if a stress is applied to a system at equilibrium, the equilibrium will shift in a tendency to reduce the stress.

light: a form of energy. Light has both wave and particle properties, with a speed in vacuum of 3.00×10^8 m/s.

limiting quantity: the reagent that will be used up before all of the other reagent(s); the substance that will be used up first in a given chemical reaction, causing the reaction to stop.

line formula: a formula for an organic compound written on one line, in which bonded groups of atoms are written together. For example, $CH_3CHClCH_3$.

liter: the basic unit of volume of the metric system (about 6% greater than a U.S. quart). The cubic meter is the basic unit of volume in SI.

m: (1) symbol for mass. (2) symbol for meter. (3) symbol for milli. (4) symbol for molal (unit of molality).

M: symbol for molar, the unit of molarity.

main group: one of the eight groups at the left and right of the periodic table that extend as high as the first or second period.

mass: a quantitative measure of the quantity of matter and energy in a body. Mass is measured by its direct proportionality to weight and/or to inertia (the resistance to change in the body's motion).

matter: anything that has mass and occupies space.

metathesis: double substitution.

meter: the basic unit of length in the metric system (about 10% greater than a yard).

metric system: a system of measurement based on the decimal system and designed for ease of use, in which prefixes are used that have the same meanings no matter what unit they are used with.

metric ton: 10^6 g = 1 000 kg = 2 200 lb.

milli: prefix meaning 0.001.

millimole: 0.001 mol.

mixture: a physical combination of elements and/or compounds.

mol: abbreviation for mole.

molality: number of moles of solute per kilogram of solute.

molar: unit of molarity.

molar heat capacity: the heat capacity per mole of substance.

molar mass: the mass of 1 mol of a substance. The molar mass is numerically equal to the formula weight, but the dimensions are grams per mole.

molarity: number of moles of solute per liter of solution.

mole: 6.02×10^{23} units; Avogadro's number.

mole fraction: the ratio of moles of a component to total number of moles in a solution:

$$X_A = \frac{\text{mol A}}{\text{total moles}}$$

mole ratio: a ratio of moles, such as given by a chemical formula or a balanced chemical equation.

molecular formula: the formula of a compound that gives the ratio of number of atoms of each element to number of molecules of compound. The molecular formula is an integral multiple of the empirical formula.

molecular weight: the formula weight of a molecular substance. The sum of the atomic weights of the atoms in a molecule.

molecule: a combination of atoms held together by covalent bonds.

monatomic: consisting of one atom. A monatomic ion contains only one atom.

Nernst equation: an equation to calculate the actual potential of a cell in which the concentrations or pressures differ from 1.00 *M* or 1.00 atm.

net ionic equation: an equation in which spectator ions (ions that begin in solution and wind up unchanged as the same ion in solution) are omitted.

neutralization: the reaction of an acid with a base.

neutron: a neutral particle in the nucleus of the atom.

noble gas configuration: an octet of electrons in the outermost shell, such as the noble gases possess when they are uncombined. A pair of electrons in the first shell if that is the only shell, such as in helium.

normal: unit of normality.

normality: the number of equivalents of solute per liter of solution.

nuclear energy: energy from reactions of nuclei.

nuclear reaction: a reaction in which at least one nucleus undergoes change.

nucleus: the tiny center of an atom containing the protons and neutrons.

octet: a set of eight electrons in the outermost shell.

octet rule: a generalization that atoms tend to form chemical bonds to get eight electrons in their outermost shells or that eight electrons in the outermost shell of an atom is a stable state. Helium is said to obey the octet rule when it has two electrons in its outermost shell, because the first shell can hold a maximum of two electrons.

orbit: the circular path of an electron about the nucleus in the Bohr theory.

orbital: a subdivision of an energy level in which an electron will have a given value for each of *n*, *l*, and *m*.

order: the exponent of a concentration in the rate law equation.

organic chemistry: chemistry of compounds with C—C and/or C—H bonds.

outermost shell: the largest shell containing electrons in an atom or ion.

overall equation: an equation with all ions present, as opposed to a net ionic equation.

oxidation: raising of oxidation number, by loss of (control of) electrons.

oxidation number: the number of outermost electrons of a free atom minus the number of electrons the atom "controls" in a compound or molecule.

oxidation number change method: a method for balancing redox equations by balancing changes in oxidation numbers first.

oxidation state: oxidation number.

oxidizing agent: reactant that causes an increase in the oxidation number of another reactant.

oxo-:   prefix meaning combined oxygen, as in oxovanadium(IV) ion: VO^{2+}.

oxyanion: a negative ion containing oxygen as well as another element, for example, ClO_4^{-}.

paraffin: a saturated hydrocarbon.

parent nucleus: a nucleus that disintegrates spontaneously yielding a small particle plus another nucleus of size reasonably equal to itself.

partial pressure: the pressure of a component of a gas mixture.

Pauli exclusion principle: no two electrons in a given atom can have the same set of four quantum numbers.

per: divided by. For example, the number of miles per hour is calculated by dividing the total number of miles by the total number of hours.

percent composition: the number of grams of each element in a compound per 100 g of the compound.

percent yield: 100 times the actual yield divided by the calculated yield:

$$100 \times \frac{\text{actual yield}}{\text{calculated yield}}$$

percentage: the number of units of an item present per 100 units total. For example, 73 g of sand in 100 g of a mixture (or 7.3 g in 10.0 g mixture, or 730 g in 1 000 g mixture) is 73% sand.

period: a horizontal array of elements in the periodic table.

periodic table: a tabulation of the elements according to atomic number with elements having similar properties in the same (vertical) group.

peroxide ion: O_2^{2-} ion.

pH: $-\log [H_3O^+]$.

phase change: change of state, as for example change of solid to liquid or solid to gas.

photon: a particle of light.

physical change: a process in which no change in composition occurs.

pipet: a piece of glassware calibrated to deliver an exact volume of liquid.

polyatomic: composed of more than one atom.

positron: a subatomic particle that may be ejected from a nucleus, with all the properties of an electron except for the sign of its charge, which is positive.

potential: the driving force for electric current.

potential energy: energy of position, as for example in a rock on top of a mountain.

precipitate: (1) form a solid from solute in solution. (Form solid or liquid from the water in air.) (2) the solid so formed.

pressure: force per unit area.

product: element or compound produced in a chemical reaction.

property: a characteristic of a substance by which the substance can be identified.

propylene: propene; $CH_3CH\!=\!CH_2$.

proton: (1) a positive particle in the nucleus of the atom. (2) the H^+ ion.

q: symbol for heat.

quadratic formula: a formula for solving for x in an equation of the general form:

$$ax^2 + bx + c = 0$$
$$x = \frac{-b \pm \sqrt{b^2 - 4ac}}{2a}$$

quantum (plural, quanta):   a particle of energy.

quantum number: one of four values that control the properties of the electron in the atom.

R: ideal gas law constant: $0.0821\ \text{L} \cdot \text{atm/mol} \cdot \text{K}$.

R: Rydberg constant (see a general chemistry text).

R—:   symbol for a radical in organic chemistry.

radical: the hydrocarbon-like portion of an organic compound with one hydrogen atom of the hydrocarbon replaced by another group.

radioactive decay: the disintegration of a sample of a naturally radioactive isotope.

radioactive series: a series of isotopes produced one from the other in a sequence of spontaneous radioactive disintegrations.

rate constant: the constant k in the rate law equation.

rate law: an equation that relates the rate of a chemical reaction to the concentrations of its reactants:

$$\text{rate} = k[A]^x[B]^y$$

rate of reaction: the number of moles per liter of product that is produced by a chemical reaction per unit time.

reactant: element or compound used up in a chemical reaction.

reacting ratio: mole ratio in which reagents react and are produced, given by the coefficients in the balanced chemical equation.

reaction: a chemical change; a process in which compositions of substances are changed.

reactivity: activity; tendency to react.

reagent: element or compound used up in a chemical reaction.

redox: oxidation-reduction.

reducing agent: reactant that causes a lowering of the oxidation number of another reactant.

reduction: lowering of oxidation number, by gain of (control of) electrons.

reduction potential: a quantitative measure of the tendency of a species to be reduced. Used in galvanic cell calculations.

rest mass: the mass of a body at rest. (According to Einstein's theory, a body's mass will increase as it is put into motion—as its energy increases—despite its having a constant quantity of matter and thus a constant rest mass.)

rounding off: reducing the number of digits in a calculated result to indicate the precision of the measurement(s).

salt: a combination of a cation and an anion (except for H^+, H_3O^+, O^{2-}, or OH^-). Salts may be produced by the reaction of acids and bases.

salt bridge: a connector between the two halves of a galvanic cell necessary to complete the circuit and prevent a charge buildup which would cause stoppage of the current.

saturated hydrocarbon: a hydrocarbon containing only single bonds.

saturated solution: a solution holding as much solute as it can stably hold at a particular temperature. Solutions in equilibrium with excess solute are saturated.

scientific notation: standard exponential notation.

SH: abbreviation for specific heat.

shell: (1) a "layer" of electrons in an atom; (2) a set of electrons all having the same principal quantum number.

shift: a change in the position of an equilibrium system. Shift to the left produces reactants from products; shift to the right produces products from reactants.

SI (Système International d'Unités): a modern version of the metric system.

significant digits: significant figures.

significant figures: the digits in a measurement or in the calculations resulting from a measurement that indicate how precisely the measurement was made; significant digits.

single bond: a covalent bond with one shared pair of electrons.

solubility: the concentration of a saturated solution at a given temperature.

solute: substance dissolved in a solvent, such as salt dissolved in water.

solution: a homogeneous mixture; a combination of a solvent and solute(s).

solvent: substance that dissolves the solute.

specific heat: the heat capacity per gram of substance; the number of joules required to raise 1 g of substance 1 °C.

spectator ions: ions that are in solution at the start of a reaction and wind up unchanged (as the same ion in solution).

stable: resistant to reaction.

standard conditions: 0 °C and 1 atm pressure; STP.

standard exponential notation: exponential notation with the coefficient having a value of 1 or more but less than 10.

standard reduction potential: reduction potential for a cell in which all solutes are 1.00 M and all gases are at 1.00 atm.

standard state: the state of a substance in which it usually occurs at 25 °C and 1 atm pressure. For example, the standard state of elementary oxygen is O_2 molecules in the gaseous state.

standard temperature and pressure: 0 °C and 1 atm pressure.

Stock system: the nomenclature system using oxidation numbers to differentiate between compounds or ions of a given element.

stoichiometry: the science of measuring how much of one substance can be produced from given quantities of others.

STP: standard temperature and pressure; 0 °C and 1 atm pressure.

stress: a change in conditions imposed on a system at equilibrium.

strong acid: an acid that reacts completely with water to form ions.

structural formula: a formula in which all atoms are shown with their covalently bonded electron pairs represented by lines; graphic formula.

subdivision: particle size of solids.

sublimation: changing directly from a solid to a gas.

subshell: the set of electrons in an atom all having the same value of n and the same value of l.

substance (pure substance): an element or a compound.

substitution reaction: the reaction of a free element plus a compound in which the free element replaces one of the elements in the compound and enters the compound itself. For example,

$$2\,Na + ZnCl_2 \longrightarrow 2\,NaCl + Zn$$

superoxide ion: O_2^- ion.

supersaturated solution: a solution holding more solute than the solution can hold stably at the particular temperature.

symbol: one- or two-letter representation of an element or an atom of an element. The first letter of a symbol is capitalized; the second letter, if any, is lowercase.

system: portion of matter under investigation, such as the contents of a particular beaker.

t: symbol for Celsius temperature.

T: symbol for absolute temperature.

temperature: the intensity of heat in an object. The property that determines the direction of spontaneous heat flow when bodies are connected thermally.

ternary compound: a compound of three elements.

theory: an accepted explanation of experimental results.

thermochemistry: the science that investigates the interaction of heat and chemical reactions.

titration: measured neutralization reaction (or other type) to determine the concentration or number of moles of one reactant from data about the other.

total bond order: the number of pairs of electrons on an atom that are shared with other atoms.

transition groups: the groups that extend only as high as the fourth period of the periodic table. That is, the groups containing the elements with atomic numbers 21–30, 39–48, 57, and 72–80, arising from the filling of the $3d$, $4d$, and $5d$ subshells with electrons.

transmutation: the change of one element into another (by a nuclear reaction).

triple bond: a covalent bond with three shared pairs of electrons.

tritium: the isotope of hydrogen with mass number 3.

unsaturated solution: a solution holding less solute than a saturated solution would hold at that temperature.

valence shell: the outermost electron shell of a neutral atom.

vaporization: changing from a liquid to a gas; evaporation.

vapor pressure: the pressure of the vapor of a liquid in equilibrium with that same substance in the liquid phase. The vapor pressure for a given substance is determined by the temperature only.

voltage: electric potential.

volumetric flask: a flask calibrated to hold an exact volume of liquid.

w: (1) symbol for weight. (2) symbol for work.

weak acid: a hydrogen-containing compound that reacts somewhat but does not react completely with water to form ions. (Weak acids react almost completely with bases, however.)

weight: the measure of the attraction of a body to the earth.

work: (1) force times the distance through which the force is applied. (2) all forms of energy except for heat.

X_A: mole fraction of A.

-yne: name ending for the alkyne series of hydrocarbons.

Z: symbol for atomic number.

INDEX

Table of the Elements

Element	Symbol	Atomic Number	Atomic Weight
Actinium	Ac	89	(227)
Aluminum	Al	13	26.9815
Americium	Am	95	(243)
Antimony	Sb	51	121.75
Argon	Ar	18	39.948
Arsenic	As	33	74.9216
Astatine	At	85	(210)
Silver	Ag	47	107.868
Gold	Au	79	196.9665
Barium	Ba	56	137.34
Berkelium	Bk	97	(249)
Beryllium	Be	4	9.01218
Bismuth	Bi	83	208.9806
Boron	B	5	10.81
Bromine	Br	35	79.904
Cadmium	Cd	48	112.40
Calcium	Ca	20	40.08
Californium	Cf	98	(251)
Carbon	C	6	12.011
Cerium	Ce	58	140.12
Cesium	Cs	55	132.9055
Chlorine	Cl	17	35.453
Chromium	Cr	24	51.996
Cobalt	Co	27	58.9332
Copper	Cu	29	63.546
Curium	Cm	96	(247)
Dysprosium	Dy	66	162.50
Einsteinium	Es	99	(254)
Erbium	Er	68	167.26
Europium	Eu	63	151.96
Fermium	Fm	100	(253)
Fluorine	F	9	18.9984
Francium	Fr	87	(223)
Iron	Fe	26	55.847
Gadolinium	Gd	64	157.25
Gallium	Ga	31	69.72
Germanium	Ge	32	72.59
Gold	Au	79	196.9665
Hafnium	Hf	72	178.49
Helium	He	2	4.00260
Holmium	Ho	67	164.9303
Hydrogen	H	1	1.0080
Mercury	Hg	80	200.59
Indium	In	49	114.82
Iodine	I	53	126.9045
Iridium	Ir	77	192.22
Iron	Fe	26	55.847
Krypton	Kr	36	83.80
Potassium	K	19	39.102
Lanthanum	La	57	138.9055
Lawrencium	Lr	103	(257)
Lead	Pb	82	207.2
Lithium	Li	3	6.941
Lutetium	Lu	71	174.97
Magnesium	Mg	12	24.305
Manganese	Mn	25	54.9380
Mendelevium	Md	101	(256)

Element	Symbol	Atomic Number	Atomic Weight
Mercury	Hg	80	200.59
Molybdenum	Mo	42	95.94
Neodymium	Nd	60	144.24
Neon	Ne	10	20.179
Neptunium	Np	93	237.0482
Nickel	Ni	28	58.71
Niobium	Nb	41	92.9064
Nitrogen	N	7	14.0067
Nobelium	No	102	(254)
Sodium	Na	11	22.9898
Osmium	Os	76	190.2
Oxygen	O	8	15.9994
Palladium	Pd	46	106.4
Phosphorus	P	15	30.9738
Platinum	Pt	78	195.09
Plutonium	Pu	94	(242)
Polonium	Po	84	(210)
Potassium	K	19	39.102
Praseodymium	Pr	59	140.9077
Promethium	Pm	61	(145)
Protactinium	Pa	91	231.0359
Lead	Pb	82	207.2
Radium	Ra	88	226.0254
Radon	Rn	86	(222)
Rhenium	Re	75	186.2
Rhodium	Rh	45	102.9055
Rubidium	Rb	37	85.4678
Ruthenium	Ru	44	101.07
Samarium	Sm	62	150.4
Scandium	Sc	21	44.9559
Selenium	Se	34	78.96
Silicon	Si	14	28.086
Silver	Ag	47	107.868
Sodium	Na	11	22.9898
Strontium	Sr	38	87.62
Sulfur	S	16	32.06
Antimony	Sb	51	121.75
Tin	Sn	50	118.69
Tantalum	Ta	73	180.9479
Technetium	Tc	43	98.9062
Tellurium	Te	52	127.60
Terbium	Tb	65	158.9254
Thallium	Tl	81	204.37
Thorium	Th	90	232.0381
Thulium	Tm	69	168.9342
Tin	Sn	50	118.69
Titanium	Ti	22	47.90
Tungsten	W	74	183.85
Uranium	U	92	238.029
Vanadium	V	23	50.9414
Tungsten	W	74	183.85
Xenon	Xe	54	131.30
Ytterbium	Yb	70	173.04
Yttrium	Y	39	88.9059
Zinc	Zn	30	65.37
Zirconium	Zr	40	91.22

Periodic Table

Classical	IA	IIA	IIIB	IVB	VB	VIB	VIIB	VIII VIII			IB	IIB	IIIA	IVA	VA	VIA	VIIA	0
Amended	IA	IIA	IIIA	IVA	VA	VIA	VIIA	VIII VIII			IB	IIB	IIIB	IVB	VB	VIB	VIIB	0
Modern	1	2	3	4	5	6	7	8	9	10	11	12	13	14	15	16	17	18
	1 H 1.0080																	2 He 4.00260
	3 Li 6.941	4 Be 9.01218											5 B 10.81	6 C 12.011	7 N 14.0067	8 O 15.9994	9 F 18.9984	10 Ne 20.179
	11 Na 22.9898	12 Mg 24.305											13 Al 26.9815	14 Si 28.086	15 P 30.9738	16 S 32.06	17 Cl 35.453	18 Ar 39.948
	19 K 39.102	20 Ca 40.08	21 Sc 44.9559	22 Ti 47.90	23 V 50.9414	24 Cr 51.996	25 Mn 54.9380	26 Fe 55.847	27 Co 58.9332	28 Ni 58.71	29 Cu 63.546	30 Zn 65.37	31 Ga 69.72	32 Ge 72.59	33 As 74.9216	34 Se 78.96	35 Br 79.904	36 Kr 83.80
	37 Rb 85.4678	38 Sr 87.62	39 Y 88.9059	40 Zr 91.22	41 Nb 92.9064	42 Mo 95.94	43 Tc 98.9062	44 Ru 101.07	45 Rh 102.9055	46 Pd 106.4	47 Ag 107.868	48 Cd 112.40	49 In 114.82	50 Sn 118.69	51 Sb 121.75	52 Te 127.60	53 I 126.9045	54 Xe 131.30
	55 Cs 132.9055	56 Ba 137.34	57 La* 138.9055	72 Hf 178.49	73 Ta 180.9479	74 W 183.85	75 Re 186.2	76 Os 190.2	77 Ir 192.22	78 Pt 195.09	79 Au 196.9665	80 Hg 200.59	81 Tl 204.37	82 Pb 207.2	83 Bi 208.9806	84 Po (210)	85 At (210)	86 Rn (222)
	87 Fr (223)	88 Ra 226.0254	89 Ac† (227)	104 Unq (261)	105 Unp (262)	106 Unh (263)												

*

58 Ce 140.12	59 Pr 140.9077	60 Nd 144.24	61 Pm (145)	62 Sm 150.4	63 Eu 151.96	64 Gd 157.25	65 Tb 158.9254	66 Dy 162.50	67 Ho 164.9303	68 Er 167.26	69 Tm 168.9342	70 Yb 173.04	71 Lu 174.97

†

90 Th 232.0381	91 Pa 231.0359	92 U 238.029	93 Np 237.0482	94 Pu (242)	95 Am (243)	96 Cm (247)	97 Bk (249)	98 Cf (251)	99 Es (254)	100 Fm (253)	101 Md (256)	102 No (254)	103 Lr (257)